Symbols		Symbol First Occurs on Page:
a_X	X-axis intercept for the least-squares regression line predicting X given Y	137
a_Y	Y-axis intercept for the least-squares regression line predicting Y given X	130
b_X	slope of the least-squares regression line for predicting X given Y	137
b_Y	slope of the least-squares regression line for predicting Y given X	130
c	number of columns in a contingency table	434
	number of columns in a two-way ANOVA data table	397
cum f	cumulative frequency	42
cum f_L	frequency of scores below the lower real limit of the interval containing the percentile point	45
cum f_P	frequency of scores below the percentile point	44
cum %	cumulative percent	43
D	difference between paired scores	319
$\bar{D}$	mean of the differences between paired scores	319
df	degrees of freedom	295
df_B	between-group degrees of freedom	358
df_C	column degrees of freedom	400
df_R	row degrees of freedom	399
df_W	within-cells degrees of freedom	396
	within-group degrees of freedom	357
df_{RC}	row $\times$ column degrees of freedom	401
F	ratio of two variance estimates	351
f	frequency	35
f_e	expected frequency	425
f_i	frequency of the interval containing the percentile point	45
f_o	observed frequency	425
H_0	null hypothesis	212
H_1	alternative hypothesis	212

continued

UNDERSTANDING STATISTICS IN THE BEHAVIORAL SCIENCES

The cover shows a detail from a five-by-five scatter plot matrix. The color clusters represent three-dimensional similarity to values taken by three additional variables. This improvement over black-only dots can be used to help conceptualize complex relationships.

UNDERSTANDING STATISTICS IN THE BEHAVIORAL SCIENCES

FIFTH EDITION

ROBERT R. PAGANO

University of Pittsburgh

BROOKS/COLE PUBLISHING COMPANY

I(T)P® An International Thomson Publishing Company

Pacific Grove • Albany • Belmont • Bonn • Boston • Cincinnati • Detroit • Johannesburg • London
Madrid • Melbourne • Mexico City • New York • Paris • Singapore • Tokyo • Toronto • Washington

Sponsoring Editor: *Denis Ralling*
Marketing Team: *Michael Campbell,*
  Christine Davis, and Deanne Brown
Editorial Assistant: *Stephanie Andersen*
Production Editor: *Jamie Sue Brooks*
Production Service: *Susan L. Reiland*
Manuscript Editor: *Christine M. Levesque*
Permissions Editor: *Elaine Jones*

Interior Design: *Image House, Inc.*
Cover Design: *Laurie Albrecht*
Illustrations: *Precision Graphics*
What Is the Truth? Illustrations: *Axelle
  Fortier*
Typesetting: *Bi-Comp, Inc.*
Cover Printing: *Phoenix Color Corp*
Printing and Binding: *Courier Kendallville*

Cover illustration by Colin Ware. From Ware, C. and Beatty, J. C., "Using Color Dimensions to Display Data Dimensions," *Human Factors, 30* (1988), 127–142.

MINITAB® is a registered trademark of MINITAB, Inc. SPSS and SPSS/PC+ are registered trademarks of SPSS, Inc. Other brand and product names are trademarks or registered trademarks of their respective holders.

For more information, contact:

BROOKS/COLE PUBLISHING COMPANY
511 Forest Lodge Road
Pacific Grove, CA 93950
USA

International Thomson Publishing Europe
Berkshire House 168-173
High Holborn
London WC1V 7AA
England

Thomas Nelson Australia
102 Dodds Street
South Melbourne, 3205
Victoria, Australia

Nelson Canada
1120 Birchmount Road
Scarborough, Ontario
Canada M1K 5G4

International Thomson Editores
Seneca 53
Col. Polanco
11560 México, D. F., México

International Thomson Publishing GmbH
Königswinterer Strasse 418
53227 Bonn
Germany

International Thomson Publishing Asia
221 Henderson Road
#05-10 Henderson Building
Singapore 0315

International Thomson Publishing Japan
Hirakawacho Kyowa Building, 3F
2-2-1 Hirakawacho
Chiyoda-ku, Tokyo 102
Japan

Printed in the United States of America

10 9 8 7 6 5 4 3 2 1

Library of Congress Cataloging-in-Publication Data

Pagano, Robert R.
   Understanding statistics in the behavioral sciences / Robert R.
   Pagano.—5th ed.
      p. cm.
   Includes bibliographic references and index.
   ISBN 0-534-35390-8
   1. Psychology—Statistical methods. 2. Psychometrics. I. Title.
BF39.P25 1998
519.5—dc21   97-40018

Do you care about reality? Does

truth matter to you? If so, then I

dedicate this fifth edition to you.

BRIEF CONTENTS

APPENDIXES

GLOSSARY

INDEX

CONTENTS

CHAPTER 11   Mann–Whitney *U* Test   241

CHAPTER 12   Sampling Distributions, Sampling Distribution of the Mean,
the Normal Deviate (*z*) Test   263

PREFACE

I have been teaching a course in introductory statistics for over twenty-five years, first within the Department of Psychology at the University of Washington and subsequently within the Department of Neuroscience at the University of Pittsburgh. This textbook is the mainstay of the course. Most of my students have been psychology majors, but many have also come from biology, business, education, neuroscience, nursing, and other fields. Because these students generally do not have high aptitude for mathematics and are not well grounded in mathematical skills, I have used an informal rather than strictly mathematical approach. This approach assumes only high school algebra for background knowledge. It relies on clarity of presentation, a particularly effective sequencing of the inferential material, detailed verbal description, interesting illustrative examples, and many fully solved practice problems to help students understand the material.

My statistics course has been quite successful. Students are able to grasp the material, even the more complicated topics such as "power," and at the same time, many even enjoy learning it. Student ratings of this course have been quite high. Students rate the textbook even higher, saying among other things that it is very clear, they like the touches of humor, and that it helps them a lot to have the material presented in such great detail.

In preparing the fifth edition, I have again been guided by feedback from students and professors. Again, I am very pleased that this feedback has been quite positive, and that for most of the textbook the advice has been not to change anything because the text works very well. However, there were two recommended changes that had strong consensus: (1) remove the computer material from the textbook and put it in separate manuals, and (2) reverse the order of correlation and regression so that correlation precedes regression. Both changes have been made. In addition to moving the computer material to manuals, because SPSS now has a student edition running on the Windows platform, and because it is used so much in the behavioral sciences, I have included material on SPSS, rather than on MYSTAT, which was used in the fourth edition.

Following a theme that tries to simplify the fifth edition without compromising depth and quality, I have also eliminated the separate chapter on power and incorporated a discussion on power into Chapter 10, the introductory hypothesis testing chapter. I have done this somewhat reluctantly, because I believe having a separate chapter on power has real merit. However, because there are so many important topics to cover in an introductory statistics course, there often is not enough time for such coverage. Moreover, discussing power in conjunction with the sign test allows briefer treatment of this complicated topic with understanding. Of course, I have kept the more extensive treatment of power that is given in Chapter 12 in conjunction with the z test. I have also made what I consider to be one other major change. I have changed the illustrative example that introduces hypothesis testing in Chapter 10. The example "Marijuana and Reaction Time" had been in the textbook since its inception in 1981, and had become dated. I have replaced it with a more current and I hope interesting example, "Marijuana and the Treatment of AIDS Patients."

The following additional changes have been made. A new section, Significance Versus Importance, has been added to Chapter 10. Proper names throughout the text have been changed to better reflect diversity and thereby enable all students to see themselves as statistical experts. Two new "What Is the Truth?" sections have been added: "What Is the Truth? Sperm Count Decline—Male or Research Inadequacy?" has been added at the end of Chapter 8; and "What Is the Truth? Statistics and Applied Social Research—Useful or 'Abuseful'?" has been added at the end of Chapter 18. Finally, in an effort to update and extend the end-of chapter Questions and Problems, there are over fifty new or revised questions and problems.

Textbook Rationale

This is an introductory textbook that covers both descriptive and inferential statistics. It is intended for students majoring in the behavioral sciences. For many behavioral sciences undergraduates, statistics is a subject that engenders considerable anxiety and that is avoided for as long as possible. Moreover, I think it is fair to say that when the usual undergraduate statistics course is completed, many students have not understood much of the inferential statistics material. This happens partly because the material is inherently difficult and the students themselves are not proficient in mathematics but also, in my opinion, partly because most textbooks do a poor job of explaining inferential statistics to this group of students. These texts usually err in one or more of the following ways: (1) They are not clearly written; (2) they are not sufficiently detailed, (3) they present the material too mathematically; (4) they present the material at too low a level; (5) they do not give a sufficient number of fully solved problems for the student to practice on; and (6) in inferential statistics, they use an inappropriate sequence of topics, beginning with the sampling distribution of the mean.

In this and the previous four editions, I have tried to correct such deficiencies through an informal writing style; a clearly written, detailed, and theoretically oriented presentation that requires only high school algebra for understanding; the inclusion of many interesting, fully solved practice problems that are located immediately following the relevant expository material; and a sequencing of the inferential material better suited to the students for whom this book is intended.

I believe a key to understanding inferential statistics is the material presented in the beginning inferential chapters and its sequencing. In my opinion, optimal learning of the material occurs when it is sequenced as follows: random sampling and probability, binomial distribution, introduction to hypothesis testing using the sign test, Mann–Whitney U test, sampling distributions (including their empirical generation), sampling distribution of the mean, z test for single samples, t test for single samples, confidence intervals, t test for correlated and independent groups, introduction to analysis of variance, multiple comparisons, two-way ANOVA, nonparametric tests, and review of inferential statistics.

At the heart of statistical inference lies the concept of sampling distribution. The first sampling distribution discussed by most texts is the sampling distribution of the mean. The difficulty here is that the sampling distribution of the mean cannot be generated from simple probability considerations, which makes it hard to understand. This problem is compounded by the fact that most texts do not attempt to generate the sampling distribution of the mean in a concrete way. Rather, they define it theoretically, as a probability distribution that would result if an infinite number of random samples were taken of size N from the population and the mean of each sample were calculated. This definition is far too abstract for students, especially when this is their initial contact with the idea of sampling distributions. And when students fail to grasp the concept of sampling distributions, they fail to grasp the rest of inferential statistics. What appears to happen is that since students do not understand the material conceptually, they are forced to memorize the equations and to solve problems by rote. Thus, students are often able to solve the problems without genuinely understanding what they are doing—all because they fail to comprehend the essence of sampling distributions.

To impart a basic understanding, I believe it is much better to begin with the sign test (Chapter 10), a simple inference test for which the binomial distribution is the appropriate sampling distribution. The binomial distribution is easy to comprehend, and it can be derived from the basic probability rules developed in an earlier chapter (Chapter 8, Random Sampling and Probability). It depends entirely on logical considerations. Hence, its generation is easily followed. Moreover, it can also be generated by the same empirical process used later on for generating the sampling distribution of the mean. It therefore serves as an important bridge to understanding all the sampling distributions discussed later in the textbook. Introducing hypothesis testing along with the sign test has other advantages: All the important concepts involving hypothesis testing can be illustrated: e.g., null hypothesis, alternative hypothesis, alpha level, Type I and Type II errors, and power. The sign test also provides an illustration of the before-after (repeated measures) design, which is a superior way to begin, as most students are familiar with this type of experiment, and the logic of the design can be followed with ease.

Chapter 11 takes up the Mann–Whitney U test. This is a practical, powerful test that also has an easily understood sampling distribution. Both the sign test and Mann–Whitney U test have sampling distributions derived from basic probability considerations. The Mann–Whitney U test is additionally useful in that it illustrates the independent groups design. By the time students finish this chapter, they should have a sound knowledge of hypothesis testing as well as having experienced excellent exposure to the two basic experimental designs. Their confidence in "getting" the fundamentals of statistics should be greatly increased.

Chapter 12 initiates a formal discussion of sampling distributions and how they can be generated. After this, the sampling distribution of the mean is introduced, and *discussion centers on how this sampling distribution can be generated empirically,* which gives students a concrete understanding of the sampling distribution of the mean. With prior experience of the binomial distribution and the sampling distribution of U, and with knowledge of the empirical approach for the sampling distribution of the mean, most conscientious students will have achieved a good grasp of why sampling distributions are essential for inferential statistics. Since the sampling distributions underlying Student's t test and the analysis of variance are also explained in terms of their empirical generation, students can conceptually comprehend the use of these tests rather than just solve problems by rote. This approach gives the insight that *all* the concepts of hypothesis testing are the same as we go from statistic to statistic—what varies from experiment to experiment is the statistic used and its accompanying sampling distribution. The stage is now set for covering the remaining inference tests.

Chapters 12, 13, 14, and 18 discuss, in a fairly conventional way, the z test and t test for single samples, the t test for correlated and independent groups, and nonparametric statistics. However, these chapters differ from those in other texts in their clarity of presentation, the quantity and interest value of the fully solved problems they contain, and the use of empirically derived sampling distributions. Then, too, there are differences specific to each test. For example: (1) the t test for correlated groups is developed as a special case of the t test for single samples, this time using difference scores rather than raw scores; (2) the sign test and the t test for correlated groups are compared to illustrate the difference in power that results from using one or the other; (3) the factors influencing the power of experiments using Student's t test are taken up; (4) the correlated and independent groups designs are contrasted with regard to utility.

Chapters 15 and 17 deal with analysis of variance. In these chapters single rather than double subscript notation is deliberately employed. The more complex double subscript notation serves to confuse students. In my view, the single subscript notation and resulting single summations work better for the undergraduate major in psychology and related fields because they are simpler and, for this audience, promote understanding of this reasonably complicated material. Here, I have followed in part the notation used by Edward Minium in *Statistical Reasoning in Psychology and Education*. I am indebted to Professor Minium for this contribution.

Other features of this textbook are worth noting. Chapter 8, on probability, does not delve deeply into probability theory. This is not necessary since the proper mathematical foundation for all the inference tests contained in this textbook can be built as is done in Chapter 8 by the use of basic probability definitions, the addition rule and the multiplication rule. Chapter 16, covering both planned and *post hoc* comparisons, contains two *post hoc* tests: the Tukey HSD test and the Newman–Keuls test. Chapter 17 is a separate chapter on two-way ANOVA for instructors wishing to cover this topic in depth. For instructors with insufficient time for in-depth handling of two-way ANOVA, at the end of Chapter 15 on one-way ANOVA, I have qualitatively described the two-way ANOVA technique, emphasizing the concept of main effects and interactions. Chapter 19 is a review chapter that brings together all of the inference tests and provides practice in determining which test to use when analyzing data from different experimental designs and data of different levels of scaling. Students especially like the tree diagram on page 473 for helping them determine the appropriate test. Finally, at various places throughout the text, there are sections titled "What Is the Truth?" These sections show students practical applications of statistics.

The inferential material in this textbook is intended to be used in the sequence presented. However, if time constrains, Chapter 11 (the Mann–Whitney U test) may be omitted without injuring the rest of the material.

Some comments about the descriptive statistics part of this book: The material is presented at a level that (1) serves as a foundation for the inference chapters and (2) enables students to adequately describe the data for its own sake. For the most part, material on descriptive statistics follows a traditional format because this works well. Chapter 1 is an exception. It discusses approaches for determining truth and establishes statistics as part of the scientific method, which is somewhat unusual for a statistics text.

Finally, I should say something about how I have handled the use of computers. I have written the textbook so that it can be used with or without computers. In the fifth edition, I have removed all computer material from the textbook and placed it in two manuals. This was done in response to suggestions from the reviewers. It was felt that incorporating the computer material into the textbook itself, as was done in the previous editions, had the potential to detract from the text material, particularly for students in classes that do not use computers, or that do so minimally. There are two computer manuals, one covering MINITAB and the other SPSS; both use the Windows platform. These manuals have been written in sufficient detail for student use with minimal recourse to additional manuals. Both manuals contain fully solved computer problems and computer end-of-chapter problems tailored for specific textbook chapters. It is important to note the following:

> *The computer material can be totally ignored or used at different levels, varying from just having students read selected, fully solved computer problems, to having students use MINITAB or SPSS software to illustrate concepts and solve computer exercises covering material throughout the textbook.*

Fifth Edition Changes

New Material   As mentioned earlier, because of positive feedback from users of the fourth edition, fifth edition changes are not extensive. However, there are several major changes as well as many additional less basic ones. These changes include:

- Reversing the presentation order of correlation and regression. In the previous editions, Linear Regression (Chapter 6) was presented first, followed by Correlation (Chapter 7). This was done because (1) I believed that the concept of correlation is better understood after first having treated regression, and (2) linear regression was rather straightforward and easily understood. This order was resisted from the beginning in the reviews for the first edition and has continued through the fourth edition. Most professors teaching statistics are used to seeing correlation presented before regression. They probably learned it in that order themselves. Moreover, it does make logical sense to present correlation first. After all, one first needs to know that a correlation exists between two variables, before attempting to do any prediction. Because the support for this ordering seems to have grown over the years, and was particularly strong amongst the reviewers for the fifth edition, I have decided to accept their wisdom and reverse the ordering, presenting Correlation first in Chapter 6 followed by Linear Regression in Chapter 7. Chapter 6 begins with background material on relationships, and then presents the material on correlation itself. Discussion of Pearson r as the slope of the least squares regression line using z scores has been moved to the chapter on linear regression.

- Removing all the computer material from the textbook, except an introductory section in Chapter 1, and placing the computer material in two new computer manuals. This was done because reviewers were almost unanimous in preferring this arrangement, regardless of whether they used computers in their courses or not. Most said that they felt inclusion in the textbook of the computer material detracted from the text material itself, and added so many pages to the textbook that it was becoming too big.

- Two new computer manuals, one for MINITAB for Windows and one for SPSS for Windows. The MINITAB manual illustrates the use of Release 11 and the SPSS manual SPSS 7.5. I have replaced MYSTAT with SPSS because SPSS is so widely used in the behavioral sciences, and because SPSS now has a student version running on the Windows platform. I have opted for the Windows platform because PCs running this operating system have become so widespread and inexpensive. Both manuals are similar, differing only in the software discussed. It is beyond the scope of these manuals to cover all of MINITAB or SPSS. However, these manuals do describe the software in sufficient detail for student use with the textbook material. Each manual contains fully solved and end-of-chapter computer problems for appropriate textbook chapters. Answers to the end-of-chapter problems are also provided.

- Abbreviated treatment of power. In the fifth edition, instead of presenting a separate chapter on power as was done in the fourth edition (Chapter 11, Power), this chapter has been eliminated, and an abbreviated discussion of power has been presented at the end of Chapter 10, Introduction to Hypothesis Testing Using the Sign Test. This has been done somewhat reluctantly because there are many advantages to having a separate chapter on power. This change has been made to streamline the textbook, and because feedback from instructors indicates that many do not use this chapter or use it only in an attenuated manner. Fortunately, discussing power in conjunction with the sign test, as is done in Chapter 10, allows briefer treatment of power with understanding. The extensive treatment of power in conjunction with the z test in Chapter 12 has been kept.

- Change in illustrative example that introduces hypothesis testing in Chapter 10. For the first four editions, hypothesis testing was introduced through the example, "Marijuana and Reaction Time." This example has become dated and has been replaced with a more current and hopefully interesting example, "Marijuana and the Treatment of AIDS Patients."

- New section in Chapter 10, Significance Versus Importance. In the past editions, this important topic was covered in the "What Is the Truth?" section at the end of Chapter 16. It was decided that it properly belongs in Chapter 10, and deserves more prominent exposure.

- Proper names changed. At various places throughout the textbook, proper names have been changed to better reflect diversity and thereby enable all students to see themselves as potential statistical experts.

- Two new "What Is the Truth?" sections. "What Is the Truth? Sperm Count Decline—Male or Research Inadequacy?" has been added at the end of Chapter 8, Random Sampling and Probability. This section is an applied illustration of the importance of random sampling to the inference process. "What Is the Truth? Statistics and Applied Social Research—Useful or 'Abuseful'?" has been added at the end of Chapter 18, Chi-Square and Other

Nonparametric Tests. This section discusses how applied socia. become biased due to the personal interests of the researchers o. group, and raises the question of its utility and ethics. It has been p. in this chapter because chi-square is one of the main inference tests use. in applied social research.

- New or revised end-of-chapter questions and problems. Over fifty new or revised end-of-chapter questions and problems have been added to update and extend this material.
- Appendices changed. Appendices B and C of the fourth edition that dealt with computers have been removed. In addition, Appendix D, Symbols, has been moved to the inside covers of the textbook. Finally, a Glossary has been added.

Ancillary Package   *The supplements consist of:*

- A **student's study guide,** intended for review and consolidation of the material contained in each chapter of the textbook. Each chapter of the study guide has a chapter outline, a programmed learning concept review, exercises and answers to exercises, true–false questions and answers, and an end-of-chapter self-quiz with answers. Many students have commented on the helpfulness of this study guide.
- An **instructor's manual** contains short answer, multiple-choice, and true–false questions for each chapter. The answers to the multiple-choice and true–false questions are given in bold type to the left of the questions. This manual also contains answers to selected end-of-chapter problems contained in the textbook. Because of requests from instructors, I have not included answers to *all* the computational end-of-chapter problems found in the text; rather, I have omitted answers from at least one problem in each chapter. The omitted answers are found near the end of the instructor's manual.
- Software: A **MINITAB manual** and an **SPSS manual** illustrate use of these statistical software packages in conjunction with the material covered in the textbook. The MINITAB manual covers MINITAB release 11 for Windows, and the SPSS manual SPSS 7.5 for Windows. Each manual discusses its respective software, and contains fully solved and end-of-chapter computer problems for appropriate textbook chapters. Each also contains answers to these problems.
- **Transparency masters** feature a selection of key figures and tables from the text.
- **WESTEST 3.2,** a computerized testing package for Windows or Macintosh computers that allows instructors to create, edit, store, and print exams.

Acknowledgments

Over the past four editions, many people have contributed to this textbook. Foremost is Clyde Perlee, who was editor-in-chief of the College Division of West Publishing Company. It is with distinct sadness and much gratitude that I bid farewell to Clyde. He has been a real driving force and inspiration for over 15 years. Thank you, Clyde.

I also wish to thank the following reviewers for their valuable comments regarding the fifth edition:

Bryan Auday
Gordon College

Michael Brown
University of Washington

Penny Fidler
California State University, Long
  Beach

Michael Gardner
California State University,
  Northridge

William Gibson
Northern Arizona University

George P. Knight
Arizona State University

Nona Phillips
University of Washington

Ladonna Rush
College of Wooster

Anna Smith
Troy State University

Paul Smith
Alverno College

Mary Tallent-Runnels
Texas Tech University

Evangeline Wheeler
Towson State University

Todd Zakrajsek
Southern Oregon State College

I am grateful to the Literary Executor of the Late Sir Ronald A. Fisher, F.R.S., to Dr. Frank Yates, F.R.S., and to the Longman Group Ltd., London, for permission to reprint Tables III, IV, and VII from their book *Statistical Tables for Biological, Agricultural and Medical Research* (6th edition, 1974).

The material covered in this textbook, study guide, instructor's manual, and MINITAB and SPSS manuals is appropriate for undergraduate students with a major in psychology or related behavioral science discipline. I believe the approach I have followed helps considerably to impart this subject matter with understanding. I am grateful to receive any comments that will improve the quality of these materials.

Robert R. Pagano

Sources

Many of the examples and problems used in this textbook are adapted from actual research. The citations for this research are given below.

Altman, et al. Trust of the stranger in the city and the small town. Unpublished research. In Kagan, J., and Havemann, E. *Psychology: An Introduction.* New York: Harcourt Brace Jovanovich, Inc. (3rd edition), 1976.

Barash, D. P. The social biology of the Olympic marmot. *Animal Behavior Monographs,* 1973, *6*(3), 171–249.

Budznyski, T. H., Stoyva, J. M., Adler, C. S., and Mullaney, D. J. EMG Biofeedback and tension headache: A controlled outcome study. *Psychosomatic Medicine,* 1973, *35,* No. 6, 484–96.

Butler, R. A. Discrimination learning by rhesus monkeys in visual-exploration motivation. *Journal of Comparative and Physiological Psychology,* 1953, *46,* 95–98.

Byrne, D., and Nelson, D. Attraction as a linear function of proportion of positive reinforcements. *Journal of Personality and Social Psychology, 1,* No. 6, 659–63.

Deutsch, M., and Collins, M. E. *Interracial housing: A psychological evaluation of a social experiment.* Minneapolis: University of Minnesota Press, 1951.

Holmes, T. H., and Rahe, R. H. The social readjustment, rating scale. *Journal of Psychosomatic Research,* 1967, *11,* 213–18.

McClelland, D. C., and Watson, R. I., Jr. Power motivation and risktaking behavior. *Journal of Personality,* 1973, *41,* 121–39.

Pagano, R. R., and Lovely, R. H. Diurnal cycle and ACTH facilitation of shuttlebox avoidance. *Physiology and Behavior,* 1972, *8,* 721–23.

U.S. Bureau of Census, *Statistical Abstract of the United States:* 1978. (99th edition) Washington, D.C., 1978.

Wallace, R. K. Physiological effects of transcendental meditation. *Science,* 1970, *167,* No. 3926, 1751–54.

Wilkinson, R. T. Sleep deprivation: Performance tests for partial and selective sleep deprivation. *Progress in Clinical Psychology,* 1968, *8,* 28–43.

Material for the "What Is the Truth?" sections has been adapted from the following sources:

What Is the Truth? (p. 12): Advertisement in *Psychology Today,* 1984, June, p. 13. Reprinted with permission.

What Is the Truth? (p. 13): Advertisement in *Cosmopolitan,* 1983, October, p. 154.

What Is the Truth? (p. 57): Puget Power awaits rate decision (1984, September 25), *Journal American,* Section D, p. 1. Graph reprinted with permission.

What Is the Truth? (p. 121): Equation for success: Good principal = good elementary school (1985, August 23), *Seattle Post-Intelligencer,* Section A, pp. 1, 11.

What Is the Truth? (p. 184): Parking tickets and missing women: Statistics and the law. Zisel, H., and Kalven, H., 139–140. In *Statistics: A guide to the unknown.*

What Is the Truth? (p. 185): 20-Year study shows sperm count decline among fertile men (1995, February 2), *Pittsburgh Post-Gazette,* Section A, p. 3.

Tanur, J. M., Mosteller, F., Kruskal, W. H., Link, R. F., Pieters, R. S., Rising, G. R., Lehmann, E. L. (Eds.) San Francisco: Holden-Day, 1978.

What Is the Truth? (p. 232): *Time,* 1976, July, p. 64. Advertisement reproduced with permission.

What Is the Truth? (p. 234): Advertisement in *Cosmopolitan;* 1983, October, p. 95.

What Is the Truth? (p. 235): When clinical studies mislead (1993, November 29), *Pittsburgh Post-Gazette,* Section A, pp. 8–9.

What Is the Truth? (p. 375): Huck, S. W., and Sandler, H. M. *Rival Hypotheses,* 51–52, New York: Harper and Row, 1979.

What Is the Truth? (p. 449): Poor results—Think tanks battle to judge the impact of welfare overhaul, J. Harwood (1997, January 30), *The Wall Street Journal,* Section A, pp. 1, 6.

PART ONE

OVERVIEW

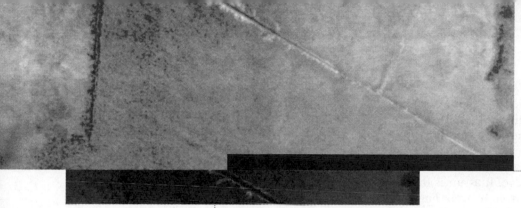

1 STATISTICS AND SCIENTIFIC METHOD

 ## INTRODUCTION

Have you ever wondered how we come to know truth? Most college students would agree that finding out what is true about the world, ourselves, and others constitutes a very important activity. A little reflection reveals that much of our time is spent in precisely this way. If we are studying geography, we want to know what is *true* about the geography of a particular region. Is the region mountainous or flat, agricultural or industrial? If our interest is in studying human beings, we want to know what is *true* about humans. Do we *truly* possess a spiritual nature, or are we *truly* reducible solely to atoms and molecules, as the reductionists would have it? How do humans think? What happens in the body to produce a sensation or a movement? When I get angry, is it *true* that there is a unique underlying physiological pattern? What is the pattern? Is my *true* purpose in life to become a teacher? Is it *true* that animals think? We could go on indefinitely with examples because so much of our lives is spent seeking and acquiring truth.

METHODS OF KNOWING

Historically, humankind has employed four methods to acquire knowledge. They are authority, rationalism, intuition, and the scientific method.

Authority   When using the method of authority, something is considered true because of tradition or because some person of distinction says it is true. Thus, we may believe in the theory of evolution because our distinguished professors tell us it is true, or we may believe that God truly exists because our parents say so. Although this method of knowing is currently in disfavor and does sometimes lead to error, it is indispensable to living our daily lives. We simply must accept a large amount of information on the basis of authority, if for no other reason than we often do not have the time or the expertise to check it out firsthand. For example, I believe, on the basis of physics authorities, that electrons exist, but I have never seen one; or, perhaps closer to home, if the Surgeon

General tells me that smoking causes cancer, I stop smoking because I have faith in the Surgeon General and do not have the time or means to investigate the matter personally.

Rationalism   The method of rationalism uses reasoning alone to arrive at knowledge. It assumes that if the premises are sound and the reasoning is carried out correctly according to the rules of logic, then the conclusions will yield truth. We are very familiar with reason because we use it so much. As an example, consider the following syllogism:

All statistics professors are interesting people.

Mr. X is a statistics professor.

Therefore, Mr. X is an interesting person.

Assuming the first statement is true (who could doubt it?), then it follows that if the second statement is true, the conclusion must be true. Joking aside, hardly anyone would question the importance of the reasoning process in yielding truth. However, there are a great number of situations in which reason alone is inadequate in determining the truth.

To illustrate, let's suppose you notice that John, a friend of yours, has been depressed for a couple months. As a psychology major, you know that psychological problems can produce depression. Therefore, it is reasonable to believe John may have psychological problems that are producing his depression. On the other hand, you also know that an inadequate diet can result in depression. It is also reasonable to believe that this may be at the root of his trouble. In this situation, there are two reasonable explanations of the phenomenon. Hence, reason alone is inadequate in distinguishing between them. We must resort to experience. Is John's diet in fact deficient? Will improved eating habits correct the situation? Or does John have serious psychological problems that, when worked through, will lift the depression? Reason alone, then, may be sufficient to yield truth in some situations, but it is clearly inadequate in others. As we shall see, the scientific method also uses reason to arrive at truth, but reasoning alone is only part of the process. Thus, the scientific method incorporates reason but is not synonymous with it.

Intuition   Knowledge is also acquired through intuition. By intuition, we mean that sudden insight, the clarifying idea that springs into consciousness all at once as a whole. It is not arrived at by reason. On the contrary, the idea often seems to occur after conscious reasoning has failed. Beveridge* gives numerous occurrences taken from prominent individuals. Here are a couple of examples:

Here is Metchnikoff's own account of the origin of the idea of phagocytosis: "One day when the whole family had gone to the circus to see some extraordinary performing apes, I remained alone with my microscope, observing the life in the mobile cells of a transparent starfish larva, when a new thought suddenly flashed across my brain. It struck me that similar cells might serve in the defense of the organism against intruders. Feeling that there was in this something of surpassing interest, I felt so

---

* W. I. B. Beveridge, *The Art of Scientific Investigation,* Vintage Books/Random House, New York, 1957, pp. 94–95.

excited that I began striding up and down the room and even went to the seashore to collect my thoughts."

Hadamard cites an experience of the mathematician Gauss, who wrote concerning a problem he had tried unsuccessfully to prove for years: "Finally two days ago I succeeded . . . like a sudden flash of lightning the riddle happened to be solved. I cannot myself say what was the conducting thread which connected what I previously knew with what made my success possible."

It is interesting to note that the intuitive idea often occurs after conscious reasoning has failed and the individual has put the problem aside for a while. Thus, Beveridge* quotes two scientists as follows:

Freeing my mind of all thoughts of the problem I walked briskly down the street, when suddenly at a definite spot which I could locate today—as if from the clear sky above me—an idea popped into my head as emphatically as if a voice had shouted it.

I decided to abandon the work and all thoughts relative to it, and then, on the following day, when occupied in work of an entirely different type, an idea came to my mind as suddenly as a flash of lightning and it was the sollution . . . the utter simplicity made me wonder why I hadn't thought of it before.

Despite the fact that intuition has probably been used as a source of knowledge for as long as humans have existed, it is still a very mysterious process about which we have only the most rudimentary understanding.

Scientific Method   Although the scientific method uses both reasoning and intuition for establishing truth, its reliance on objective assessment is what differentiates this method from the others. At the heart of science lies the *scientific experiment*. The method of science is rather straightforward. By some means, usually by reasoning deductively from existing theory or inductively from existing facts or through intuition, the scientist arrives at a hypothesis about some feature of reality. He or she then designs an experiment to objectively test the hypothesis. The data from the experiment are then analyzed statistically, and the hypothesis is either supported or rejected. The feature of overriding importance in this methodology is that no matter what the scientist believes is true regarding the hypothesis under study, the experiment provides the basis for an *objective* evaluation of the hypothesis. The data from the experiment force a conclusion consonant with reality. Thus, scientific methodology has a built-in safeguard for assuring that truth assertions of any sort about reality must conform to what is demonstrated to be objectively true about the phenomena before the assertions are given the status of scientific truth.

An important aspect of this methodology is that the experimenter can hold incorrect hunches, and the data will expose them. The hunches can then be revised in light of the data and retested. This methodology, although sometimes painstakingly slow, has a self-correcting feature that, over the long run, has a high probability of yielding truth. Since in this textbook we emphasize statistical analysis rather than experimental design, we cannot spend a great deal of time discussing the design of experiments. Nevertheless, some experimental design will be covered because it is so intertwined with statistical analysis.

---

* Ibid., p. 92.

In discussing this and other material throughout the book, we shall be using certain technical terms. The terms and their definitions follow:

- **Population** *A population is the complete set of individuals, objects, or scores that the investigator is interested in studying.* In an actual experiment, the population is the larger group of individuals from which the subjects run in the experiment have been taken.

- **Sample** *A sample is a subset of the population.* In an experiment, for economical reasons, the investigator usually collects data on a smaller group of subjects than the entire population. This smaller group is called the sample.

- **Variable** *A variable is any property or characteristic of some event, object, or person that may have different values at different times, depending on the conditions.* Height, weight, reaction time, and drug dosage are examples of variables. A variable should be contrasted with a *constant,* which, of course, does not have different values at different times. An example of a constant is the mathematical constant ʄ; it always has the same value (3.14 to two-decimal-place accuracy).

- **Independent variable** *The independent variable in an experiment is the variable that is systematically manipulated by the investigator.* In most experiments, the investigator is interested in determining the effect that one variable, say, variable A, has on one or more other variables. To do so, the investigator manipulates the levels of variable A and measures the effect on the other variables. Variable A is called the *independent* variable because its levels are controlled by the experimenter, independent of any change in the other variables. To illustrate, an investigator might be interested in the effect of alcohol on social behavior. To investigate this, he or she would probably vary the amount of alcohol consumed by the subjects and measure its effect on their social behavior. In this example, the experimenter is manipulating the amount of alcohol and measuring its consequences on social behavior. Alcohol amount is the independent variable. In another experiment, the effect of sleep deprivation on aggressive behavior is studied. Subjects are deprived of various amounts of sleep, and the consequences on aggressiveness are observed. Here, the amount of sleep deprivation is being manipulated. Hence, it is the independent variable.

- **Dependent variable** *The dependent variable in an experiment is the variable that the investigator measures to determine the effect of the independent variable.* For example, in the experiment studying the effects of alcohol on social behavior, amount of alcohol is the independent variable. The social behavior of the subjects is measured to see whether it is affected by the amount of alcohol consumed. Thus, social behavior is the dependent variable. It is called *dependent* because it may depend on the amount of alcohol consumed. In the investigation of sleep deprivation and aggressive behavior, the amount of sleep deprivation is being manipulated, and the subjects' aggressive behavior is being measured. Amount of sleep deprivation is the independent variable, and aggressive behavior is the dependent variable.

- **Data** *The measurements that are made on the subjects of an experiment are called data.* Usually data consist of the measurements of the dependent variable or of other subject characteristics, such as age, gender, number of subjects, etc. The data as originally measured are often referred to as *raw* or *original* scores.

- **Statistic** *A statistic is a number calculated on sample data that quantifies a characteristic of the sample.* Thus, the average value of a sample set of scores would be called a statistic.

- **Parameter** *A parameter is a number calculated on population data that quantifies a characteristic of the population.* For example, the average value of a population set of scores is called a parameter. It should be noted that a statistic and a parameter are very similar concepts. The only difference is that a statistic is calculated on a sample and a parameter is calculated on a population.

An Overall Example: Mode of Presentation and Retention

Let's now consider an illustrative experiment and apply the previously discussed terms to it.

An educator conducts an experiment to determine whether the mode of presentation affects how well prose material is remembered. For this experiment, the educator uses several prose passages that are presented visually or auditorily. Fifty students are selected from the undergraduates attending the university at which the educator works. The students are divided into two groups of 25 students per group. The first group receives a visual presentation of the prose passages, and the second group hears the passages through an auditory presentation. At the end of their respective presentations, the subjects are asked to write down as much of the material as they can remember. The average number of words remembered by each group is calculated, and the two group averages are compared to see whether the mode of presentation had an effect.

In this experiment, the independent variable is the mode of presentation of the prose passages, i.e., auditory or visual presentation. The dependent variable is the number of words remembered. The sample is the 50 students who participated in the experiment. The population is the larger group of individuals from which the sample was taken, namely, the undergraduates attending the university. The data are the number of words recalled by each student in the sample. The average number of words recalled by each group is a statistic because it measures a characteristic of the sample scores. Since there was no measurement made of any population characteristic, there was no parameter calculated in this experiment. However, for illustrative purposes, suppose the entire population had been given a visual presentation of the passages. If we calculate the average number of words remembered by the population, the average number would be called a parameter, because it measures a characteristic of the population scores.

Now, let's do a problem to practice identifying these terms.

For the experiment described below, specify the following: the independent variable, the dependent variable(s), the sample, the population, the data, the statistic(s), and the parameter(s).

A professor of gynecology at a prominent medical school wants to determine whether an experimental birth control implant has side effects on body weight and depression. A group of 1000 adult women living in a nearby city volunteers for the experiment. The gynecologist selects 100 of these women to participate in the study. Fifty of the women are assigned to group 1 and the other fifty to group 2 such that the mean body weight and the mean depression scores of each group are equal at the beginning of the experiment. Treatment conditions are the same for both groups, except that the women in group 1 are surgically implanted with the experimental birth control device, whereas the women in group 2 receive a placebo implant. Body weight and depressed mood state are measured at the beginning and end of the experiment. A standardized questionnaire designed to measure degree of depression is used for the mood state measurement. The higher the score on this questionnaire, the more depressed is the individual. The mean body weight and the mean depression scores of each group at the end of the experiment are compared to determine whether the experimental birth control implant had an effect on these variables. To safeguard the women from unwanted pregnancy, another method of birth control that does not interact with the implant was used for the duration of the experiment.

SOLUTION

Independent variable: The experimental birth control implant versus the placebo

Dependent variables: Body weight and depressed mood state

Sample: 100 women who participated in the experiment

Population: 1000 women who volunteered for the experiment

Data: The individual body weight and depression scores of the 100 women at the beginning and end of the experiment

Statistics: Mean body weight of group 1 at the beginning of the experiment, mean body weight of group 1 at the end of the experiment, mean depression score of group 1 at the beginning of the experiment, mean depression score of group 1 at the end of the experiment, plus the same four statistics for group 2

Parameter: No parameters were given or computed in this experiment. If the gynecologist had measured the body weights of all 1000 volunteers at the beginning of the experiment, the mean of these 1000 weights would be a parameter.

SCIENTIFIC RESEARCH AND STATISTICS

Scientific research may be divided into two categories, observational studies and true experiments. Statistical techniques are important in both kinds of research.

Observational Studies In this type of research, no variables are actively manipulated by the investigator. Included within this category of research are (1) naturalistic observation, (2) parameter estimation, and (3) correlational studies. With naturalistic observation research, a major goal is to obtain an accurate description of the situation being studied. Much anthropological and etiological research is of this type. Parameter estimation research is conducted on samples to estimate the level of one or more population characteristics; e.g., the population average or percentage. Surveys, public opinion polls, and much market research fall into this category. In correlational research, the investigator focuses attention on two or more variables to determine whether they are related. For example, to determine whether in adults over 30 years old, obesity and high blood pressure are related, an investigator might measure the fat level and blood pressure of individuals in a sample of adults over 30. The investigator would then analyze the results to see whether a relationship exists between these variables; i.e., do individuals with low fat levels also have low blood pressure, individuals with moderate fat levels also have moderate blood pressure, and individuals with high fat levels have high blood pressure?

True Experiments In this type of research, an attempt is made to determine whether changes in one variable produce changes in another variable. With this type of research, an independent variable is manipulated, and its effect on some dependent variable is studied. However, there can be more than one independent variable and more than one dependent variable. In the simplest case, there is only one independent and one dependent variable. One example of this case is the experiment mentioned previously that investigated the effect of alcohol on social behavior. In this experiment, you will recall, alcohol level was manipulated by the experimenter, and its effect on social behavior was measured.

 ## RANDOM SAMPLING

In all of the research described previously, data are usually collected on a sample of subjects, rather than on the entire population to which the results are intended to apply. Ideally, of course, the experiment would be performed on the whole population, but usually it is far too costly, and so a sample is taken. Note that not just any sample will do. The sample should be a *random* sample. Random sampling is discussed in Chapter 8. For now, it is sufficient to know that random sampling allows the laws of probability to apply to the data and at the same time helps achieve a sample that is representative of the population. Thus, the results obtained from the sample should also apply to the population. Once the data are collected, they are statistically analyzed, and the appropriate conclusions about the population are drawn.

DESCRIPTIVE AND INFERENTIAL STATISTICS

Statistical analysis, of course, is the main theme of this textbook. It has been divided into two areas: (1) *descriptive statistics* and (2) *inferential statistics*. Both involve analyzing data. If an analysis is done for the purpose of describing or characterizing the data that have been collected, then we are in the area of descriptive statistics. To illustrate, suppose your biology professor has just recorded the scores from an exam he has recently given you. He hands back the tests and now wants to describe the scores. He might decide to calculate the average of the distribution so as to describe its *central tendency*. Perhaps he will also determine its range so as to characterize its *variability*. He might also plot the scores on a graph so as to show the *shape* of the distribution. Since all of these procedures are for the purpose of describing or characterizing the data already collected, they fall within the realm of descriptive statistics.

Inferential statistics, on the other hand, is not concerned with just describing the obtained data. Rather, it embraces techniques that allow one to use obtained sample data to infer to or draw conclusions about populations. This is the more complicated part of statistical analysis. It involves probability and various inference tests, such as *Student's* t *test* and the *analysis of variance*. To illustrate the difference between descriptive and inferential statistics, suppose we were interested in determining the average IQ of the entire freshman class at your university. Since it would be too costly and time-consuming to measure the IQ of every student in the population, we would take a random sample of, say, 100 students and give each an IQ test. We would then have 100 *sample* IQ scores, which we want to use to determine the average IQ in the *population*. Although we can't determine the exact value of the population average, we can estimate it, using the sample data in conjunction with an inference test called Student's *t* test. The results would allow us to make a statement such as "We are 95% confident that the interval of 115–120 contains the mean IQ of the population." Here, we are not just describing the obtained scores, as was the case with the biology exam. Rather, we are using the sample scores to infer to a population value. We are therefore in the domain of inferential statistics. Thus, *descriptive* and *inferential statistics* can be defined as follows:

DEFINITIONS

- **Descriptive statistics** *is concerned with techniques that are used to describe or characterize the obtained data.*

- **Inferential statistics** *involves techniques that use the obtained sample data to infer to populations.*

 ## USING COMPUTERS IN STATISTICS

The use of computers in statistics has increased greatly over the past decade. In fact, today almost all research data in the behavioral sciences are analyzed using statistical computer programs rather than by "hand" with a calculator. This is good news for students, who often like the ideas, concepts, and results of statistics but hate the drudgery of hand computation. The fact is that researchers hate computational drudgery too and therefore almost always use a computer to analyze data sets of any appreciable size. Computers have the advantages of saving time and labor, minimizing chances of computational error, allowing easy graphical display of the data, and providing better management of large data sets. As useful as computers are, there is often not enough time in a basic statistics course to include them. Therefore, I have written this edition of the textbook so that you can learn the statistical content with or without the computer material.

Several computer programs are available to do statistical analysis. The most popular are Statistical Package for the Social Sciences (SPSS), Biomedical Computer Programs-P series (BMDP), Statistical Analysis System (SAS), SYSTAT, and MINITAB. Versions of SPSS, SAS, SYSTAT, and MINITAB are available for both mainframes and microcomputers. It is worth taking the extra time to learn one or more of these programs. In conjuction with this textbook, I have chosen to illustrate the use of two of them: MINITAB and SPSS. The computer material is presented in two separate manuals.

As you begin solving problems using MINITAB or SPSS, I believe you will begin to experience the fun and power that computers can bring to your study of statistics. In fact, once you have used a computer to analyze data, you will probably wonder, "Why do I have to do any of these complicated calculations by hand?" Unfortunately, when you are using a computer to calculate the value of a statistic, it does not help you understand that statistic. Understanding the statistic and its proper use is best achieved by doing hand calculations primarily and using computers as a supporting medium. The computer is especially effective in promoting understanding when used for graphical presentations. Of course, once you have learned everything you can from the hand calculations, using the computer to grind out correct values of the statistic seems eminently reasonable.

STATISTICS AND THE "REAL WORLD"

As I mentioned previously, one major purpose of statistics is to aid in the scientific evaluation of truth assertions. Although you may view this as rather esoteric and far removed from everyday life, I believe you will be convinced, by the time you have finished this text, that understanding statistics has very important practical aspects for your success in life. As you go through this text, I hope you will become increasingly aware of how frequently in ordinary life we are bombarded with "authorities" telling us, based on "truth assertions," what we should do, how we should live, what we should buy, what we should value, and so on. In areas of real importance to you, I hope you will begin to ask questions like: "Are these truth assertions supported by data?"; "How good are the data?"; "Is chance a reasonable explanation of the data?" If there are no data presented, or if the data presented are of the form, "My experience is that . . . ," rather than from well-controlled experiments, you will begin to question how seriously you should take the authority's advice.

To help develop this aspect of your statistical decision making, I have included (at the end of certain chapters) applications taken from everyday life. These are titled, "What Is the Truth?" To begin, let's consider the following material.

DATA, DATA, WHERE ARE THE DATA?

The accompanying advertisement was printed in a recent issue of *Psychology Today*. From a scientific point of view, what's missing?

ANSWER This ad is similar to a great many that appear these days. It promises a lot, but offers no experimental data to back up its claims. The ad puts forth a truth assertion: **"Think And Be Thin."** It further claims **"Here's a tape program that really works ... and permanently!"** The program consists of listening to a subliminal tape that is supposed to program your mind so as to produce thinness. The glaring lack is that there are no controlled experiments, no data offered to substantiate the claim. This is the kind of claim that cries out for empirical verifica-

tion. Apparently, the authors of the ad do not believe the readers of *Psychology Today* are very sophisticated, statistically. I certainly hope the readers of this textbook would ask for the data before they spent 6 months of their time listening to a tape, the message of which they can't even hear!

AUTHORITIES ARE NICE, BUT ...

An advertisement promoting Anacin-3 appeared in a recent issue of *Cosmopolitan.* The heading of the advertisement was "3 Good Reasons to Try Anacin-3." The advertisement pictured a doctor, a nurse, and a pharmacist making the following three statements:

1. "*Doctors* are recommending acetaminophen, the aspirin-free pain reliever in Anacin-3, more than any other aspirin-free pain reliever."
2. "*Hospitals* use acetaminophen, the aspirin-free pain reliever in Anacin-3, more than any other aspirin-free pain reliever."
3. "*Pharmacists* recommend acetaminophen, the aspirin-free pain reliever in Anacin-3, more than any other aspirin-free pain reliever."

From a scientific point of view, is anything missing?

ANSWER This is somewhat better than the previous ad. At least relevant authorities are invoked in support of the product. How-ever, the ad is misleading and again fails to present the appropriate data. Much better than the "3 Good Reasons to Try Anacin-3" given in the ad, would be reason 4, data from well-conducted experiments showing that (a) acetaminophen is a better pain reliever than any other aspirin-free pain reliever and (b) Anacin-3 relieves pain better than any competitor. Any guesses about why these data haven't been presented? As a budding statistician, are you satisfied with the case made by this ad?

2 GOOD REASONS TO TRY PYRO-Z®

1. Doctors are recommending PYRO-Z® for pain relief over other products.

2. Hospitals use PYRO-Z, the pain reliever, more than any other pain reliever.

SUMMARY

In this chapter, we have discussed how truth is established. Traditionally, four methods have been used: authority, reason, intuition, and science. At the heart of science is the scientific experiment. By reasoning or through intuition, the scientist forms a hypothesis about some feature of reality. He or she then designs an experiment to objectively test the hypothesis. The data from the experiment are then analyzed statistically, and the hypothesis is either confirmed or rejected. Most scientific research falls into two categories: observational studies and true experiments. Natural observation, parameter estimation, and correlational studies are included within the observational category. Their major goal is to give an accurate description of the situation, estimate population parameters, or determine whether two or more of the variables are related. Since there is no systematic manipulation of any variable by the experimenter when doing an observational study, this type of research cannot determine whether changes in one variable will produce changes in another variable. This kind of result can be determined only from true experiments. In true experiments, the investigator systematically manipulates the independent variable and observes its effect on one or more dependent variables. Due to practical considerations, data are collected on only a sample of subjects rather than on the whole population. It is important that the sample be a random sample. The obtained data are then analyzed statistically. The statistical analysis may be descriptive or inferential. If the analysis just describes or characterizes the obtained data, we are in the domain of descriptive statistics. If the analysis uses the obtained data to infer to populations, we are in the domain of inferential statistics. Today, computers are used extensively in analyzing data. This textbook uses the programs MINITAB and SPSS to teach computer use in statistics. Understanding statistical analysis has important practical consequences in life.

IMPORTANT TERMS

Constant (p. 6)
Correlational research (p. 9)
Data (p. 7)
Dependent variable (p. 6)
Descriptive statistics (p. 10)
Independent variable (p. 6)
Inferential statistics (p. 10)
Intuition (p. 4)

Method of authority (p. 3)
MINITAB (p. 11)
Natural observation research
   (p. 9)
Parameter (p. 7)
Parameter estimation research
   (p. 9)
Population (p. 6)

Rationalism (p. 4)
Sample (p. 6)
Scientific method (p. 5)
SPSS (p. 11)
Statistic (p. 7)
True experiment (p. 9)
Variable (p. 6)

QUESTIONS AND PROBLEMS

1. Define each of the following terms:

Population	Dependent variable
Sample	Constant
Data	Statistic
Variable	Parameter
Independent variable	

2. What are four methods of acquiring knowledge? Write a short paragraph describing the essential characteristics of each.

3. How does the scientific method differ from each of the methods listed here?
   a. Method of authority
   b. Method of rationalism
   c. Method of intuition

4. Write a short paragraph comparing naturalistic observation and true experiments.

5. Distinguish between descriptive and inferential statistics. Use examples to illustrate the points you make.

6. In each of the experiments described here, specify (1) the independent variable, (2) the dependent

variable, (3) the sample, (4) the population, (5) the data, and (6) the statistic:

a. A health psychologist is interested in whether fear motivation is effective in reducing the incidence of smoking. Forty smoking adults are selected from individuals residing in the city in which the psychologist works. Twenty are asked to smoke a cigarette, after which they see a gruesome film about how smoking causes cancer. Vivid pictures of the diseased lungs and other internal organs of deceased smokers are shown in an effort to instill fear of smoking in these subjects. The other group receives the same treatment, except they see a neutral film that is unrelated to smoking. For 2 months after showing the film, the experimenter keeps records on the number of cigarettes smoked daily by the participants. A mean for each group is then computed of the number of cigarettes smoked daily since seeing the film, and these means are compared to determine whether the fear-inducing film had an effect on smoking.

b. A physiologist wants to know whether a particular region of the brain (the hypothalamus) is involved in the regulation of eating. An experiment is performed in which 30 rats are selected from the university vivarium and divided into two groups. One of the groups receives lesions in the hypothalamus, whereas the other group is lesioned in a neutral area. After recovery from the operations, all animals are given free access to food for 2 weeks, and a record is kept of the daily food intake of each animal. At the end of the 2-week period, the mean daily food intake for each group is determined. Finally, these means are compared to see whether the lesions in the hypothalamus have affected the amount eaten.

c. A typing teacher believes that a different arrangement of the typing keys will promote faster typing. Twenty secretarial trainees, selected from a large business school, participate in an experiment designed to test this belief. Ten of the trainees learn to type on the conventional keyboard. The other ten are trained using the new arrangement of keys. At the end of the training period, the typing speed in words per minute of each trainee is measured. The mean typing speeds are then calculated for both groups and compared to determine whether the new arrangement has had an effect.

d. A clinical psychologist is interested in evaluating three methods of treating depression: medication, cognitive restructuring, and exercise. A fourth treatment condition, a waiting-only treatment group, is included to provide a baseline control group. Sixty depressed students are recruited from the undergraduate student body at a large state university and 15 are assigned to each treatment method. Treatments are administered for 6 months, after which each student is given a questionnaire designed to measure the degree of depression. The questionnaire is scaled from 0 to 100, with higher scores indicating a higher degree of depression. The mean depression values are then computed for the four treatments and compared to determine the relative effectiveness of each treatment.

7. Indicate which of the following represent a variable and which a constant:
   a. The number of letters in the alphabet
   b. The number of hours in a day
   c. The time at which you eat dinner
   d. The number of students who major in psychology at your university each year
   e. The number of centimeters in a meter
   f. The amount of sleep you get each night
   g. The amount you weigh
   h. The volume of a liter

8. Indicate which of the following situations involve descriptive statistics and which involve inferential statistics:
   a. An annual stockholders' report details the assets of the corporation.
   b. A history instructor tells his class the number of students who received an A on a recent exam.
   c. Calculating the mean of a sample set of scores to characterize the sample
   d. Using the sample data from a poll to estimate the opinion of the population
   e. Conducting a correlational study on a sample to determine whether educational level and income in the population are related
   f. A newspaper article reports the average salaries of federal employees from data collected on all federal employees.

9. For each of the following, identify the sample and population scores:
   a. A liquor board official investigates the number of drinks served in bars in a particular city on a Friday during "happy hour." In the city there

are 213 bars. Since there are too many bars to monitor all of them, she selects 20 and records the number of drinks served in them. The following are the data:

50	82	47	65
40	76	61	72
35	43	65	76
63	66	83	82
57	72	71	58

b. To make a profit from a restaurant that specializes in low-cost ¼-pound hamburgers, it is necessary that each hamburger served weigh very close to 0.25 pound. Accordingly, the manager of the restaurant is interested in the variability among the weights of the hamburgers served each day. On a particular day, there are 450 hamburgers served. Since it would take too much time to weigh all 450, the manager decides instead to weigh just 15. The following weights in pounds were obtained:

0.25	0.27	0.25
0.26	0.35	0.27
0.22	0.32	0.38
0.29	0.22	0.28
0.27	0.40	0.31

c. A machine that cuts steel blanks (used for making bolts) to their proper length is suspected of being unreliable. The shop foreman decides to check the output of the machine. On the day of checking, the machine is set to produce 2-centimeter blanks. The acceptable tolerance is ±0.05 centimeter. Since it would take too much time to measure all 600 blanks produced in 1 day, a representative group of 25 is selected. The following lengths in centimeters were obtained:

2.01	1.99	2.05	1.94	2.05
2.01	2.02	2.04	1.93	1.95
2.03	1.97	2.00	1.98	1.96
2.05	1.96	2.00	2.01	1.99
1.98	1.95	1.97	2.04	2.02

PART TWO

DESCRIPTIVE STATISTICS

2

Basic Mathematical and Measurement Concepts

Study Hints for the Student

Statistics is not an easy subject. It requires learning difficult concepts as well as doing mathematics. There is, however, some advice that I would like to pass on, which I believe will help you greatly in learning this material. This advice is based on many years of teaching the subject; I hope you will take it seriously.

Most students in the behavioral sciences have a great deal of anxiety about taking a course on mathematics or statistics. Without minimizing the difficulty of the subject, a good deal of this anxiety is unnecessary. To learn the material contained in this textbook, you do not have to be a whiz in calculus or differential equations. I have tried hard to present the material so that nonmathematically inclined students can understand it. I cannot, however, totally do away with mathematics. To be successful, you must be able to do elementary algebra and a few other mathematical operations. To help you review, I have included Appendix A, which covers prerequisite mathematics. You should study that material and be sure you can do the problems it contains. If you have difficulty with these problems, it will help to review the topic in a basic textbook on elementary algebra.

Another factor of which you should be aware is that a lot of symbols are used in statistics. For example, to designate the mean of a sample set of scores, we shall use the symbol $\bar{X}$ (read "X bar"). Students often make the material more difficult than necessary by failing to thoroughly learn what the symbols stand for. You can save yourself much grief by taking the symbols seriously. Treat them as though they were foreign vocabulary. Memorize them and be able to deal with them conceptually; e.g., if the text says $\bar{X}$, the concept "the mean of the sample" should immediately come into your mind.

It is also important to realize that the material in statistics is cumulative. Do not let yourself fall behind. If you do, you will not understand the current material either. The situation can then snowball, and before you know it, you may seem hopelessly behind. Remember, do all you can to keep up with the material.

Finally, our experience indicates that a good deal of the understanding of statistics comes from working lots of problems. Very often, one problem is worth a thousand words. Frequently, although the text is clearly worded, the material won't come into focus until you have worked the problems associated with the topic. Therefore, do lots of problems, and afterward reread the textual material to be sure you understand it.

In sum, I believe that if you can handle elementary algebra, work diligently on learning the symbols and studying the text, keep up with the material, and work lots of problems, you will do quite well. Believe it or not, as you begin to experience the elegance and fun that are inherent in statistics, you may even come to enjoy it.

MATHEMATICAL NOTATION

In statistics, we usually deal with group data that result from measuring one or more variables. The data are most often derived from samples, occasionally from populations. For mathematical purposes, it is useful to let symbols stand for the variables measured in the study. Throughout this text, we shall use the Roman capital letter X, and sometimes Y, to stand for the variable(s) measured. Thus, if we were measuring the age of subjects, we would let X stand for the variable "age." When there are many values of the variable, it is important to distinguish among them. We do this by subscripting the symbol X. This process is illustrated in the following table, which contains the ages of six subjects:

Subject Number	Score Symbol	Score Value, Age (yr)
1	X_1	8
2	X_2	10
3	X_3	7
4	X_4	6
5	X_5	10
6	X_6	12

In this example, we are letting the variable "age" be represented by the symbol X. We shall also let N represent the number of scores in the distribution. In this example, $N = 6$. Each of the six scores represents a specific value of X. We distinguish among the six scores by assigning a subscript to X that corresponds to the number of the subject that had the specific value. Thus, the score symbol X_1 corresponds to the score value 8, X_2 to the score value 10, X_3 to the value 7, X_4 to 6, X_5 to 10, and X_6 to 12. In general, we can refer to a single score in the X distribution as X_i, where i can take on any value from 1 to N, depending on which score we wish to designate. To summarize,

- X or Y stands for the variable measured.
- N stands for the total number of subjects or scores.
- X_i is the ith score, where i can vary from 1 to N.

SUMMATION*

One of the most frequent operations performed in statistics is to sum all or part of the scores in the distribution. Since it is awkward to write out "sum of all the scores" each time this operation is required, particularly in equations, a symbolic abbreviation is used instead. The capital Greek letter sigma (Σ) is used to indicate the operation of summation. The algebraic phrase employed for summation is

$$\sum_{i=1}^{N} X_i$$

This is read as "sum of the X variable from $i = 1$ to N." The notations above and below the summation sign designate which scores to include in the summation. The term below the summation sign tells us the first score in the summation, and the term above the summation sign designates the last score. This phrase, then, indicates that we are to add the X scores, beginning with the first score and ending with the Nth score. Thus,

$$\sum_{i=1}^{N} X_i = X_1 + X_2 + X_3 + \cdots + X_N \qquad \textit{summation equation}$$

Applied to the age data of the previous table,

$$\sum_{i=1}^{N} X_i = X_1 + X_2 + X_3 + X_4 + X_5 + X_6$$

$$= 8 + 10 + 7 + 6 + 10 + 12 = 53$$

When the summation is over all the scores (from 1 to N), the summation phrase itself is often abbreviated by omitting the notations above and below the summation sign and by omitting the subscript i. Thus,

$$\sum_{i=1}^{N} X_i \text{ is often written as } \Sigma X.$$

In the previous example,

$$\Sigma X = 53$$

This says that the sum of all the X scores is 53.

Note that it is not necessary for the summation to be from 1 to N. For example, we might desire to sum only the second, third, fourth, and fifth scores. Remember, the notation below the summation sign tells us where to begin the summation, and the term above the summation sign tells us where to stop the summation.

---

* After finishing this section, see Note 2.1 at the end of Chapter 2 for additional summation rules, if desired.

Thus, to indicate the operation of summing the second, third, fourth, and fifth scores, we would use the symbol $\sum_{i=2}^{5} X_i$. For the preceding age data,

$$\sum_{i=2}^{5} X_i = X_2 + X_3 + X_4 + X_5 = 10 + 7 + 6 + 10 = 33$$

Let's do some practice problems in summation.

PRACTICE PROBLEM 2.1

a. For the following scores, find $\sum_{i=1}^{N} X_i$:

X: 6, 8, 13, 15 $\Sigma X = 6 + 8 + 13 + 15 = 42$

X: 4, −10, −2, 20, 25, 8 $\Sigma X = 4 - 10 - 2 + 20 + 25 + 8 = 45$

X: 1.2, 3.5, 0.8, 4.5, 6.1 $\Sigma X = 1.2 + 3.5 + 0.8 + 4.5 + 6.1 = 16.1$

b. For the following scores, find $\sum_{i=1}^{3} X_i$:

$$X_1 = 10, X_2 = 12, X_3 = 13, X_4 = 18$$

$$\sum_{i=1}^{3} X_i = 10 + 12 + 13 = 35$$

c. For the following scores, find $\sum_{i=2}^{4} X_i + 3$:

$$X_1 = 20, X_2 = 24, X_3 = 25, X_4 = 28, X_5 = 30, X_6 = 31$$

$$\sum_{i=2}^{4} X_i + 3 = (24 + 25 + 28) + 3 = 80$$

d. For the following scores, find $\sum_{i=2}^{4} (X_i + 3)$:

$$X_1 = 20, X_2 = 24, X_3 = 25, X_4 = 28, X_5 = 30, X_6 = 31$$

$$\sum_{i=2}^{4} (X_i + 3) = (24 + 3) + (25 + 3) + (28 + 3) = 86$$

There are two more summations that we shall frequently encounter later in the textbook. They are ΣX^2 and $(\Sigma X)^2$. Although they look alike, they are different and generally will yield different answers. The symbol ΣX^2

(sum of the squared X scores) indicates that we are first to square the X scores and then sum them. Thus,

$$\Sigma X^2 = X_1^2 + X_2^2 + X_3^2 + \cdots + X_N^2$$

Given the scores $X_1 = 3$, $X_2 = 5$, $X_3 = 8$, and $X_4 = 9$,

$$\Sigma X^2 = 3^2 + 5^2 + 8^2 + 9^2 = 179$$

The symbol $(\Sigma X)^2$ (or sum of the X scores, quantity squared) indicates that we are first to sum the X scores and then square the resulting sum. Thus,

$$(\Sigma X)^2 = (X_1 + X_2 + X_3 + \cdots + X_N)^2$$

For the previous scores, namely, $X_1 = 3$, $X_2 = 5$, $X_3 = 8$, and $X_4 = 9$,

$$(\Sigma X)^2 = (3 + 5 + 8 + 9)^2 = (25)^2 = 625$$

Note that $\Sigma X^2 \neq (\Sigma X)^2$ ($179 \neq 625$). Confusing ΣX^2 and $(\Sigma X)^2$ is a common error made by students, particularly when calculating the standard deviation. We shall return to this point in Chapter 4.

MEASUREMENT SCALES

Since statistics deals with data and data are the result of measurement, we need to spend some time discussing measuring scales. This subject is particularly important because the type of measuring scale employed in collecting the data helps determine which statistical inference test is used to analyze the data. Theoretically, a measuring scale can have one or more of the following mathematical attributes: magnitude, an equal interval between adjacent units, and an absolute zero point. There are four types of scales commonly encountered in the behavioral sciences: *nominal, ordinal, interval,* and *ratio.* They differ in the number of mathematical attributes that they possess.

Nominal Scales

A *nominal scale* is the lowest level of measurement and is most often used with variables that are qualitative in nature, rather than quantitative. Examples of qualitative variables are brands of jogging shoes, kinds of fruit, types of music, days of the month, nationality, religious preference, and eye color. When using a nominal scale, the variable is divided into its several categories. These categories comprise the "units" of the scale, and objects are "measured" by determining the category to which they belong. Thus, measurement with a nominal scale really amounts to classifying the objects and giving them the name (hence *nominal scale*) of the category to which they belong.

To illustrate: If you are a jogger, you are probably interested in the different brands of jogging shoes available for use; e.g., Brooks, Nike, Adidas, Saucony, and New Balance, to name a few. Jogging shoes are important because, in jogging, each shoe hits the ground about 800 times a mile. In a 5-mile run, that's 4000 times. If you weigh 125 pounds, you have a total impact of 300 tons on each foot during each 5-mile jog. That's quite a pounding. No wonder joggers are extremely careful about selecting shoes.

The variable "brand of jogging shoes" is a qualitative variable. It is measured on a nominal scale. The different brands mentioned here represent some of the possible categories (units) of this scale. If we had a group of jogging shoes and wished to measure them using this scale, we would take each one and determine to which brand it belonged. It is important to note that since the units of a nominal scale are categories, there is no magnitude relationship between the units of a nominal scale. Thus, there is no quantitative relationship between the categories of Nike and Brooks. The Nike is no more or less a brand of jogging shoe than is the Brooks. They are just different brands. The point becomes even clearer if we were to call the categories jogging shoes 1 and jogging shoes 2, instead of Nike and Brooks. Here the numbers 1 and 2 are really just names and bear no magnitude relationship to each other.

A fundamental property of nominal scales is that of *equivalence*. By this we mean that all members of a given class are the same from the standpoint of the classification variable. Thus, all pairs of Nike jogging shoes are considered the same from the standpoint of "brand of jogging shoes"—this despite the fact that there may be different types of Nike jogging shoes present.

An operation often performed in conjunction with nominal measurement is that of counting the instances within each class. For example, if we had a bunch of jogging shoes and we determined the brand of each shoe, we would be doing nominal measurement. In addition, we might want to count the number of shoes in each category. Thus, we might find that there are 20 Nike, 19 Saucony, and 6 New Balance shoes in the bunch. These frequencies allow us to compare the number of shoes within each category. This quantitative comparison of numbers within each category should not be confused with the statement made earlier that there is no magnitude relationship between the units of a nominal scale. We can compare quantitatively the numbers of Nike with the numbers of Saucony shoes, but Nike is no more or less a brand of jogging shoe than is Saucony. Thus, a nominal scale does not possess any of the mathematical attributes of magnitude, equal interval, or absolute zero point. It merely allows categorization of objects into mutually exclusive categories.

Ordinal Scales

An *ordinal scale* represents the next higher level of measurement. It possesses a relatively low level of the property of magnitude. With an ordinal scale, we rank-order the objects being measured according to whether they possess more, less, or the same amount of the variable being measured. Thus, an ordinal scale allows determination of whether $A > B$, $A = B$, or $A < B$.

An example of an ordinal scale is the rank ordering of the top five contestants in a speech contest according to speaking ability. Among these speakers, the individual ranked 1 was judged a better speaker than the individual ranked 2, who in turn was judged better than the individual ranked 3. The individual ranked 3 was judged a better speaker than the individual ranked 4, who in turn was judged better than the individual ranked 5. It is important to note that although this scale allows better than, equal to, or less than comparisons, it says nothing about the magnitude of the difference between adjacent units on the scale. In the present example, the difference in speaking ability between the individuals ranked 1 and 2 might be large, and the difference between individuals ranked 2 and 3 might be small. Thus, an ordinal scale does not have the property of equal intervals between adjacent units. Further, since all we have is relative rankings, the scale doesn't tell the absolute level of the variable. Thus, all five of the top-ranked speakers could have a very high level of speaking ability or a low level. This information can't be obtained from an ordinal scale. Other examples of ordinal scaling are the ranking of runners in the Boston Marathon according to their finishing order, the rank ordering of college football teams according to merit by the Associated Press, the rank ordering of teachers according to teaching ability, and the rank ordering of students according to motivation level.

Interval Scales

The *interval scale* represents a higher level of measurement than the ordinal scale. It possesses the properties of magnitude and equal interval between adjacent units but doesn't have an absolute zero point. Thus, the interval scale possesses the properties of the ordinal scale and has equal intervals between adjacent units. Equal intervals between adjacent units means that there are equal amounts of the variable being measured between adjacent units on the scale.

The Celsius scale of temperature measurement is a good example of the interval scale. It has the property of equal intervals between adjacent units but does not have an absolute zero point. The property of equal intervals is shown by the fact that a given change of heat will cause the same change in temperature reading on the scale no matter where on the scale the change occurs. Thus, the additional amount of heat that will cause a temperature reading to change from 2° to 3° Celsius will also cause a reading to change from 51° to 52° or from 105° to 106° Celsius. This illustrates the fact that equal amounts of heat are indicated between adjacent units throughout the scale.

Since with an interval scale there are equal amounts of the variable between adjacent units on the scale, equal differences between the numbers on the scale represent equal differences in the magnitude of the variable. Thus, we can say the difference in heat is the same between 78° and 75° Celsius as between 24° and 21° Celsius. It also follows logically that greater differences between the numbers on the scale represent greater differences in the magnitude of the variable being measured, and smaller differences between the numbers on the scale represent smaller differences in the magnitude of the variable being measured. Thus, the difference in heat between 80° and 65° Celsius is greater than between 18° and 15° Celsius, and the difference in heat between 93° and 91° Celsius is less than between 48° and 40° Celsius. In light of the preceding discussion, we can see that in addition to being able to determine whether $A = B$, $A > B$, or $A < B$, an interval scale allows us to determine whether $A - B = C - D$, $A - B > C - D$, or $A - B < C - D$.

Ratio Scales

The next, and highest, level of measurement is called a *ratio scale*. It has all the properties of an interval scale and, in addition, has an absolute zero point. Without an absolute zero point, it is not legitimate to compute ratios with the scale readings. Since the ratio scale has an absolute zero point, ratios are permissible (hence the name *ratio* scale).

A good example to illustrate the difference between interval and ratio scales is to compare the Celsius scale of temperature with the Kelvin scale. Zero on the Kelvin scale is absolute zero (the complete absence of heat). Zero on the Celsius scale is the temperature at which water freezes. It is an arbitrary zero point that actually occurs at 273° Kelvin. The Celsius scale is an interval scale, and the Kelvin scale is a ratio scale. The difference in heat between 8° and 9° is the same as between 99° and 100° whether the scale is Celsius or Kelvin. However, we cannot compute ratios with the Celsius scale. A reading of 20° Celsius is not twice as hot as 10° Celsius. This can be seen by converting the Celsius readings to the actual heat they represent. In terms of actual heat, 20° Celsius is really 293° (273° + 20°), and 10° Celsius is really 283° (273° + 10°). It is obvious that 293° is not twice 283°. Since

FIGURE 2.1   Scales of measurement and their characteristics

Nominal
Units of the scale are categories. Objects are measured by determining the category to which they belong. There is no magnitude relationship between the categories.

Nike   New Balance   Saucony

Ordinal
Possesses the property of magnitude. Can rank-order the objects according to whether they possess more, less, or the same amount of the variable being measured. Thus, can determine whether $A > B$, $A = B$, or $A < B$. Cannot determine how much greater or less A is than B in the attribute being measured.

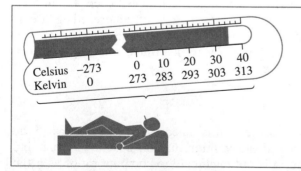

Celsius	−273	0	10	20	30	40
Kelvin	0	273	283	293	303	313

Interval and Ratio
Interval: Possesses the properties of magnitude and equal intervals between adjacent units. Can do same determinations as with an ordinal scale, plus can determine whether $A - B = C - D$, $A - B > C - D$, or $A - B < C - D$.
Ratio: Possesses the properties of magnitude, equal interval between adjacent units, and an absolute zero point. Can do all the mathematical operations usually associated with numbers, including ratios.

the Kelvin scale has an absolute zero, a reading of 20° on it is twice as hot as 10°. Thus, ratios are permissible with the Kelvin scale. Other examples of variables measured with ratio scales include reaction time, length, weight, age, and frequency of any event, such as the number of Nike shoes contained in the bunch of jogging shoes discussed previously. With a ratio scale, you can construct ratios and perform all the other mathematical operations usually associated with numbers—e.g., addition, subtraction, multiplication, and division. The four scales of measurement and their characteristics are summarized in Figure 2.1.

MEASURING SCALES IN THE BEHAVIORAL SCIENCES

In the behavioral sciences, many of the scales used are frequently treated as though they were of interval scaling without clearly establishing that the scale really does possess equal intervals between adjacent units. Measurement of IQ; emotional variables such as anxiety and depression; personality variables like self-sufficiency, introversion, extroversion, and dominance; end-of-course proficiency or achievement variables; attitudinal variables; and so forth fall into this category. With all of these variables, it is clear that the scales are not ratio. For example, with IQ, if an individual actually scored a zero on the Wechsler Adult Intelligence Scale (WAIS), we would not say that he had zero intelligence. Presumably, some questions could be found that he could answer which would indicate an IQ over zero. Thus, the WAIS does not have an absolute zero point, and ratios are not appropriate. Hence, it is not correct to say that a person with an IQ of 160 is twice as smart as someone with an IQ of 80. On the other hand, it seems that we can do more than just specify a rank ordering of individuals. An individual with an IQ of 100 is closer in intelligence to someone with an IQ of 110 than to someone with an IQ of 60. This appears to be interval scaling, but it is difficult to establish that the scale actually possesses *equal* intervals between adjacent units. Many researchers treat those variables as though they were measured on interval scales, particularly when the measuring instrument is well standardized, as is the WAIS. It is more controversial to treat poorly standardized scales measuring psychological variables as interval scales. This issue arises particularly in inferential statistics, where the level of scaling can influence the selection of the test to be used for data analysis. There are two schools of thought. The first claims that certain tests such as Student's *t* test and the analysis of variance should be limited in use to data that are interval or ratio in scaling. The second disagrees, claiming that these tests can also be used with nominal and ordinal data. The issue, however, is too complex to be treated here.*

---

* The interested reader should consult N. H. Anderson, "Scales and Statistics: Parametric and Nonparametric," *Psychological Bulletin, 58* (1961), 305–316; F. M. Lord, "On the Statistical Treatment of Football Numbers," *American Psychologist, 8* (1953), 750–751; W. L. Hays, *Statistics for the Social Sciences,* 2nd ed., Holt, Rinehart and Winston, New York, 1973, pp. 87–90; S. Siegel, *Nonparametric Statistics for the Behavioral Sciences,* McGraw-Hill, New York, 1956, pp. 18–20; and S. S. Stevens, "Mathematics, Measurement, and Psychophysics," in *Handbook of Experimental Psychology,* S. S. Stevens, ed., Wiley, New York, 1951, pp. 23–30.

CONTINUOUS AND DISCRETE VARIABLES

In Chapter 1, we defined a variable as a property or characteristic of something that can take on more than one value. We also distinguished between independent and dependent variables. In addition, variables may be continuous or discrete:

DEFINITIONS

- *A **continuous variable** is one that theoretically can have an infinite number of values between adjacent units on the scale.*

- *A **discrete variable** is one in which there are no possible values between adjacent units on the scale.*

Weight, height, and time are examples of continuous variables. With each of these variables, there are potentially an infinite number of values between adjacent units. If we are measuring time and the smallest unit on the clock that we are using is 1 second, between 1 and 2 seconds there are an infinite number of possible values: 1.1 seconds, 1.01 seconds, 1.001 seconds, etc. The same argument could be made for weight and height. This is not the case with a discrete variable. "Number of children in a family" is an example of a discrete variable. Here the smallest unit is one child, and there are no possible values beween one or two children, two or three children, etc. The characteristic of a discrete variable is that the variable changes in fixed amounts, with no intermediate values possible. Other examples include "number of students in your class," "number of professors at your university," and "number of dates you had last month."

Real Limits of a Continuous Variable

Since a continuous variable may have an infinite number of values between adjacent units on the scale, all measurements made on a continuous variable are *approximate.* Let's use weight to illustrate this point. Suppose you began dieting yesterday. Assume it is spring, heading into summer, and bathing suit weather is just around the corner. Anyway, you weighed yourself yesterday morning, and your weight was shown by the solid needle in Figure 2.2. Assume that the scale shown in the figure has accuracy only to the nearest pound. The weight you record is 180 pounds. This morning, after a day of starvation, when you weighed yourself the pointer was shown by the dashed needle. What weight do you report this time? We know as a humanist that you would like to record 179 pounds, but as a budding scientist, it is *truth* at all costs. You again record 180 pounds. When would you be justified in reporting 179 pounds? When the needle is below the halfway point between 179 and 180 pounds. Similarly, you would still record 180 pounds if the needle was above 180 but below the halfway point between 180 and 181 pounds. Thus, anytime the weight 180 pounds is recorded, we don't necessarily mean exactly 180 pounds, but rather that the weight was between 180 ± 0.5 pounds. We don't know the exact value of the weight, but we are sure it is in the range 179.5–180.5. This range specifies the *real limits* of the weight 180 pounds. The value 179.5 is called the *lower real limit,* and 180.5 is the *upper real limit.*

FIGURE 2.2

Real limits of a continuous
variable

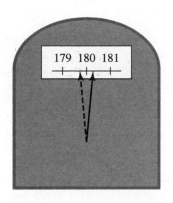

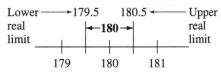

DEFINITION

- *The **real limits of a continuous variable** are those values that are above and below the recorded value by one-half of the smallest measuring unit of the scale.*

To illustrate, if the variable is weight, the smallest unit is 1 pound, and we record 180 pounds, the real limits are above and below 180 pounds by $\frac{1}{2}$ pound. Thus, the real limits are 179.5 and 180.5 pounds.* If the smallest unit were 0.1 pound rather than 1 pound and we recorded 180.0 pounds, then the real limits would be $180 \pm \frac{1}{2}(0.1)$, or 179.95 and 180.05.

Significant Figures

In statistics, we analyze data, and data analysis involves lots of mathematical calculations. Very often, we wind up with a decimal remainder, e.g., after doing a division. When this happens, we need to decide to how many decimal places we should carry the remainder.

In the physical sciences, we usually follow the practice of carrying the same number of significant figures as are in the raw data. For example, if we measured the weights of five subjects to three significant figures (173, 156, 162, 165, and 175 pounds) and we were calculating the average of these weights, our answer should also contain only three significant figures. Thus,

$$\overline{X} = \frac{\Sigma X}{N} = \frac{173 + 156 + 162 + 165 + 175}{5} = \frac{831}{5} = 166.2 = 166$$

---

* Actually, the real limits are 179.500000 . . . and 180.499999 . . . , but it is not necessary to be that accurate here.

The answer of 166.2 would be rounded to three significant figures, giving a final answer of 166 pounds. For various reasons, this procedure has not been followed in the behavioral sciences. Instead, a tradition has evolved in which most final values are reported to two or three decimal places, regardless of the number of significant figures in the raw data. Since this is a text for use in the behavioral sciences, we have chosen to follow this tradition. Thus, in this text, we shall report most of our final answers to two decimal places. Occasionally there will be exceptions. For example, correlation and regression coefficients have three decimal places, and probability values are often given to four places, as is consistent with tradition. *It is standard practice to carry all intermediate calculations to two or more decimal places than will be reported in the final answer.* Thus, when the final answer is required to have two decimal places, you should carry intermediate calculations to at least four decimal places and round the final answer to two places.

Rounding

Given that we shall be reporting our final answers to two and sometimes three or four decimal places, we need to decide how we determine what value the last digit should have. Happily, the rules to be followed are rather simple and straightforward, as follows:

1. Divide the number you wish to round into two parts: the potential answer and the remainder.
2. Place a decimal point in front of the first digit of the remainder, creating a decimal remainder.
3. If the decimal remainder is greater than $\frac{1}{2}$, add 1 to the last digit of the answer.
4. If the decimal remainder is less than $\frac{1}{2}$, leave the last digit of the answer unchanged.
5. If the decimal remainder is equal to $\frac{1}{2}$, add 1 to the last digit of the answer if it is an odd digit but, if it is even, leave it unchanged.

Let's try a few examples. Round the numbers in the left-hand column in the following table to two decimal places.

Number	Answer;Remainder	Decimal Remainder	Final Answer	Reason
34.01350	34.01;350	.350	34.01	Decimal remainder is below $\frac{1}{2}$.
34.01761	34.01;761	.761	34.02	Decimal remainder is above $\frac{1}{2}$.
45.04500	45.04;500	.500	45.04	Decimal remainder equals $\frac{1}{2}$, and last digit is even.
45.05500	45.05;500	.500	45.06	Decimal remainder equals $\frac{1}{2}$, and last digit is odd.

To accomplish the rounding, the number is divided into two parts: the potential answer and the remainder. Since we are rounding to two decimal places, the potential answer ends at the second decimal place. The rest of the number constitutes the remainder. For the first number, 34.01350, 34.01 constitutes the potential answer and .350 the remainder. Since .350 is below $\frac{1}{2}$, the last digit of the potential answer remains unchanged, and the final answer is 34.01. For the second number, 34.01761, the decimal remainder (.761) is above $\frac{1}{2}$. Therefore, we must add 1 to the last digit, making the correct answer 34.02. For the next two numbers, the decimal remainder equals $\frac{1}{2}$. The number 45.04500 becomes 45.04 because the last digit in the potential answer is even. The number 45.05500 becomes 45.06 because the last digit is odd.

SUMMARY

In this chapter, we have discussed basic mathematical and measurement concepts. The topics covered were notation, summation, measuring scales, discrete and continuous variables, and rounding. In addition, we pointed out that to do well in statistics, you do not need to be a mathematical giant. If you have a sound knowledge of elementary algebra, do lots of problems, pay special attention to the symbols, and keep up, you should achieve a thorough understanding of the material.

IMPORTANT TERMS

Continuous variable (p. 28)
Discrete variable (p. 28)
Interval scale (p. 25)

Nominal scale (p. 23)
Ordinal scale (p. 24)
Ratio scale (p. 25)

Real limits of a variable (p. 29)
Summation (p. 21)

QUESTIONS AND PROBLEMS

1. Define and give an example of each of the terms in the "Important Terms" section.
2. Identify which of the following represent continuous variables and which discrete variables:
   a. Time of day
   b. Number of women in your class
   c. Number of bar presses
   d. Age of subjects in an experiment
   e. Number of words remembered
   f. Weight of food eaten
   g. Percentage of students in your class who are women
   h. Speed of runners in a race
3. Identify the scaling of each of the following variables:

a. Number of bicycles ridden by students in the freshman class
b. Types of bicycles ridden by students in the freshman class
c. The IQ of your teachers (assume equal interval scaling)
d. Proficiency in mathematics graded in the categories of poor, fair, and good
e. Anxiety over public speaking is scored on a scale of 0–100. (Assume the difference in anxiety between adjacent units throughout the scale is not the same.)
f. The weight of a group of dieters
g. The time it takes to react to the sound of a tone
h. Proficiency in mathematics is scored on a scale

of 0–100. The scale is well standardized and can be thought of as having equal intervals between adjacent units.

i. Ratings of professors by students on a 50-point scale. There is insufficient basis for assuming equal intervals between adjacent units.

4. A student is measuring assertiveness with an interval scale. Is it correct to say that a score of 30 on the scale represents half as much assertiveness as a score of 60? Explain.

5. For each of the following sets of scores, find $\sum_{i=1}^{N} X_i$:

a. 2, 4, 5, 7
b. 2.1, 3.2, 3.6, 5.0, 7.2
c. 11, 14, 18, 22, 25, 28, 30
d. 110, 112, 115, 120 133

6. Round the following numbers to two decimal places:

a. 14.53670   b. 25.26231
c. 37.83500   d. 46.50499
e. 52.46500   f. 25.48501

7. Determine the real limits of the following values:

a. 10 pounds (assume the smallest unit of measurement is 1 pound)
b. 2.5 seconds (assume the smallest unit of measurement is 0.1 second)
c. 100 grams (assume the smallest unit of measurement is 10 grams)
d. 2.01 centimeters (assume the smallest unit of measurement is 0.01 centimeter)
e. 5.232 seconds (assume the smallest unit of measurement is 1 millisecond)

8. Find the values of the expressions listed here:

a. Find $\sum_{i=1}^{4} X_i$ for the scores $X_1 = 3$, $X_2 = 5$, $X_3 = 7$, $X_4 = 10$.

b. Find $\sum_{i=1}^{6} X_i$ for the scores $X_1 = 2$, $X_2 = 3$, $X_3 = 4$, $X_4 = 6$, $X_5 = 9$, $X_6 = 11$, $X_7 = 14$.

c. Find $\sum_{i=2}^{N} X_i$ for the scores $X_1 = 10$, $X_2 = 12$, $X_3 = 13$, $X_4 = 15$, $X_5 = 18$.

d. Find $\sum_{i=3}^{N-1} X_i$ for the scores $X_1 = 22$, $X_2 = 24$, $X_3 = 28$, $X_4 = 35$, $X_5 = 38$, $X_6 = 40$.

9. In an experiment measuring the reaction times of eight subjects, the following scores in milliseconds were obtained:

Subject	Reaction Time
1	250
2	378
3	451
4	275
5	225
6	430
7	325
8	334

a. If X represents the variable of reaction time, assign each of the scores its appropriate X_i symbol.
b. Compute ΣX for these data.

10. Represent each of the following with summation notation. Assume the number of scores is 10.

a. $X_1 + X_2 + X_3 + X_4 + \cdots + X_{10}$
b. $X_1 + X_2 + X_3$
c. $X_2 + X_3 + X_4$
d. $X_2^2 + X_3^2 + X_4^2 + X_5^2$

11. Round the following numbers to one decimal place:

a. 1.423
b. 23.250
c. 100.750
d. 41.652
e. 35.348

12. For each of the sets of scores given in Problems 5b and 5c, show that $\Sigma X^2 \neq (\Sigma X)^2$.

13. Given the scores $X_1 = 3$, $X_2 = 4$, $X_3 = 7$, and $X_4 = 12$, find the values of the following expressions. (This question pertains to Note 2.1.)

a. $\sum_{i=1}^{N} (X_i + 2)$ b. $\sum_{i=1}^{N} (X_i - 3)$

c. $\sum_{i=1}^{N} (2X_i)$ d. $\sum_{i=1}^{N} (X_i/4)$

NOTES

2.1 Many textbooks present a discussion of additional summation rules, such as the summation of a variable plus a constant, summation of a variable times a constant, and so forth. Since this knowledge is not necessary for understanding any of the material in this textbook, we have not included it in the main body but have presented the material here. Knowledge of summation rules may come in handy as background for statistics courses taught at the graduate level.

Rule 1

The sum of the values of a variable plus a constant is equal to the sum of the values of the variable plus N times the constant. In equation form,

$$\sum_{i=1}^{N} (X_i + a) = \sum_{i=1}^{N} X_i + Na$$

The validity of this equation can be seen from the following simple algebraic proof:

$$\sum_{i=1}^{N} (X_i + a) = (X_1 + a) + (X_2 + a) + (X_3 + a)$$
$$+ \cdots + (X_N + a)$$
$$= (X_1 + X_2 + X_3 + \cdots + X_N)$$
$$+ (a + a + a + \cdots + a)$$
$$= \sum_{i=1}^{N} X_i + Na$$

To illustrate the use of this equation, suppose we wish to find the sum of the following scores with a constant of 3 added to each score:

$$X: \quad 4, 6, 8, 9$$
$$\sum_{i=1}^{N} (X_i + 3) = \sum_{i=1}^{N} X_i + Na = 27 + 4(3) = 39$$

Rule 2

The sum of the values of a variable minus a constant is equal to the sum of the values of the variable minus N times the constant. In equation form,

$$\sum_{i=1}^{N} (X_i - a) = \sum_{i=1}^{N} X_i - Na$$

The algebraic proof of this equation is as follows:

$$\sum_{i=1}^{N} (X_i - a) = (X_1 - a) + (X_2 - a) + (X_3 - a)$$
$$+ \cdots + (X_N - a)$$
$$= (X_1 + X_2 + X_3 + \cdots + X_N)$$
$$+ (-a - a - a - \cdots - a)$$
$$= \sum_{i=1}^{N} X_i - Na$$

To illustrate the use of this equation, suppose we wish to find the sum of the following scores with a constant of 2 subtracted from each score:

$$X: \quad 3, 5, 6, 10$$
$$\sum_{i=1}^{N} (X_i - 2) = \sum_{i-1}^{N} X_i - Na = 24 - 4(2) = 16$$

Rule 3

The sum of a constant times the value of a variable is equal to the constant times the sum of the values of the variable. In equation form,

$$\sum_{i=1}^{N} aX_i = a \sum_{i=1}^{N} X_i$$

The validity of this equation is shown here:

$$\sum_{i=1}^{N} aX_i = aX_1 + aX_2 + aX_3 + \cdots + aX_N$$
$$= a(X_1 + X_2 + X_3 + \cdots + X_N)$$
$$= a \sum_{i=1}^{N} X_i$$

To illustrate the use of this equation, suppose we wish to determine the sum of 4 times each of the following scores:

$$X: \quad 2, 5, 7, 8, 12$$
$$\sum_{i=1}^{N} 4X_i = 4 \sum_{i=1}^{N} X_i = 4(34) = 136$$

Rule 4

The sum of a constant divided into the values of a variable is equal to the constant divided into the sum of the values of the variable. In equation form,

$$\sum_{i=1}^{N} \frac{X_i}{a} = \frac{\sum_{i=1}^{N} X_i}{a}$$

The validity of this equation is shown here:

$$\sum_{i=1}^{N} \frac{X_i}{a} = \frac{X_1}{a} + \frac{X_2}{a} + \frac{X_3}{a} + \cdots + \frac{X_N}{a}$$

$$= \frac{X_1 + X_2 + X_3 + \cdots + X_N}{a}$$

$$= \frac{\sum_{i=1}^{N} X_i}{a}$$

Again, let's do an example to illustrate the use of this equation. Suppose we want to find the sum of 4 divided into the following scores:

$$X: \quad 3, 4, 7, 10, 11$$

$$\sum_{i=1}^{N} \frac{X_i}{4} = \frac{\sum_{i=1}^{N} X_i}{4} = \frac{35}{4} = 8.75$$

3 | FREQUENCY DISTRIBUTIONS

INTRODUCTION: UNGROUPED FREQUENCY DISTRIBUTIONS

Let's suppose you have just been handed back your first exam in statistics. You received an 86. Naturally, you are interested in how well you did relative to the other students. You have lots of questions: How many other students received an 86? Were there many scores higher than yours? How many scores were lower? The raw scores from the exam are presented haphazardly in Table 3.1. Although all the scores are shown, it is difficult to make much sense out of them the way they are arranged in the table. A more efficient arrangement, and one that conveys more meaning, would be to list the scores with their frequency of occurrence. This listing is called a *frequency distribution.*

DEFINITION

• *A* **frequency distribution** *presents the score values and their frequency of occurrence.* When presented in a table, the score values are listed in rank order, with the lowest score value usually at the bottom of the table.

The scores in Table 3.1 have been arranged into a frequency distribution and presented in Table 3.2. The data now are more meaningful. First, it is easy to see that there are 2 scores of 86. Further, by summing the appropriate frequencies (f), we can determine the number of scores higher and lower than 86. It turns out that there were 15 scores higher and 53 scores lower than your score. It is also quite easy to determine the range of the scores when they are displayed as a frequency distribution. For the statistics test, the scores ranged from 46 to 99. From this illustration, it can be seen that the major purpose of a frequency distribution is to present the scores in such a way as to facilitate ease of understanding and interpreting the scores.

TABLE 3.1   Scores from Statistics Exam ($N = 70$)

95	57	76	93	86	80	89
76	76	63	74	94	96	77
65	79	60	56	72	82	70
67	79	71	77	52	76	68
72	88	84	70	83	93	76
82	96	87	69	89	77	81
87	65	77	72	56	78	78
58	54	82	82	66	73	79
86	81	63	46	62	99	93
82	92	75	76	90	74	67

TABLE 3.2   Scores from Table 3.1 Organized into a Frequency Distribution

Score	f	Score	f	Score	f	Score	f
99	1	85	0	71	1	57	1
98	0	84	1	70	2	56	2
97	0	83	1	69	1	55	0
96	2	82	5	68	1	54	1
95	1	81	2	67	2	53	0
94	1	80	1	66	1	52	1
93	3	79	3	65	2	51	0
92	1	78	2	64	0	50	0
91	0	77	4	63	2	49	0
90	1	76	6	62	1	48	0
89	2	75	1	61	0	47	0
88	1	74	2	60	1	46	1
87	2	73	1	59	0		
86	2	72	3	58	1		

GROUPING SCORES

When there are many scores and the scores range widely, as they do on the statistics exam we have been considering, listing individual scores results in many values with a frequency of zero and a display from which it is difficult to visualize the shape of the distribution and its central tendency. Under these conditions, the individual scores are usually grouped into class intervals and presented as a *frequency distribution of grouped scores*. Table 3.3 shows the statistics exam scores grouped into two frequency distributions, one with each interval being 2 units wide and the other having intervals 19 units wide.

When grouping data, one of the important issues is how wide each interval should be. Whenever data are grouped, some information is lost. The wider the interval, the more information is lost. For example, consider the distribution shown in Table 3.3 with intervals 19 units wide. Although this large an interval does result in a smooth display (there are no zero frequencies), a lot of information has been lost. For instance, how are the 38 scores distributed in the interval

TABLE 3.3   Scores from Table 3.1 Grouped into Class Intervals of Different Widths

Class Interval (width = 2)	f	Class Interval (width = 19)	f
98–99	1	95–113	4
96–97	2	76–94	38
94–95	2	57–75	23
92–93	4	38–56	5
90–91	1		$N = \overline{70}$
88–89	3		
86–87	4		
84–85	1		
82–83	6		
80–81	3		
78–79	5		
76–77	10		
74–75	3		
72–73	4		
70–71	3		
68–69	2		
66–67	3		
64–65	2		
62–63	3		
60–61	1		
58–59	1		
56–57	3		
54–55	1		
52–53	1		
50–51	0		
48–49	0		
46–47	1		
	$N = \overline{70}$		

from 76 to 94? Do they fall at 94? Or at 76? Or are they evenly distributed throughout the interval? The point is that we do not know how they are distributed in the interval. We have lost that information by the grouping. Note that the larger the interval, the more the ambiguity.

It should be obvious that the more narrow the interval, the more faithfully the original data are preserved. (The extreme case is where the interval is 1 unit wide and we are back to the individual scores.) Unfortunately, when the interval is made too narrow, we encounter the same problems as with individual scores—namely, values with zero frequency and an unclear display of the shape of the distribution and its central tendency. The frequency distribution with intervals 2 units wide, shown in Table 3.3, is an example where the intervals are too narrow.

From the preceding discussion we can see that in grouping scores there is a trade-off between losing information and presenting a meaningful visual display. To have the best of both worlds, we must choose an interval width neither too wide nor too narrow. In practice, we usually determine interval width by dividing the distribution into 10 to 20 intervals. Over the years, this range of intervals has been shown to work well with most distributions. Within this range, the specific number of intervals used depends on the number and range of the raw scores. Note that the more intervals employed, the narrower each interval becomes.

Constructing a Frequency Distribution of Grouped Scores

The steps for constructing a frequency distribution of grouped scores are as follows:

1. Find the range of the scores.
2. Determine the width of each class interval (i).
3. List the limits of each class interval, placing the interval containing the lowest score value at the bottom.
4. Tally the raw scores into the appropriate class intervals.
5. Add the tallies for each interval to obtain the interval frequency.

Let's apply these steps to the data of Table 3.1.

1. *Finding the range.*

$$\text{Range} = \text{Highest score minus lowest score} = 99 - 46 = 53$$

2. *Determining interval width (i).*   Let's assume we wish to group the data into approximately 10 class intervals.

$$i = \frac{\text{Range}}{\text{Number of class intervals}} = \frac{53}{10} = 5.3 \quad (\textit{round to 5})$$

When i has a decimal remainder, we'll follow the rule of rounding i to the same number of decimal places as in the raw scores. Thus, i rounds to 5.

3. *Listing the intervals.*   We begin with the lowest interval. The first step is to determine the lower limit of this interval. There are two requirements:
   a. The lower limit of this interval must be such that the interval contains the lowest score.
   b. It is customary to make the lower limit of this interval evenly divisible by i.

   Given these two requirements, the lower limit is assigned the value of the lowest score in the distribution, if it is evenly divisible by i. If not, then the lower limit is assigned the next lower value that is evenly divisible by i. In the present example, the lower limit of the lowest interval begins with 45, because the lowest score (46) is not evenly divisible by 5.

   Once the lower limit of the lowest interval has been found, we can list all of the intervals. Since each interval is 5 units wide, the lowest interval ranges from 45 to 49. Although it may seem as though this interval is only 4 units wide, it really is 5. If in doubt, count the units (45, 46, 47, 48, 49). In listing the other intervals, we must be sure that the intervals are continuous and mutually exclusive. By mutually exclusive, we mean that the intervals must be such that no score can be legitimately included in more than one interval. Following these rules, we wind up with the intervals shown in Table 3.4. Note that, consistent with our discussion of real limits in Chapter 2, the class intervals shown in the first column represent apparent limits. The real limits are shown in the second column. The usual practice is to list just the apparent limits of each interval and omit listing the real limits. We have followed this practice in the remaining examples.

4. *Tallying the scores.* Next, the raw scores are tallied into the appropriate class intervals. Tallying is a procedure whereby one systematically goes through the distribution and for each raw score enters a tally mark next to the interval that contains the score. Thus, for 95 (the first score in Table 3.1), a tally mark is placed in the interval 95–99. This procedure has been followed for all the scores, and the results are shown in Table 3.4.

5. *Summing into frequencies.* Finally, the tally marks are converted into frequencies by adding the tallies within each interval. These frequencies are also shown in Table 3.4.

TABLE 3.4   Construction of Frequency Distribution for Grouped Scores

Class Interval	Real Limits	Tally	f
95–99	94.5–99.5	IIII	4
90–94	89.5–94.5	IN I	6
85–89	84.5–89.5	IN II	7
80–84	79.5–84.5	IN IN	10
75–79	74.5–79.5	IN IN IN I	16
70–74	69.5–74.5	IN IIII	9
65–69	64.5–69.5	IN II	7
60–64	59.5–64.5	IIII	4
55–59	54.5–59.5	IIII	4
50–54	49.5–54.5	II	2
45–49	44.5–49.5	I	1
			$N = \overline{70}$

PRACTICE PROBLEM 3.1

Let's try a practice problem. Given these 90 scores, construct a frequency distribution of grouped scores having approximately 12 intervals.

112	68	55	33	72	80	35	55	62
102	65	104	51	100	74	45	60	58
92	44	122	73	65	78	49	61	65
83	76	95	55	50	82	51	138	73
83	72	89	37	63	95	109	93	65
75	24	60	43	130	107	72	86	71
128	90	48	22	67	76	57	86	114
33	54	64	82	47	81	28	79	85
42	62	86	94	52	106	30	117	98
58	32	68	77	28	69	46	53	38

SOLUTION

The solution is shown in Table 3.5.

TABLE 3.5

1. *Find the range.*   Range = highest score − lowest score = 138 − 22 = 116.
2. *Determine the interval width (i):*

$$i = \frac{Range}{Number\ of\ intervals} = \frac{116}{12} = 9.7 \qquad i\ rounds\ to\ 10$$

3. *List the limits of each class interval.*   Since the lowest score in the distribution (22) is not evenly divisible by *i*, the lower limit of the lowest interval is 20. Why 20? Because it is the next lowest scale value evenly divisible by 10. The limits of each class interval have been listed here.
4. *Tally the raw scores into the appropriate class intervals.*   This has been done in the following table.
5. *Add the tallies for each interval to obtain the interval frequency.*   This has been done in the table.

Class Interval	Tally	f	Class Interval	Tally	f
130–139	//	2	70–79	/// /// ///	13
120–129	//	2	60–69	/// /// ///	15
110–119	///	3	50–59	/// /// //	12
100–109	/// /	6	40–49	/// ///	8
90–99	/// //	7	30–39	/// //	7
80–89	/// /// /	11	20–29	////	4
					N = 90

PRACTICE PROBLEM 3.2

Given the 130 scores shown here, construct a frequency distribution of grouped scores having approximately 15 intervals.

1.4	2.9	3.1	3.2	2.8	3.2	3.8	1.9	2.5	4.7
1.8	3.5	2.7	2.9	3.4	1.9	3.2	2.4	1.5	1.6
2.5	3.5	1.8	2.2	4.2	2.4	4.0	1.3	3.9	2.7
2.5	3.1	3.1	4.6	3.4	2.6	4.4	1.7	4.0	3.3
1.9	0.6	1.7	5.0	4.0	1.0	1.5	2.8	3.7	4.2
2.8	1.3	3.6	2.2	3.5	3.5	3.1	3.2	3.5	2.7
3.8	2.9	3.4	0.9	0.8	1.8	2.6	3.7	1.6	4.8
3.5	1.9	2.2	2.8	3.8	3.7	1.8	1.1	2.5	1.4
3.7	3.5	4.0	1.9	3.3	2.2	4.6	2.5	2.1	3.4
1.7	4.6	3.1	2.1	4.2	4.2	1.2	4.7	4.3	3.7
1.6	2.8	2.8	2.8	3.5	3.7	2.9	3.5	1.0	4.1
3.0	3.1	2.7	2.2	3.1	1.4	3.0	4.4	3.3	2.9
3.2	0.8	3.2	3.2	2.9	2.6	2.2	3.6	4.4	2.2

SOLUTION

The solution is shown in Table 3.6.

TABLE 3.6

1. *Find the range.*   Range = highest score − lowest score = 5.0 − 0.6 = 4.4.
2. *Determine the interval width (i):*

$$i = \frac{\text{Range}}{\text{Number of intervals}} = \frac{4.4}{15} = 0.29 \qquad i \; rounds \; to \; 0.3$$

3. *List the limits of each class interval.*   Since the lowest score in the distribution (0.6) is evenly divisible by *i*, it becomes the lower limit of the lowest interval. The limits of each class interval are listed in the accompanying table.
4. *Tally the raw scores into the appropriate class intervals.*   This has been done here.
5. *Add the tallies for each interval to obtain the interval frequency.*   This has been done below. Note that since the smallest unit of measurement in the raw scores is 0.1, the real limits for any score are ±0.05 away from the score. Thus, the real limits for the interval 4.8–5.0 are 4.75–5.05.

Class Interval	Tally	f	Class Interval	Tally	f
4.8–5.0	//	2	2.4–2.6	//// ////	10
4.5–4.7	////	5	2.1–2.3	//// ////	9
4.2–4.4	//// ///	8	1.8–2.0	//// ////	9
3.9–4.1	//// /	6	1.5–1.7	//// ///	8
3.6–3.8	//// //// /	11	1.2–1.4	//// /	6
3.3–3.5	//// //// //// /	16	0.9–1.1	////	4
3.0–3.2	//// //// //// /	16	0.6–0.8	///	3
2.7–2.9	//// //// //// //	17		N =	130

Relative Frequency, Cumulative Frequency, and Cumulative Percentage Distributions

It is often desirable to express the data from a frequency distribution as a relative frequency, a cumulative frequency, or a cumulative percentage distribution.

DEFINITIONS

• *A relative frequency distribution indicates the proportion of the total number of scores that occurred in each interval.*

DEFINITIONS

- *A cumulative frequency distribution* indicates the number of scores that fell below the upper real limit of each interval.

- *A cumulative percentage distribution* indicates the percentage of scores that fell below the upper real limit of each interval.

Table 3.7 shows the frequency distribution of statistics exam scores expressed as relative frequency, cumulative frequency, and cumulative percentage distributions. To convert a frequency distribution into a relative frequency distribution, the frequency for each interval is divided by the total number of scores. Thus,

$$\text{Relative } f = \frac{f}{N}$$

For example, the relative frequency for the interval 45–49 is found by dividing its frequency (1) by the total number of scores (70). Thus, the relative frequency for this interval $= \frac{1}{70} = 0.01$. The relative frequency is useful because it tells us the proportion of scores contained in the interval.

The cumulative frequency for each interval is found by adding the frequency of that interval to the frequencies of all the class intervals below it. Thus, the cumulative frequency for the interval $60 - 64 = 4 + 4 + 2 + 1 = 11$.

The cumulative percentage for each interval is found by converting cumulative frequencies to cumulative percentages. The equation for doing this is:

$$\text{cum } \% = \frac{\text{cum } f}{N} \times 100$$

TABLE 3.7   Relative Frequency, Cumulative Frequency, and Cumulative Percentage Distributions for the Grouped Scores in Table 3.4

Class Interval	f	Relative f	Cumulative f	Cumulative %
95–99	4	0.06	70	100
90–94	6	0.09	66	94.29
85–89	7	0.10	60	85.71
80–84	10	0.14	53	75.71
75–79	16	0.23	43	61.43
70–74	9	0.13	27	38.57
65–69	7	0.10	18	25.71
60–64	4	0.06	11	15.71
55–59	4	0.06	7	10.00
50–54	2	0.03	3	4.29
45–49	1	0.01	1	1.43
	70	1.00		

For the interval 60–64,

$$\text{cum \%} = \frac{\text{cum } f}{N} \times 100 = \frac{11}{70} \times 100 = 15.71\%$$

Cumulative frequency and cumulative percentage distributions are especially useful for finding percentiles and percentile ranks.

 ## PERCENTILES

Percentiles are measures of relative standing. They are used extensively in education to compare the performance of an individual to that of a reference group.

DEFINITION

> • A **percentile** *or* **percentile point** *is the value on the measurement scale below which a specified percentage of the scores in the distribution fall.*

Thus, the 60th percentile point is the value on the measurement scale below which 60% of the scores in the distribution fall.

Computation of Percentile Points

Let's assume we are interested in computing the 50th percentile point for the statistics exam scores. The scores have been presented in Table 3.8 as cumulative frequency and cumulative percentage distributions. We shall use the symbol P_{50} to stand for the 50th percentile point. What do we mean by the 50th percentile point? From the definition of percentile point, P_{50} is the scale value below which 50% of the scores fall. Since there are 70 scores in the distribution, P_{50} must be

TABLE 3.8   Computation of Percentile Points for the Scores of Table 3.1

Class Interval	f	Cum f	Cum %	Percentile Computation
95–99	4	70	100	Percentile point $= X_L + (i/f_i)(\text{cum } f_P - \text{cum } f_L)$
90–94	6	66	94.29	
85–89	7	60	85.71	
80–84	10	53	75.71	
75–79	16	43	61.43	$P_{50} = 74.5 + (\frac{5}{16})(35 - 27) = 77.00$
70–74	9	27	38.57	
65–69	7	18	25.71	$P_{20} = 64.5 + (\frac{5}{7})(14 - 11) = 66.64$
60–64	4	11	15.71	
55–59	4	7	10.00	
50–54	2	3	4.29	
45–49	1	1	1.43	

the value below which 35 scores fall (50% of 70 = 35). Looking at the cumulative frequency column and moving up from the bottom, we see that P_{50} falls in the interval 75–79. At this stage, however, we do not know what scale value to assign P_{50}. All we know is that it falls somewhere between the real limits of the interval 75–79, which are 74.5 to 79.5. To find where in the interval P_{50} falls, we make the assumption that all the scores in the interval are equally distributed throughout the interval.

Since 27 of the scores fall below a value of 74.5, we need to move into the interval until we acquire 8 more scores (see Figure 3.1). Since there are 16 scores in the interval and the interval is 5 scale units wide, each score in the interval takes up $\frac{5}{16}$ of a unit. To acquire 8 more scores, we need to move into the interval $\frac{5}{16} \times 8 = 2.5$ units. Adding 2.5 to the lower limit of 74.5, we arrive at P_{50}. Thus,

$$P_{50} = 74.5 + 2.5 = 77.0$$

To find any percentile point follow these steps:

1. *Determine the frequency of scores below the percentile point.* We will symbolize this frequency as "cum f_P."

$$\text{cum } f_P = (\% \text{ of scores below the percentile point}) \times N$$

$$\text{cum } f_P \text{ for } P_{50} = (50\%) \times N = (0.50) \times 70 = 35$$

2. *Determine the lower real limit of the interval containing the percentile point.* We will call the real lower limit X_L. Knowing the number of scores below the percentile point, we can locate the interval containing the percentile

FIGURE 3.1

....

Determining the scale value of P_{50} for the statistics exam scores

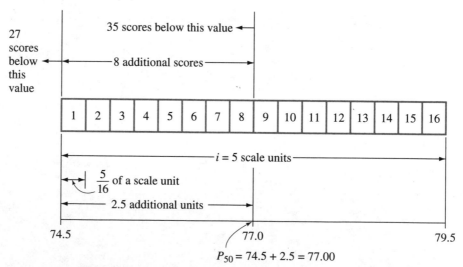

From *Statistical Reasoning in Psychology and Education* by E. W. Minium. Copyright © 1978 John Wiley & Sons, Inc. Adapted by permission.

point by comparing cum f_P with the cumulative frequency for each interval. Once the interval containing the percentile point is located, we can immediately ascertain its lower real limit, X_L. For this example, the interval containing P_{50} is 75–79 and its real lower limit, $X_L = 74.5$.

3. *Determine the number of additional scores we must acquire in the interval to reach the percentile point.*

$$\text{Number of additional scores} = \text{cum } f_P - \text{cum } f_L$$

where   cum f_L = frequency of scores below the lower real limit of the interval containing the percentile point.

For the above example,

$$\text{Number of additional scores} = \text{cum } f_P - \text{cum } f_L$$
$$= 35 - 27$$
$$= 8$$

4. *Determine the number of additional units into the interval we must go to acquire the additional number of scores.*

$$\text{Additional units} = (\text{Number of units per score}) \times \text{Number of additional scores}$$
$$= (i/f_i) \times \text{Number of additional scores}$$
$$= \left(\tfrac{5}{16}\right) \times 8$$
$$= 2.5$$

Note that,

f_i *is the number of scores in the interval and*
i/f_i *gives us the number of units per score for the interval*

5. *Determine percentile point.* This is accomplished by adding the additional units to the lower real limit of the interval containing the percentile point.

$$\text{Percentile point} = X_L + \text{Additional units}$$
$$P_{50} = 74.5 + 2.5 = 77.00$$

These steps can be put into equation form. Thus,

$$\text{Percentile point} = X_L + (i/f_i)(\text{cum } f_P - \text{cum } f_L)^*$$

equation for computing percentile point

---

* I am indebted to LeAnn Wilson for suggesting this form of the equation.

where X_L = value of the lower real limit of the interval containing the percentile point

cum f_P = frequency of scores below the percentile point

cum f_L = frequency of scores below the lower real limit of the interval containing the percentile point

f_i = frequency of the interval containing the percentile point

i = width of the interval

Using this equation to calculate P_{50}, we obtain

$$\text{Percentile point} = X_L + (i/f_i)(\text{cum } f_P - \text{cum } f_L)$$

$$P_{50} = 74.5 + (\tfrac{5}{16})(35 - 27)$$

$$= 74.5 + 2.5 = 77.00$$

PRACTICE PROBLEM 3.3

Let's try another problem. This time we'll calculate P_{20}, the value below which 20% of the scores fall. In terms of cumulative frequency, P_{20} is the value below which 14 scores fall (20% of 70 = 14). From Table 3.8 (p. 43), we see that P_{20} lies in the interval 65–69. Since 11 scores fall below a value of 64.5, we need 3 more scores. Given there are 7 scores in the interval and the interval is 5 units wide, we must move $\tfrac{5}{7} \times 3 = 2.14$ units into the interval. Thus,

$$P_{20} = 64.5 + 2.14 = 66.64$$

P_{20} could also have been found directly by using the equation for percentile point. Thus,

$$\text{Percentile point} = X_L + (i/f_i)(\text{cum } f_P - \text{cum } f_L)$$

$$P_{20} = 64.5 + (\tfrac{5}{7})(14 - 11)$$

$$= 64.5 + 2.14 = 66.64$$

PRACTICE PROBLEM 3.4

Let's try one more problem. This time, let's compute P_{75}. P_{75} is the scale value below which 75% of the scores fall. In terms of cumulative frequency, P_{75} is the scale value below which 52.5 scores fall (cum f_P = 75% of 70 = 52.5). From Table 3.8 (p. 43), we see that P_{75} falls in the interval 80–84. Since 43 scores fall below this interval's lower limit of 79.5, we need to add to 79.5 the number of scale units appropriate for 52.5 − 43 = 9.5 more scores. Since there are 10 scores in this interval and the interval is 5 units wide, we need to move into the interval $\frac{5}{10} \times 9.5 = 4.75$ units. Thus,

$$P_{75} = 79.5 + 4.75 = 84.25$$

P_{75} also could have been found directly by using the equation for percentile point. Thus,

$$\text{Percentile point} = X_L + (i/f_i)(\text{cum } f_P - \text{cum } f_L)$$

$$P_{75} = 79.5 + (\tfrac{5}{10})(52.5 - 43)$$

$$= 79.5 + 4.75$$

$$= 84.25$$

PERCENTILE RANK

Sometimes we want to know the percentile rank of a raw score. For example, since your score on the statistics exam was 86, it would be useful to you to know the percentile rank of 86.

DEFINITION

- *The **percentile rank** of a score is the percentage of scores with values lower than the score in question.*

Computation of Percentile Rank

This situation is just the reverse of the one where we were calculating percentile point. Now, we are given the score and must calculate the percentage of scores below it. Again, we must assume the scores within any interval are evenly distributed throughout the interval. From the class interval column of Table 3.9, we see that the score of 86 falls in the interval 85–89. There are 53 scores below 84.5, the lower limit of this interval. Since there are 7 scores in the interval, and the interval is 5 scale units wide, there are $\frac{7}{5}$ scores per scale unit. Between a score of 86 and 84.5, there are $(\frac{7}{5})(86 - 84.5) = 2.1$ additional scores. There are, therefore, a total of $53 + 2.1 = 55.1$ scores below 86. Since there are 70 scores in the distribution, the percentile rank of 86 = $(\frac{55.1}{70}) \times 100 = 78.71$.

These operations are summarized in the following equation:

$$\text{Percentile rank} = \frac{\text{cum } f_L + (f_i/i)(X - X_L)}{N} \times 100 \qquad \textit{equation for computing percentile rank}$$

where cum f_L = frequency of scores below the lower real limit of the interval containing the score X
X = score whose percentile rank is being determined
X_L = scale value of the lower real limit of the interval containing the score X
i = interval width
f_i = frequency of the interval containing the score X
N = total number of raw scores

Using this equation to find the percentile rank of 86, we obtain

$$\text{Percentile rank} = \frac{\text{cum} f_L + (f_i/i)(X - X_L)}{N} \times 100$$

$$= \frac{53 + (\frac{7}{5})(86 - 84.5)}{70} \times 100$$

$$= \frac{53 + 2.1}{70} \times 100$$

$$= \frac{55.1}{70} \times 100$$

$$= 78.71$$

PRACTICE PROBLEM 3.5

Let's do one more problem for practice. Find the percentile rank of 59. The score of 59 falls in the interval 55–59. There are 3 scores below 54.5. Since there are 4 scores within the interval, there are $(\frac{4}{5})(59 - 54.5) = 3.6$ scores within the interval below 59. In all, there are $3 + 3.6 = 6.6$ scores below 59. Thus, the percentile rank of $59 = (\frac{6.6}{70}) \times 100 = 9.43$.

SOLUTION

The solution is presented in equation form in Table 3.9.

TABLE 3.9   Computation of Percentile Rank for the Scores of Table 3.1

Class Interval	f	Cum f	Cum %	Percentile Rank Computation
95–99	4	70	100	$\text{Percentile rank} = \dfrac{\text{cum} f_L + (f_i/i)(X - X_L)}{N} \times 100$
90–94	6	66	94.29	
85–89	7	60	85.71	
80–84	10	53	75.71	
75–79	16	43	61.43	$\text{Percentile rank of } 86 = \dfrac{53 + (\frac{7}{5})(86 - 84.5)}{70} \times 100$
70–74	9	27	38.57	
65–69	7	18	25.71	$= 78.71$
60–64	4	11	15.71	
55–59	4	7	10.00	$\text{Percentile rank of } 59 = \dfrac{3 + (\frac{4}{5})(59 - 54.5)}{70} \times 100$
50–54	2	3	4.29	
45–49	1	1	1.43	$= 9.43$

PRACTICE PROBLEM 3.6

Let's do one more practice problem. Using the frequency distribution of grouped scores shown in Table 3.5 (p. 40), determine the percentile rank of a score of 117. The score of 117 falls in the interval 110–119. The lower limit of this interval is 109.5. There are 6 + 7 + 11 + 13 + 15 + 12 + 8 + 7 + 4 = 83 scores below 109.5. Since there are 3 scores within the interval and the interval is 10 units wide, there are $(\frac{3}{10})(117 - 109.5) =$ 2.25 scores within the interval that are below a score of 117. In all, there are 83 + 2.25 = 85.25 scores below a score of 117. Thus, the percentile rank of 117 = $(\frac{85.25}{90}) \times 100 = 94.72$.

This problem could also have been solved by using the equation for percentile rank. Thus,

$$\text{Percentile Rank} = \frac{\text{cum} f_L + (f_i/i)(X - X_L)}{N} \times 100$$

$$= \frac{83 + (\frac{3}{10})(117 - 109.5)}{90} \times 100$$

$$= 94.72$$

GRAPHING FREQUENCY DISTRIBUTIONS

Frequency distributions are often displayed as graphs rather than tables. Since a graph is based completely on the tabled scores, the graph does not contain any new information. However, a graph presents the data pictorially, which often makes it easier to see important features of the data. We have assumed, in writing this section, that you are already familiar with constructing graphs. Even so, it is worthwhile to review a few of the important points.

1. A graph has two axes: vertical and horizontal. The vertical axis is called the *ordinate*, or *Y* axis, and the horizontal axis, the *abscissa*, or *X* axis.
2. The scores are plotted along the horizontal axis, and some characteristic of the scores is plotted on the vertical axis. In a frequency distribution, the frequency of the scores is plotted on the vertical axis.
3. Suitable units for plotting scores should be chosen along the axes.
4. To avoid distorting the data, it is customary to set the intersection of the two axes at zero and then choose scales for the axes such that the height of the graphed data is about $\frac{3}{4}$ of the width. Figure 3.2 shows how violation of this rule can bias the impression conveyed by the graph. The figure shows two graphs plotted from the same data, namely, enrollment at a large university during the years 1984–1996. Part (a) follows the rule we have just elaborated. In part (b), the scale on the ordinate does not begin

FIGURE 3.2
....

Enrollment at a large
university from 1984 to
1996

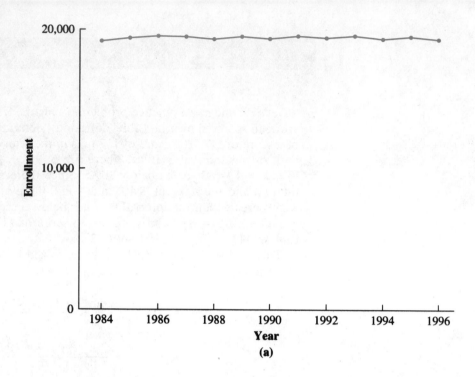

(a)

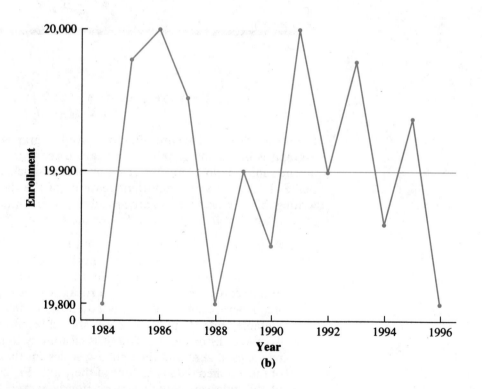

(b)

at zero and is greatly expanded from that of part (a). The impressions conveyed by the two graphs are very different. Part (a) gives the correct impression of a very stable enrollment, whereas part (b) greatly distorts the data, making them seem as though there were large enrollment fluctuations.

5. Ordinarily, the intersection of the two axes is at zero for both scales. When it is not, this is indicated by breaking the relevant axis near the intersection. For example, in Figure 3.4 the horizontal axis is broken to indicate that a part of the scale has been left off.

6. Each axis should be labeled, and the title of the graph should be both short and explicit.

When graphing frequency distributions, four main types of graphs are encountered: the *bar graph,* the *histogram,* the *frequency polygon,* and the *cumulative percentage* curve.

The Bar Graph

Frequency distributions of nominal or ordinal data are customarily plotted using a bar graph. This type of graph is shown in Figure 3.3. A bar is drawn for each category, where the height of the bar represents the frequency or number of members of that category. Since there is no numerical relationship between the categories in nominal data, the various groups can be arranged along the horizontal axis in any order. In Figure 3.3, they are arranged from left to right according to the magnitude of frequency in each category. Note that the bars for each category in a bar graph do not touch each other. This further emphasizes the lack of a quantitative relationship between the categories.

FIGURE 3.3

Bar graph: Number of students enrolled in various undergraduate majors in a college of arts and sciences

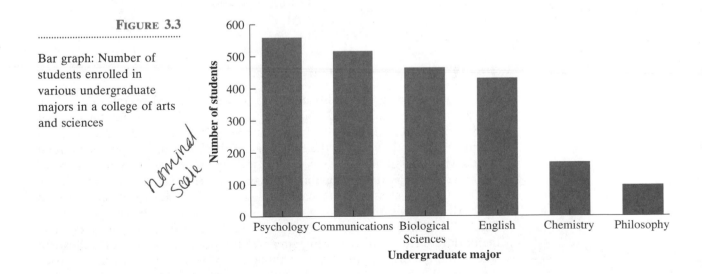

The Histogram

The histogram is used to represent frequency distributions composed of interval or ratio data. It resembles the bar graph, except that with the histogram a bar is drawn for each class interval. The class intervals are plotted on the horizontal axis such that each class bar begins and terminates at the real limits of the interval. The height of the bar corresponds to the frequency of the class interval. Since the intervals are continuous, the vertical bars must touch each other rather than being spaced apart as is done with the bar graph. Figure 3.4 shows the statistics exam scores (Table 3.4, p. 39) displayed as a histogram. Note that it is customary to plot the midpoint of each class interval on the abscissa. The grouped scores have been presented again in the figure for your convenience.

FIGURE 3.4

Histogram: Statistics exam scores of Table 3.4

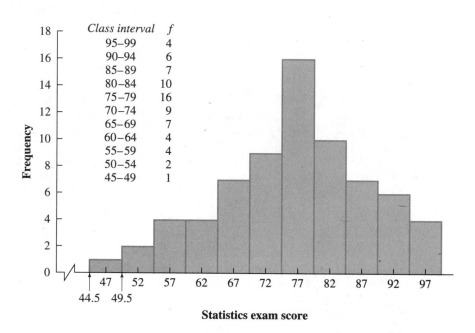

Class interval	f
95–99	4
90–94	6
85–89	7
80–84	10
75–79	16
70–74	9
65–69	7
60–64	4
55–59	4
50–54	2
45–49	1

The Frequency Polygon

The frequency polygon is also used to represent interval or ratio data. The horizontal axis is identical to that of the histogram. However, for this type of graph, instead of using bars, a point is plotted over the midpoint of each interval at a height corresponding to the frequency of the interval. The points are then joined with straight lines. Finally, the line joining the points is extended to meet the horizontal axis at the midpoint of the two class intervals falling immediately beyond the end class intervals containing scores. This closing of the line with the horizontal axis forms a polygon from which the name of this graph is taken. Figure 3.5 displays the scores listed in Table 3.4 as a frequency polygon. The major difference between a histogram and a frequency polygon is the following: The histogram displays the scores as though they were equally distributed over the interval, whereas the frequency polygon displays the scores as though they

FIGURE 3.5

Frequency polygon: Statistics exam scores of Table 3.4

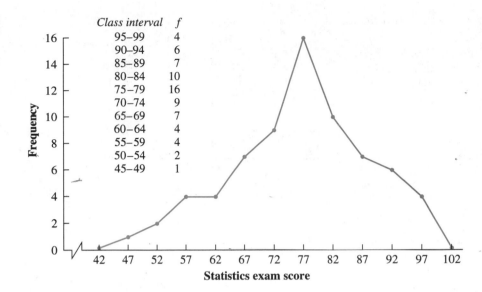

Class interval	f
95–99	4
90–94	6
85–89	7
80–84	10
75–79	16
70–74	9
65–69	7
60–64	4
55–59	4
50–54	2
45–49	1

were all concentrated at the midpoint of the interval. Some investigators prefer to use the frequency polygon when they are comparing the shapes of two or more distributions. The frequency polygon also has the effect of displaying the scores as though they were continuously distributed, which in many instances is actually the case.

The Cumulative Percentage Curve

Cumulative frequency and cumulative percentage distributions may also be presented in graphical form. We shall illustrate only the latter, since the graphs are basically the same, and cumulative percentage distributions are more often encountered. You will recall that the cumulative percentage for a class interval indicates the percentage of scores that fall below the upper real limit of the interval. Thus, the vertical axis for the cumulative percentage curve is plotted in cumulative percentage units. On the horizontal axis, instead of plotting points at the midpoint of each class interval, we plot them at the upper real limit of the interval. Figure 3.6 shows the scores of Table 3.7 (p. 42) displayed as a cumulative percentage curve. It should be obvious that the cumulative frequency curve would have the same shape, the only difference being that the vertical axis would be plotted in cumulative frequency rather than in cumulative percentage units. Both percentiles and percentile ranks can be read directly off the cumulative percentage curve. The cumulative percentage curve is also called an *ogive*, implying an S shape.

Shapes of Frequency Curves

Frequency distributions can take many different shapes. Some of the more commonly encountered shapes are shown in Figure 3.7. Curves are generally classified as *symmetrical* or *skewed*.

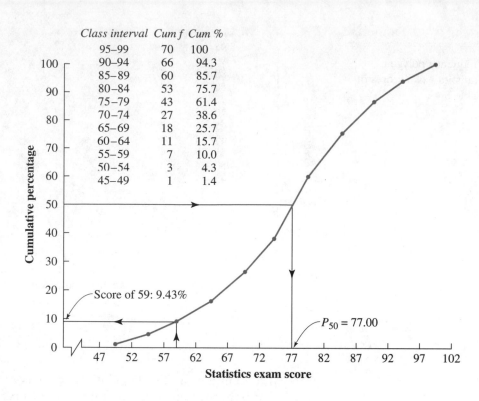

FIGURE 3.6

Cumulative percentage curve: Statistics exam scores of Table 3.7

Class interval	Cum f	Cum %
95–99	70	100
90–94	66	94.3
85–89	60	85.7
80–84	53	75.7
75–79	43	61.4
70–74	27	38.6
65–69	18	25.7
60–64	11	15.7
55–59	7	10.0
50–54	3	4.3
45–49	1	1.4

Score of 59: 9.43%

$P_{50} = 77.00$

FIGURE 3.7   Shapes of frequency curves

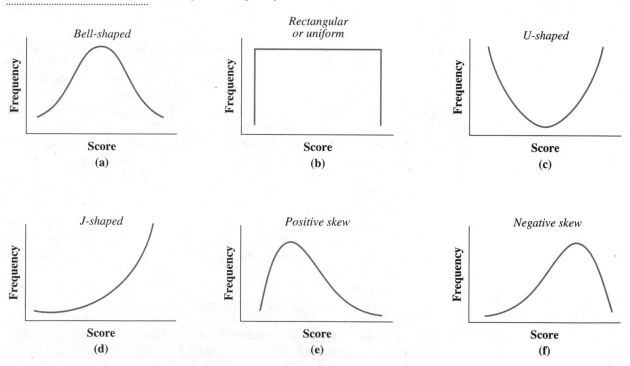

(a) Bell-shaped

(b) Rectangular or uniform

(c) U-shaped

(d) J-shaped

(e) Positive skew

(f) Negative skew

• *A curve is* **symmetrical** *if when folded in half the two sides coincide. If a curve is not symmetrical, it is* **skewed**.

The curves shown in Figure 3.7(a), (b), (c) are symmetrical. The curves shown in parts (d), (e), and (f) are skewed. If a curve is skewed, it may be *positively* or *negatively skewed*.

• *When a curve is* **positively skewed,** *most of the scores occur at the lower values of the horizontal axis, and the curve tails off toward the higher end. When a curve is* **negatively skewed,** *most of the scores occur at the higher values, and the curve tails off toward the lower end of the horizontal axis.*

The curve in part (e) is positively skewed, and the curve in part (f) is negatively skewed.

Frequency curves are often referred to according to their shape. Thus, the curves shown in parts (a), (b), (c), and (d) are, respectively, called bell-shaped, rectangular or uniform, U-shaped, and J-shaped curves.

EXPLORATORY DATA ANALYSIS

Exploratory data analysis is a recently developed procedure. It employs easy-to-construct diagrams that are quite useful in summarizing and describing sample data. One of the most popular of these is the *stem and leaf diagram.*

Stem and Leaf Diagrams

Stem and leaf diagrams were first developed in 1977 by John Tukey, working at Princeton University. They are a simple alternative to the histogram and are most useful for summarizing and describing data when the data set is under 100 scores. Unlike what happens with a histogram, however, a stem and leaf diagram does not lose any of the original data. A stem and leaf diagram for the statistics exam scores of Table 3.1 is shown in Figure 3.8.

In constructing a stem and leaf diagram, each score is represented by a *stem* and a *leaf.* The stem is placed to the left of the vertical line and the leaf to the right. For example, the stems and leafs for the first and last original scores are:

stem	leaf		stem	leaf
9	5		6	7

In a stem and leaf diagram, stems are placed in order, vertically down the page, and the leafs are placed in order, horizontally across the page. The leaf for each score is usually the last digit, and the stem, the remaining digits. Occasionally, the leaf is the last two digits, depending on the range of the scores.

Note that in stem and leaf diagrams, stem values can be repeated. In Figure 3.8, the stem values are repeated twice. This has the effect of stretching the stem,

FIGURE 3.8

Stem and leaf diagram:
Statistics exam scores of
Table 3.1

ORIGINAL SCORES

95	57	76	93	86	80	89
76	76	63	74	94	96	77
65	79	60	56	72	82	70
67	79	71	77	52	76	68
72	88	84	70	83	93	76
82	96	87	69	89	77	81
87	65	77	72	56	78	78
58	54	82	82	66	73	79
86	81	63	46	62	99	93
82	92	75	76	90	74	67

STEM AND LEAF DIAGRAM

```
4 | 6
5 | 2 4
5 | 6 6 7 8
6 | 0 2 3 3
6 | 5 5 6 7 7 8 9
7 | 0 0 1 2 2 2 3 4 4
7 | 5 6 6 6 6 6 6 7 7 7 7 8 8 9 9 9
8 | 0 1 1 2 2 2 2 2 3 4
8 | 6 6 7 7 8 9 9
9 | 0 2 3 3 3 4
9 | 5 6 6 9
```

i.e., creating more intervals and spreading the scores out. A stem and leaf diagram for the statistics scores with stem values listed only once is shown here.

```
4 | 6
5 | 2 4 6 6 7 8
6 | 0 2 3 3 5 5 6 7 7 8 9
7 | 0 0 1 2 2 2 3 4 4 5 6 6 6 6 6 6 7 7 7 7 8 8 9 9 9
8 | 0 1 1 2 2 2 2 2 3 4 6 6 7 7 8 9 9
9 | 0 2 3 3 3 4 5 6 6 9
```

Listing stem values only once results in fewer, wider intervals, with each interval generally containing more scores. This makes the display appear more crowded. Whether stem values should be listed once, twice, or even more than twice, depends on the range of the scores.

You should observe that rotating the stem and leaf diagram of Figure 3.8 counterclockwise 90°, such that the stems are at the bottom, results in a diagram very similar to the histogram shown in Figure 3.4. With the histogram, however, we have lost the original scores; with the stem and leaf diagram, the original scores are preserved.

WHAT IS THE TRUTH

STRETCH THE SCALE, CHANGE THE TALE

An article appeared in the business section of a newspaper discussing the rate increases of Puget Power & Light Company. The company was in poor financial condition and had proposed still another rate increase in 1984 to try to help it get out of trouble. The issue was particularly sensitive, as rate increases had plagued the region recently to pay for huge losses in nuclear power plant construction. The following graph appeared in the article along with the caption, "Puget Power rates have climbed steadily during the past 14 years." Do you notice anything peculiar about the graph?

ANSWER Take a close look at the X axis. From 1970 to 1980, the scale is calibrated in 2-year intervals. After 1980, the same distance on the X axis represents 1 year rather than 2 years. Given the data, stretching this part of the scale gives the false impression that costs have risen "steadily." When plotted properly, as is done here, the graph shows

that rates have not risen steadily, but instead have greatly accelerated over the last 3 years (including the proposed rate increase). Labeling the rise as a "steady" rise, rather than a greatly accelerating increase, obviously is in the company's interest. It is unclear whether the company furnished the graph or whether the newspaper constructed its own. In any case, when the axes of graphs are not uniform, reader beware!

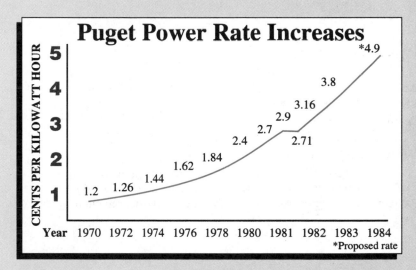

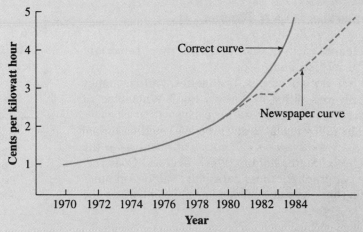

SUMMARY

In this chapter, I have discussed frequency distributions and how to present them in tables and graphs. In descriptive statistics, we are interested in characterizing a set of scores in the most meaningful manner. When faced with a large number of scores, it is easier to understand, interpret, and discuss the scores when they are presented as a frequency distribution. A frequency distribution is a listing of the score values in rank order along with their frequency of occurrence. If there are many scores existing over a wide range, the scores are usually grouped together in equal intervals to allow a more meaningful interpretation. The scores can be presented as an ordinary frequency distribution, a relative frequency distribution, a cumulative frequency distribution, or a cumulative percentage distribution. I discussed each of

these and how to construct them. I also presented the concepts of percentile point and percentile rank and discussed how to compute each.

When graphing frequency distributions, frequency is plotted on the vertical axis and the score value on the horizontal axis. Four main types of graphs are used: the bar graph, the histogram, the frequency polygon, and the cumulative percentage curve. I discussed the use of each type and how to construct them. Frequency curves can also take on various shapes. I illustrated some of the common shapes encountered, e.g., bell-shaped, U-shaped, and J-shaped, and discussed the difference between symmetrical and skewed curves. Finally, I discussed the use of an exploratory data analysis technique, stem and leaf diagrams.

IMPORTANT TERMS

Bar graph (p. 51)
Bell-shaped curve (p. 55)
Cumulative frequency
  distribution (p. 42)
Cumulative percentage
  distribution (p. 42)
Exploratory data analysis (p. 55)
Frequency distribution (p. 35)

Frequency distribution of grouped
  scores (p. 36)
Frequency polygon (p. 52)
Histogram (p. 52)
J-shaped curve (p. 55)
Negatively skewed curve (p. 55)
Percentile point (p. 43)
Percentile rank (p. 47)
Positively skewed curve (p. 55)

Relative frequency distribution
  (p. 41)
Skewed curve (p. 55)
Stem and leaf diagrams (p. 55)
Symmetrical curve (p. 55)
U-shaped curve (p. 55)
X axis (abscissa) (p. 49)
Y axis (ordinate) (p. 49)

QUESTIONS AND PROBLEMS

1. Define each of the terms in the "Important Terms" section.
2. How do bar graphs, histograms, and frequency polygons differ in construction? What type of scaling is appropriate for each?
3. The following table gives the 1993 median annual salaries of various categories of scientists in the United States holding Ph.D. degrees. Construct a bar graph for these data with "Salary Amount" on the Y axis and "Category of Scientist" on the

X axis. Arrange the categories so that the salaries decrease from left to right.

Category of Scientist	Annual Salary ($)
Biological and Health Sciences	55,100
Chemistry	64,100
Computer and Math Sciences	60,000
Psychology	51,700
Sociology and Anthropology	48,100

4. The following scores were obtained by a college sophomore class on an English exam:

60	94	75	82	72	57	92	75	85	77	91
72	85	64	78	75	62	49	70	94	72	84
55	90	88	81	64	91	79	66	68	67	74
45	76	73	68	85	73	83	85	71	87	57
82	78	68	70	71	78	69	98	65	61	83
84	69	77	81	87	79	64	72	55	76	68
93	56	67	71	83	72	82	78	62	82	49
63	73	89	78	81	93	72	76	73	90	76

a. Construct a frequency distribution of the ungrouped scores ($i = 1$).

b. Construct a frequency distribution of grouped scores having approximately 15 intervals. List both the apparent and real limits of each interval.

c. Construct a histogram of the frequency distribution constructed in part **b**.

d. Is the distribution skewed or symmetrical? If it is skewed, is it skewed positively or negatively?

e. Construct a stem and leaf diagram with the last digit being a leaf and the first digit a stem. Repeat stem values twice.

f. Which diagram do you like better, the histogram of part **c** or the stem and leaf diagram of part **e**? Explain.

5. Express the grouped frequency distribution of part **b** of Problem 4 as a relative frequency, a cumulative frequency, and a cumulative percentage distribution.

6. Using the cumulative frequency arrived at in Problem 5, determine

a. P_{75}

b. P_{40}

7. Again, using the cumulative distribution and grouped scores arrived at in Problem 5, determine

a. The percentile rank of a score of 81

b. The percentile rank of a score of 66

c. The percentile rank of a score of 87

8. Construct a histogram of the distribution of grouped English exam scores determined in Problem 4b.

9. The following scores show the amount of weight lost (in pounds) by each client of a weight control clinic during the last year:

10	13	22	26	16	23	35	53	17	32
41	35	24	23	27	16	20	60	48	43
52	31	17	20	33	18	23	8	24	15
26	46	30	19	22	13	22	14	21	39
28	43	37	15	20	11	25	9	15	21
21	25	34	10	23	29	28	18	17	24
16	26	7	12	28	20	36	16	14	
18	16	57	31	34	28	42	19	26	

a. Construct a frequency distribution of grouped scores with approximately 10 intervals.

b. Construct a histogram of the frequency distribution constructed in part **a**.

c. Is the distribution skewed or symmetrical? If it is skewed, is it skewed positively or negatively?

d. Construct a stem and leaf diagram with the last digit being a leaf and the first digit a stem. Repeat stem values twice.

e. Which diagram do you like better, the histogram of part **b** or the stem and leaf diagram of part **d**? Explain.

10. Convert the grouped frequency distribution of weight losses determined in Problem 9 to a relative frequency and a cumulative frequency distribution.

11. Using the cumulative frequency distribution arrived at in Problem 10, determine

a. P_{50}

b. P_{25}

12. Again, using the cumulative frequency distribution of Problem 10, determine

a. The percentile rank of a score of 41

b. The percentile rank of a score of 28

13. Construct a frequency polygon using the grouped frequency distribution determined in Problem 9. Is the curve symmetrical? If not, is it positively or negatively skewed?

14. A small eastern college uses the grading system of 0–4.0, with 4.0 being the highest possible grade. The scores shown on page 60 are the grade point averages of the students currently enrolled as psychology majors at the college.

2.7	1.9	1.0	3.3	1.3	1.8	2.6	3.7
3.1	2.2	3.0	3.4	3.1	2.2	1.9	3.1
3.4	3.0	3.5	3.0	2.4	3.0	3.4	2.4
2.4	3.2	3.3	2.7	3.5	3.2	3.1	3.3
2.1	1.5	2.7	2.4	3.4	3.3	3.0	3.8
1.4	2.6	2.9	2.1	2.6	1.5	2.8	2.3
3.3	3.1	1.6	2.8	2.3	2.8	3.2	2.8
2.8	3.8	1.4	1.9	3.3	2.9	2.0	3.2

a. Construct a frequency distribution of grouped scores with approximately 10 intervals.

b. Construct a histogram of the frequency distribution constructed in part **a**.

c. Is the distribution skewed or symmetrical? If skewed, is it skewed positively or negatively?

d. Construct a stem and leaf diagram with the last digit being a leaf and the first digit a stem. Repeat stem values five times.

e. Which diagram do you like better, the histogram of part **b** or the stem and leaf diagram of part **d**? Explain.

15. For the grouped scores in Problem 14, determine
   a. P_{80}
   b. P_{20}

16. Sarah's grade point average is 3.1. Based on the frequency distribution of grouped scores constructed in Problem 14, part **a**, what is the percentile rank of Sarah's grade point average?

17. The policy of the school in Problem 14 is that to graduate with a major in psychology, a student must have a grade point average of 2.5 or higher.
   a. Based on the ungrouped scores shown in Problem 14, what percentage of current psychology majors needs to raise its grades?
   b. Based on the frequency distribution of grouped scores, what percentage needs to raise its grades?
   c. Explain the difference between the answers to parts **a** and **b**.

18. Construct a frequency polygon using the distribution of grouped scores constructed in Problem 14. Is the curve symmetrical, or positively or negatively skewed?

4 | MEASURES OF CENTRAL TENDENCY AND VARIABILITY

INTRODUCTION

In Chapter 3, we discussed how to organize and present data in meaningful ways. The frequency distribution and its many derivatives are useful in this regard, but, by themselves, they do not allow quantitative statements that characterize the distribution as a whole to be made; nor do they allow quantitative comparisons to be made between two or more distributions. It is often desirable to describe the characteristics of distributions quantitatively. For example, suppose a psychologist has conducted an experiment to determine whether men and women differ in mathematical aptitude. She has two sets of scores, one from the men and one from the women in the experiment. How can she compare the distributions? To do so, she needs to quantify them. The way this is most often done is to compute the average score for each group and then compare the averages. The measure computed is a measure of the *central tendency* of each distribution.

A second characteristic of distributions that is very useful to quantify is the *variability* of the distribution. Variability deals with the extent to which scores are different from each other, are dispersed, or spread out. It is important for two reasons. First, quantifying the variability of the data is required by many of the statistical inference tests that we shall be discussing later in the book. In addition, the variability of a distribution can be useful in its own right. For example, suppose you were hired to design and evaluate an educational program for disadvantaged youngsters. When evaluating the program, you would not only be interested in the average value of the end-of-program scores but also in how variable the scores were. The variability of the scores is important, because you need to know whether the effect of the program is uniform or varies over the youngsters. If it varies, as it almost assuredly will, how large is the variability? Is the program doing a good job with some students and a poor job with others? If so, the program may need to be redesigned to do a better job with those youngsters who have not been adequately benefiting from it.

Central tendency and variability are the two characteristics of distributions that are most often quantified. In this chapter, we shall discuss the most important measures of these two characteristics.

 ## MEASURES OF CENTRAL TENDENCY

The three most often used measures of central tendency are the arithmetic mean, the median, and the mode.

The Arithmetic Mean

You are probably already familiar with the arithmetic mean. It is the value you ordinarily calculate when you average something. For example, if you wanted to know the average number of hours you studied per day for the past 5 days, you would add the hours you studied each day and divide by 5. In so doing, you would be calculating the arithmetic mean.

DEFINITION

• *The **arithmetic mean** is defined as the sum of the scores divided by the number of scores. In equation form,*

$$\overline{X} = \frac{\sum X_i}{N} = \frac{X_1 + X_2 + X_3 + \cdots + X_N}{N} \qquad mean\ of\ a\ sample$$

or

$$\mu = \frac{\sum X_i}{N} = \frac{X_1 + X_2 + X_3 + \cdots + X_N}{N} \qquad \begin{array}{l} mean\ of\ a\ population \\ set\ of\ scores \end{array}$$

where $\quad X_1, \ldots, X_N$ = raw scores
$\overline{X}$ (read "X bar") = mean of a sample set of scores
μ (read "mew") = mean of a population set of scores
Σ (read "sigma") = summation sign
N = number of scores

Note that we use two symbols for the mean, $\overline{X}$ if the scores are sample scores and μ (the Greek letter mu) if the scores are population scores. The computations, however, are the same regardless of whether the scores are sample or population scores. We shall use μ without any subscript to indicate that this is the mean of a population of raw scores. Later on in the text, we shall calculate population means of other kinds of scores for which we shall add the appropriate subscript.

Let's try a few problems for practice.

Properties of the Mean   The mean has many important properties or characteristics. First,

The mean is sensitive to the exact value of all the scores in the distribution.

Since to calculate the mean you have to add *all* the scores, a change in any of the scores will cause a change in the mean. This is not true of the median or the mode.

PRACTICE PROBLEM 4.1

Calculate the mean for each of the following sample sets of scores:

a. X: 3, 5, 6, 8, 14

$$\overline{X} = \frac{\Sigma X_i}{N} = \frac{3+5+6+8+14}{5}$$

$$= \frac{36}{5} = 7.20$$

b. X: 20, 22, 28, 30, 37, 38

$$\overline{X} = \frac{\Sigma X_i}{N} = \frac{20+22+28+30+37+38}{6}$$

$$= \frac{175}{6} = 29.17$$

c. X: 2.2, 2.4, 3.1, 3.1

$$\overline{X} = \frac{\Sigma X_i}{N} = \frac{2.2+2.4+3.1+3.1}{4}$$

$$= \frac{10.8}{4} = 2.70$$

A second property is the following:

The sum of the deviations about the mean equals zero. Written algebraically, this property becomes $\Sigma (X_i - \overline{X}) = 0$.

This property says that if the mean is subtracted from each score, the sum of the differences will equal zero. The algebraic proof is presented in Note 4.1 at the end of this chapter. A demonstration of its validity is shown in Table 4.1. This property results from the fact that the mean is the balance point of the

TABLE 4.1   Demonstration that $\Sigma (X_i - \overline{X}) = 0$

X_i	$X_i - \overline{X}$	Calculation of $\overline{X}$
2	−4	
4	−2	$\overline{X} = \dfrac{\Sigma X_i}{N} = \dfrac{30}{5}$
6	0	
8	+2	$= 6.00$
10	+4	
$\Sigma X_i = 30$	$\Sigma (X_i - \overline{X}) = 0$	

distribution. The mean can be thought of as the fulcrum of a seesaw, to use a mechanical analogy. The analogy is shown in Figure 4.1, using the scores of Table 4.1. When the scores are distributed along the seesaw according to their values, the mean of the distribution occupies the position where the scores are in balance.

FIGURE 4.1

The mean as the balance point in the distribution

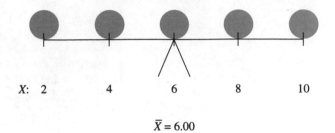

X: 2   4   6   8   10

$$\bar{X} = 6.00$$

A third property of the mean also derives from the fact that the mean is the balance point of the distribution:

The mean is very sensitive to extreme scores.

A glance at Figure 4.1 should convince you that, if we added an extreme score (one far from the mean), it would greatly disrupt the balance. The mean would have to shift a considerable distance to reestablish balance. The mean is more sensitive to extreme scores than is the median or the mode. We shall discuss this more fully when we take up the median.

A fourth property of the mean has to do with the variability of the scores about the mean. This property states the following:

The sum of the squared deviations of all the scores about their mean is a minimum. Stated algebraically, $\Sigma (X_i - \bar{X})^2$ is a minimum.

This is an important characteristic used in many areas of statistics, particularly in regression. Elaborated a little more fully, this property states that although the sum of the squared deviations about the mean does not usually equal zero, it is smaller than if the squared deviations were taken about any other value. The validity of this property is demonstrated in Table 4.2. The scores of

TABLE 4.2   Demonstration that $\Sigma (X_i - \bar{X})^2$ Is a Minimum

X_i	$(X_i - 3)^2$	$(X_i - 4)^2$	$(X_i - \bar{X})^2$ $(X_i - 5)^2$	$(X_i - 6)^2$	$(X_i - 7)^2$	Calculation of $\bar{X}$
2	1	4	9	16	25	$\bar{X} = \dfrac{\Sigma X_i}{N}$
4	1	0	1	4	9	
6	9	4	1	0	1	$= \dfrac{20}{4}$
8	25	16	9	4	1	
	36	24	20	24	36	$= 5.00$

the distribution are given in the first column. Their mean equals 5.00. The fourth column shows the squared deviations of the X_i scores about their mean $(X_i - 5)^2$. The sum of these squared deviations is 20. The other columns show the squared deviations of the X_i scores about values other than the mean. In the third column, the value is $4 (X_i - 4)^2$, the second column 3, the fifth column 6, and the last column 7. Note that the sum of the squared deviations about each of these values is larger than the sum of the squared deviations about the mean of the distribution. Not only is the sum larger, but the farther the value gets from the mean, the larger the sum becomes. This implies that although we've compared only four other values, it holds true for all other values. Thus, although the sum of the squared deviations about the mean does not usually equal zero, it is smaller than if the squared deviations are taken about any other value.

The last property has to do with the use of the mean for statistical inference. This property states the following:

Under most circumstances, of the measures used for central tendency, the mean is least subject to sampling variation.

If we were repeatedly to take samples from a population on a random basis, the mean would vary from sample to sample. The same is true for the median and the mode. However, the mean varies less than these other measures of central tendency. This is very important in inferential statistics and is a major reason why the mean is used in inferential statistics whenever possible.

The Overall Mean

Occasionally the situation arises where we know the mean of several groups of scores and we want to calculate the mean of all the scores combined. Of course, we could start from the beginning again and just sum all the raw scores and divide by the total number of scores. However, there is a shortcut available if we already know the mean of the groups and the number of scores in each group. The equation for this method derives from the basic definition of the mean. Suppose we have several groups of scores that we wish to combine to calculate the overall mean. We'll let k equal the number of groups. Then,

$$\overline{X}_{\text{overall}} = \frac{\text{Sum of all scores}}{N}$$

$$= \frac{\Sigma X_i(\text{first group}) + \Sigma X_i(\text{second group}) + \cdots + \Sigma X_i(\text{last group})}{n_1 + n_2 + \cdots + n_k}$$

where N = total number of scores
n_1 = number of scores in the first group
n_2 = number of scores in the second group
n_k = number of scores in the last group

Since $\overline{X}_1 = \Sigma X_i(\text{first group})/n_1$, multiplying by n_1, we have $\Sigma X_i(\text{first group}) = n_1\overline{X}_1$. Similarly, $\Sigma X_i(\text{second group}) = n_2\overline{X}_2$, and $\Sigma X_i(\text{last group}) = n_k\overline{X}_k$, where $\overline{X}_k$ is the mean of the last group. Substituting these values in the numerator of the above equation, we arrive at

$$\overline{X}_{\text{overall}} = \frac{n_1\overline{X}_1 + n_2\overline{X}_2 + \cdots + n_k\overline{X}_k}{n_1 + n_2 + \cdots + n_k} \qquad \textit{overall mean of several groups}$$

In words, this equation states that the overall mean is equal to the sum of the mean of each group times the number of scores in the group, divided by the sum of the number of scores in each group.

To illustrate how this equation is used, suppose a sociology professor gave a final exam to two classes. The mean of one of the classes was 90, and the number of scores was 20. The mean of the other class was 70, and 40 students took the exam. Calculate the mean of the two classes combined.

The solution is as follows: Given that $\overline{X}_1 = 90$ and $n_1 = 20$ and that $\overline{X}_2 = 70$ and $n_2 = 40$,

$$\overline{X}_{overall} = \frac{n_1\overline{X}_1 + n_2\overline{X}_2}{n_1 + n_2} = \frac{20(90) + 40(70)}{20 + 40} = 76.67$$

The overall mean is much closer to the average of the class with 40 scores than the class with 20 scores. In this context, we can see that each of the means is being *weighted* by its number of scores. We are counting the mean of 70 forty times and the mean of 90 only twenty times. Thus, the overall mean really is a weighted mean, where the weights are the number of scores used in determining each mean. Let's do one more problem for practice.

PRACTICE PROBLEM 4.2

A researcher conducted an experiment involving three groups of subjects. The mean of the first group was 75, and there were 50 subjects in the group. The mean of the second group was 80, and there were 40 subjects. The third group had 25 subjects and a mean of 70. Calculate the overall mean of the three groups combined.

SOLUTION

The solution is as follows: Given that $\overline{X}_1 = 75$, $n_1 = 50$; $\overline{X}_2 = 80$, $n_2 = 40$; and $\overline{X}_3 = 70$, $n_3 = 25$,

$$\overline{X}_{overall} = \frac{n_1\overline{X}_1 + n_2\overline{X}_2 + n_3\overline{X}_3}{n_1 + n_2 + n_3} = \frac{50(75) + 40(80) + 25(70)}{50 + 40 + 25}$$

$$= \frac{8700}{115} = 75.65$$

The Median

The second most frequently encountered measure of central tendency is the median.

DEFINITION

• *The median* (*symbol Mdn*) *is defined as the scale value below which 50%* *of the scores fall.* It is therefore the same thing as P_{50}.

In Chapter 3, we discussed how to calculate P_{50}, and therefore you already know how to calculate the median for grouped scores. For practice, however, Practice Problem 4.3 contains another problem and its solution. You should try this problem and be sure you can solve it before going on.

PRACTICE PROBLEM 4.3

Calculate the median of the grouped scores listed in Table 4.3:

TABLE 4.3   Calculating the Median from Grouped Scores

Class Interval	f	Cum f	Cum %	Calculation of Median
3.6–4.0	4	52	100.00	
3.1–3.5	6	48	92.31	
2.6–3.0	8	42	80.77	
2.1–2.5	10	34	65.38	$\text{Mdn} = P_{50}$
1.6–2.0	9	24	46.15	$= X_L + (i/f_i)(\text{cum } f_P - \text{cum } f_L)$
1.1–1.5	7	15	28.85	$= 2.05 + (0.5/10)(26 - 24)$
0.6–1.0	5	8	15.38	$= 2.05 + 0.10 = 2.15$
0.1–0.5	3	3	5.77	

SOLUTION

The median is the value below which 50% of the scores fall. Since $N = 52$, the median is the value below which 26 of the scores fall (50% of 52 = 26). From Table 4.3, we see that the median lies in the interval 2.1–2.5. Since 24 scores fall below a value of 2.05, we need two more scores to make up the 26. Given there are 10 scores in the interval and the interval is 0.5 unit wide, we must move $0.5/10 \times 2 = 0.10$ unit into the interval. Thus,

$$\text{Median} = 2.05 + 0.10 = 2.15$$

The median could also have been found by using the equation for percentile point. This solution is shown in Table 4.3.

When dealing with raw (ungrouped) scores, it is quite easy to find the median. First, arrange the scores in rank order.

The median is the centermost score if the number of scores is odd. If the number is even, the median is taken as the average of the two centermost scores.

To illustrate, suppose we have the scores 5, 2, 3, 7, and 8 and want to determine their median. First, we rank-order the scores: 2, 3, 5, 7, 8. Since the number of scores is odd, the median is the centermost score. In this example, the median is 5. It may seem that 5 is not really P_{50} for the set of scores. However, consider the score of 5 to be evenly distributed over the interval 4.5–5.5. Now it becomes obvious that half of the scores fall below 5.0. Thus, 5.0 is P_{50}. Let's try another example, this time with an even number of scores. Given the scores 2, 8, 6, 4, 12, and 10, determine their median. First, we rank-order the scores: 2, 4, 6, 8, 10, 12. Since the number of scores is even, the median is the average of the two centermost scores. The median for this example is $(6 + 8)/2 = 7$. For additional practice, Practice Problem 4.4 presents a few problems dealing with raw scores.

PRACTICE PROBLEM 4.4

Calculate the median for the following sets of scores:

a. 8, 10, 4, 3, 1, 15   Rank order: 1, 3, 4, 8, 10, 15   Mdn = (4 + 8)/2 = 6
b. 100, 102, 108, 104, 112   Rank order: 100, 102, 104, 108, 112   Mdn = 104
c. 2.5, 1.8, 1.2, 2.4, 2.0   Rank order: 1.2, 1.8, 2.0, 2.4, 2.5   Mdn = 2.0
d. 10, 11, 14, 14, 16, 14, 12   Rank order: 10, 11, 12, 14, 14, 14, 16   Mdn = 14

In the last set of scores in Practice Problem 4.4, the median occurs at 14, where there are three scores. Technically, we should consider the three scores equally spread out over the interval 13.5–14.5. Then we would find the median by using the equation shown in Table 4.3 (p. 67), with $i = 1$ (Mdn = 13.67). However, when dealing with raw scores, this refinement is often not made. Rather, the median is taken at 14. We shall follow this procedure. Thus, if the median occurs at a value where there are tied scores, we shall use the tied score as the median.

Properties of the Median   There are two properties of the median worth noting. First,

The median is less sensitive than the mean to extreme scores.

To illustrate this property, consider the scores shown in the first column of Table 4.4. The three distributions shown are the same except for the last score. In the

TABLE 4.4   Effect of Extreme Scores on the Mean and Median

Scores	Mean	Median
3, 4, 6, 7, 10	6	6
3, 4, 6, 7, 100	24	6
3, 4, 6, 7, 1000	204	6

second distribution, the score of 100 is very different in value from the other scores. In the third distribution, the score of 1000 is even more extreme. Note what happens to the mean in the second and third distributions. Since the mean is sensitive to extreme scores, it changes considerably with the extreme scores. How about the median? Does it change too? As we see from the third column, the answer is no! The median stays the same. Since the median is not responsive to each individual score, but rather divides the distribution in half, it is not as sensitive to extreme scores as is the mean. For this reason, when the distribution is strongly skewed, it is probably better to represent the central tendency with the median rather than the mean. Certainly in the third distribution of Table 4.4, the median of 6 does a better job representing most of the scores than does the mean of 204.

The second property of the median involves its sampling stability. It states that

Under usual circumstances, the median is more subject to sampling variability than the mean but less subject to sampling variability than the mode.

Because the median is usually less stable than the mean from sample to sample, it is not as useful in inferential statistics.

The Mode

The third and last measure of central tendency that we shall discuss is the mode.

 • *The* **mode** *is defined as the most frequent score in the distribution.**

Clearly, this is the easiest of the three measures to determine. The mode is found by inspection of the scores; there isn't any calculation necessary. For instance, to find the mode of the data in Table 3.2 (p. 36) all we need to do is search the frequency column. The mode for these data is 76. With grouped scores, the mode is designated as the midpoint of the interval with the highest frequency. The mode of the grouped scores in Table 3.4 (p. 39) is 77.

Usually distributions are *unimodal;* that is, they have only one mode. However, it is possible for a distribution to have many modes. When a distribution has two modes, as is the case with the scores 1, 2, 3, 3, 3, 3, 4, 5, 7, 7, 7, 7, 8, 9, the distribution is called *bimodal.* Histograms of a unimodal and bimodal distribution are shown in Figure 4.2. Although the mode is the easiest measure of central tendency to determine, it is not used very much in the behavioral sciences, because it is not very stable from sample to sample and often there is more than one mode for a given set of scores.

Measures of Central Tendency and Symmetry

If the distribution is unimodal and symmetrical, then the mean, median, and mode will all be equal. An example of this is the bell-shaped curve mentioned in Chapter 3 and shown in Figure 4.3. When the distribution is skewed, the mean and median will not be equal. Since the mean is most affected by extreme scores, it will have a value closer to the extreme scores than will the median. Thus, with

---

* When all the scores in the distribution have the same frequency, it is customary to say that the distribution has no mode.

FIGURE 4.2

Unimodal and bimodal
histograms

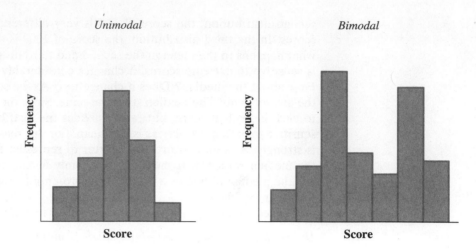

FIGURE 4.3

Symmetry and measures of
central tendency

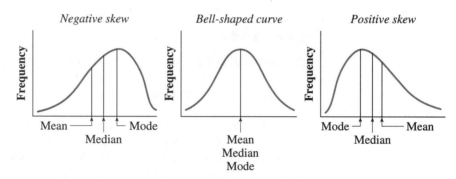

From *Statistical Reasoning in Psychology and Education* by E. W. Minium. Copyright © 1978
John Wiley & Sons, Inc. Adapted by permission.

a negatively skewed distribution, the mean will be lower than the median. With
a positively skewed curve, the mean will be larger than the median. Figure 4.3
shows these relationships.

MEASURES OF VARIABILITY

Previously in this chapter we pointed out that variability has to do with how
far apart the scores are spread. Whereas measures of central tendency are a
quantification of the average value of the distribution, measures of variability
quantify the extent of dispersion. There are three measures of variability commonly used in the behavioral sciences: the range, the standard deviation, and
the variance.

The Range

We have already used the range when we were constructing frequency distributions of grouped scores.

DEFINITION

• *The range is defined as the difference between the highest and lowest scores in the distribution. In equation form,*

$$\text{Range} = \text{Highest score} - \text{Lowest score}$$

The range is easy to calculate but gives us only a relatively crude measure of dispersion, because the range really measures the spread of only the extreme scores and not the spread of any of the scores in between. Although the range is easy to calculate, we've included below some problems for you to practice on. Better to be sure than sorry.

PRACTICE PROBLEM 4.5

Calculate the range for the following distributions:

a. 2, 3, 5, 8, 10   Range = 10 − 2 = 8
b. 18, 12, 28, 15, 20   Range = 28 − 12 = 16
c. 115, 107, 105, 109, 101   Range = 115 − 101 = 14
d. 1.2, 1.3, 1.5, 1.8, 2.3   Range = 2.3 − 1.2 = 1.1

The Standard Deviation

Before discussing the standard deviation, it is necessary to introduce the concept of a deviation score.

Deviation Scores   So far, we've been dealing mainly with raw scores. You will recall that a raw score is the score as originally measured. For example, if we are interested in IQ and we measure an IQ of 126, then 126 is a raw score.

DEFINITION

• *A deviation score tells how far away the raw score is from the mean of its distribution.*

In equation form, a deviation score is defined as

$$X - \overline{X} \qquad \textit{deviation score for sample data}$$
$$X - \mu \qquad \textit{deviation score for population data}$$

As an illustration, consider the sample scores in Table 4.5. The raw scores are shown in the first column and their transformed deviation scores in the second column. The deviation score tells how far the raw score lies above or below the mean. Thus, the raw score of 2 ($X = 2$) lies 4 units below the mean ($X - \overline{X} = -4$). The raw scores and their deviation scores are also shown pictorially in Figure 4.4.

TABLE 4.5   Calculating Deviation Scores

X	$X - \bar{X}$	Calculation of $\bar{X}$
2	$2 - 6 = -4$	$\bar{X} = \dfrac{\sum X}{N} = \dfrac{30}{5}$
4	$4 - 6 = -2$	
6	$6 - 6 = 0$	$= 6.00$
8	$8 - 6 = +2$	
10	$10 - 6 = +4$	

FIGURE 4.4

Raw scores and their corresponding deviation scores

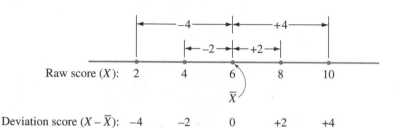

Let's suppose that you are a budding mathematician (use your imagination if necessary). You have been assigned the task of deriving a measure of dispersion that gives the average deviation of the scores about the mean. After some reflection, you say, "That's easy. Just calculate the deviation from the mean of each score and average the deviation scores." Your logic is impeccable. There is only one stumbling block. Consider the scores in Table 4.6. For the sake of this example, we will assume this is a population set of scores. The first column contains the population raw scores, and the second column the deviation scores. We want to calculate the average deviation of the raw scores about their mean. According to your method, we would first compute the deviation scores (second column) and average them by dividing the sum of the deviation scores $[\sum (X - \mu)]$ by N. The stumbling block is that $\sum (X - \mu) = 0$. Remember this

TABLE 4.6   Calculation of the Standard Deviation of a Population Set of Scores Using the Deviation Method

X	$X - \mu$	$(X - \mu)^2$	Calculation of μ and σ
3	-2	4	$\mu = \dfrac{\sum X}{N} = \dfrac{25}{5} = 5.00$
4	-1	1	
5	0	0	
6	$+1$	1	$\sigma = \sqrt{\dfrac{SS_{pop}}{N}} = \sqrt{\dfrac{\sum (X - \mu)^2}{N}} = \sqrt{\dfrac{10}{5}} = 1.41$
7	$+2$	4	
	$\sum (X - \mu) = 0$	$\sum (X - \mu)^2 = 10$	

is a general property of the mean. The sum of the deviations about the mean always equals zero. Thus, if we follow your suggestion, the average of the deviations would always equal zero, no matter how dispersed the scores were $[\Sigma\,(X - \mu)/N = 0/N = 0]$.

You are momentarily stunned by this unexpected low blow. However, you don't give up. You look at the deviation scores and you see that the negative scores are canceling the positive ones. Suddenly, you have a flash of insight. Why not square each deviation score? Then all the scores would be positive, and their sum would no longer be zero. Eureka, you have solved the problem. Now you can divide the sum of the squared scores by N to get the average value $[\Sigma\,(X - \mu)^2/N]$, and the average won't equal zero. You should note that the numerator of this formula $[\Sigma\,(X - \mu)^2]$ is called the *sum of squares* or, more accurately, *sum of squared deviations,* and is symbolized as SS_{pop}. The only trouble at this point is that you have now calculated the average *squared deviation,* not the average deviation. What you need to do is "unsquare" the answer. This is done by taking the square root of SS_{pop}/N.

Your reputation as a mathematician is vindicated! You have come up with the equation for standard deviation used by many statisticians. The symbol for the standard deviation of population scores is σ (the lowercase Greek letter sigma) and for samples is s. Your derived equation for population scores is as follows:

$$\sigma = \sqrt{\frac{SS_{pop}}{N}} = \sqrt{\frac{\Sigma\,(X - \mu)^2}{N}} \qquad \text{\textit{standard deviation of a population set of raw scores—deviation method}}$$

where $\quad SS_{pop} = \Sigma\,(X - \mu)^2 \qquad$ *sum of squares—population data*

Calculation of the standard deviation of a population set of scores using the deviation method is shown in Table 4.6.

Technically, the equation is the same for calculating the standard deviation of sample scores. However, when we calculate the standard deviation of sample data, we usually want to use our calculation to estimate the population standard deviation. It can be shown algebraically that the equation with N in the denominator gives an estimate, which on the average is too small. Dividing by $N - 1$, instead of N, gives a more accurate estimate of σ. Since estimation of the population standard deviation is an important use of the sample standard deviation, and since it saves confusion later on in the textbook when we cover Student's t test and the F test, we have chosen to adopt the equation with $N - 1$ in the denominator for calculating the standard deviation of sample scores. Thus,

$$s = \text{Estimated } \sigma = \sqrt{\frac{SS}{N-1}} = \sqrt{\frac{\Sigma\,(X - \bar{X})^2}{N-1}} \qquad \text{\textit{standard deviation of a sample set of raw scores—deviation method}}$$

where $\quad SS = \Sigma\,(X - \bar{X})^2 \qquad$ *sum of squares—sample data*

In most practical situations, the data are from samples rather than populations. Calculation of the standard deviation of a sample using the above equation for samples is shown in Table 4.7. Although this equation gives the best conceptual

TABLE 4.7   Calculation of the Standard Deviation of Sample Scores Using the Deviation Method

X	$X - \bar{X}$	$(X - \bar{X})^2$	Calculation of $\bar{X}$ and s
2	−4	16	$\bar{X} = \dfrac{\Sigma X}{N} = \dfrac{30}{5} = 6.00$
4	−2	4	
6	0	0	
8	+2	4	$s = \sqrt{\dfrac{SS}{N-1}} = \sqrt{\dfrac{\Sigma (X - \bar{X})^2}{N-1}} = \sqrt{\dfrac{40}{5-1}}$
10	+4	16	
	$\Sigma (X - \bar{X}) = 0$	$SS = 40$	$= \sqrt{10} = 3.16$

understanding of the standard deviation and it does yield the correct answer, it is quite cumbersome to use in practice. This is especially true if the mean is not a whole number. Table 4.8 shows an illustration, using the previous equation with a mean that has a decimal remainder. Note that each deviation score has a decimal remainder that must be squared to get $(X - \bar{X})^2$. A great deal of rounding is necessary, which may contribute to inaccuracy. In addition, we are dealing with adding 5-digit numbers, which increases the possibility of error. You can see how cumbersome using this equation becomes when the mean is not an integer, and in most practical problems, the mean is not an integer!

TABLE 4.8   Calculation of the Standard Deviation Using Deviation Scores When the Mean Is Not a Whole Number

X	$X - \bar{X}$	$(X - \bar{X})^2$	Calculation of $\bar{X}$ and s
10	−6.875	47.2656	$\bar{X} = \dfrac{\Sigma X}{N} = \dfrac{135}{8} = 16.875$
12	−4.875	23.7656	
13	−3.875	15.0156	
15	−1.875	3.5156	$s = \sqrt{\dfrac{SS}{N-1}} = \sqrt{\dfrac{\Sigma (X - \bar{X})^2}{N-1}}$
18	1.125	1.2656	
20	3.125	9.7656	
22	5.125	26.2656	$= \sqrt{\dfrac{192.8748}{7}}$
25	8.125	66.0156	
$\Sigma X = 135$	$\Sigma (X - \bar{X}) = 0.000$	$SS = 192.8748$	$= \sqrt{27.5535}$
$N = 8$			$= 5.25$

Calculating the Standard Deviation of a Sample Using the Raw Scores Method
It can be shown algebraically that

$$SS = \Sigma X^2 - \frac{(\Sigma X)^2}{N} \qquad \textit{sum of squares}$$

The derivation is presented in Note 4.2. Using this equation to find SS allows us to use the raw scores without the necessity of calculating deviation scores. This, in turn, avoids the decimal remainder difficulties described previously. We shall call this method of computing SS "the raw score method" to distinguish it from the "deviation method." Since the raw score method is generally easier to use and avoids potential errors, it is the method of choice in computing SS, and will be used throughout the remainder of this text. When using the raw score method, *you must be sure not to confuse* ΣX^2 *and* $(\Sigma X)^2$. ΣX^2 *is read "sum X square," and* $(\Sigma X)^2$ *is read "sum X quantity squared."* To find ΣX^2, we square each score and then sum the squares. To find $(\Sigma X)^2$, we sum each score and then square the sum. The result is different for the two procedures. In addition, SS must be positive. If your calculation turns out negative, you have probably confused ΣX^2 and $(\Sigma X)^2$.

Table 4.9 shows the calculation of the standard deviation of the data presented in Table 4.8, using the raw score method. When using this method, we first calculate SS from the raw score equation and then substitute the obtained value in the equation for the standard deviation.

TABLE 4.9   Calculation of the Standard Deviation Using the Raw Score Method

X	X^2	Calculation of SS	Calculation of s
10	100	$SS = \Sigma X^2 - \dfrac{(\Sigma X)^2}{N}$	$s = \sqrt{\dfrac{SS}{N-1}}$
12	144		
13	169		
15	225	$= 2471 - \dfrac{(135)^2}{8}$	$= \sqrt{\dfrac{192.875}{7}}$
18	324		
20	400	$= 2471 - 2278.125$	$= \sqrt{27.5536}$
22	484		
25	625	$= 192.875$	$= 5.25$
$\Sigma X = 135$	$\Sigma X^2 = 2471$		
$N = 8$			

Properties of the Standard Deviation   The standard deviation has many important characteristics. First, the *standard deviation gives us a measure of dispersion relative to the mean.* This differs from the range, which tells us directly the spread of the two most extreme scores. Second, *the standard deviation is sensitive to each score in the distribution.* If a score is moved closer to the mean, then the standard deviation will become smaller. Conversely, if a score shifts away from the mean, then the standard deviation will increase. Third, *like the mean, the standard deviation is stable with regard to sampling fluctuations.* If samples were taken repeatedly from populations of the type usually encountered in the behavioral sciences, the standard deviation of the samples would vary much less from sample to sample than the range. This property is one of the main reasons why the standard deviation is used so much more often than the range for reporting variability. Finally, both the mean and the standard deviation can be manipulated algebraically. This allows mathematics to be done with them for use in inferential statistics.

Now let's do Practice Problems 4.6 and 4.7.

Calculate the standard deviation of the scores contained in the first column of the following table:

X	X^2	Calculation of SS	Calculation of s
25	625	$SS = \sum X^2 - \dfrac{(\sum X)^2}{N}$	$s = \sqrt{\dfrac{SS}{N-1}}$
28	784		
35	1,225		
37	1,369	$= 15{,}545 - \dfrac{(387)^2}{10}$	$= \sqrt{\dfrac{568.1}{9}}$
38	1,444		
40	1,600	$= 15{,}545 - 14{,}976.9$	$= \sqrt{63.1222}$
42	1,764		
45	2,025	$= 568.1$	$= 7.94$
47	2,209		
50	2,500		
$\sum X = 387$	$\sum X^2 = 15{,}545$		
$N = 10$			

Calculate the standard deviation of the scores contained in the first column of the following table:

X	X^2	Calculation of SS	Calculation of s
1.2	1.44	$SS = \sum X^2 - \dfrac{(\sum X)^2}{N}$	$s = \sqrt{\dfrac{SS}{N-1}}$
1.4	1.96		
1.5	2.25		
1.7	2.89	$= 60.73 - \dfrac{(25.9)^2}{12}$	$= \sqrt{\dfrac{4.8292}{11}}$
1.9	3.61		
2.0	4.00	$= 60.73 - 55.9008$	$= \sqrt{0.4390}$
2.2	4.84		
2.4	5.76	$= 4.8292$	$= 0.66$
2.5	6.25		
2.8	7.84		
3.0	9.00		
3.3	10.89		
$\sum X = 25.9$	$\sum X^2 = 60.73$		
$N = 12$			

The Variance

The variance of a set of scores is just the square of the standard deviation. For sample scores, the variance equals

$$s^2 = \text{Estimated } \sigma^2 = \frac{SS}{N-1} \qquad \textit{variance of a sample}$$

For population scores, the variance equals

$$\sigma^2 = \frac{SS_{\text{pop}}}{N} \qquad \textit{variance of a population}$$

The variance is not used much in descriptive statistics because it gives us squared units of measurement. It is used, however, quite frequently in inferential statistics.

SUMMARY

In this chapter, we have discussed the central tendency and variability of distributions. The most common measures of central tendency are the arithmetic mean, the median, and the mode. The arithmetic mean gives the average of the scores and is computed by summing the scores and dividing by N. The median divides the distribution in half and, hence, is the scale value that is at the 50th percentile point of the distribution. The mode is the most frequent score in the distribution. The mean possesses special properties that make it by far the most commonly used measure of central tendency. However, if the distribution is quite skewed, the median should be used instead of the mean, since it is less affected by extreme scores. In addition to presenting these measures, we showed how to calculate each and elaborated their most important properties. We also showed how to obtain the overall mean when the average of several means is desired. Finally, we discussed the relationship between the mean, median, and mode of a distribution and its symmetry.

The most common measures of variability are the range, the standard deviation, and the variance. The range is a crude measure that tells the dispersion between the two most extreme scores. The standard deviation is the most frequently encountered measure of variability. It gives the average dispersion about the mean of the distribution. The variance is just the square of the standard deviation. As with the measures of central tendency, our discussion of variability included how to calculate each measure. Finally, since the standard deviation is the most important measure of variability, we also presented its properties.

IMPORTANT TERMS

Arithmetic mean (p. 62)
Central tendency (p. 61)
Deviation score (p. 71)
Dispersion (p. 61)

Median (p. 66)
Mode (p. 69)
Overall mean (p. 65)
Range (p. 70)

Standard deviation (p. 71)
Sum of squares (p. 73)
Variability (p. 70)
Variance (p. 77)

78   CHAPTER 4   MEASURES OF CENTRAL TENDENCY AND VARIABILITY

QUESTIONS AND PROBLEMS

1. Define or identify the terms in the "Important Terms" section.
2. State four properties of the mean and illustrate each with an example.
3. Under what condition might you prefer to use the median rather than the mean as the best measure of central tendency?
4. Why is the mode not used very much as a measure of central tendency?
5. The overall mean ($\overline{X}_{overall}$) is a weighted mean. Is this statement correct? Explain.
6. Discuss the relationship between the mean and median for distributions that are symmetrical and skewed.
7. Why is the range not as useful a measure of dispersion as the standard deviation?
8. The standard deviation is a relative measure of average dispersion. Is this statement correct? Explain.
9. Why do we use $N - 1$ in the denominator for computing s, whereas we use N in the denominator for determining σ?
10. What is the raw score equation for SS? When is it useful?
11. Give three properties of the standard deviation.
12. How are the variance and standard deviation related?
13. Give the symbol for each of the following:
    a. Mean of a sample
    b. Mean of a population
    c. Standard deviation of a sample
    d. Standard deviation of a population
    e. A raw score
    f. Variance of a sample
    g. Variance of a population
14. Calculate the mean, median, and mode for the following scores:
    a. 5, 2, 8, 2, 3, 2, 4, 0, 6
    b. 30, 20, 17, 12, 30, 30, 14, 29
    c. 1.5, 4.5, 3.2, 1.8, 5.0, 2.2
15. Calculate the mean of the following set of sample scores: 1, 3, 4, 6, 6.
    a. Add a constant of 2 to each score. Calculate the mean for the new values. Generalize to answer the question, "What is the effect on the mean of adding a constant to each score?"
    b. Substract a constant of 2 from each score. Calculate the mean for the new values. Generalize to answer the question, "What is the

effect on the mean of subtracting a constant from each score?"
    c. Multiply each score by a constant of 2. Calculate the mean for the new values. Generalize to answer the question, "What is the effect on the mean of multiplying each score by a constant?"
    d. Divide each score by a constant of 2. Calculate the mean for the new values. Generalize to answer the question, "What is the effect on the mean of dividing each score by a constant?"
16. The following scores resulted from a biology exam:

Scores	f	Scores	f
95–99	3	65–69	7
90–94	3	60–64	6
85–89	5	55–59	5
80–84	6	50–54	3
75–79	6	45–49	2
70–74	8		

    a. What is the median for this exam?
    b. What is the mode?
17. Using the scores shown in Table 3.5 (p. 40),
    a. Determine the median.
    b. Determine the mode.
18. Using the scores shown in Table 3.6 (p. 41),
    a. Determine the median.
    b. Determine the mode.
19. For the following distributions, state whether you would use the mean or the median to represent the central tendency of the distribution. Explain why.
    a. 2, 3, 8, 5, 7, 8
    b. 10, 12, 15, 13, 19, 22
    c. 1.2, 0.8, 1.1, 0.6, 25
20. Given the following values of central tendency for each distribution, determine whether the distribution is symmetrical, positively skewed, or negatively skewed:
    a. Mean = 14, median = 12, mode = 10
    b. Mean = 14, median = 16, mode = 18
    c. Mean = 14, median = 14, mode = 14
21. A student kept track of the number of hours she studied each day for a 2-week period. The

following daily scores were recorded (scores are in hours): 2.5, 3.2, 3.8, 1.3, 1.4, 0, 0, 2.6, 5.2, 4.8, 0, 4.6, 2.8, 3.3. Calculate
a. The mean number of hours studied per day
b. The median number of hours studied per day
c. The modal number of hours studied per day

22. Two salesmen working for the same company are having an argument. Each claims that the average number of items he sold, averaged over the last month, was the highest in the company. Can they both be right? Explain.

23. An ornithologist studying the glaucous-winged gull on Puget Sound counts the number of aggressive interactions per minute among a group of sea gulls during 9 consecutive minutes. The following scores resulted: 24, 9, 12, 15, 10, 13, 22, 20, 14. Calculate
a. The mean number of aggressive interactions per minute
b. The median number of aggressive interactions per minute
c. The model number of aggressive interactions per minute

24. A reading specialist tests the reading speed of children in four ninth-grade English classes. There are 42 students in class A, 35 in class B, 33 in class C, and 39 in class D. The mean reading speed in words per minute for the classes were as follows: class A, 220; class B, 185; class C, 212; and class D, 172. What is the mean reading speed for all classes combined?

25. For the following sample sets of scores, calculate the range, the standard deviation, and the variance:
a. 6, 2, 8, 5, 4, 4, 7
b. 24, 32, 27, 45, 48
c. 2.1, 2.5, 6.6, 0.2, 7.8, 9.3

26. In a particular statistics course, three exams were given. Each student's grade was based on a weighted average of his exam scores. The first test had a weight of 1, the second test had a weight of 2, and the third test had a weight of 2. The exam scores for one student are listed here. What was the student's overall average?

Exam	1	2	3
Score	83	97	92

27. The timekeeper for a particular mile race uses a stopwatch to determine the finishing times of the racers. He then calculates that the mean time for the first three finishers was 4.25 minutes. After checking his stopwatch, he notices to his horror that the stopwatch begins timing at 15 seconds rather than at 0, resulting in scores each of which is 15 seconds too long. What is the correct mean time for the first three finishers?

28. The manufacturer of brand A jogging shoes wants to determine how long the shoes last before resoling is necessary. She randomly samples from users in Chicago, New York, and Seattle. In Chicago, the sample size was 28, and the mean duration before resoling was 7.2 months. In New York, the sample size was 35, and the mean duration before resoling was 6.3 months. In Seattle, the sample size was 22, and the mean duration before resoling was 8.5 months. What is the overall mean duration before resoling is necessary for brand A jogging shoes?

29. Calculate the standard deviation of the following set of sample scores: 1, 3, 4, 6, 6.
a. Add a constant of 2 to each score. Calculate the standard deviation for the new values. Generalize to answer the question, "What is the effect on the standard deviation of adding a constant to each score?"
b. Subtract a constant of 2 from each score. Calculate the standard deviation for the new values. Generalize to answer the question, "What is the effect on the standard deviation of subtracting a constant from each score?"
c. Multiply each score by a constant of 2. Calculate the standard deviation for the new values. Generalize to answer the question, "What is the effect on the standard deviation of multiplying each score by a constant?"
d. Divide each score by a constant of 2. Calculate the standard deviation for the new values. Generalize to answer the question, "What is the effect on the standard deviation of dividing each score by a constant?"

30. An industrial psychologist observed eight drill-press operators for 3 working days. She recorded the number of times each operator pressed the "faster" button instead of the "stop" button to determine whether the design of the control panel was contributing to the high rate of accidents in the plant. Given the scores 4, 7, 0, 2, 7, 3, 6, 7, compute the following:
a. Mean   b. Median
c. Mode   d. Range
e. Standard deviation   f. Variance

31. Without actually calculating the variability, study the following sample distributions:
    Distribution a: 21, 24, 28, 22, 20
    Distribution b: 21, 32, 38, 15, 11
    Distribution c: 22, 22, 22, 22, 22
    a. Rank-order them according to your best guess of their relative variability.
    b. Calculate the standard deviation of each to verify your rank ordering.
32. Compute the standard deviation for the following sample scores. Why is s so high in part **b**, relative to part **a**?
    a. 6, 8, 7, 3, 6, 4
    b. 6, 8, 7, 3, 6, 35
33. A psychologist interested in the dating habits of college undergraduates samples 10 students and determines the number of dates they have had in the last month. Given the scores 1, 8, 12, 3, 8, 14, 4, 5, 8, 16, compute the following:
    a. Mean   b. Median
    c. Mode   d. Range
    e. Standard deviation   f. Variance
34. What happens to the mean of a set of scores if
    a. A constant a is added to each score in the set?
    b. A constant a is subtracted from each score in the set?
    c. Each score is multiplied by a constant a?
    d. Each score is divided by a constant a?
    Illustrate each of these with a numerical example.
35. What happens to the standard deviation of a set of scores if
    a. A constant a is added to each score in the set?

b. A constant a is subtracted from each score in the set?
c. Each score is multiplied by a constant a?
d. Each score is divided by a constant a?
Illustrate each of these with a numerical example.
36. Suppose that, as is done in some lotteries, we sample balls from a big vessel. The vessel contains a large number of balls, each labeled with a single number, 0–9. There are an equal number of balls for each number, and the balls are continually being mixed. For this example, let's collect 10 samples of three balls each. Each sample is formed by selecting balls one at a time, and replacing each ball back in the vessel prior to selecting the next ball. The selection process used assures that every ball in the vessel has an equal chance of being chosen on each selection. Assume the following samples are collected.

1, 3, 4	2, 2, 6	3, 8, 8	1, 6, 7	5, 6, 9
3, 4, 7	1, 2, 6	2, 3, 7	6, 8, 9	4, 7, 9

a. Calculate the mean of each sample.
b. Calculate the median of each sample.
c. Based on the properties of the mean and median discussed previously in the chapter, do you expect more variability in the means or medians? Verify this by calculating the standard deviation of the means and medians.
37. If $s = 0$, what must be true about the scores in the distribution? Verify your answer using an example.

NOTES

4.1 To show that $\Sigma (X_i - \bar{X}) = 0$,

$$\Sigma (X_i - \bar{X}) = \Sigma X_i - \Sigma \bar{X}$$
$$= \Sigma X_i - N\bar{X}$$
$$= \Sigma X_i - N\left(\frac{\Sigma X_i}{N}\right)$$
$$= \Sigma X_i - \Sigma X_i$$
$$= 0$$

4.2 To show that $SS = \Sigma X^2 - [(\Sigma X)^2/N]$,

$$SS = \Sigma (X - \bar{X})^2$$
$$= \Sigma (X^2 - 2X\bar{X} + \bar{X}^2)$$
$$= \Sigma X^2 - \Sigma 2X\bar{X} + \Sigma \bar{X}^2$$
$$= \Sigma X^2 - 2\bar{X} \Sigma X + N\bar{X}^2$$
$$= \Sigma X^2 - 2\left(\frac{\Sigma X}{N}\right) \Sigma X + \frac{N(\Sigma X)^2}{N^2}$$
$$= \Sigma X^2 - \frac{2(\Sigma X)^2}{N} + \frac{(\Sigma X)^2}{N}$$
$$= \Sigma X^2 - \frac{(\Sigma X)^2}{N}$$

5 | The Normal Curve and Standard Scores

 ## INTRODUCTION

The normal curve is a very important distribution in the behavioral sciences. There are three principal reasons why. First, many of the variables measured in behavioral science research have distributions that quite closely approximate the normal curve. Height, weight, intelligence, and achievement are a few examples. Second, many of the inference tests used in analyzing experiments have sampling distributions that become normally distributed with increasing sample size. The sign test and Mann–Whitney U test are two such tests, which we shall cover later in the text. Finally, many inference tests require sampling distributions that are normally distributed (we shall discuss sampling distributions in Chapter 12). The z test, Student's t test, and the F test are examples of inference tests that depend on this point. Thus, much of the importance of the normal curve occurs in conjunction with inferential statistics.

 ## THE NORMAL CURVE

The normal curve is a theoretical distribution of population scores. It is a bell-shaped curve that is described by the following equation:

$$Y = \frac{N}{\sqrt{2\pi}\sigma} e^{-(X-\mu)^2/2\sigma^2} \qquad \textit{equation of the normal curve}$$

where

Y = frequency of a given value of X
X = any score in the distribution
μ = mean of the distribution
σ = standard deviation of the distribution

N = total frequency of the distribution
π = a constant of 3.1416
e = a constant of 2.7183

Most of us will never need to know the exact equation for the normal curve. It has been given here primarily to make the point that the normal curve is a theoretical curve that is mathematically generated. An example of the normal curve is shown in Figure 5.1.

Normal curve

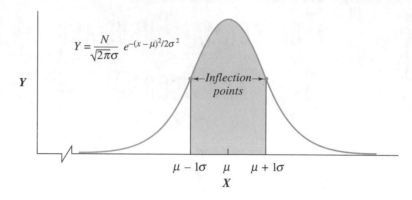

$$Y = \frac{N}{\sqrt{2\pi}\sigma} \, e^{-(x-\mu)^2/2\sigma^2}$$

Note that the curve has two inflection points, one on each side of the mean. Inflection points are located where the curvature changes direction. In Figure 5.1, the inflection points are located where the curve changes from being convex downward to being convex upward. If the bell-shaped curve is a normal curve, the inflection points are at one standard deviation from the mean ($\mu + 1\sigma$ and $\mu - 1\sigma$). Note also that as the curve approaches the horizontal axis it is slowly changing its Y value. Theoretically, the curve never quite reaches the axis. It approaches the horizontal axis and gets closer and closer to it, but it never quite touches it. The curve is said to be *asymptotic* to the horizontal axis.

Area Contained Under the Normal Curve

In distributions that are normally shaped, there is a special relationship between the mean and the standard deviation with regard to the area contained under the curve. When a set of scores is normally distributed, 34.13% of the area under the curve is contained between the mean (μ) and a score that is equal to $\mu + 1\sigma$; 13.59% of the area is contained between a score equal to $\mu + 1\sigma$ and a score of $\mu + 2\sigma$; 2.15% of the area is contained between scores of $\mu + 2\sigma$ and $\mu + 3\sigma$; and 0.13% of the area exists beyond $\mu + 3\sigma$. This accounts for 50% of the area. Since the curve is symmetrical, the same percentages hold for scores below the mean. These relationships are shown in Figure 5.2. Since frequency is plotted on the vertical axis, these percentages represent the *percentage of scores* contained within the area.

To illustrate, suppose we have a population of 10,000 IQ scores. The distribution is normally shaped, with $\mu = 100$ and $\sigma = 16$. Since the scores are normally distributed, 34.13% of the scores are contained between scores of 100 and 116 ($\mu + 1\sigma = 100 + 16 = 116$), 13.59% between 116 and 132 ($\mu + 2\sigma = $

Areas under the normal
curve for selected scores

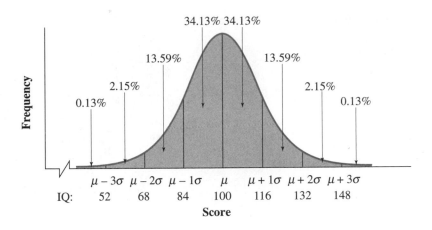

100 + 32 = 132), 2.15% between 132 and 148, and 0.13% are greater than 148. Similarly, 34.13% of the scores fall between 84 and 100, 13.59% between 68 and 84, 2.15% between 52 and 68, and 0.13% are lower than 52. These relationships are also shown in Figure 5.2.

To calculate the number of scores in each area, all we must do is multiply the relevant percentage by the total number of scores. Thus, there are 34.13% × 10,000 = 3413 scores between 100 and 116, 13.59% × 10,000 = 1359 scores between 116 and 132, and 215 scores between 132 and 148; 13 scores are greater than 148. For the other half of the distribution, there are 3413 scores between 84 and 100, 1359 scores between 68 and 84, and 215 scores between 52 and 68; there are 13 scores below 52. Note that these frequencies would be true only if the distribution is exactly normally distributed. In actual practice, the frequencies would vary slightly depending on how close the distribution is to this theoretical model.

 ## STANDARD SCORES (*z* SCORES)

Suppose someone told you your IQ was 132. Would you be happy or sad? In the absence of additional information, it is difficult to say. An IQ of 132 is meaningless unless you have a reference group to compare against. Without such a group, you can't tell whether the score is high, average, or low. For the sake of this illustration, let's assume your score is one of the 10,000 scores of the distribution described above. Now we can begin to give your IQ score of 132 some meaning. For example, we can determine the percentage of scores in the distribution that are lower than 132. You will recognize this as determining the percentile rank of the score of 132. Referring to Figure 5.2, we can see that 132 is two standard deviations above the mean. In a normal curve, there are 34.13 + 13.59 = 47.72% of the scores between the mean and a score that is two standard deviations above the mean. To find the percentile rank of 132, we need to add to this percentage the 50.00% that lie below the mean. Thus, 97.72% (47.72 + 50.00) of the scores fall below your IQ score of 132. You should be quite happy to be so intelligent. The solution is shown in Figure 5.3.

FIGURE 5.3

Percentile rank of an IQ
of 132

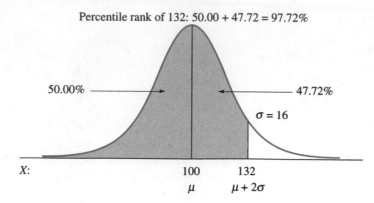

To solve this problem, we had to determine how many standard deviations
the raw score of 132 was above or below the mean. In so doing, we transformed
the raw score into a *standard score,* also called a *z score.*

DEFINITION

• *A z* **score** *is a transformed score that designates how many standard
deviation units the corresponding raw score is above or below the mean.*

In equation form,

$$z = \frac{X - \mu}{\sigma} \qquad \text{z score for population data}$$

$$z = \frac{X - \overline{X}}{s} \qquad \text{z score for sample data}$$

For the previous example,

$$z = \frac{X - \mu}{\sigma} = \frac{132 - 100}{16} = 2.00$$

The process by which the raw score is altered is called a *score transformation.*
We shall see later that the *z* transformation results in a distribution having a
mean of 0 and a standard deviation of 1. The reason *z* scores are called standard
scores is that they are expressed relative to a distribution mean of 0 and a
standard deviation of 1.

In conjunction with a normal curve, *z* scores allow us to determine the number
or percentage of scores that fall above or below any score in the distribution.
In addition, *z* scores allow comparison between scores in different distributions,
even when the units of the distributions are different. To illustrate this point,
let's consider another population set of scores that are normally distributed.
Suppose that the weights of all the rats housed in a university vivarium are
normally distributed, with $\mu = 300$ and $\sigma = 20$ grams. What is the percentile
rank of a rat weighing 340 grams?

The solution is shown in Figure 5.4. First, we need to convert the raw score

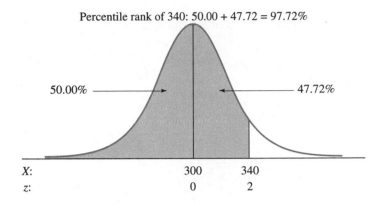

FIGURE 5.4

Percentile rank of a rat
weighing 340 grams

Percentile rank of 340: 50.00 + 47.72 = 97.72%

50.00%   47.72%

X: 300   340
z: 0   2

of 340 grams to its corresponding *z* score:

$$z = \frac{X - \mu}{\sigma} = \frac{340 - 300}{20} = 2.00$$

Since the scores are normally distributed, 34.13 + 13.59 = 47.72% of the scores are between the score and the mean. Adding the remaining 50.00% that lie below the mean, we arrive at a percentile rank of 47.72 + 50.00 = 97.72% for the weight of 340 grams. Thus, the IQ score of 132 and the rat's weight of 340 grams have something in common. They both occupy the same relative position in their respective distributions. The rat is as heavy as you are smart.

This example, although somewhat facetious, illustrates an important use of *z* scores—namely, to compare scores that are not otherwise directly comparable. Ordinarily we would not be able to compare intelligence and weight. They are measured on different scales and have different units. But by converting the scores to their *z*-transformed scores, we eliminate the original units and replace them with a universal unit, the standard deviation. Thus, your score of 132 IQ units becomes a score of two standard deviation units above the mean, and the rat's weight of 340 grams also becomes a score of two standard deviation units above the mean. In this way, it is possible to compare "anything with anything" as long as the measuring scales allow computation of the mean and standard deviation. The ability to compare scores that are measured on different scales is of fundamental importance to the topic of correlation. We shall discuss this in more detail when we take up that topic in Chapter 6.

So far, the examples we've been considering have dealt with populations. It might be useful to practice computing *z* scores using sample data. Let's do this in the next practice problem.

PRACTICE PROBLEM 5.1

For the set of sample raw scores *X* = 1, 4, 5, 7, 8, determine the *z* score for each raw score.

Step 1: Determine the mean of the raw scores.

$$\bar{X} = \frac{\sum X_i}{N} = \frac{25}{5} = 5.00$$

Step 2: Determine the standard deviation of the scores.

$$s = \sqrt{\frac{SS}{N-1}} \qquad SS = \sum X^2 - \frac{(\sum X)^2}{N}$$

$$= \sqrt{\frac{30}{4}} \qquad\qquad = 155 - \frac{(25)^2}{5}$$

$$= 2.7386 \qquad\qquad = 30$$

Step 3: Compute the z score for each raw score.

X	z
1	$z = \dfrac{X - \bar{X}}{s} = \dfrac{1-5}{2.7386} = -1.46$
4	$z = \dfrac{X - \bar{X}}{s} = \dfrac{4-5}{2.7386} = -0.37$
5	$z = \dfrac{X - \bar{X}}{s} = \dfrac{5-5}{2.7386} = 0.00$
7	$z = \dfrac{X - \bar{X}}{s} = \dfrac{7-5}{2.7386} = 0.73$
8	$z = \dfrac{X - \bar{X}}{s} = \dfrac{8-5}{2.7386} = 1.10$

Characteristics of z Scores

There are three characteristics of z scores worth noting. First, *the z scores have the same shape as the set of raw scores.* Transforming the raw scores into their corresponding z scores does not change the shape of the distribution. Nor do the scores change their relative positions. All that is changed are the score values. Figure 5.5 illustrates this point by showing the IQ scores and their corresponding z scores. You should note that although we have used z scores in conjunction with the normal distribution, all z distributions are not normally shaped. Using the z equation given previously, z scores can be calculated for distributions of any shape. The resulting z scores will take on the shape of the raw scores.

Second, *the mean of the z scores always equals zero* ($\mu_z = 0$). This follows

FIGURE 5.5

Raw IQ scores and corresponding *z* scores

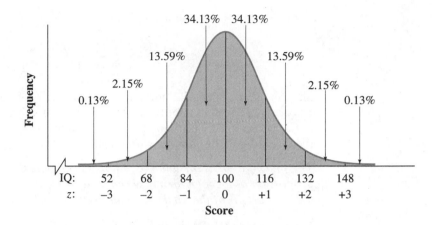

from the observation that the scores located at the mean of the raw scores will also be at the mean of the *z* scores (see Figure 5.5). The *z* value for raw scores at the mean equals zero. For example, the *z* transformation for a score at the mean of the IQ distribution is given by $z = (X - \mu)/\sigma = (100 - 100)/16 = 0$. Thus, the mean of the *z* distribution equals zero. The last characteristic of importance is that the *standard deviation of z scores always equals* 1 ($\sigma_z = 1$). This follows because a raw score that is one standard deviation above the mean has a *z* score of +1:

$$z = \frac{(\mu + 1\sigma) - \mu}{\sigma} = 1$$

Finding Areas Corresponding to Any Raw Score

In the previous examples with IQ and weight, the *z* score was carefully chosen so that the solution could be found from Figure 5.2. However, suppose instead of an IQ of 132, we desire to find the percentile rank of an IQ of 142. Assume the same population parameters. The solution is shown in Figure 5.6. First, draw a curve showing the population and locate the relevant area by entering the

FIGURE 5.6

Percentile rank of an IQ of 142 in a normal distribution with $\mu = 100$ and $\sigma = 16$

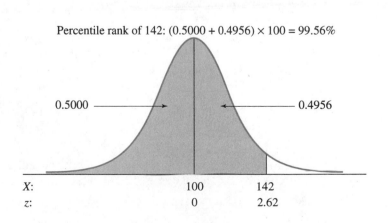

Percentile rank of 142: $(0.5000 + 0.4956) \times 100 = 99.56\%$

score 142 on the horizontal axis. Then shade in the area desired. Next, calculate z:

$$z = \frac{X - \mu}{\sigma} = \frac{142 - 100}{16} = \frac{42}{16} = 2.62$$

Since neither Figure 5.2 nor Figure 5.5 shows a percentage corresponding to a z score of 2.62, we cannot use these figures to solve the problem. Fortunately, the areas under the normal curve for various z scores have been computed, and the resulting values are shown in Table A of Appendix D.

The first column of the table (column A) contains the z score. Column B lists the proportion of the total area between a given z score and the mean. Column C lists the proportion of the total area that exists beyond the z score.

We can use Table A to find the percentile rank of 142. First, we locate the z score of 2.62 in column A. Next, we determine from column B the proportion of the total area between the z score and the mean. For a z score of 2.62, this area equals 0.4956. To this value we must add 0.5000 to take into account the scores lying below the mean (the picture helps remind us to do this). Thus, the proportion of scores that lie below an IQ of 142 is 0.4956 + 0.5000 = 0.9956. To convert this proportion to a percentage, we must multiply by 100. Thus, the percentile rank of 142 is 99.56. Table A can be used to find the area for any z score, provided the scores are normally distributed. When using Table A, it is usually sufficient to round z values to two-decimal-place accuracy. Let's do a few more illustrative problems for practice.

PRACTICE PROBLEM 5.2

The scores on a nationwide mathematics aptitude exam are normally distributed, with $\mu = 80$ and $\sigma = 12$. What is the percentile rank of a score of 84?

SOLUTION

In solving problems involving areas under the normal curve, it is wise, at the outset, to draw a picture of the curve and locate the relevant areas on the curve. Figure 5.7 shows such a picture. The shaded area contains all the scores lower than 84. To find the percentile rank of 84, we must first convert 84 to its corresponding z score:

$$z = \frac{X - \mu}{\sigma} = \frac{84 - 80}{12} = \frac{4}{12} = 0.33$$

To find the area between the mean and a z score of 0.33, we enter Table A, locate the z value in column A, and read off the corresponding entry in column B. This value is 0.1293. Thus, the proportion of the total area between the mean and a *z* score of 0.33 is 0.1293. From Figure 5.7, we can see that the remaining scores below the mean occupy 0.5000 proportion of the total area. If we add these two areas together, we shall have the proportion of scores lower than 84. Thus, the proportion of scores lower than 84 is 0.1293 + 0.5000 = 0.6293. The percentile rank of 84 is then 0.6293 × 100 = 62.93.

FIGURE 5.7   Percentile rank of a mathematics exam score of 84 in a normal distribution with $\mu = 80$ and $\sigma = 12$

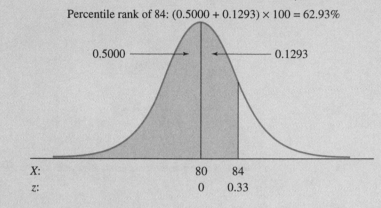

Percentile rank of 84: (0.5000 + 0.1293) × 100 = 62.93%

0.5000 ⟶   ← 0.1293

X:	80	84
z:	0	0.33

PRACTICE PROBLEM 5.3

What percentage of aptitude scores are below a score of 66?

SOLUTION

Again, the first step is to draw the appropriate diagram. This is shown in Figure 5.8. From this diagram, we can see that the relevant area (shaded) lies beyond the score of 66. To find the percentage of scores contained in this area, we must first convert 66 to its corresponding *z* score. Thus,

$$z = \frac{X - \mu}{\sigma} = \frac{66 - 80}{12} = \frac{-14}{12} = -1.17$$

From Table A, column C, we find that the area beyond a *z* score of 1.17 is 0.1210. Thus, the percentage of scores below 66 is 0.1210 × 100 = 12.10%. Table A does not show any negative *z* scores. However, this does not cause a problem because the normal curve is symmetrical, and negative *z* scores have the same proportion of area as positive z scores of the same magnitude. Thus, the proportion of total area lying beyond a z score of $+1.17$ is the same as the proportion lying beyond a z score of -1.17.

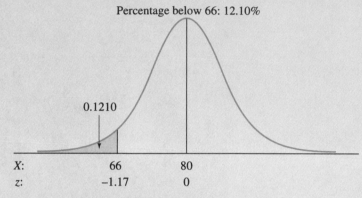

FIGURE 5.8   Percentage of scores below a score of 66 in a normal distribution with $\mu = 80$ and $\sigma = 12$

Percentage below 66: 12.10%

0.1210

X: 66   80
z: −1.17   0

PRACTICE PROBLEM 5.4

Using the same population as in Practice Problem 5.3, what percentage of scores fall between 64 and 90?

SOLUTION

Figure 5.9 shows the relevant diagram. This time, the shaded areas are on either side of the mean. To solve this problem, we must find the area between 64 and 80 and add it to the area between 80 and 90. As before, to determine area, we must calculate the appropriate z score. This time, however, we must compute two z scores. For the area to the left of the mean,

$$z = \frac{64 - 80}{12} = \frac{-16}{12} = -1.33$$

For the area to the right of the mean,

$$z = \frac{90 - 80}{12} = \frac{10}{12} = 0.83$$

Since the areas we want to determine are between the mean and the z score, we shall use column B of Table A. The area corresponding to a z

score of −1.33 is 0.4082, and the area corresponding to a *z* score of 0.83 is 0.2967. The total area equals the sum of these two areas. Thus, the proportion of scores falling between 64 and 90 is 0.4082 + 0.2967 = 0.7049. The percentage of scores between 64 and 90 is 0.7049 × 100 = 70.49%. Note that in this problem we cannot just subtract 64 from 90 and divide by 12. The areas in Table A are designated with the mean as a reference point. Therefore, to solve this problem we must relate the scores of 64 and 90 to the mean of the distribution. You should also note that you cannot just subtract one *z* value from the other, because the curve is not rectangular but has differing amounts of area under various points of the curve.

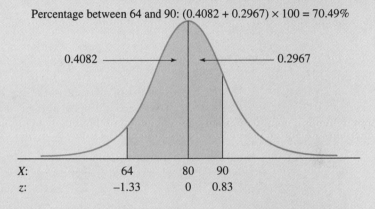

FIGURE 5.9   Percentage of scores falling between 64 and 90 in a normal distribution with $\mu = 80$ and $\sigma = 12$

Percentage between 64 and 90: (0.4082 + 0.2967) × 100 = 70.49%

0.4082   0.2967

| X: | 64 | 80 | 90 |
| z: | −1.33 | 0 | 0.83 |

PRACTICE PROBLEM 5.5

Another type of problem arises when we want to determine th[e]
between two scores and both scores are either above or below th[e]
Let's try a problem of this sort. Find the percentage of aptit[ude]
falling between the scores of 95 and 110.

SOLUTION

Figure 5.10 shows the distribution and the relevant
Problem 5.4, we can't just subtract 95 from 110 an[d]
the appropriate *z* score. Rather we must use the
point. In this problem, we must find (1) the area b
and (2) the area between 95 and the mean. By s
we shall arrive at the area between 95 and 110.
two *z* scores:

z transformation of 110:

$$z = \frac{110 - 80}{12} = \frac{30}{12} = 2.50$$

z transformation of 95:

$$z = \frac{95 - 80}{12} = \frac{15}{12} = 1.25$$

From column B of Table A,

$$\text{Area } (z = 2.50) = 0.4938$$

and

$$\text{Area } (z = 1.25) = 0.3944$$

Thus, the proportion of scores falling between 95 and 110 is 0.4938 − 0.3944 = 0.0994. The percentage of scores is 0.0994 × 100 = 9.94%.

FIGURE 5.10   Percentage of scores falling between 95 and 110 in a normal distribution with $\mu = 80$ and $\sigma = 12$

Percentage between 95 and 110: (0.4938 − 0.3944) × 100 = 9.94%

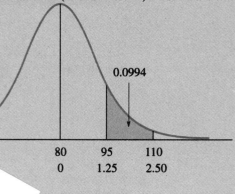

Finding the Raw Score Corresponding to a Given Area

Sometimes we know the area and want to determine the corresponding score. The following problem is of this kind. Find the raw score that divides the distribution of aptitude scores such that 70% of the scores are below it.

This problem is just the reverse of the previous one. Here, we are given the area and need to determine the score. Figure 5.11 shows the appropriate diagram.

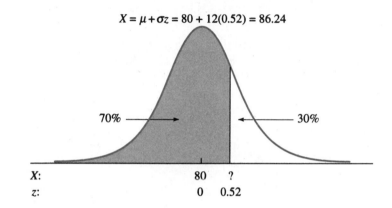

$X = \mu + \sigma z = 80 + 12(0.52) = 86.24$

70% ⟶   ⟵ 30%

X: 80   ?
z: 0   0.52

Although we don't know what the raw score value is, we can determine its corresponding z score from Table A. Once we know the z score, we can solve for the raw score using the z equation. If 70% of the scores lie below the raw score, then 30% must lie above it. We can find the z score by searching in Table A, column C, until we locate the area closest to 0.3000 (30%) and then noting that the z score corresponding to this area is 0.52. To find the raw score, all we need to do is substitute the relevant values in the z equation and solve for X. Thus,

$$z = \frac{X - \mu}{\sigma}$$

Substituting and solving for X,

$$0.52 = \frac{X - 80}{12}$$

$$X = 80 + 12(0.52) = 86.24$$

PRACTICE PROBLEM 5.6

Let's try another problem of this type. What is the score that divides the distribution such that 99% of the area is below it?

SOLUTION

The diagram is shown in Figure 5.12. If 99% of the area is below the score, 1% must be above it. To solve this problem, we locate the area in column C of Table A that is closest to 0.0100 (1%) and note that $z = 2.33$. We convert the z score to its corresponding raw score by substituting the relevant values in the z equation and solving for X. Thus,

$$z = \frac{X - \mu}{\sigma}$$

$$2.33 = \frac{X - 80}{12}$$

$$X = 80 + 12(2.33) = 107.96$$

FIGURE 5.12   Determining the score below which 99% of the distribution falls in a normal distribution with $\mu = 80$ and $\sigma = 12$

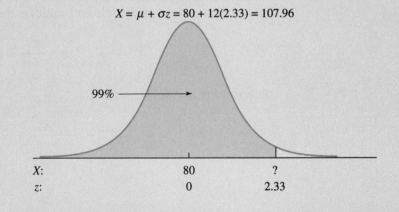

$$X = \mu + \sigma z = 80 + 12(2.33) = 107.96$$

99%		

X:	80	?
z:	0	2.33

PRACTICE PROBLEM 5.7

Let's do one more problem. What are the scores that bound the middle 95% of the distribution?

SOLUTION

The diagram is shown in Figure 5.13. There is an area of 2.5% above and below the middle 95%. To determine the scores that bound the middle 95% of the distribution, we must first find the *z* values and then convert these values to raw scores. The *z* scores are found in Table A by locating the area in column C closest to 0.0250 (2.5%) and reading the associated *z* score in column A. In this case, $z = \pm 1.96$. The raw scores are found by substituting the relevant values in the *z* equation and solving for *X*. Thus,

$$z = \frac{X - \mu}{\sigma}$$

$$-1.96 = \frac{X - 80}{12} \qquad\qquad +1.96 = \frac{X - 80}{12}$$

$$X = 80 + 12(-1.96) \qquad\qquad X = 80 + 12(1.96)$$

$$= 56.48 \qquad\qquad\qquad = 103.52$$

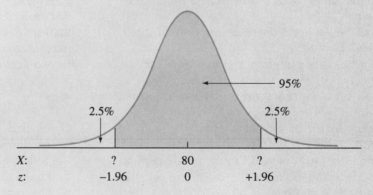

FIGURE 5.13   Determining the scores that bound the middle 95% of the distribution in a normal distribution with $\mu = 80$ and $\sigma = 12$

SUMMARY

In this chapter, we have discussed the normal curve and standard scores. We pointed out that the normal curve is a bell-shaped curve and gave the equation describing it. Next, we discussed the area contained under the normal curve and its relation to z scores. A z score is a transformation of a raw score. It designates how many standard deviation units the corresponding raw score is above or below the mean. A z distribution has the following characteristics:

(1) the z scores have the same shape as the set of raw scores; (2) the mean of z scores always equals 0; and (3) the standard deviation of z scores always equals 1. Finally, we showed how to use z scores in conjunction with a normal distribution to find (1) the percentage or frequency of scores corresponding to any raw score in the distribution and (2) the raw score corresponding to any frequency or percentage of scores in the distribution.

IMPORTANT TERMS

Asymptotic (p. 82)   Normal curve (p. 81)   Standard scores (z scores) (p. 83)

QUESTIONS AND PROBLEMS

1. Define
   a. The normal curve
   b. z scores
   c. Standard scores
2. What is a score transformation? Provide an example.
3. What are the values of the mean and standard deviation of the z distribution?
4. Must the shape of a z distribution be normal? Explain.
5. Are all bell-shaped distributions normal distributions? Explain.
6. Given the set of sample raw scores 10, 12, 16, 18, 19, 21,
   a. Convert each raw score to its z-transformed value.
   b. Compute the mean and standard deviation of the z scores.
7. Assume the raw scores in Problem 6 are population scores and perform the calculations called for in parts **a** and **b**.
8. A population of raw scores is normally distributed with $\mu = 60$ and $\sigma = 14$. Determine the z scores for the following raw scores taken from that population:
   a. 76   b. 48
   c. 86   d. 60
   e. 74   f. 46
9. For the following z scores, determine the percentage of scores that lie beyond z:
   a. 0   b. 1

   c. 1.54   d. −2.05
   e. 3.21   f. −0.45
10. For the following z scores, determine the percentage of scores that lie between the mean and the z score:
    a. 1   b. −1
    c. 2.34   d. −3.01
    e. 0   f. 0.68
    g. −0.73
11. For each of the following, determine the z score that divides the distribution such that the given percentage of scores lies above the z score (round to two decimal places):
    a. 50%   b. 2.50%
    c. 5%   d. 30%
    e. 80%   f. 90%
12. Given that a population of scores is normally distributed with $\mu = 110$ and $\sigma = 8$, determine the following:
    a. The percentile rank of a score of 120
    b. The percentage of scores that are below a score of 99
    c. The percentage of scores that are between a score of 101 and 122
    d. The percentage of scores that are between a score of 114 and 124
    e. The score in the population above which 5% of the scores lie
13. At the end of a particular quarter, Carol took four final exams. The mean and standard deviation for each exam along with Carol's grade on

each exam are listed here. Assume that the grades on each exam are normally distributed.

Exam	Mean	Standard Deviation	Carol's Grade
French	75.4	6.3	78.2
History	85.6	4.1	83.4
Psychology	88.2	3.5	89.2
Statistics	70.4	8.6	82.5

a. On which exam did Carol do best, relative to the other students taking the exam?
b. What was her percentile rank on this exam?

14. A hospital in a large city records the weight of every infant born at the hospital. The distribution of weights is normally shaped, has a mean $\mu = 2.9$ kilograms, and has a standard deviation $\sigma = 0.45$. Determine the following:
a. The percentage of infants who weighed under 2.1 kilograms
b. The percentile rank of a weight of 4.2 kilograms
c. The percentage of infants who weighed between 1.8 and 4.0 kilograms
d. The percentage of infants who weighed between 3.4 and 4.1 kilograms
e. The weight that divides the distribution such that 1% of the weights are above it
f. Beyond what weights do the most extreme 5% of the scores lie?
g. If 15,000 infants have been born at the hospital, how many weighed less than 3.5 kilograms?

15. A statistician studied the records of monthly rainfall for a particular geographical locale. She found that the average monthly rainfall was normally distributed with a mean $\mu = 8.2$ centimeters and a standard deviation $\sigma = 2.4$. What is the percentile rank of the following scores?
a. 12.4   b. 14.3
c. 5.8   d. 4.1
e. 8.2

16. Using the same population parameters as in Problem 15, what percentage of scores are above the following scores?
a. 10.5   b. 13.8
c. 7.6   d. 3.5
e. 8.2

17. Using the same population parameters as in Problem 15, what percentage of scores are between the following scores?
a. 6.8 and 10.2
b. 5.4 and 8.0
c. 8.8 and 10.5

18. A jogging enthusiast keeps track of how many miles he jogs each week. The following scores are sampled from his 1996 records:

Week	Distance	Week	Distance
5	32*	30	36
8	35	32	38
10	30	38	35
14	38	43	31
15	37	48	33
19	36	49	34
24	38	52	37

* Scores are miles run.

a. Determine the z scores for the distances shown in the table. Note that the distances are sample scores.
b. Plot a frequency polygon for the raw scores.
c. On the same graph, plot a frequency polygon for the z scores.
d. Is the z distribution normally shaped? If not, explain why.
e. Compute the mean and standard deviation of the z distribution.

19. A stock market analyst has kept records for the past several years of the daily selling price of a particular blue-chip stock. The resulting distribution of scores is normally shaped with a mean $\mu = \$84.10$ and a standard deviation $\sigma = \$7.62$.
a. Determine the percentage of selling prices that were below a price of \$95.00.
b. What percentage of selling prices were between \$76.00 and \$88.00?
c. What percentage of selling prices were above \$70.00?
d. What selling price divides the distribution such that 2.5% of the scores are above it?

20. Anthony is deciding whether to go to graduate school in business or law. He has taken nationally administered aptitude tests for both fields. Anthony's scores along with the national norms are shown here. Based solely on Anthony's relative

standing on these tests, which field should he enter? Assume that the distributions of both tests are normally distributed.

Field	National Norms		Anthony's Score
	μ	σ	
Business	68	4.2	80.4
Law	85	3.6	89.8

21. On which of the following exams did Rebecca do better? How about Maurice? Assume the scores on each exam were normally distributed.

	μ	σ	Rebecca's Scores	Maurice's Scores
Exam 1	120	6.8	130	132
Exam 2	50	2.4	56	52

6 | CORRELATION

 ## INTRODUCTION

In the previous chapters, we were mainly concerned with single distributions and how to best characterize them. In addition to describing individual distributions, it is often desirable to determine whether the scores of one distribution are related to the scores of another distribution. For example, the person in charge of hiring employees for a large corporation might be very interested in knowing whether there was a relationship between the college grades that were earned by their employees and their success in the company. If a strong relationship between these two variables did exist, college grades could be used to predict success in the company and hence would be very useful in screening prospective employees.

Aside from the practical utility of using a relationship for prediction, why would anyone be interested in determining whether two variables are related? One important reason is that if the variables are related, it is possible that one of them is the cause* of the other. As we shall see later in this chapter, the fact that two variables are related is not sufficient basis for proving causality. Nevertheless, because correlational studies are among the easiest to carry out, showing that a correlation exists between the variables is often the first step toward proving that they are causally related. Conversely, if a correlation does not exist between the two variables, a causal relationship can be ruled out.

Another very important use of correlation is to assess the "test–retest reliability" of testing instruments. Test–retest reliability means consistency in scores over repeated administrations of the test. For example, assuming an individual's IQ is stable from month to month, we would expect a good test of IQ to show a strong relationship between the scores of two administrations of the test 1 month apart to the same people. Correlational techniques allow us to measure the relationship between the scores derived on the two administrations and, hence, to measure the test–retest reliability of the instrument.

---

* We recognize that the topic of cause and effect has engendered much philosophical debate. However, we cannot consider the intricacies of this topic here. When we use the term *cause,* we mean it in the commonsense way it is used by nonphilosophers. That is, when we say that A caused B, we mean that a variation in A produced a variation in B with all other variables appropriately controlled.

Correlation and regression are very much related. They both involve the relationship between two or more variables. Correlation is concerned primarily with finding out whether a relationship exists and with determining its magnitude and direction, whereas regression is primarily concerned with using the relationship for prediction. In this chapter, we discuss correlation, and in Chapter 7 we will take up the topic of linear regression.

 ## RELATIONSHIPS

Correlation is a topic that deals primarily with the magnitude and direction of relationships. Before delving into these special aspects of relationships, we will discuss some general features of relationships. With these in hand, we can better understand the material specific to correlation.

Linear Relationships

To begin our discussion of relationships, let's illustrate a linear relationship between two variables. The following table shows one month's salary for five salesmen and the dollar value of the merchandise each sold that month.

Salesman	X Variable Merchandise Sold (\$)	Y Variable Salary (\$)
1	0	500
2	1000	900
3	2000	1300
4	3000	1700
5	4000	2100

The relationship between these variables can best be seen by plotting a graph using the paired X and Y values for each salesman as the points on the graph. Such a graph is called a *scatter plot*.

DEFINITION   • *A scatter plot is a graph of paired X and Y values.*

The scatter plot for the salesman data is shown in Figure 6.1. Referring to this figure, we see that all of the points fall on a straight line. When a straight line describes the relationship between two variables, the relationship is called *linear*.

DEFINITION   • *A linear relationship between two variables is one in which the relationship can be most accurately represented by a straight line.*

Note that not all relationships are linear. Some relationships are *curvilinear*. In these cases, when a scatter plot of the X and Y variables is drawn, a *curved* line fits the points better than a straight line.

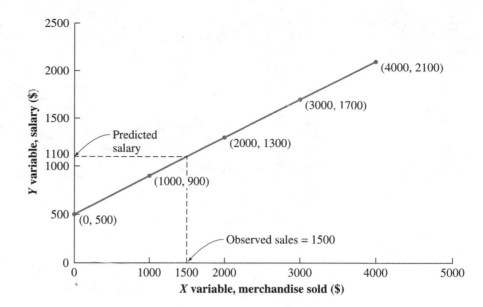

FIGURE 6.1

Scatter plot of the relationship between salary and merchandise sold

Deriving the Equation of the Straight Line   The relationship between "salary" and "merchandise sold" shown in Figure 6.1 can be described by an equation. Of course, this equation is the equation of the line joining all of the points. The general form of the equation is given by

$$Y = bX + a \qquad \textit{equation of a straight line}$$

where $a = Y$ intercept (value of Y when $X = 0$)
b = slope of the line

Finding the Y Intercept a   The Y intercept is the value of Y where the line intersects the Y axis. Thus, it is the Y value when $X = 0$. In this problem, we can see from Figure 6.1 that

$$a = Y \text{ intercept} = 500$$

Finding the Slope b   The slope of a line is a measure of its rate of change. It tells us how much the Y score changes for each unit change in the X score. In equation form,

$$b = \text{slope} = \frac{\Delta Y}{\Delta X} = \frac{Y_2 - Y_1}{X_2 - X_1} \qquad \textit{slope of a straight line}$$

Since we are dealing with a straight line, its slope is constant. This means it doesn't matter what values we pick for X_2 and X_1; the corresponding Y_2 and Y_1 scores will yield the same value of slope. To calculate the slope, let's vary X from 2000 to 3000. If $X_1 = 2000$, then $Y_1 - 1300$. If $X_2 = 3000$, then $Y_2 = 1700$.

FIGURE 6.2

Graph of salary and
amount of merchandise
sold

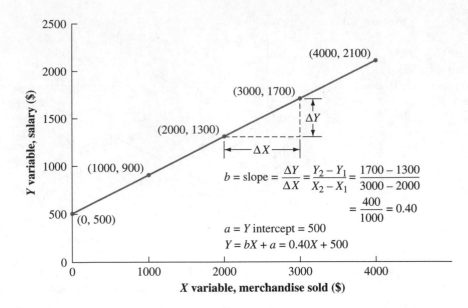

Substituting these values into the slope equation,

$$b = \text{slope} = \frac{\Delta Y}{\Delta X} = \frac{Y_2 - Y_1}{X_2 - X_1} = \frac{1700 - 1300}{3000 - 2000} = \frac{400}{1000} = 0.40$$

Thus, the slope is 0.40. This means that the Y value increases 0.40 unit for every 1-unit increase in X. The slope and Y intercept determinations are also shown in Figure 6.2. Note that the same slope would occur if we had chosen other values for X_1 and X_2. For example, if $X_1 = 1000$ and $X_2 = 4000$, then $Y_1 = 900$ and $Y_2 = 2100$. Solving for the slope,

$$b = \text{slope} = \frac{\Delta Y}{\Delta X} = \frac{Y_2 - Y_1}{X_2 - X_1} = \frac{2100 - 900}{4000 - 1000} = \frac{1200}{3000} = 0.40$$

Again, the slope is 0.40.

The full equation for the linear relationship that exists between salary and merchandise sold can now be written:

$$Y = bX + a$$

Substituting for a and b,

$$Y = 0.40X + 500$$

The equation $Y = 0.40X + 500$ describes the relationship between the Y variable, "salary," and the X variable, "merchandise sold." It tells us that Y increases by 1 unit for every 0.40 increase in X. Moreover, as long as the relationship holds, this equation lets us compute an appropriate value for Y, given any value of X. That makes the equation very useful for prediction.

Predicting Y given X When used for prediction, the equation becomes

$$Y' = 0.40X + 500$$

where Y' is the predicted value of the Y variable.

With this equation we can predict any Y value just by knowing the corresponding X value. For example, if $X = 1500$ as in our previous problem, then

$$Y' = 0.40X + 500$$
$$= 0.40(1500) + 500$$
$$= 600 + 500$$
$$= 1100$$

Thus, if a salesman sells $1500 worth of merchandise, his salary would equal $1100.

Of course, prediction could also have been done graphically, as shown in Figure 6.1. By vertically projecting the X value of $1500 until it intersects with the straight line, we can read the predicted Y value from the Y axis. The predicted value is $1100, which is the same value we arrived at using the equation.

Positive and Negative Relationships

In addition to being linear or curvilinear, the relationship between two variables may be *positive* or *negative*.

DEFINITION

• *A* **positive relationship** *indicates that there is a direct relationship between the variables. A* **negative relationship** *exists when there is an inverse relationship between X and Y.*

The slope of the line tells us whether the relationship is positive or negative. When the relationship is positive, the slope is positive. The previous example had a positive slope; i.e., higher values of X were associated with higher values of Y, and lower values of X were associated with lower values of Y. When the slope is positive, the line tends to run upward from left to right, indicating that as X increases, Y increases. Thus, a direct relationship exists between the two variables.

When the relationship is negative, there is an inverse relationship between the variables, making the slope negative. An example of a negative relationship is shown in Figure 6.3. Note that with a negative slope, the curve runs downward from left to right. Low values of X are associated with high values of Y, and high values of X are associated with low values of Y. Another way of saying this is that as X increases, Y decreases.

Perfect and Imperfect Relationships

In the relationships we have graphed so far, all of the points have fallen on the straight line. When this is the case, the relationship is a *perfect* one. Unfortunately, in the behavioral sciences, perfect relationships are rare. It is much more common to find imperfect relationships.

FIGURE 6.3

Example of a negative
relationship

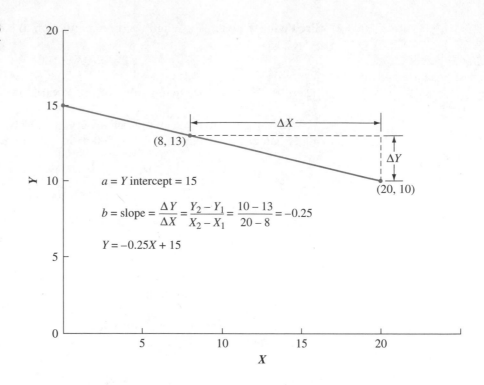

DEFINITION

• *A **perfect relationship** is one in which a positive or negative relationship exists and all of the points fall on the line. An **imperfect relationship** is one in which a relationship exists, but all of the points do not fall on the line.*

TABLE 6.1

IQ and Grade Point
Average of 12 College
Students

Student No.	IQ	Grade Point Average
1	110	1.0
2	112	1.6
3	118	1.2
4	119	2.1
5	122	2.6
6	125	1.8
7	127	2.6
8	130	2.0
9	132	3.2
10	134	2.6
11	136	3.0
12	138	3.6

As an example, Table 6.1 shows the IQ scores and grade point averages of a sample of 12 college students. Suppose we wanted to determine the relationship between these hypothetical data. The scatter plot is shown in Figure 6.4. From the scatter plot, it is obvious that the relationship between IQ and college grades is imperfect. The imperfect relationship is positive because lower values of IQ are associated with lower values of grade point average and higher values of IQ are associated with higher values of grade point average. In addition, the relationship appears linear.

To describe this relationship with a straight line, the best we can do is to draw the line that *best* fits the data. Another way of saying this is that, when the relationship is imperfect, we cannot draw a single straight line through all of the points. We can, however, construct a straight line that most accurately fits the data.* This line has been drawn in Figure 6.4. This best-fitting line is often used for prediction; when so used, it is called a *regression* line.

A recent *USA Today* article reported that there is an inverse relationship between the amount of television watched by primary school students and their reading skills. Suppose the sixth-grade data for the article appeared as shown in Figure 6.5. This is an example of a negative, imperfect, linear relationship. The relationship is negative because higher values of television watching are associated with lower values of reading skill, and lower values of television

---

* The details on how to construct this line will be discussed in Chapter 7.

FIGURE 6.4

Scatter plot of IQ and grade point average

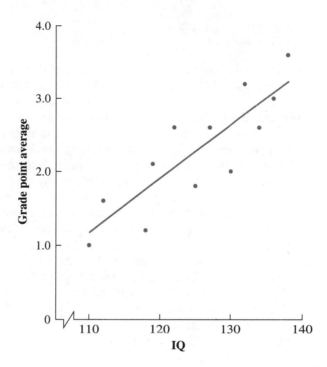

watching are associated with higher values of reading skill. The linear relationship is imperfect, because not all of the points fall on a single straight line. The regression line for these data is also shown in Figure 6.5.

Having completed our background discussion of relationships, we can now move on to the topic of correlation.

FIGURE 6.5

Scatter plot of reading skill and amount of television watched by sixth graders

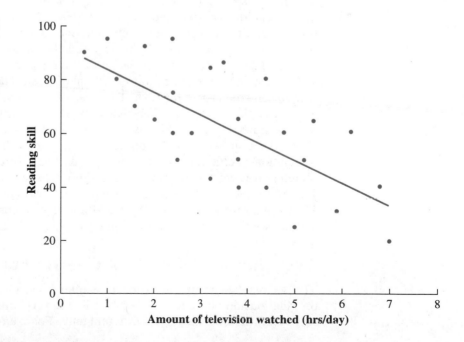

CORRELATION

Correlation is a topic that focuses on the *direction* and *degree* of the relationship. The direction of the relationship refers to whether the relationship is positive or negative. The degree of relationship refers to the magnitude or strength of the relationship. The degree of relationship can vary from nonexistent to perfect. When the relationship is perfect, correlation is at its highest, and we can exactly predict from one variable to the other. In this situation, as X changes, so does Y. Moreover, the same value of X always leads to the same value of Y. Alternatively, the same value of Y always leads to the same value of X. The points all fall on a straight line, assuming the relationship is linear. When the relationship is nonexistent, correlation is at its lowest, and knowing the value of one of the variables doesn't help at all in predicting the other. Imperfect relationships have intermediate levels of correlation, and prediction is approximate. Here, the same value of X doesn't always lead to the same value of Y. Nevertheless, on the average, Y changes systematically with X, and we can do a better job of predicting Y with knowledge of X than without it.

Although it suffices for some purposes to talk rather loosely about "high" or "low" correlations, it is much more often desirable to know the exact magnitude and direction of the correlation. A *correlation coefficient* gives us this information.

DEFINITION

• *A **correlation coefficient** expresses quantitatively the magnitude and direction of the relationship.*

A correlation coefficient can vary from $+1$ to -1. The *sign* of the coefficient tells us whether the relationship is *positive* or *negative*. The numerical part of the correlation coefficient describes the magnitude of the correlation. The higher the number, the greater the correlation. Since 1 is the highest number possible, it represents a perfect correlation. A correlation coefficient of $+1$ means the correlation is perfect and the relationship is positive. A coefficient of -1 means the correlation is perfect and the relationship is negative. When the relationship is nonexistent, the correlation coefficient equals 0. Imperfect relationships have correlation coefficients varying in magnitude between 0 and 1. They will be plus or minus, depending on the direction of the relationship.

Figure 6.6 shows scatter plots of several different linear relationships and the correlation coefficients for each. The Pearson r correlation coefficient has been used since the relationships are linear. We shall discuss Pearson r in the next section. Each scatter plot is made up of paired X and Y values. Note that the closer the points are to the regression line, the higher the magnitude of the correlation coefficient and the more accurate the prediction. Also, when the correlation is zero, there is no relationship between X and Y. This means that Y does not increase or decrease systematically with increases or decreases in X. Thus, with zero correlation, the regression line for predicting Y is horizontal, and knowledge of X does not aid in predicting Y.

The Linear Correlation Coefficient Pearson r

You will recall from our discussion in Chapter 5 that a basic problem in measuring the relationship between two variables is that very often the variables are measured on different scales and in different units. For example, if we were interested

FIGURE 6.6   Scatter plots of several linear relationships

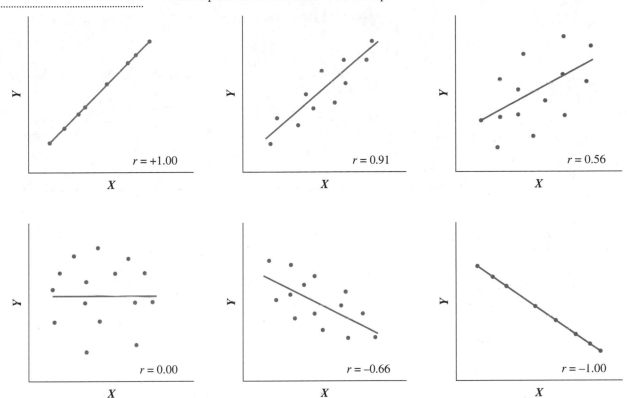

in measuring the correlation between IQ and grade point average for the data presented in Table 6.1, we are faced with the problem that IQ and grade point average have very different scaling. As was mentioned in Chapter 5, this problem is resolved by converting each score to its z-transformed value, in effect putting both variables on the same scale, a z scale.

To appreciate how useful z scores are for determining correlation, consider the following example. Suppose your neighborhood supermarket is having a sale on oranges. The oranges are prebagged, and each bag has the total price marked on it. You want to know whether there is a relationship between the weight of the oranges in each bag and their cost. Being a natural-born researcher, you randomly sample six bags and weigh each one. The cost and weight in pounds of the six bags are shown in Table 6.2. A scattergram of the data is plotted in Figure 6.7. Are these two variables related? Yes; in fact, all the points fall on a straight line. There is a perfect positive correlation between the cost and weight of the oranges. Thus, the correlation coefficient must equal +1.

Next, let's see what happens when we convert these raw scores to z scores. The raw scores for weight (X) and cost (Y) have been expressed as standard scores in the fourth and fifth columns of Table 6.2. Something quite interesting has happened. *The paired raw scores for each bag of oranges have the same z value.* For example, the paired raw scores for bag A are 2.25 and 0.75. However, their respective z scores are both −1.34. The raw score of 2.25 is as many standard deviation units below the mean of the X distribution as the raw score of 0.75 is below the mean of the Y distribution. The same is true for the other paired

TABLE 6.2   Cost and Weight in Pounds of Six Bags of Oranges

Bag	Weight (lb) X	Cost ($) Y	z_X	z_Y
A	2.25	0.75	−1.34	−1.34
B	3.00	1.00	−0.80	−0.80
C	3.75	1.25	−0.27	−0.27
D	4.50	1.50	0.27	0.27
E	5.25	1.75	0.80	0.80
F	6.00	2.00	1.34	1.34

FIGURE 6.7

Cost of oranges versus their weight in pounds

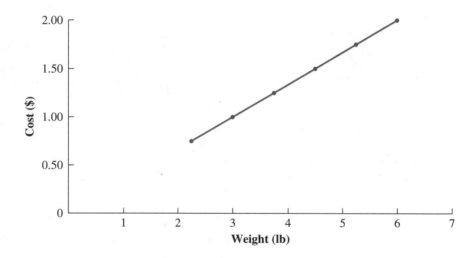

scores. All of the paired raw scores occupy the same relative position within their own distributions. That is, they have the same z values. When using raw scores, this relationship is obscured because of differences in scaling between the two variables. If the paired scores occupy the same relative position within their own distributions, then the correlation must be perfect ($r = 1$), because knowing one of the paired values will allow us to exactly predict the other value. If prediction is perfect, the relationship must be perfect.

This brings us to the definition of Pearson r.

DEFINITION

• **Pearson r is a measure of the extent to which paired scores occupy the same or opposite positions within their own distributions.**

Note that this definition also includes the paired scores occupying opposite positions. If the paired z scores have the same magnitude, but opposite signs, the correlation would again be perfect, and r would equal −1.

This example highlights a very important point. Since correlation is concerned with the relationship between two variables and since the variables are often measured in different units and scaling, the magnitude and direction of the correlation coefficient must be independent of the differences in units and scaling

TABLE 6.3   Cost and Weight in Kilograms of Six Bags of Oranges

Bag	Weight (kg) X	Cost ($) Y	z_X	z_Y
A	1.02	0.75	−1.34	−1.34
B	1.36	1.00	−0.80	−0.80
C	1.70	1.25	−0.27	−0.27
D	2.04	1.50	0.27	0.27
E	2.38	1.75	0.80	0.80
F	2.72	2.00	1.34	1.34

that exist between the two variables. Pearson r achieves this by using z scores. Thus, we can correlate such diverse variables as time of day and position of the sun, percent body fat and caloric intake, test anxiety and examination grades, etc.

Since this is such an important point, we would like to illustrate it once more by taking the previous example one more step. In the example involving the relationship between the cost of oranges and their weight, suppose you weighed the oranges in kilograms rather than in pounds. Should this change the degree of relationship between the cost and weight of the oranges? In light of what we have just presented, the answer is surely "no." Correlation must be independent of the units used in measuring the two variables. If the correlation is 1 between the cost of the oranges and their weight in pounds, the correlation should also be 1 between the cost of the oranges and their weight in kilograms. We've converted the weight of each bag of oranges from pounds to kilograms. The data are presented in Table 6.3, and the raw scores are plotted in Figure 6.8. Again all the scores fall on a straight line, so the correlation equals 1.00. Notice the values of the paired z scores in the fourth and fifth columns of Table 6.3. Again they have the same values. Thus, using z scores allows a measurement of

FIGURE 6.8

Cost of oranges versus their weight in kilograms

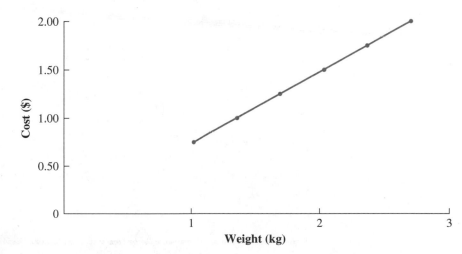

the relationship between the two variables that is independent of differences in scaling and of the units used in measuring the variables.

Calculating Pearson r   The equation for calculating Pearson r using z scores is

$$r = \frac{\sum z_X z_Y}{N - 1}$$

where $\sum z_X z_Y$ is the sum of the product of each z score pair.

To use this equation, you must first convert each raw score into its z-transformed value. This can take a considerable amount of time and possibly create rounding errors. Using some algebra, this equation can be transformed into a calculation equation that uses the raw scores:

$$r = \frac{\sum XY - \frac{(\sum X)(\sum Y)}{N}}{\sqrt{\left[\sum X^2 - \frac{(\sum X)^2}{N}\right]\left[\sum Y^2 - \frac{(\sum Y)^2}{N}\right]}}$$

computational equation for Pearson r

where $\sum XY$ is the sum of the product of each X and Y pair. $\sum XY$ is also called the sum of the cross products.

Table 6.4 contains some hypothetical data collected from five subjects.

TABLE 6.4   Hypothetical Data for Computing Pearson r

Subject	X	Y	X^2	Y^2	XY
A	1	2	1	4	2
B	3	5	9	25	15
C	4	3	16	9	12
D	6	7	36	49	42
E	7	5	49	25	35
Total	21	22	111	112	106

$$r = \frac{\sum XY - \frac{(\sum X)(\sum Y)}{N}}{\sqrt{\left[\sum X^2 - \frac{(\sum X)^2}{N}\right]\left[\sum Y^2 - \frac{(\sum Y)^2}{N}\right]}}$$

$$= \frac{106 - \frac{21(22)}{5}}{\sqrt{\left[111 - \frac{(21)^2}{5}\right]\left[112 - \frac{(22)^2}{5}\right]}}$$

$$= \frac{13.6}{18.616} = 0.731 = 0.73$$

Let's use these data to calculate Pearson r:

$$r = \frac{\sum XY - \dfrac{(\sum X)(\sum Y)}{N}}{\sqrt{\left[\sum X^2 - \dfrac{(\sum X)^2}{N}\right]\left[\sum Y^2 - \dfrac{(\sum Y)^2}{N}\right]}}$$

$\sum XY$ is called the sum of the cross products. It is found by multiplying the X and Y scores for each subject and then summing the resulting products. Calculation of $\sum XY$ and the other terms is illustrated in Table 6.4. Substituting these values in the previous equation, we obtain

$$r = \frac{106 - \dfrac{21(22)}{5}}{\sqrt{\left[111 - \dfrac{(21)^2}{5}\right]\left[112 - \dfrac{(22)^2}{5}\right]}} = \frac{13.6}{\sqrt{22.8(15.2)}} = \frac{13.6}{18.616} = 0.731 = 0.73$$

PRACTICE PROBLEM 6.1

Let's try another problem. This time we shall use data given in Table 6.1. For your convenience, these data are reproduced in the first three columns of Table 6.5. In this example, we have an imperfect linear relationship, and we are interested in computing the magnitude and direction of the relationship, using Pearson r. The solution is also shown in Table 6.5.

TABLE 6.5   IQ and Grade Point Average: Computing Pearson r

Student No.	IQ X	Grade Point Average Y	X^2	Y^2	XY
1	110	1.0	12,100	1.00	110.0
2	112	1.6	12,544	2.56	179.2
3	118	1.2	13,924	1.44	141.6
4	119	2.1	14,161	4.41	249.9
5	122	2.6	14,884	6.76	317.2
6	125	1.8	15,625	3.24	225.0
7	127	2.6	16,129	6.76	330.2
8	130	2.0	16,900	4.00	260.0
9	132	3.2	17,424	10.24	422.4
10	134	2.6	17,956	6.76	348.4
11	136	3.0	18,496	9.00	408.0
12	138	3.6	19,044	12.96	496.8
Total	1503	27.3	189,187	69.13	3488.7

$$r = \frac{\sum XY - \dfrac{(\sum X)(\sum Y)}{N}}{\sqrt{\left[\sum X^2 - \dfrac{(\sum X)^2}{N}\right]\left[\sum Y^2 - \dfrac{(\sum Y)^2}{N}\right]}}$$

$$= \frac{3488.7 - \dfrac{1503(27.3)}{12}}{\sqrt{\left[189{,}187 - \dfrac{(1503)^2}{12}\right]\left[69.13 - \dfrac{(27.3)^2}{12}\right]}} = \frac{69.375}{81.088} = 0.856 \approx 0.86$$

PRACTICE PROBLEM 6.2

Let's try one more problem. Have you ever wondered whether it is true that opposites attract? We've all been with couples in which the two individuals seem so different from each other. But is this the usual experience? Does similarity or dissimilarity foster attraction? A social psychologist investigating this problem asked 15 college students to fill out a questionnaire concerning their attitudes toward a wide variety of topics. Some time later, they were shown the "attitudes" of a stranger to the same items and were asked to rate the stranger as to probable liking for the stranger and probable enjoyment of working with him. The "attitudes" of the stranger were really made up by the experimenter and varied over subjects regarding the proportion of attitudes held by the stranger that were similar to those held by the rater. Thus, for each subject, data were collected concerning his attitudes and the attraction of a stranger based on the stranger's attitudes to the same items. If similarities attract, then there should be a direct relationship between the attraction of the stranger and the proportion of his similar attitudes. The data are presented in Table 6.6. The higher the attraction, the higher the score. The maximum possible attraction score is 14. Compute the Pearson r correlation coefficient* to determine whether there is a direct relationship between similarity of attitudes and attraction.

SOLUTION

The solution is shown in Table 6.6

---

* As will be pointed out later in the chapter, it is legitimate to calculate Pearson r only where the data are of interval or ratio scaling. Therefore, to calculate Pearson r for this problem, we must assume the data are at least of interval scaling.

TABLE 6.6 Data and Solution to Practice Problem 6.2

Student No.	Proportion of Similar Attitudes X	Attraction Y	X^2	Y^2	XY
1	0.30	8.9	0.090	79.21	2.670
2	0.44	9.3	0.194	86.49	4.092
3	0.67	9.6	0.449	92.16	6.432
4	0.00	6.2	0.000	38.44	0.000
5	0.50	8.8	0.250	77.44	4.400
6	0.15	8.1	0.022	65.61	1.215
7	0.58	9.5	0.336	90.25	5.510
8	0.32	7.1	0.102	50.41	2.272
9	0.72	11.0	0.518	121.00	7.920
10	1.00	11.7	1.000	136.89	11.700
11	0.87	11.5	0.757	132.25	10.005
12	0.09	7.3	0.008	53.29	0.657
13	0.82	10.0	0.672	100.00	8.200
14	0.64	10.0	0.410	100.00	6.400
15	0.24	7.5	0.058	56.25	1.800
Total	7.34	136.5	4.866	1279.69	73.273

$$r = \frac{\sum XY - \frac{(\sum X)(\sum Y)}{N}}{\sqrt{\left[\sum X^2 - \frac{(\sum X)^2}{N}\right]\left[\sum Y^2 - \frac{(\sum Y)^2}{N}\right]}}$$

$$= \frac{73.273 - \frac{7.34(136.5)}{15}}{\sqrt{\left[4.866 - \frac{(7.34)^2}{15}\right]\left[1279.69 - \frac{(136.5)^2}{15}\right]}}$$

$$= \frac{6.479}{\sqrt{1.274(37.54)}} = \frac{6.479}{6.916} = 0.937 = 0.94$$

Therefore, based on these students, there is a very strong relationship between similarity and attractiveness.

A Second Interpretation for Pearson r Pearson r can also be interpreted in terms of the variability of Y accounted for by X. This approach leads to important additional information about r and the relationship between X and Y. Consider Figure 6.9 in which an imperfect relationship is shown between X and Y. In this example, the X variable represents spelling competence and the Y variable, writing ability of six students in the third grade. Suppose we are interested in predicting the writing score for Maria, the student whose spelling score is 88. If there were no relationship between writing and spelling, we would predict a

FIGURE 6.9

Relationship between
spelling and writing

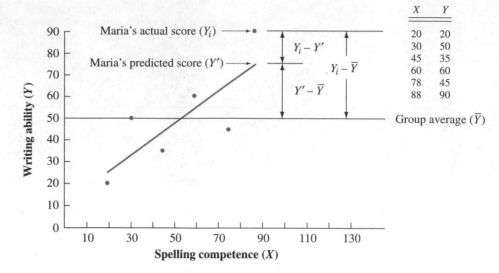

X	Y
20	20
30	50
45	35
60	60
78	45
88	90

score of 50, which is the overall mean of all the writing scores. In the absence of a relationship between X and Y, the overall mean is the best predictor, because the sum of the squared deviations from it is a minimum. You will recognize this as the fourth property of the mean, discussed in Chapter 4. Since Maria's actual writing score is 90, our estimate of 50 is in error by 40 points. Thus,

$$\text{Maria's actual writing score} - \text{Group average} = Y_i - \overline{Y} = 90 - 50 = 40$$

However, in this example, the relationship between X and Y is not zero. Although it is not perfect, a relationship greater than zero exists between X and Y. Therefore, the overall mean of the writing scores is not the best predictor. Rather, as discussed previously in the chapter, we can use the regression line for these data as the basis of our prediction. The regression line for the writing and spelling scores is shown in Figure 6.9. Using this line, we would predict a writing score of 75 for Maria. Now the error is only 15 points. Thus,

$$\left(\begin{array}{c}\text{Maria's actual}\\\text{score}\end{array}\right) - \left(\begin{array}{c}\text{Maria's predicted}\\\text{score using } X\end{array}\right) = Y_i - Y' = 90 - 75 = 15$$

It can be observed in Figure 6.9 that the distance between Maria's score and the mean of the Y scores is divisible into two segments. Thus,

$$Y_i - \overline{Y} = (Y_i - Y') + (Y' - \overline{Y})$$

Deviation of Y_i =   Error in   +   Deviation
prediction   of Y_i
using the   accounted
relationship   for by the
between X relationship
and Y between X
and Y

The segment $Y_i - Y'$ represents the error in prediction. The remaining segment $Y' - \overline{Y}$ represents that part of the deviation of Y_i that is accounted for by the relationship between X and Y. You should note that "accounted for by the relationship between X and Y" is often abbreviated as "accounted for by X."

Suppose we now determine the predicted Y score (Y') for each X score using the regression line. We could then construct $Y_i - \overline{Y}$ for each score. If we squared each $Y_i - \overline{Y}$ and summed over all the scores, we would obtain

$$\Sigma\,(Y_i - \overline{Y})^2 = \Sigma\,(Y_i - Y')^2 + \Sigma\,(Y' - \overline{Y})^2$$

| *Total variability of Y* | = *Variability of prediction errors* | + *Variability of Y accounted for by X* |

Note that $\Sigma\,(Y_i - \overline{Y})^2$ is the sum of squares of the Y scores. It represents the total variability of the Y scores. Thus, this equation states that the total variability of the Y scores can be divided into two parts: the variability of the prediction errors and the variability of Y accounted for by X.

We know that, as the relationship gets stronger, the prediction gets more accurate. In the previous equation, as the relationship gets stronger, the prediction errors get smaller, also causing the variability of prediction errors $\Sigma\,(Y_i - Y')^2$ to decrease. Since the total variability $\Sigma\,(Y_i - \overline{Y})^2$ hasn't changed, the variability of Y accounted for by X, namely, $\Sigma\,(Y' - \overline{Y})^2$, must increase. Thus, the proportion of the total variability of the Y scores that is accounted for by X, namely, $\Sigma\,(Y' - \overline{Y})^2 / \Sigma\,(Y_i - \overline{Y})^2$, is a measure of the strength of relationship. It turns out that if we take the square root of this ratio and substitute for Y' the appropriate values, we obtain the computational formula for Pearson r. We previously defined Pearson r as a measure of the extent to which paired scores occupy the same or opposite positions within their own distributions. From what we have just said, it is also the case that Pearson r equals the square root of the proportion of the variability of Y accounted for by X. In equation form,

$$r = \sqrt{\frac{\Sigma\,(Y' - \overline{Y})^2}{\Sigma\,(Y_i - \overline{Y})^2}} = \sqrt{\frac{\text{Variability of } Y \text{ accounted for by } X}{\text{Total variability of } Y}}$$

$$r = \sqrt{\text{Proportion of the variability of } Y \text{ accounted for by } X}$$

It follows from this equation that the higher r is, the greater the proportion of the variability of Y that is accounted for by X.

Relationship of r^2 and Explained Variability   If we square the previous equation, we obtain

$$r^2 = \text{Proportion of the variability of } Y \text{ accounted for by } X$$

Thus, r^2 is called the *coefficient of determination*. As shown in the equation, r^2 equals the proportion of the total variability of Y that is accounted for or explained by X. In the problem dealing with grade point average and IQ, the correlation was 0.86. If we square r, we obtain

$$r^2 = (0.86)^2 = 0.74$$

TABLE 6.7

Relationship Between r and Explained Variability

r	Explained Variability, r^2 (%)
0.10	1
0.20	4
0.30	9
0.40	16
0.50	25
0.60	36
0.70	49
0.80	64
0.90	81
1.00	100

This means that 74% of the variability in Y can be accounted for by IQ. If it turns out that IQ is a causal factor in determining grade point average, then r^2 tells us that IQ accounts for 74% of the variability in grade point average. What about the remaining 26%? Other factors that can account for the remaining 26% must be influencing grade point average. The important point here is that one can be misled by using r into thinking that X may be a major cause of Y when really it is r^2 that tells us how much of the change in Y can be accounted for by X. The error isn't so serious when you have a correlation coefficient as high as 0.86. However, in the behavioral sciences, such high correlations are rare. Correlation coefficients of $r = 0.50$ or 0.60 are considered fairly high, and yet correlations of this magnitude account for only 25 to 36% of the variability in Y ($r^2 = 0.25$ to 0.36). Table 6.7 shows the relationship between r and the explained variability expressed as a percentage.

Other Correlation Coefficients

So far, we have discussed correlation and described in some detail the linear correlation coefficient Pearson r. We have chosen Pearson r because it is the most frequently encountered correlation coefficient in behavioral science research. However, you should be aware that there are many different correlation coefficients one might employ, each of which is appropriate under different conditions. In deciding which correlation coefficient to calculate, the shape of the relationship and the measuring scale of the data are the two most important considerations.

Shape of the Relationship   The choice of which correlation coefficient to calculate depends on whether the relationship is linear or curvilinear. If the data are curvilinear, using a linear correlation coefficient such as Pearson r can seriously underestimate the degree of relationship that exists between X and Y. Accordingly, another correlation coefficient η (eta) is used for curvilinear relationships. Since η is not frequently encountered in behavioral science research, we have not presented a discussion of it.* This does, however, emphasize the importance of doing a scatter plot to determine whether the relationship is linear before just routinely going ahead and calculating a linear correlation coefficient.

Measuring Scale   The choice of correlation coefficient also depends on the type of measuring scale underlying the data. We've already discussed the linear correlation coefficient Pearson r. It assumes the data are measured on an interval or ratio scale. Some examples of other linear correlation coefficients are the Spearman rank order correlation coefficient rho (r_s), the biserial correlation coefficient (r_b), and the phi (ϕ) coefficient. In actuality, each of these coefficients is really the equation for Pearson r simplified to apply to the lower-order scaling. Rho is used when one or both of the variables are of ordinal scaling, r_b is used when one of the variables is at least interval and the other is dichotomous, and phi is used when each of the variables is dichotomous. Although it is beyond the scope of this textbook to present each of these correlation coefficients in detail, the Spearman rank order correlation coefficient rho occurs frequently enough to warrant discussion here.

---

* A discussion of η as well as the other coefficients presented in this section is contained in N. Downie and R. Heath, *Basic Statistical Methods,* 4th ed., Harper & Row, New York, 1974, pp. 102–114.

The Spearman Rank Order Correlation Coefficient Rho (r_s)   As mentioned, the Spearman rank order correlation coefficient rho is used when one or both of the variables are only of ordinal scaling. Spearman rho is really the linear correlation coefficient Pearson r applied to data that meet the requirements of ordinal scaling. The easiest equation for calculating rho when there are no ties or just a few ties relative to the number of paired scores is

$$r_s = 1 - \frac{6 \sum D_i^2}{N^3 - N} \qquad \textit{computational equation for rho}$$

where
D_i = difference between the ith pair of ranks = $R(X_i) - R(Y_i)$
$R(X_i)$ = rank of the ith X score
$R(Y_i)$ = rank of the ith Y score
N = number of pairs of ranks

It can be shown that, with ordinal data having no ties, Pearson r reduces algebraically to the previous equation.

To illustrate the use of rho, let's consider an example. Assume that a large corporation is interested in rating a current class of 12 management trainees on their leadership ability. Two psychologists are hired to do the job. As a result of their tests and interviews, the psychologists each independently rank-order the students according to leadership ability. The rankings are from 1 to 12, with 1 representing the highest level of leadership. The data are given in Table 6.8. What is the correlation between the rankings of the two psychologists?

Since the data are of ordinal scaling, we should compute rho. The solution is shown in Table 6.8. Note that subjects 5 and 6 were tied in the rankings of psychologist A. When ties occur, the rule is to give each subject the average of the tied ranks. For example, subjects 5 and 6 were tied for ranks 2 and 3.

TABLE 6.8   Calculation of r_s for Leadership Example

Subject	Rank Order of Psychologist A $R(X_i)$	Rank Order of Psychologist B $R(Y_i)$	$D_i = R(X_i) - R(Y_i)$	D_i^2
1	6	5	1	1
2	5	3	2	4
3	7	4	3	9
4	10	8	2	2.25
5	2.5	1	1.5	12.25
6	2.5	6	−3.5	1
7	9	10	−1	1
8	1	2	−1	4
9	11	9	2	9
10	4	7	−3	9
11	8	11	−3	0
12	12	12	0	$\sum D_i^2 = 56.5$

$N - 12$

$$r_s = 1 - \frac{6 \sum D_i^2}{N^3 - N} = 1 - \frac{6(56.5)}{(12)^3 - 12} = 1 - \frac{339}{1716} = 0.80$$

Therefore, they each received a ranking of 2.5 $[(2 + 3)/2 = 2.5]$. In giving the two subjects a rank of 2.5, we have effectively used up ranks 2 and 3. The next rank is 4. D_i is the difference between the paired rankings for the ith subject. Thus, $D_i = 1$ for subject 1. It doesn't matter whether you subtract $R(X_i)$ from $R(Y_i)$ or $R(Y_i)$ from $R(X_i)$ to get D_i because we square each D_i value. The squared D_i values are then summed ($\sum D_i^2 = 56.5$). This value is then entered in the equation along with N ($N = 12$), and r_s is computed. For this problem,

$$r_s = 1 - \frac{6\sum D_i^2}{N^3 - N} = 1 - \frac{6(56.5)}{12^3 - 12} = 1 - \frac{339}{1716} = 0.80$$

PRACTICE PROBLEM 6.3

To illustrate computation of r_s, let's assume that the raters' attitude and attraction scores given in Practice Problem 6.2 were only of ordinal scaling. Given this assumption, determine the value of the lienar correlation coefficient rho for these data and compare the value with the value of Pearson r determined in Practice Problem 6.2.

SOLUTION

The data and solution are shown in Table 6.9.

TABLE 6.9   Calculation of r_s for Attitude Example

Subject	Proportion of Similar Attitudes X_i	Attraction Y_i	Rank of X_i $R(X_i)$	Rank of Y_i $R(Y_i)$	$D_i = R(X_i) - R(Y_i)$	D_i^2
1	0.30	8.9	5	7	−2	4
2	0.44	9.3	7	8	−1	1
3	0.67	9.6	11	10	1	1
4	0.00	6.2	1	1	0	0
5	0.50	8.8	8	6	2	4
6	0.15	8.1	3	5	−2	4
7	0.58	9.5	9	9	0	0
8	0.32	7.1	6	2	4	16
9	0.72	11.0	12	13	−1	1
10	1.00	11.7	15	15	0	0
11	0.87	11.5	14	14	0	0
12	0.09	7.3	2	3	−1	1
13	0.82	10.0	13	11.5	1.5	2.25
14	0.64	10.0	10	11.5	−1.5	2.25
15	0.24	7.5	4	4	0	0

$N = 15$ $\sum D_i^2 = 36.5$

$$r_s = 1 - \frac{6\sum D_i^2}{N^3 - N} = 1 - \frac{6(36.5)}{(15)^3 - 15} = 1 - \frac{219}{3360} = 0.93$$

Note that $r_s = 0.93$ and $r = 0.94$. The values are not identical but quite close. In general, when Pearson r is calculated using the interval or ratio properties of data, its values will be close but not exactly the same as when calculated on only the ordinal properties of those data.

Effect of Range on Correlation

If a correlation exists between X and Y, restricting the range of either of the variables will have the effect of *lowering* the correlation. This can be seen in Figure 6.10, where we have drawn a scatter plot of freshman grade point average and College Entrance Examination Board (CEEB) scores. The figure has been subdivided into low, medium, and high CEEB scores. Taking the figure as a whole, i.e., considering the full range of the CEEB scores, there is a high correlation between the two variables. However, if we were to consider the three sections separately, the correlation for each section would be much lower. Within each

FIGURE 6.10
....

Freshman grades and
CEEB scores

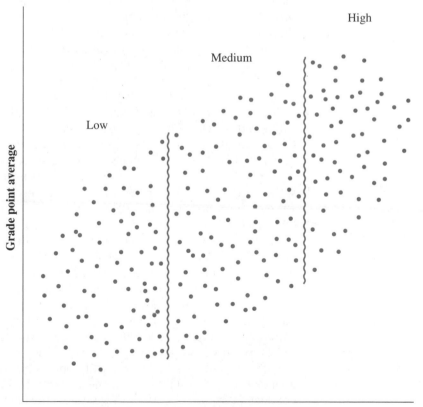

College Entrance Examination Board (CEEB) scores

section, the points show much less systematic change in Y with changes in X. This, of course, indicates a lower correlation between X and Y. The effect of range restriction on correlation is often encountered in education or industry. For instance, suppose that on the basis of the high correlation between freshman grades and CEEB scores as shown in Figure 6.10, a college decided to admit only high school graduates who have scored in the high range of the CEEB scores. If the subsequent freshman grades of these students were correlated with their CEEB scores, we would expect a much lower correlation because of the range restriction of the CEEB scores for these freshmen. In a similar vein, if one is doing a correlational study and obtains a low correlation coefficient, one should check to be sure that range restriction is not responsible for the low value.

Correlation Does Not Imply Causation

When two variables (X and Y) are correlated, it is tempting to conclude that one of them is the cause of the other. However, to do so without further experimentation would be a serious error, because whenever two variables are correlated, there are four possible explanations of the correlation: (1) the correlation between X and Y is spurious; (2) X is the cause of Y; (3) Y is the cause of X; or (4) a third variable is the cause of the correlation between X and Y. The first possibility asserts that it was just due to accidents of sampling unusual people or unusual behavior that the sample showed a correlation; that is, if the experiment were repeated or more samples were taken, the correlation would disappear. If the correlation is really spurious, it is obviously wrong to conclude that there is a causal relation between X and Y.

It is also erroneous to assume causality between X and Y if the fourth alternative is correct. Quite often when X and Y are correlated, they are not causally related to each other, but rather a third variable is responsible for the correlation. For example, do you know that there is a close relationship between the salaries of university professors and the price of a fifth of scotch whiskey? Which is the cause and which the effect? Do the salaries of university professors dominate the scotch whiskey market such that when the professors get a raise and thereby can afford to buy more scotch, the price of scotch is raised accordingly? Or perhaps the university professors are paid from the profits of scotch whiskey sales, so that when the professors need a raise, the price of a fifth of scotch whiskey goes up? Actually, neither of these explanations is correct. Rather, a third factor is responsible for this correlation. What is that factor? Inflation! Recently, a newspaper article reported a positive correlation between obesity and female crime. Does this mean that if a woman gains 20 pounds, she will become a criminal? Or does it mean that if she is a criminal, she is doomed to being obese? Neither of these explanations seems satisfactory. Frankly, we are not sure how to interpret this correlation. One possibility is that it is a spurious correlation. If not, it could be due to a third factor, namely, socioeconomic status. Both obesity and crime are related to lower socioeconomic status.

The point we are trying to make is that a correlation between two variables is not sufficient to establish causality between them. There are other possible explanations. To establish that one variable is the cause of another, we must conduct an experiment in which we systematically vary *only* the suspected causal variable and measure its effect on the other variable.

WHAT IS THE TRUTH?

"GOOD PRINCIPAL = GOOD ELEMENTARY SCHOOL," OR DOES IT?

A major newspaper of a large city recently carried as front page headlines, printed in large bold letters, **"Equation for success: Good principal = good elementary school."** The article that followed described a study in which elementary school principals were rated by their teachers on a series of questions indicating whether the principals were strong, average, or weak leaders. Students in these schools were evaluated in reading and mathematics on the annual California Achievement Tests. As far as we can tell from the newspaper article, the ratings and test scores were obtained from ongoing principal assignments, with no attempt in the study to randomly assign principals to schools. The results showed that: (1) in 11 elementary schools that had strong principals, students were making big academic strides, (2) in 11 schools where principals were weak leaders, students were showing less improvement than average or even falling behind, and (3) in 39 schools where principals were rated as average, the students' test scores were showing just average improvement.

The newspaper reporter interpreted these data as indicated by the headline, "Equation for success: Good principal = good elementary school." In the article,

an elementary school principal was quoted, "I've always said, 'Show me a good school, and I'll show you a good principal,' but now we have powerful, incontrovertible data that corroborates that." The article further quoted the president of the principals' association as saying: "It's exciting information that carries an enormous responsibility. It shows we can make a real difference in our students' lives." In your view, do the data warrant these conclusions?

ANSWER Although, personally, I believe that school principals are important to educational quality, the study seems to be strictly a correlational one: paired measurements on two variables, without random assignment to groups. From what we said pre-

viously, it is impossible to determine causality from such a study. The individuals quoted herein have taken a study that shows there is a *correlation* between "strong" leadership by elementary school principals and educational gain, and concluded that the principals *caused* the educational gain. The conclusion is too strong. The correlation could be spurious or due to a third variable.

It is truly amazing how often this error is made in real life. Stay on the lookout, and I believe you will be surprised how frequently you encounter individuals concluding causation when the data are only correlational. Of course, now that you are so well informed on this point, you will never make this mistake yourself!

SUMMARY

In this chapter, we have discussed the topic of correlation. Correlation is a measure of the relationship that exists between two variables. The magnitude and direction of the relationship are given by a correlation coefficient. The correlation coefficient can vary from +1 to −1. The sign of the coefficient tells us whether the relationship is positive or negative. The numerical part describes the magnitude of the correlation. When the relationship is perfect, the magnitude is 1. If the relationship is nonexistent, the magnitude is 0. Magnitudes between 0 and 1 indicate imperfect relationships.

There are many correlation coefficients that can be computed, depending on the scaling of the data and the shape of the relationship. In this chapter, we emphasized Pearson r and Spearman rho. Pearson r is defined as a measure of the extent to which paired scores occupy the same or opposite positions within their own distributions. Using standard scores allows measurement of the relationship that is independent of the differences in scaling and of the units used in measuring the variables. Pearson r is also equal to the square root of the proportion of the variability in Y that is accounted for by X. In addition to these concepts, we presented a computational equation for r and practice calculating r.

Spearman rho is used for linear relationships when one or both of the variables are only of ordinal scaling. The computational equation for rho was presented and several practice problems worked out. Next, we discussed the effect of range on correlation and pointed out that truncated range will result in a lower correlation coefficient.

As the last topic of correlation, we discussed correlation and causation. We pointed out that if a correlation exists between two variables in an experiment, we cannot conclude they are causally related on the basis of the correlation alone, because there are other possible explanations. The correlation may be spurious, or a third variable may be responsible for the correlation between the first two variables. To establish causation, one of the variables must be independently manipulated and its effect on the other variable measured. All other variables should be held constant or vary unsystematically. Even if the two variables are causally related, it is important to keep in mind that r^2, rather than r, indicates the magnitude of the effect of one variable on the other.

IMPORTANT TERMS

Biserial coefficient (p. 116)
Coefficient of determination
  (p. 115)
Correlation (p. 99)
Correlation coefficient (p. 106)
Curvilinear relationship (p. 100)
Direct relationship (p. 103)

Imperfect relationship (p. 103)
Inverse relationship (p. 103)
Linear relationship (p. 100)
Negative relationship (p. 103)
Pearson r (p. 106)
Perfect relationship (p. 103)
Phi coefficient (p. 116)

Positive relationship (p. 103)
Scatter plot (p. 100)
Slope (p. 101)
Spearman rho (p. 117)
Variability accounted for by X
  (p. 115)
Y intercept (p. 101)

QUESTIONS AND PROBLEMS

1. Define or identify each of the terms in the "Important Terms" section.
2. Discuss the different kinds of relationships that are possible between two variables.
3. For each scatter plot in the accompanying figure (parts **a**–**f**), determine whether the relationship is
   a. Linear or curvilinear. If linear, further determine whether it is positive or negative.
   b. Perfect or imperfect

Scatter plots for
Question 3

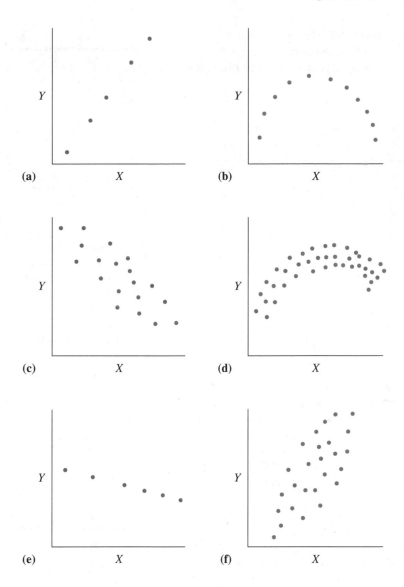

(a)   X

(b)   X

(c)   X

(d)   X

(e)   X

(f)   X

4. Professor Taylor does an experiment and establishes that a correlation exists between variables
A and *B*. Based on this correlation, she asserts
that *A* is the cause of *B*. Is this assertion correct? Explain.

5. Give two meanings of Pearson *r*.

6. Why are *z* scores used as the basis for determining Pearson *r*?

7. What is the range of values that a correlation
coefficient may take?

8. A recent study has shown that the correlation
between fatigue and irritability is 0.53. On the
basis of this correlation, the author concludes
that fatigue is an important factor in producing
irritability. Is this conclusion justified? Explain.

9. What factors influence the choice of whether to
use a particular correlation coefficient? Give
some examples.

10. The Pearson *r* and Spearman rho correlation coefficients are related. Is this statement correct?
Explain.

11. When two variables are correlated, there are four
possible explanations of the correlation. What
are they?

12. What effect does decreasing the range of the paired scores have on the correlation coefficient?

13. Given the following sets of paired sample scores:

A		B		C	
X	Y	X	Y	X	Y
1	1	4	2	1	5
4	2	5	4	4	4
7	3	8	5	7	3
10	4	9	1	10	2
13	5	10	4	13	1

a. Use the equation

$$r = \Sigma\, z_X z_Y / (N - 1)$$

to compute the value of Pearson r for each set. Note that in set B, where the correlation is lowest, some of the $z_X z_Y$ values are positive and some are negative. These tend to cancel each other, causing r to have a low magnitude. However, in both sets A and C, all the products have the same sign, causing r to be large in magnitude. When the paired scores occupy the same or opposite positions within their own distributions, the $z_X z_Y$ products have the same sign, resulting in high magnitudes for r.

b. Compute r for set B, using the raw score equation. Which do you prefer, using the raw score or the z score equation?

c. Add the constant 5 to the X scores in set A and compute r again, using the raw score equation. Has the value changed?

d. Multiply the X scores in set A by 5 and compute r again. Has the value changed?

e. Generalize the results obtained in parts c and d to subtracting and dividing the scores by a constant. What does this tell you about r?

14. In a large introductory sociology course, a professor gives two exams. The professor wants to determine whether the scores students receive on the second exam are correlated with their scores on the first exam. To make the calculations easier, a sample of eight students is selected. Their scores are shown in the accompanying table.

Student	Exam 1	Exam 2
1	60	60
2	75	100
3	70	80
4	72	68
5	54	73
6	83	97
7	80	85
8	65	90

a. Construct a scatter plot of the data, using exam 1 score as the X variable. Does the relationship look linear?

b. Assuming a linear relationship exists between scores on the two exams, compute the value for Pearson r.

c. How well does the relationship account for the scores on exam 2?

15. A researcher conducts a study to investigate the relationship between cigarette smoking and illness. The number of cigarettes smoked daily and the number of days absent from work in the last year due to illness is determined for 12 individuals employed at the company where the researcher works. The scores are given in the table.

Subject	Cigarettes Smoked	Days Absent
1	0	1
2	0	3
3	0	8
4	10	10
5	13	4
6	20	14
7	27	5
8	35	6
9	35	12
10	44	16
11	53	10
12	60	16

a. Construct a scatter plot for these data. Does the relationship look linear?

b. Calculate the value of Pearson r.

c. Eliminate the data from subjects 1, 2, 3, 10, 11, and 12. This decreases the range of both variables. Recalculate r for the remaining subjects. What effect does decreasing the range have on r?

d. Using the full set of scores, what percentage of the variability in the number of days absent is accounted for by the number of cigarettes smoked daily? Of what use is this value?

16. An educator has constructed a test for mechanical aptitude. He wants to determine how reliable the test is over two administrations spaced by 1 month. A study is conducted in which 10 students are given two administrations of the test, with the second administration being given 1 month after the first. The data are given in the table.

Student	Administration 1	Administration 2
1	10	10
2	12	15
3	20	17
4	25	25
5	27	32
6	35	37
7	43	40
8	40	38
9	32	30
10	47	49

a. Construct a scatter plot of the paired scores.
b. Determine the value of r.
c. Would it be fair to say that this is a reliable test? Explain using r^2.

17. A group of researchers has devised a stress questionnaire consisting of 15 life events. They are interested in determining whether there is cross-cultural agreement on the relative amount of adjustment each event entails. The questionnaire is given to 300 Americans and 300 Italians. Each individual is instructed to use the event of "marriage" as the standard and to judge each of the other life events in relation to the adjustment required in marriage. Marriage is arbitrarily given a value of 50 points. If an event is judged to require greater adjustment than marriage, the event should receive more than 50 points. How many more points depends on how much more adjustment is required. After each subject within each culture has assigned points to the 15 life events, the points for each event are averaged. The results are shown in the table that follows.

Life Event	Americans	Italians
Death of spouse	100	80
Divorce	73	95
Marital separation	65	85
Jail term	63	52
Personal injury	53	72
Marriage	50	50
Fired from work	47	40
Retirement	45	30
Pregnancy	40	28
Sex difficulties	39	42
Business readjustment	39	36
Trouble with in-laws	29	41
Trouble with boss	23	35
Vacation	13	16
Christmas	12	10

a. Assume the data are at least of interval scaling and compute the correlation between the American and Italian ratings.
b. Assume the data are only of ordinal scaling and compute the correlation between ratings of the two cultures.

18. A psychologist has constructed a paper and pencil test purported to measure depression. To see how the test compares to the ratings of experts, 12 "emotionally disturbed" individuals are given the paper and pencil test. The individuals are also independently rank-ordered by two psychiatrists according to the degree of depression each psychiatrist finds as a result of detailed interviews. The scores are given here. Higher scores represent greater depression.

Individual	Paper & Pencil Test	Psychiatrist A	Psychiatrist B
1	48	12	9
2	37	11	12
3	30	4	5
4	45	7	8
5	31	10	11
6	24	8	7
7	28	3	4
8	18	1	1
9	35	9	6
10	15	2	2
11	42	6	10
12	22	5	3

a. What is the correlation between the rankings of the two psychiatrists?
b. What is the correlation between the scores on the paper and pencil test and the rankings of each psychiatrist?

19. For this problem, let's suppose that you are a psychologist employed in the human resources department of a large corporation. The corporation president has just finished talking with you about the importance of hiring productive personnel in the manufacturing section of the corporation and has asked you to help improve the corporation's ability to do so. There are 300 employees in this section, with each employee making the same item. Until now, the corporation has been depending solely on interviews for selecting these employees. You search the literature and discover two well-standardized paper and pencil performance tests that you think might be related to the performance requirements of this section. To determine whether either might be used as a screening device, you select 10 representative employees from the manufacturing section, making sure that a wide range of performance is represented in the sample, and administer the two tests to each employee. The data are shown in the table. The higher the score, the better the performance. The work performance scores are the actual number of items completed by each employee per week, averaged over the past 6 months.

a. Construct a scatter plot of work performance and test 1, using test 1 as the X variable. Does the relationship look linear?
b. Assuming it is linear, compute the value of Pearson r.
c. Construct a scatter plot of work performance and test 2, using test 2 as the X variable. Is the relationship linear?
d. Assuming it is linear, compute the value of Pearson r.
e. If you could use only one of the two tests for screening prospective employees, would you use either test? If yes, which one? Explain.

| | **Employee** | | | | | | | | | |
	1	2	3	4	5	6	7	8	9	10
Work Performance	50	74	62	90	98	52	68	80	88	76
Test 1	10	19	20	20	21	14	10	24	16	14
Test 2	25	35	40	49	50	29	32	44	46	35

7 | LINEAR REGRESSION

INTRODUCTION

Regression and correlation are closely related. At the most basic level, they both involve the relationship between two variables; and they both utilize the same set of basic data, paired scores taken from the same or matched subjects. As we saw in Chapter 6, correlation is concerned with the magnitude and direction of the relationship. Regression focuses on using the relationship for prediction. Prediction is quite easy when the relationship is perfect. If the relationship is perfect, all the points fall on a straight line and all we must do is derive the equation of the straight line and use it for prediction. Prediction is perfect. The situation is more complicated when the relationship is imperfect.

DEFINITION

- **Regression** *is a topic that considers using the relationship between two or more variables for prediction.*

PREDICTION AND IMPERFECT RELATIONSHIPS

Let's return to the data involving grade point average and IQ that were presented in Chapter 6. For convenience, the data have been reproduced in Table 7.1. Figure 7.1 shows a scatter plot of the data. The relationship is imperfect, positive, and linear. The problem we face is how to determine the single straight line that best describes the data. The solution most often used is to construct the line that minimizes errors of prediction according to a *least-squares* criterion. This line is called the *least-squares regression line.*

The least-squares regression line for the data in Table 7.1 is shown in Figure 7.2(a). The vertical distance between each point and the line represents the error in prediction. If we let Y' = the predicted Y value and Y = the actual value, then $Y - Y'$ equals the error for each point. It might seem that the total error in prediction should be the simple algebraic sum of $Y - Y'$, summed over all of the points. If this were true, since we are interested in minimizing the error, we would construct the line that minimizes $\Sigma (Y - Y')$. However, the total error in prediction does not equal $\Sigma (Y - Y')$, because some of the Y' values will be

TABLE 7.1   IQ and Grade Point Average of 12 College Students

Student No.	IQ	Grade Point Average
1	110	1.0
2	112	1.6
3	118	1.2
4	119	2.1
5	122	2.6
6	125	1.8
7	127	2.6
8	130	2.0
9	132	3.2
10	134	2.6
11	136	3.0
12	138	3.6

FIGURE 7.1

Scatter plot of IQ and grade point average

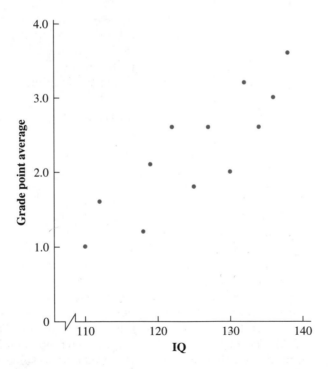

FIGURE 7.2   Two regression lines and prediction error

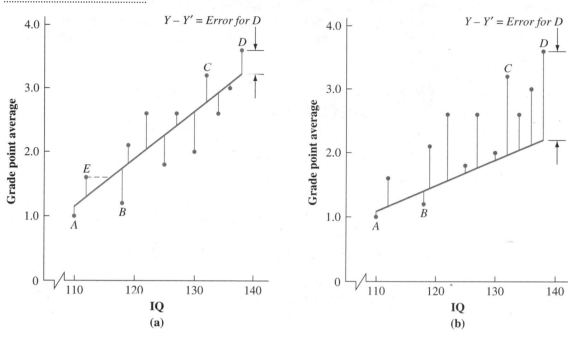

greater than Y and some will be less. Thus, there will be both positive and negative error scores, and the simple algebraic sums of these would cancel each other. We encountered a similar situation when considering measures of the average dispersion. In deriving the equation for the standard deviation, we squared $X - \overline{X}$ to overcome the fact that there were positive and negative deviation scores that canceled each other. The same solution works here, too. Instead of just summing $Y - Y'$, we first compute $(Y - Y')^2$ for each score. This removes the negative values and eliminates the cancellation problem. Now, if we minimize $\Sum (Y - Y')^2$, we minimize the total error of prediction.

DEFINITION

- *The **least-squares regression line** is the prediction line that minimizes* $\Sigma (Y - Y')^2$.

For any linear relationship, there is only one line that will minimize $\Sigma (Y - Y')^2$. Thus, there is only one least-squares regression line for each linear relationship.

We said before that there are many "possible" prediction lines we could construct when the relationship is imperfect. Why should we use the least-squares regression line? We use the least-squares regression line because it gives the greatest overall accuracy in prediction. To illustrate this point, another prediction line has been drawn in Figure 7.2(b). This line has been picked arbitrarily and is just one of an infinite number that could have been drawn. How does it compare in prediction accuracy with the least-squares regression line? We can see that it actually does better for some of the points; e.g., points A and B. However, it also misses horribly on others; e.g., points C and D. If we consider all of the points, it is clear that the line of Figure 7.2(a) fits the points better than the line of Figure 7.2(b). The total error in prediction, represented by

$\Sigma (Y - Y')^2$, is less for the least-squares regression line than for the line in Figure 7.2(b). In fact, the total error in prediction is less for the least-squares regression line than for any other possible prediction line. Thus, the least-squares regression line is used because it gives greater overall accuracy in prediction than any other regression line.

CONSTRUCTING THE LEAST-SQUARES REGRESSION LINE

The equation for the least-squares regression line is given by

$$Y' = b_Y X + a_Y \qquad \textit{linear regression equation for predicting Y given X}$$

where Y' = predicted or estimated value of Y
b_Y = slope of the line for minimizing errors in predicting Y
a_Y = Y axis intercept for minimizing errors in predicting Y

This is, of course, the general equation for a straight line that we have been using all along. In this context, however, a_Y and b_Y are called *regression constants*.
The b_Y regression constant is equal to

$$b_Y = \frac{\Sigma XY - \dfrac{(\Sigma X)(\Sigma Y)}{N}}{SS_X}$$

where SS_X = sum of squares of X scores

$$= \Sigma X^2 - \frac{(\Sigma X)^2}{N}$$

N = number of *paired* scores

The equation for computing b_Y from the raw scores is

$$b_Y = \frac{\Sigma XY - \dfrac{(\Sigma X)(\Sigma Y)}{N}}{\Sigma X^2 - \dfrac{(\Sigma X)^2}{N}} \qquad \begin{array}{l}\textit{computational equation for determining the b}\\ \textit{regression constant for predicting Y given X}\end{array}$$

where ΣXY = sum of the product of each X and Y pair (also called the sum of the cross products)

The a_Y regression constant is given by

$$a_Y = \overline{Y} - b_Y \overline{X} \qquad \begin{array}{l}\textit{computational equation for determining the a regression}\\ \textit{constant for predicting Y given X}\end{array}$$

Since we need the b_Y constant to determine the a_Y constant, the procedure is to first find b_Y and then a_Y. Once both are found, they are substituted into the regression equation. Let's construct the least-squares regression line for the IQ and grade point data presented previously. For convenience, the data have been presented again in Table 7.2.

TABLE 7.2 IQ and Grade Point Average of 12 College Students: Predicting Y from X

Student No.	IQ X	Grade Point Average Y	XY	X^2
1	110	1.0	110.0	12,100
2	112	1.6	179.2	12,544
3	118	1.2	141.6	13,924
4	119	2.1	249.9	14,161
5	122	2.6	317.2	14,884
6	125	1.8	225.0	15,625
7	127	2.6	330.2	16,129
8	130	2.0	260.0	16,900
9	132	3.2	422.4	17,424
10	134	2.6	348.4	17,956
11	136	3.0	408.0	18,496
12	138	3.6	496.8	19,044
Total	1503	27.3	3488.7	189,187

$$b_Y = \frac{\sum XY - \frac{(\sum X)(\sum Y)}{N}}{\sum X^2 - \frac{(\sum X)^2}{N}} = \frac{3488.7 - \frac{1503(27.3)}{12}}{189,187 - \frac{(1503)^2}{12}}$$

$$= \frac{69.375}{936.25} = 0.0\overline{741} = 0.074$$

$$a_Y = \overline{Y} - b_Y\overline{X} = 2.275 - 0.0741(125.25) = -7.006$$

$$Y' = b_Y X + a_Y = 0.074X - 7.006$$

$$b_Y = \frac{\sum XY - \frac{(\sum X)(\sum Y)}{N}}{\sum X^2 - \frac{(\sum X)^2}{N}}$$

$$= \frac{3488.7 - \frac{1503(27.3)}{12}}{189,187 - \frac{(1503)^2}{12}}$$

$$= \frac{69.375}{936.25} = 0.0741 = 0.074$$

$$a_Y = \overline{Y} - b_Y\overline{X}$$

$$= 2.275 - 0.0741(125.25)$$

$$= -7.006$$

FIGURE 7.3

Regression line for grade point average and IQ

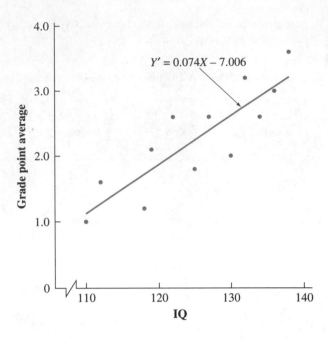

and

$$Y' = 0.074X - 7.006$$

The full solution is also shown in Table 7.2. The regression line has been plotted in Figure 7.3. The equation for Y' can now be used to predict the grade point average knowing only the student's IQ score. For example, suppose a student's IQ score is 124; using this regression line, what is the student's predicted grade point average?

$$Y' = 0.074X - 7.006$$
$$= 0.074(124) - 7.006$$
$$= 2.17$$

Let's try a couple of practice problems.

PRACTICE PROBLEM 7.1

A developmental psychologist is interested in determining whether it is possible to use the heights of young boys to predict their eventual height at maturity. To answer this question, she collects the data shown in Table 7.3.

a. Draw a scatter plot of the data.
b. If the data are linearly related, derive the least-squares regression line.
c. Based on these data, what height would you predict for a 20-year-old if at 3 years his height were 42 inches?

TABLE 7.3 Height of Males at Ages 3 and 20 Years

Individual No.	Height at Age 3 X (in.)	Height at Age 20 Y (in.)	XY	X^2
1	30	59	1,770	900
2	30	63	1,890	900
3	32	62	1,984	1,024
4	33	67	2,211	1,089
5	34	65	2,210	1,156
6	35	61	2,135	1,225
7	36	69	2,484	1,296
8	38	66	2,508	1,444
9	40	68	2,720	1,600
10	41	65	2,665	1,681
11	41	73	2,993	1,681
12	43	68	2,924	1,849
13	45	71	3,195	2,025
14	45	74	3,330	2,025
15	47	71	3,337	2,209
16	48	75	3,600	2,304
Total	618	1077	41,956	24,408

$$b_Y = \frac{\sum XY - \frac{(\sum X)(\sum Y)}{N}}{\sum X^2 - \frac{(\sum X)^2}{N}} = \frac{41,956 - \frac{618(1077)}{16}}{24,408 - \frac{(618)^2}{16}} = 0.6636 = 0.664$$

$$a_Y = \bar{Y} - b_Y\bar{X} = 67.3125 - 0.6636(38.625) = 41.681$$

$$Y' = b_Y X + a_Y = 0.664X + 41.681$$

SOLUTION

a. The scatter plot is shown in Figure 7.4. It is clear that an imperfect relationship that is linear and positive exists between the heights at ages 3 and 20.

b. Derive the least-squares regression line.

$$Y' = b_Y X + a_Y$$

$$b_Y = \frac{\sum XY - \frac{(\sum X)(\sum Y)}{N}}{\sum X^2 - \frac{(\sum X)^2}{N}} = \frac{41,956 - \frac{618(1077)}{16}}{24,408 - \frac{(618)^2}{16}}$$

$$= 0.6636 = 0.664$$

$$a_Y = \bar{Y} - b_Y\bar{X} = 67.3125 - 0.6636(38.625)$$

$$= 41.681$$

Therefore,

$$Y' = b_Y X + a_Y = 0.664X + 41.681$$

The solution is shown in Table 7.3. The least-squares regression line is drawn on the scatter plot of Figure 7.4.

FIGURE 7.4   Scatter plot of height at ages 3 and 20

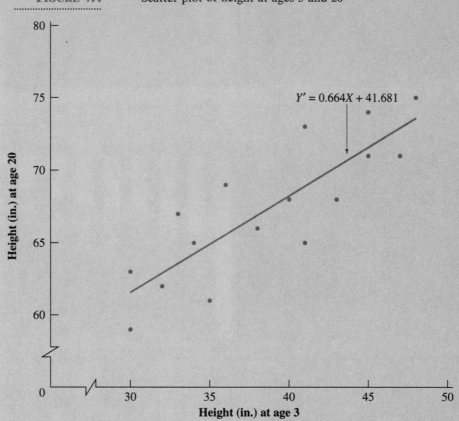

c. Predicted height for the 3-year-old of 42 inches:

$$Y' = 0.664X + 41.681$$
$$= 0.664(42) + 41.681$$
$$= 69.57 \text{ inches}$$

PRACTICE PROBLEM 7.2

A neuroscientist suspects that low levels of the brain neurotransmitter serotonin may be causally related to aggressive behavior. As a first step in investigating this hunch, she decides to do a correlative study involving nine rhesus monkeys. The monkeys are observed daily for 6 months, and the number of aggressive acts recorded. Serotonin levels in the striatum (a brain region associated with aggressive behavior) are also measured once per day for each animal. The resulting data are shown in Table 7.4. The number of aggressive acts for each animal is the average for the 6 months, given on a per day basis. Serotonin levels are also average values over the 6-month period.

a. Draw a scatter plot of the data.
b. If the data are linearly related, derive the least-squares regression line for predicting the number of aggressive acts from serotonin level.
c. Based on these data, what is the number of aggressive acts you would predict if a rhesus monkey had a serotonin level of 0.46 microgm/gm?

TABLE 7.4   Serotonin Levels and Aggression in Rhesus Monkeys

Subject No.	Serotonin Level (microgm/gm) X	Number of Aggressive Acts/day Y	XY	X^2
1	0.32	6.0	1.920	0.1024
2	0.35	3.8	1.330	0.1225
3	0.38	3.0	1.140	0.1444
4	0.41	5.1	2.091	0.1681
5	0.43	3.0	1.290	0.1849
6	0.51	3.8	1.938	0.2601
7	0.53	2.4	1.272	0.2809
8	0.60	3.5	2.100	0.3600
9	0.63	2.2	1.386	0.3969
Total	4.16	32.8	14.467	2.0202

$$b_Y = \frac{\Sigma XY - \frac{(\Sigma X)(\Sigma Y)}{N}}{\Sigma X^2 - \frac{(\Sigma X)^2}{N}} = \frac{14.467 - \frac{(4.16)(32.8)}{9}}{2.0202 - \frac{(4.16)^2}{9}} = -7.127$$

$$a_Y = \bar{Y} - b_Y\bar{X} = 3.6444 - (-7.1274)(0.4622) = 6.939$$

$$Y' = b_Y X + a_Y = -7.127X + 6.939$$

SOLUTION

a. The scatter plot follows. It is clear that an imperfect, linear, negative relationship exists between the two variables.

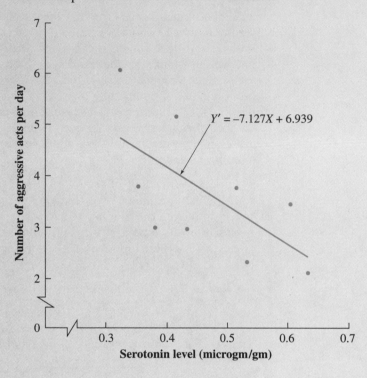

b. Derive the least-squares regression line. The solution is shown at the bottom of Table 7.4 and the regression line has been plotted on the scatter plot of part **a**.

c. Predicted number of aggressive acts:

$$Y' = -7.127X + 6.939$$

$$= -7.127(0.46) + 6.939$$

$$= 3.7 \text{ aggressive acts per day}$$

REGRESSION OF X ON Y

So far we have been concerned with predicting Y scores from the X scores. To do so, we derived a regression line that enabled us to predict Y given X. This is sometimes referred to as the *regression line of Y on X*. It is also possible to predict X given Y. However, to predict X given Y we must derive a new regression line. We cannot use the regression equation for predicting Y given X. For example, in the problem involving IQ (X) and grade point average (Y), we derived the following regression line:

$$Y' = 0.074X - 7.006$$

This line was appropriate for predicting grade point average given IQ, i.e., for predicting Y given X. However, if we want to predict IQ (X) given grade point average (Y), we cannot use this regression line. We must derive new regression constants, because the old regression line was derived to minimize errors in the Y variable. The errors we minimized were represented by vertical lines parallel to the Y axis [see Figure 7.2(a) on p. 129]. Now we want to minimize errors in the X variable. These errors would be represented by horizontal lines parallel to the X axis. An example is shown in Figure 7.2(a) by the dashed line connecting point E to the regression line.

In general, minimizing Y' errors and minimizing X' errors will not lead to the same regression lines. The exception occurs when the relationship is perfect rather than imperfect. In that case, both regression lines coincide, forming the single line that hits all of the points. The regression line for predicting X from Y is sometimes called the *regression line of X on Y* or simply the *regression of X on Y*. To illustrate the computation of the line, let us use the IQ and grade point data again. This time we shall predict IQ (X) from the grade point average (Y). For convenience, the data are shown in Table 7.5. The linear regression equation for predicting X given Y is

$$X' = b_X Y + a_X \qquad \textit{linear regression equation for predicting X given Y}$$

where X' = predicted value of X
b_X = slope of the line for minimizing X' errors
a_X = X intercept for minimizing X' errors

The equations for b_X and a_X are

$$b_X = \frac{\sum XY - \dfrac{(\sum X)(\sum Y)}{N}}{SS_Y} = \frac{(\sum XY) - \dfrac{(\sum X)(\sum Y)}{N}}{\sum Y^2 - \dfrac{(\sum Y)^2}{N}} \qquad \begin{array}{l} \textit{b regression constant} \\ \textit{for predicting X given} \\ \textit{Y—computational} \\ \textit{equation} \end{array}$$

where SS_Y = sum of squares of Y scores

$$= \sum Y^2 - \frac{(\sum Y)^2}{N}$$

$$a_X = \bar{X} - b_X \bar{Y} \qquad \textit{regression constant for predicting X given Y}$$

Solving for b_X and a_X,

$$b_X = \frac{\sum XY - \dfrac{(\sum X)(\sum Y)}{N}}{\sum Y^2 - \dfrac{(\sum Y)^2}{N}}$$

TABLE 7.5   IQ and Grade Point Average of 12 Students: Predicting X from Y

Student No.	IQ X	Grade Point Average Y	XY	Y^2
1	110	1.0	110.0	1.00
2	112	1.6	179.2	2.56
3	118	1.2	141.6	1.44
4	119	2.1	249.9	4.41
5	122	2.6	317.2	6.76
6	125	1.8	225.0	3.24
7	127	2.6	330.2	6.76
8	130	2.0	260.0	4.00
9	132	3.2	422.4	10.24
10	134	2.6	348.4	6.76
11	136	3.0	408.0	9.00
12	138	3.6	496.8	12.96
Total	1503	27.3	3488.7	69.13

$$b_X = \frac{\sum XY - \frac{(\sum X)(\sum Y)}{N}}{\sum Y^2 - \frac{(\sum Y)^2}{N}} = \frac{3488.7 - \frac{1503(27.3)}{12}}{69.13 - \frac{(27.3)^2}{12}}$$

$$= \frac{69.375}{7.0225} = 9.879$$

$$a_X = \bar{X} - b_X\bar{Y} = 125.25 - 9.8790(2.275) = 102.775$$

$$X' = b_X Y + a_X = 9.879Y + 102.775$$

$$b_X = \frac{3488.7 - \frac{1503(27.3)}{12}}{69.13 - \frac{(27.3)^2}{12}} = \frac{69.375}{7.0225} = 9.879$$

$$a_X = \bar{X} - b_X\bar{Y}$$

$$= 125.25 - 9.8790(2.275)$$

$$= 102.775$$

The linear regression equation for predicting X given Y is

$$X' = b_X Y + a_X$$

$$= 9.879Y + 102.775$$

This line along with the line predicting Y given X is shown in Figure 7.5. Note that the two lines are different, as would be expected when the relationship is imperfect. The solution is summarized in Table 7.5.

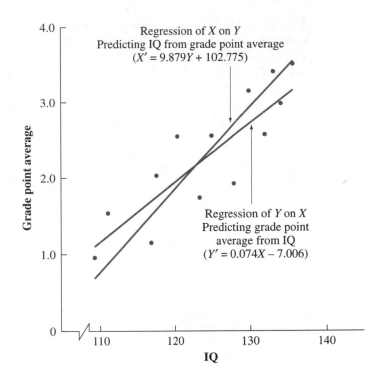

FIGURE 7.5

Regression of *X* on *Y* and regression of *Y* on *X*

Although different equations do exist for computing the second regression line, they are seldom used. Instead, it is common practice to designate the predicted variable as *Y'* and the given variable as *X*. Thus, if we wanted to predict IQ from grade point average, we would designate IQ as the *Y'* variable and grade point average as the *X* variable and then use the regression equation for predicting *Y* given *X*.

MEASURING PREDICTION ERRORS: THE STANDARD ERROR OF ESTIMATE

The regression line represents our best estimate of the *Y* scores given their corresponding *X* values. However, unless the relationship between *X* and *Y* is perfect, most of the actual *Y* values will not fall on the regression line. Thus, when the relationship is imperfect, there will necessarily be prediction errors. It is useful to know the magnitude of the errors. For example, it sounds nice to say that, on the basis of the relationship between IQ and grade point average given previously, we predict that John's grade point average will be 3.2 when he is a senior. However, since the relationship is imperfect, it is unlikely that our prediction is exactly correct. Well, if it is not exactly correct, then how far off is it? If it is likely to be very far off, we can't put much reliance on the prediction. However, if the error is likely to be small, the prediction can be taken seriously and decisions made accordingly.

Quantifying prediction errors involves computing the *standard error of estimate*. The standard error of estimate is much like the standard deviation. You will recall that the standard deviation gave us a measure of the average deviation about the mean. The standard error of estimate gives us a measure of the average deviation of the prediction errors about the regression line. In this context, the regression line can be considered an estimate of the mean of the Y values for each of the X values. It is like a "floating" mean of the Y values, which changes with the X values. With the standard deviation, the sum of the deviations, $\Sigma\,(X - \overline{X})$, equaled 0. We had to square the deviations to obtain a meaningful average. The situation is the same with the standard error of estimate. Since the sum of the prediction errors, $\Sigma\,(Y - Y')$, equals 0, we must square them also. The average is then obtained by summing the squared values, dividing by $N - 2$, and taking the square root of the quotient (very much like with the standard deviation). The equation for the standard error of estimate for predicting Y given X is

$$s_{Y|X} = \sqrt{\frac{\Sigma\,(Y - Y')^2}{N - 2}} \qquad \textit{standard error of estimate when predicting Y given X}$$

Note that we have divided by $N - 2$ rather than $N - 1$, as was done with the sample standard deviation.* The calculations involved in using this equation are quite laborious. The computational equation, which is given here, is much easier to use. In determining the b_Y regression coefficient, we have already calculated the values for SS_X and SS_Y.

$$s_{Y|X} = \sqrt{\frac{SS_Y - \dfrac{[\Sigma\,XY - (\Sigma\,X)(\Sigma\,Y)/N]^2}{SS_X}}{N - 2}}$$

To illustrate the use of these equations, let's calculate the standard error of estimate for the grade point and IQ data shown in Tables 7.1 and 7.2. As before, we shall let grade point be the Y variable and IQ the X variable, and we shall calculate the standard error of estimate for predicting grade point given IQ. As computed in the tables, $SS_X = 936.25$, $SS_Y = 7.022$, $\Sigma\,XY - (\Sigma\,X)(\Sigma\,Y)/N = 69.375$, and $N = 12$. Substituting these values in the equation for the standard error of estimate for predicting Y given X, we obtain

$$s_{Y|X} = \sqrt{\frac{SS_Y - \dfrac{[\Sigma\,XY - (\Sigma\,X)(\Sigma\,Y)/N]^2}{SS_X}}{N - 2}}$$

$$= \sqrt{\frac{7.022 - \dfrac{(69.375)^2}{936.25}}{12 - 2}}$$

$$= \sqrt{0.188} = 0.43$$

---

* We divide by $N - 2$ because calculation of the standard error of estimate involves fitting the data to a straight line. To do so requires estimation of two parameters, slope and intercept, leaving the deviations about the line with $N - 2$ degrees of freedom. We shall discuss degrees of freedom in Chapter 13.

FIGURE 7.6

Scatter plots showing the
variability of Y as a
function of X

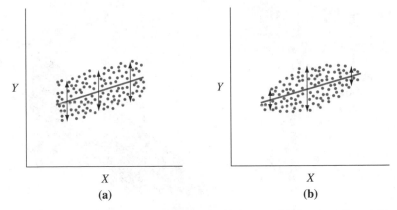

Thus, the standard error of estimate = 0.43. This measure has been computed
over all the Y scores. For it to be meaningful, we must assume that the variability
of Y remains constant as we go from one X score to the next. This assumption
is called the assumption of *homoscedasticity*. Figure 7.6(a) shows an illustration
where the homoscedasticity assumption is met. Figure 7.6(b) shows an illustration
where the assumption is violated. The homoscedasticity assumption implies that
if we divided the X scores into columns, the variability of Y would not change
from column to column. We can see how this is true for Figure 7.6(a) but not
for 7.6(b).

What meaning does the standard error of estimate have? Certainly, it is a
quantification of the errors of prediction. The larger its value, the less confidence
we have in the prediction. Conversely, the smaller its value, the more likely the
prediction will be accurate. We can still be more quantitative. We can assume
the points are normally distributed about the regression line (Figure 7.7). If the
assumption is valid and we were to construct two lines parallel to the regression
line at distances of $\pm 1 s_{Y|X}$, $\pm 2 s_{Y|X}$, and $\pm 3 s_{Y|X}$, we would find that approximately
68% of the scores fall between the lines at $\pm 1 s_{Y|X}$, approximately 95% would lie
between $\pm 2 s_{Y|X}$, and approximately 99% would lie between $\pm 3 s_{Y|X}$. To illustrate
this point, in Figure 7.8, we have drawn two dashed lines parallel to the regression
line for the grade point and IQ data at a distance of $\pm 1 s_{Y|X}$. We have also entered

FIGURE 7.7

Normal distribution of Y
scores about the
regression line

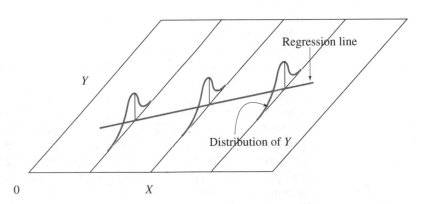

FIGURE 7.8

Regression line for grade
point and IQ data with
parallel lines $1s_{Y|X}$ above
and below the regression
line

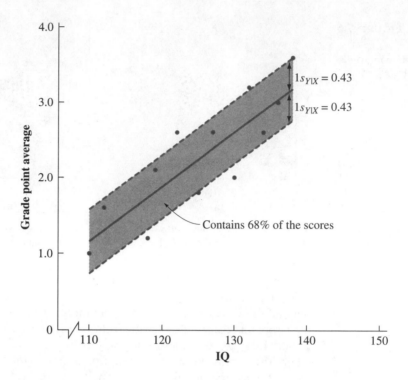

the scores in the figure. According to what we said previously, approximately 68% of the scores should lie between these lines. Since there are 12 points, we would expect 0.68(12) = 8 of the scores to be contained within the lines. In fact, there are 8. The agreement isn't always this good, particularly when there are only 12 scores in the sample. As N increases, the agreement usually increases also.

CONSIDERATIONS IN USING LINEAR REGRESSION FOR PREDICTION

The procedures we have described are appropriate for predicting scores based on presumption of a linear relationship existing between the X and Y variables. If the relationship is nonlinear, the prediction will not be very accurate. It follows, then, that the first assumption for successful use of this technique is that the relationship between X and Y must be *linear*. Second, we are not ordinarily interested in using the regression line to predict scores of the individuals that were in the group used for calculating the regression line. After all, why predict their scores when we already know them? Generally, a regression line is determined for use with subjects where one of the variables is unknown. For instance, in the IQ and grade point average problem, a university admissions officer might want to use the regression line to predict the grade point averages of *prospective* students, knowing their IQ scores. It doesn't make any sense to predict the grade point averages of the 12 students whose data were used in the problem. He already knows their grade point averages. If we are going to use data collected on one group to predict scores of another group, it is important that the basic computation group is representative of the prediction group. Often this require-

FIGURE 7.9

Limiting prediction to
range of base data

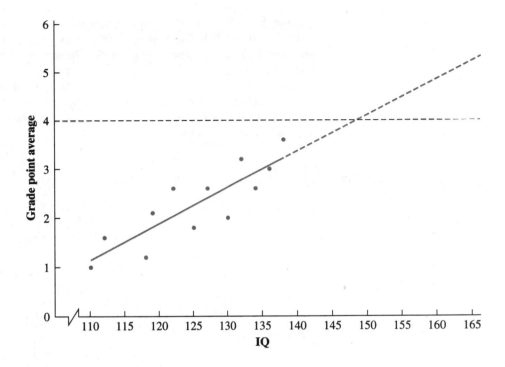

ment is handled by randomly sampling from the prediction population and using
the sample for deriving the regression equation. Random sampling is discussed
in Chapter 8. Finally, the linear regression equation is properly used just for the
range of the variable on which it is based. For example, when we were predicting
grade point average from IQ, we should have limited our predictions to IQ
scores ranging from 110 to 138. Since we do not have any data beyond this range,
we do not know whether the relationship continues to be linear for more extreme
values of IQ.

To illustrate this point, consider Figure 7.9, where we have extended the
regression line to include IQ values up to 165. At the university from which
these data were sampled, the highest possible grade point average is 4.0. If we
used the extended regression line to predict the grade point average for an IQ
of 165, we would predict a grade point average of 5.2, a value that is obviously
very wrong. Prediction for IQs over 165 would be even worse. Looking at Figure
7.9, we can see that if the relationship does extend beyond an IQ of 138, it can't
extend beyond an IQ of about 149 (the IQ value where the regression line hits
a grade point average of 4.0). Of course, there is no reason to believe the
relationship exists beyond the base data point of IQ = 138, and hence predictions
using this relationship should not be made for IQ values greater than 138.

RELATION BETWEEN REGRESSION CONSTANTS AND PEARSON *r*

Although we haven't presented this aspect of Pearson *r* before, it can be shown
that Pearson *r* is the slope of the least-squares regression line when the scores
are plotted as *z* scores. As an example, let's use the data given in Table 6.2 on

the weight and cost of six bags of oranges. For convenience, the data have been reproduced in Table 7.6. Figure 7.10(a) shows the scatter plot of the raw scores and the least-squares regression line for these raw scores. Since this is a perfect, linear relationship, $r = 1.00$. Figure 7.10(b) shows the scatter plot of the paired z scores and the least-squares regression line for these z scores. The slope of the regression line for the raw scores is b, and the slope of the regression line for the z scores is r. Note that the slope of this latter regression line is 1.00, as it should since $r = 1.00$.

TABLE 7.6   Cost and Weight in Pounds of Six Bags of Oranges

Bag	Weight (lb) X	Cost ($) Y	z_X	z_Y
A	2.25	0.75	−1.34	−1.34
B	3.00	1.00	−0.80	−0.80
C	3.75	1.25	−0.27	−0.27
D	4.50	1.50	0.27	0.27
E	5.25	1.75	0.80	0.80
F	6.00	2.00	1.34	1.34

Since Pearson r is a slope, it is related to b_Y and b_X. It can be shown algebraically that

$$b_Y = r \frac{s_Y}{s_X}$$

and

$$b_X = r \frac{s_X}{s_Y}$$

FIGURE 7.10

Relationship between b and r

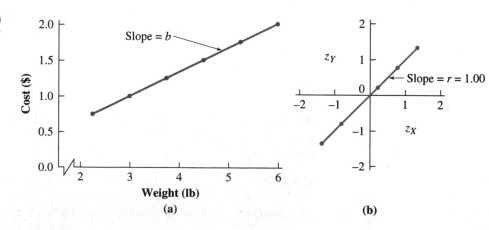

(a)   (b)

These equations are useful if we have already calculated r, s_X, and s_Y and want to determine the least-squares regression line. For example, in the problem involving IQ and grade point average, $r = 0.8556$, $s_Y = 0.7990$, and $s_X = 9.2257$. Suppose we desire to find b_Y and a_Y, having already calculated r, s_Y, and s_X. The simplest way is to use the equation

$$b_Y = r\frac{s_Y}{s_X} = 0.8556 \left(\frac{0.7990}{9.2257}\right) = 0.074$$

Note that this is the same value arrived at previously in the chapter on p. 131. Having found b_Y, we would calculate a_Y in the usual way.

MULTIPLE REGRESSION AND MULTIPLE CORRELATION

Thus far we have discussed regression and correlation using examples that have involved only two variables. When we were discussing the relationship between grade point average (GPA) and IQ, we determined that $r = 0.856$ and that the equation of the regression line for predicting grade point average from IQ was

$$Y' = .074X - 7.006$$

where Y' = predicted value of grade point average
       X = IQ score

This equation gave us reasonably accurate prediction. Although we didn't compute it, total prediction error squared $[\Sigma (Y - Y')^2]$ was 1.88 and the amount of variability accounted for was 73.2%. Of course there are other variables besides IQ that might affect grade point average. The amount of time that students spend studying, motivation to achieve high grades, and interest in the courses taken are a few that come to mind. Even though we have reasonably good prediction accuracy using just IQ alone, we might be able to do better if we also had data relating GPA to one or more of these other variables.

Multiple regression is an extension of simple regression to situations that involve two or more predictor variables. To illustrate, let's assume we had data from the 12 college students that include a second predictor variable called "study time," as well as the original grade point average and IQ scores. The data for these three variables are shown in columns 2, 3, and 4 of Table 7.7. Now we can derive a regression equation for predicting grade point average using the two predictor variables, IQ and study time. The general form of the multiple regression equation for two predictor variables is

$$Y' = b_1X_1 + b_2X_2 + a$$

where Y' = predicted value of Y
       b_1 = coefficient of the first predictor variable
       X_1 = first predictor variable
       b_2 = coefficient of the second predictor variable
       X_2 = second predictor variable
        a = prediction constant

TABLE 7.7　A Comparison of Prediction Accuracy Using One or Two Predictor Variables

Student No.	IQ (X_1)	Study Time (Hr/Wk) (X_2)	Grade Point Average (GPA) (Y)	Predicted GPA Using IQ (Y')	Predicted GPA Using IQ + Study Time (Y')	Error Using Only IQ	Error Using IQ + Study Time
1	110	8	1.0	1.14	1.13	−0.14	−0.13
2	112	10	1.6	1.29	1.46	0.31	0.13
3	118	6	1.2	1.74	1.29	−0.54	−0.09
4	119	13	2.1	1.81	2.16	0.29	−0.06
5	122	14	2.6	2.03	2.43	0.57	0.17
6	125	6	1.8	2.26	1.63	−0.46	0.17
7	127	13	2.6	2.40	2.56	0.20	0.04
8	130	12	2.0	2.63	2.59	−0.63	−0.59
9	132	13	3.2	2.77	2.81	0.42	0.39
10	134	11	2.6	2.92	2.67	−0.32	−0.07
11	136	12	3.0	3.07	2.88	−0.07	0.12
12	138	18	3.6	3.21	3.69	0.38	0.09

Total error squared $= \Sigma\,(Y - Y')^2 = 1.88$ $= 0.63$

This equation is very similar to the one we used in simple regression except that we have added another predictor variable and its coefficient. As before, the coefficient and constant values are determined according to the least-squares criterion that $\Sigma\,(Y - Y')^2$ is a minimum. However, this time the mathematics is rather formidable and the actual calculations are almost always done on a computer, using statistical software. For the data of our example, the multiple regression equation that minimizes errors in Y is given by

$$Y' = 0.049X_1 + 0.118X_2 - 5.249$$

where　　Y' = predicted value of grade point average
　　　　　$b_1 = 0.049$
　　　　　X_1 = IQ score
　　　　　$b_2 = 0.118$
　　　　　X_2 = study time score
　　　　　$a = -5.249$

To determine whether prediction accuracy is increased by using the multiple regression equation, we have listed in column 5 of Table 7.7 the predicted GPA scores using only IQ for prediction, in column 6 the predicted GPA scores using both IQ and study time as predictor variables, and prediction errors from using each in columns 7 and 8, respectively. We have also plotted in Figure 7.11(a) the actual Y value and the two predicted Y' values for each student. Students have been ordered from left to right on the X axis according to the increased prediction accuracy that results for each by using the multiple regression equation. In Figure 7.11(b), we have plotted the percent improvement in prediction accuracy for each student that results from using IQ + study time rather than just

FIGURE 7.11

Comparison of prediction accuracy using one or two predictor variables

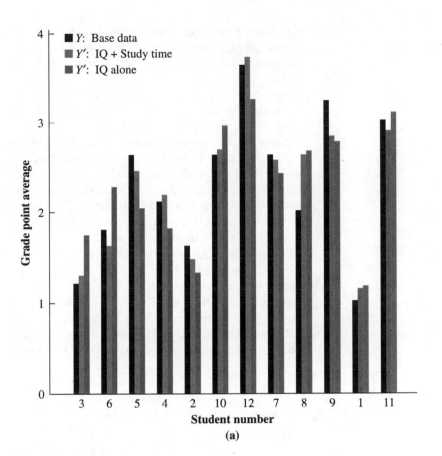

(a)

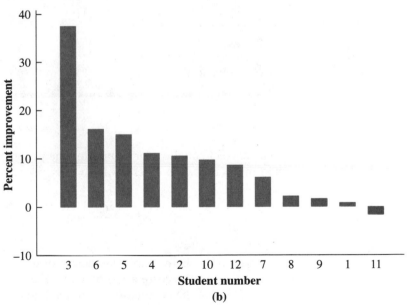

(b)

IQ alone. It is clear from Table 7.7 and Figure 7.11 that using the multiple regression equation has greatly improved overall prediction accuracy. For example, prediction accuracy was increased in all students except student number 11, and for student number 3, accuracy was increased by almost 40%. We have also shown $\Sigma (Y - Y')^2$ for each regression line at the bottom of Table 7.7. Adding the second predictor variable reduced the total prediction error squared from 1.88 to 0.63, an improvement of more than 66%.

Since, in the present example, prediction accuracy was increased by using two predictors rather than one, it follows that the proportion of the variability of Y accounted for has also increased. In trying to determine this proportion, you might be tempted, through extension of the concept of r^2 from our previous discussion of correlation, to compute r^2 between grade point average and each predictor and then simply add the resulting values. Table 7.8 shows a Pearson r correlation matrix involving grade point average, IQ, and study time. If we followed this procedure, the proportion of variability accounted for would be greater than 1.00 $[(0.856)^2 + (0.829)^2 = 1.42]$, which is clearly impossible. One cannot account for more than 100% of the variability. The error occurs because there is overlap in variability accounted for between IQ and study time. Students with higher IQ also tend to study more. Therefore, part of the variability in GPA that is explained by IQ is also explained by study time. To correct for this, we must take the correlation between IQ and study time into account.

TABLE 7.8   Pearson Correlation Matrix Between IQ, Study Time, and Grade Point Average

	IQ (X_1)	Study Time (X_2)	Grade Point Average (Y)
IQ (X_1)	1.000		
Study Time (X_2)	0.560	1.000	
Grade Point Average (Y)	0.856	0.829	1.000

The correct equation for computing the proportion of variance accounted for when there are two predictor variables is given by

$$R^2 = \frac{r_{YX_1}^2 + r_{YX_2}^2 - 2r_{YX_1}r_{YX_2}r_{X_1X_2}}{1 - r_{X_1X_2}^2}$$

where R^2 = the multiple coefficient of determination
r_{YX_1} = the correlation between Y and predictor variable X_1
r_{YX_2} = the correlation between Y and predictor variable X_2
$r_{X_1X_2}$ = the correlation between predictor variables X_1 and X_2

R^2 is also often called the *squared multiple correlation*. Based on the data of the present study, r_{YX_1} = the correlation between grade point average and IQ = 0.856, r_{YX_2} = the correlation between Y and study time = 0.829, and $r_{X_1X_2}$ = the correlation between IQ and study time = 0.560. For these data,

$$R^2 = \frac{(0.856)^2 + (0.829)^2 - 2(0.856)(0.829)(0.560)}{1 - (0.560)^2}$$

$$= 0.910$$

Thus the proportion of variance accounted for has increased from 0.73 to 0.91 by using IQ and study time.

Of course, just adding another predictor variable per se will not necessarily increase prediction accuracy or the amount of variance accounted for. Whether prediction accuracy and the amount of variance acounted for are increased depends on the strength of the relationship between the variable being predicted and the additional predictor variable, and also on the strength of the relationship between the predictor variables themselves. For example, notice what happens to R^2 when $r_{x_1 x_2} = 0$. This topic is taken up in more detail in advanced textbooks.

SUMMARY

In this chapter, we have discussed how to use the relationship between two variables for prediction. When the line that best fits the points is used for prediction, it is called a regression line. The regression line most used for linear imperfect relationships fits the points according to a least-squares criterion. Next, we presented the equations for determining the least-squares regression line when predicting Y given X and the regression line when predicting X given Y. The two lines are not the same unless the relationship is perfect. We then used these equations to construct regression lines for various sets of data and showed how to use these lines for prediction. Next, we discussed how to quantify the errors in prediction by computing the standard error of estimate. We presented the conditions under which the use of the linear regression line was appropriate: The relationship must be linear; the regression line must have been derived from data representative of the group to which prediction is desired; and prediction must be limited to the range of the base data. Next, we discussed the relationship between b and r. Finally, we introduced the topic of multiple regression and multiple correlation; discussed the multiple coefficient of determination, R^2; and showed how using two predictor variables can increase the accuracy of prediction.

IMPORTANT TERMS

Homoscedasticity (p. 141)
Least-squares regression line (p. 127)
Multiple coefficient of determination (p. 148)

Multiple correlation (p. 145)
Multiple regression (p. 145)
Regression (p. 127)
Regression constant (p. 130)

Regression line (p. 127)
Regression of X on Y (p. 136)
Regression of Y on X (p. 136)
Standard error of estimate (p. 139)

QUESTIONS AND PROBLEMS

1. Define or identify each of the terms in the "Important Terms" section.
2. List some situations in which it would be useful to have accurate prediction.

3. The least-squares regression line minimizes $\Sigma (Y - Y')^2$, rather than $\Sigma (Y - Y')$. Is this statement correct? Explain.
4. The least-squares regression line is the prediction

line that results in the most direct "hits." Is this statement correct? Explain.

5. In general, the regression line of Y on X is not the same as the regression line of X on Y. Is this statement correct? Explain.

6. Of what value is it to know the standard error of estimate for a set of paired X and Y scores?

7. Why are there usually two regression lines but only one correlation coefficient for any set of paired scores?

8. What is R^2 called? Is it true that conceptually R^2 is analogous to r^2, except that R^2 applies to situations where there are two or more predictor variables? Explain. Will using a second predictor variable always increase the precision of prediction? Explain.

9. Given the set of paired X and Y scores,

X	7	10	9	13	7	11	13
Y	1	2	4	3	3	4	5

a. Construct a scatter plot of the paired scores. Does the relationship appear linear?

b. Determine the least-squares regression line for predicting Y given X.

c. Determine the least-squares regression line for predicting X given Y. Are they the same? Explain.

d. Draw both regression lines on the scatter plot.

e. Using the relationship between X and Y, what value would you predict for Y if $X = 12$ (round to two decimal places)?

10. A popular attraction at a carnival recently arrived in town is the booth where Mr. Clairvoyant (a bright statistics student of somewhat questionable moral character) claims that he can guess the weight of females to within 1 kilogram by merely studying the lines in their hands and fingers. He offers a standing bet that if he guesses incorrectly the woman can pick out any stuffed animal in the booth. However, if he guesses correctly, as a reward for his special powers, she must pay him $2. Unknown to the women who make bets, Mr. Clairvoyant is able to surreptitiously measure the length of their left index fingers while "studying" their hands. Also unknown to the bettors, but known to Mr. Clairvoyant, is the following relationship between the weight of

females and the length of their left index fingers:

Length of Left Index Finger (cm)	5.6	6.2	6.0	5.4
Weight (kg)	79.0	83.5	82.0	77.5

a. If you were a prospective bettor, having all this information before you, would you make the bet with Mr. Clairvoyant? Explain.

b. Using the data in the accompanying table, what is the least-squares regression line for predicting a woman's weight, given the length of her index finger?

c. Using the least-squares regression line determined in part **b**, if a woman's index finger is 5.7 centimeters, what would be her predicted weight (round to two decimal places)?

11. A statistics professor conducts a study to investigate the relationship between the performance of his students on exams and their anxiety. Ten students from his class are selected for the experiment. Just prior to taking the final exam, the 10 students are given an anxiety questionnaire. Here are final exam and anxiety scores for the 10 students:

Anxiety	28	41	35	39	31	42	50	46	45	37
Final exam	82	58	63	89	92	64	55	70	51	72

a. On a piece of graph paper, construct a scatter plot of the paired scores. Use anxiety as the X variable.

b. Describe the relationship shown in the graph.

c. Assuming the relationship is linear, compute the value of Pearson r.

d. Determine the least-squares regression line for predicting the final exam score given the anxiety level. Should b_Y be positive or negative? Why?

e. Draw the least-squares regression line of part **d** on the scatter plot of part **a**.

f. Based on the data of the 10 students, if a student has an anxiety score of 38, what value would you predict for her final exam score (round to two decimal places)?

g. Calculate the standard error of estimate for predicting final exam scores from anxiety scores.

12. The sales manager of a large sporting goods store has recently started a national advertising campaign. He has kept a record of the monthly costs of the advertising and the monthly profits. These are shown here. The entries are in thousands of dollars.

Month	Jan.	Feb.	Mar.	Apr.	May	Jun.	Jul.
Monthly Advertising Cost	10.0	14.0	11.4	15.6	16.8	11.2	13.2
Monthly Profit	125	200	160	150	210	110	125

a. Assuming a linear relationship exists, derive the least-squares regression line for predicting monthly profits from monthly advertising costs.
b. In August, the manager plans to spend $17,000 on advertising. Based on the data, how much profit should he expect that month (round to the nearest $1000)?
c. Given the relationship shown by the paired scores, can you think of a reason why the manager doesn't spend a lot more money on advertising?

13. During inflationary times, Mr. Chevez has become budget conscious. Since his house is heated electrically, he has kept a record for the past year of his monthly electric bills and of the average monthly outdoor temperature. The data are shown in the table. Temperature is in degrees Celsius, and the electric bills are in dollars.
a. Assuming there is a linear relationship between the average monthly temperature and the monthly electric bill, determine the least-squares regression line for predicting the monthly electric bill from the average monthly temperature.

Month	Average Temp.	Elec. Bill
Jan.	10	120
Feb.	18	90
Mar.	35	118
Apr.	39	60
May	50	81
Jun.	65	64
Jul.	75	26
Aug.	84	38
Sep.	52	50
Oct.	40	80
Nov.	25	100
Dec.	21	124

b. Based on the almanac forecast for this year, Mr. Chevez expects a colder winter. If February is 8 degrees colder this year, how much should Mr. Chevez allow in his budget for February's electric bill? In calculating your answer, assume that the costs of electricity will rise 10% from last year's costs because of inflation.
c. Calculate the standard error of estimate for predicting the monthly electric bill from average monthly temperature.

14. Refer to Chapter 6, Problem 19 on page 126. In this problem, you were asked to recommend one test to be used as a screening device for prospective employees in the manufacturing section of a large corporation. You recommended using test 2. Now the question is, Would it be better to use both test 1 and test 2, rather than just test 2 alone? Explain your answer, using R^2 and r^2. Use a computer and statistical software to solve this problem, if you have access to them.

PART THREE

INFERENTIAL STATISTICS

8 RANDOM SAMPLING AND PROBABILITY

INTRODUCTION

We have now completed our discussion of descriptive statistics and are ready to begin considering the fascinating area of inferential statistics. With descriptive statistics, we were concerned primarily with presenting and describing sets of scores in the most meaningful and efficient way. With inferential statistics, we go beyond mere description of the scores. A basic aim of inferential statistics is to use the sample scores to make a statement about a characteristic of the population. There are two kinds of statements made. One has to do with *hypothesis testing* and the other with *parameter estimation*.

In hypothesis testing, the experimenter is collecting data in an experiment on a sample set of subjects in an attempt to validate some hypothesis involving a population. For example, suppose an educational psychologist believes a new method of teaching mathematics to the third graders in her school district (population) is superior to the usual way of teaching the subject. In her experiment, she employs two samples of third graders, one of which receives the new teaching method and the other the old one. Each group is tested on the same final exam. In doing this experiment, the psychologist is not satisfied with just reporting that the mean of the group that received the new method was higher than the mean of the other group. She wants to make a statement such as "The improvement in final exam scores was due to the new teaching method and not chance factors. Further, the improvement does not apply just to the particular sample tested. Rather, the improvement would be found in the whole population of third graders if they were taught by the new method." The techniques used in inferential statistics make these statements possible.

In parameter estimation experiments, the experimenter is interested in determining the magnitude of a population characteristic. For example, an economist might be interested in determining the average monthly amount of money spent last year on food by single college students. Using sample data, with the techniques of inferential statistics, he can estimate the mean amount spent by the population. He would conclude with a statement like "The probability is 0.95 that the interval of $140–$160 contains the population mean."

The topics of *random sampling* and *probability* are central to the methodology of inferential statistics. In the next section, we shall consider random sampling. In the remainder of the chapter we shall be concerned with presenting the basic principles of probability.

RANDOM SAMPLING

To generalize validly from the sample to the population, both in hypothesis testing and parameter estimation experiments, the sample cannot be *just any subset* of the population. Rather, it is crucial that the sample is a *random* sample.

DEFINITION

- *A **random sample** is defined as a sample selected from the population by a process that assures that (1) each possible sample of a given size has an equal chance of being selected and (2) all the members of the population have an equal chance of being selected into the sample.**

To illustrate, consider the situation where we have a population comprising the scores 2, 3, 4, 5, and 6, and we want to randomly draw a sample of size 2 from the population. Note that normally the population would have a great many more scores in it. We've restricted the population to five scores for ease in understanding the points we wish to make. Let's assume we shall be sampling from the population one score at a time and then placing it back into the population before drawing again. This is called *sampling with replacement* and is discussed later in this chapter. The following comprise all the samples of size 2 we could get from the population using this method of sampling:

2, 2	3, 2	4, 2	5, 2	6, 2
2, 3	3, 3	4, 3	5, 3	6, 3
2, 4	3, 4	4, 4	5, 4	6, 4
2, 5	3, 5	4, 5	5, 5	6, 5
2, 6	3, 6	4, 6	5, 6	6, 6

There are 25 samples of size 2 we might get when sampling one score at a time with replacement. To achieve random sampling, the process must be such that (1) all of the 25 possible samples have an equally likely chance of being selected and (2) all of the population scores (2, 3, 4, 5, and 6) have an equal chance of being selected into the sample.

The sample should be a random sample for two reasons. First, to generalize from a sample to a population, it is necessary to apply the laws of probability to the sample. If the sample has not been generated by a process assuring that each possible sample of that size has an equal chance of being selected, we can't apply the laws of probability to the sample. The importance of this aspect of randomness and of probability to statistical inference will become apparent when we have covered the chapters on hypothesis testing and sampling distributions.

The second reason for random sampling is that, to generalize from a sample to a population, it is necessary that the sample be representative of the population. One way to achieve representativeness is to choose the sample by a process that assures that all the members of the population have an equal chance of being

* See Note 8.1, p. 189.

selected into the sample. Thus, requiring the sample to be random allows the laws of probability to be used on the sample, while at the same time results in a sample that should be representative of the population.

It is tempting to think that we can achieve representativeness by using methods other than random sampling. Very often, however, the procedure used results in a biased (unrepresentative) sample. An example of this was the famous *Literary Digest* poll of 1936, which predicted a landslide victory for Landon (57 to 43%). In fact, Roosevelt won, gaining 62% of the ballots. The *Literary Digest* prediction was grossly in error. Why? Later analysis showed that the error occurred because the sample was not representative of the voting population. It was a *biased* sample. The individuals selected were chosen from sources like the telephone book, club lists, and lists of registered automobile owners. These lists systematically excluded the poor, who were unlikely to have telephones or automobiles. It turned out that the poor voted overwhelmingly for Roosevelt. Even if other methods of sampling do on occasion result in a representative sample, the methods would not be useful for inference because we could not apply the laws of probability necessary to go from the sample to the population.

Techniques for Random Sampling

It is beyond the scope of this textbook to delve deeply into the ways of generating random samples. This topic can be complex, particularly when dealing with surveys. We shall, however, present a few of the more commonly used techniques in conjunction with some simple situations so that you can get a feel for what is involved. Suppose we have a population of 100 people and wish to randomly sample 20 for an experiment. One way to do this would be to number the individuals in the population from 1 to 100, then take 100 slips of paper and write one of the numbers on each slip, and put the slips into a hat, shake them around a lot, and pick out one. We would repeat the shaking and pick out another. Then we would continue this process until 20 slips have been picked. The numbers contained on the slips of paper would identify the individuals to be used in the sample. With this method of random sampling, it is crucial that the population be thoroughly mixed to assure randomness.

A common way to produce random samples is to use a table of random numbers, such as Table J in Appendix D. These tables are most often constructed by a computer using a program that guarantees that all the digits (0–9) have an equal chance of occurring each time a digit is printed.

The table may be used as successive single digits, as successive two-digit numbers, as successive three-digit numbers, and so forth. For example, in Table J, if we begin at row 1 and move horizontally across the page, the random order of single digits would be 3, 2, 9, 4, 2, ... If we wish to use two-digit numbers, the random order would be 32, 94, 29, 54, 16, ...

Since the digits in the table are random, they may be used vertically in both directions and horizontally in both directions. The direction to be used should be specified prior to entering the table. To use the table properly, it should be entered randomly. One way would be to make cards with row and column numbers and place the cards in a box, mix them up, and then pick a row number and a column number. The intersection of the row and column would be the location of the first random number. The remaining numbers would be located by moving from the first number in the direction specified prior to entering the table. To illustrate, suppose we wanted to form a random sample of 3 subjects from a population of 10 subjects.* For this example, we have decided to move horizontally to the right in the table. To choose the sample, we would first assign each individual in the population a number from 0 to 9. Next the table would be entered randomly to locate the first number. Let's assume the entry turns out to be the first number of row 7, which is 3. This number designates the first subject in the sample. Thus, the first subject in the sample would be the subject bearing the number 3. Since we have decided to move to the right in the table, the next two numbers are 5 and 6. Thus, the individuals bearing the numbers 5 and 6 would complete the sample.

Next, let's do a problem where there are more individuals in the population. For the purposes of illustration, we shall assume that a random sample of 15 subjects is desired from a population of 100. To vary things a bit, we have decided to move vertically down in the table for this problem, rather than horizontally to the right. As before, we need to assign a number to each member of the population. This time, the numbers assigned are from 00 to 99 instead of from 0 to 9. Again the table is entered randomly. This time, let's assume the entry occurs at the intersection of the first two-digit number of column 3 with row 12. The two-digit number located at this intersection is 70. Thus, the first subject in the sample is the individual bearing the number 70. The next subject would be located by moving vertically down from 70. Thus, the second subject in the sample would be the individual bearing the number 33. This process would be continued until 15 subjects have been selected. The complete set of subject numbers would be 70, 33, 82, 22, 96, 35, 14, 12, 13, 59, 97, 37, 54, 42, and 89. In arriving at this set of numbers, the number 82 appeared twice in the table. Since the same individual cannot be in the sample more than once, the repeated number was not included.

Sampling With or Without Replacement

So far we have defined a random sample, discussed the importance of random sampling, and presented some techniques for producing random samples. To complete our discussion, we need to distinguish between "sampling with replacement" and "sampling without replacement." To illustrate the difference between these two methods of sampling, let's assume we wish to form a sample of two scores from a population composed of the scores 4, 5, 8, and 10. One way would be to randomly draw one score from the population, record its value, and then place it back in the population before drawing the second score. Thus, the first score would be eligible to be selected again on the second draw. This method of sampling is called *sampling with replacement*. A second method would be to randomly draw one score from the population and not replace it before drawing the second one. Thus, the same number of the population could appear in the sample only once. This method of sampling is called *sampling without replacement*.

DEFINITIONS

....

- **Sampling with replacement** *is defined as a method of sampling in which each member of the population selected for the sample is returned to the population before the next member is selected.*

---

* Of course, in real experiments the number of elements in the population is much greater than 10. We are using 10 in the first example to help us understand how to use the table.

• **Sampling without replacement** *is defined as a method of sampling in which the members of the sample are not returned to the population prior to selecting subsequent members.*

When selecting subjects to participate in an experiment, sampling without replacement must be used, since the same individual can't be in the sample more than once. You will probably recognize this as the method we used in the preceding section. Sampling with replacement forms the mathematical basis for many of the inference tests discussed later in the textbook. Although the two methods do not yield identical results, when sample size is small relative to population size, the differences are negligible and "with-replacement" techniques are much easier to use in providing the mathematical basis for inference. Let's now move on to the topic of probability.

PROBABILITY

Probability may be approached in two ways: (1) from an *a priori* or classical viewpoint and (2) from an *a posteriori* or empirical viewpoint. *A priori* means that which can be deduced from reason alone, without experience. From the *a priori* or classical viewpoint, probability is defined as

$$p(A) = \frac{\text{Number of events classifiable as } A}{\text{Total number of possible events}} \qquad \textit{a priori probability}$$

The symbol $p(A)$ is read "the probability of occurrence of event A." Thus, the equation states that the probability of occurrence of event A is equal to the number of events classifiable as A divided by the number of possible events. To illustrate how this equation is used, let's look at an example involving dice. Figure 8.1 shows a picture of a pair of dice. Each die (the singular of dice is die) has six sides with a different number of spots painted on each side. The spots vary from one to six. These innocent-looking cubes are used for gambling in a game called *craps*. They have been the basis of many tears and much happiness, depending on the "luck" of the gambler.

Returning to *a priori* probability, suppose we are going to roll a die once. What is the probability it will come to rest with a "two" (the side with two spots on it) facing upward? Since there are six possible numbers that might occur, and only one of these is a two, the probability of a two in one roll of one die is

FIGURE 8.1

A pair of dice

$$p(A) = p(2) = \frac{\text{Number of events classifiable as 2}}{\text{Total number of possible events}} = \frac{1}{6} = 0.1667*$$

Let's try one more problem using the *a priori* approach. What is the probability of getting a number greater than 4 in one roll of one die? This time there are

---

* In this and all other problems involving dice, we shall assume that the dice will not come to rest on any of their edges.

two events classifiable as A (rolling a 5 or 6). Thus,

$$p(A) = p(5 \text{ or } 6) = \frac{\text{Number of events classifiable as 5 or 6}}{\text{Total number of possible events}} = \frac{2}{6} = 0.3333$$

Note that the previous two problems were solved by reason alone, without recourse to any data collection. This approach is to be contrasted with the *a posteriori* or empirical approach to probability. *A posteriori* means "after the fact," and, in the context of probability, it means after some data have been collected. From the *a posteriori* or empirical viewpoint, probability is defined as

$$p(A) = \frac{\text{Number of times } A \text{ has occurred}}{\text{Total number of occurrences}} \qquad \text{\textit{a posteriori probability}}$$

To determine the probability of a two in one roll of one die using the empirical approach, we would have to take the actual die, roll it many times, and count the number of times a two has occurred. The more times we roll the die, the better. Let's assume for this problem that we roll the die 100,000 times and that a two occurs 16,000 times. The probability of a two occurring in one roll of the die is found by

$$p(2) = \frac{\text{Number of times 2 has occurred}}{\text{Total number of occurrences}} = \frac{16,000}{100,000} = 0.1600$$

Note that, with this approach, it is necessary to have the actual die and to collect some data before determining the probability. The interesting thing is that if the die is evenly balanced (spoken of as a fair die), when we roll the die many, many times, the *a posteriori* probability approaches the *a priori* probability. If we roll an infinite number of times, the two probabilities will equal each other. Note also that, if the die is loaded (weighted so that one side comes up more often than the others), the *a posteriori* probability will differ from the *a priori* determination. For example, if the die is heavily weighted for a six to come up, a two might never appear. We can see now that the *a priori* equation assumes that each possible outcome has an equal chance of occurrence. For most of the problems in this chapter and the next, we shall use the *a priori* approach to probability.

Some Basic Points Concerning Probability Values

Since probability is fundamentally a proportion, it ranges in value from 0.00 to 1.00. If the probability of an event occurring equals 1.00, then the event is certain to occur. If the probability equals 0.00, then the event is certain *not* to occur. For example, an ordinary die does not have a side with seven dots on it. Therefore, the probability of rolling a seven with it equals 0.00. Rolling a seven is certain not to occur. On the other hand, the probability that a number from one to six will occur equals 1.00. It is certain that one of the numbers one, two, three, four, five, or six will occur.

The probability of occurrence of an event is expressed as a fraction or a decimal number. For example, the probability of randomly picking the ace of spades in one draw from a deck of ordinary playing cards is $\frac{1}{52}$ or 0.0192. The answer may be left as a fraction ($\frac{1}{52}$) but usually is converted to its decimal equivalent (0.0192).

Sometimes probability is expressed as "chances in a hundred." For example, someone might say the probability that event A will occur is 5 chances in 100. What he really means is $p(A) = 0.05$. Occasionally, probability is also expressed as the odds for or against an event occurring. For example, a betting man might say that the odds are 3 to 1 favoring Fred to win the race. In probability terms, p(Fred's winning) $= \frac{3}{4} = 0.75$. If the odds were 3 to 1 against Fred's winning, p(Fred's winning) $= \frac{1}{4} = 0.25$.

Computing Probability

Determining the probability of events can be complex. In fact, there are whole courses devoted to this topic, and they are quite difficult. Fortunately, for our purposes, there are only two major probability rules we need to learn, the addition rule and the multiplication rule. These rules provide the foundation necessary for understanding the statistical inference tests that follow in the textbook.

The Addition Rule

The addition rule is concerned with determining the probability of occurrence of any one of several possible events. To begin our discussion, let's assume there are only two possible events, A and B. When there are two events, the addition rule states

DEFINITION

• *The probability of occurrence of A or B is equal to the probability of occurrence of A plus the probability of occurrence of B minus the probability of occurrence of both A and B.*

In equation form, the addition rule states

$$p(A \text{ or } B) = p(A) + p(B) - p(A \text{ and } B)$$

addition rule for two events— general equation

Let's illustrate how this rule is used. Suppose we want to determine the probability of picking an ace or a club in one draw from a deck of ordinary playing cards. The problem has been solved in two ways in Figure 8.2. Please refer to the figure as you read this paragraph. The first way is by enumerating all the events classifiable as an ace or a club and using the basic equation for probability. Since there are 16 ways to get an ace or a club, the probability of getting an ace or a club $= \frac{16}{52} = 0.3077$. The second method uses the addition rule. The probability of getting an ace $= \frac{4}{52}$, and the probability of getting a club $= \frac{13}{52}$. The probability of getting both an ace and a club $= \frac{1}{52}$. By the addition rule, the probability of getting an ace or a club $= \frac{4}{52} + \frac{13}{52} - \frac{1}{52} = \frac{16}{52} = 0.3077$. Why do we need to subtract the probability of getting both an ace and a club? Because we have already counted the ace of clubs twice. Without subtracting it, we would be misled into thinking there are 17 favorable events rather than just 16.

In this course, we shall be using the addition rule almost entirely in situations where the events are *mutually exclusive*.

FIGURE 8.2   Determining the probability of randomly picking an ace or a club in one draw from a deck of ordinary playing cards

(a) By enumeration using the basic definition of probability

Ace or club

$$p(A) = \frac{\text{Number of events favorable to } A}{\text{Total number of possible events}}$$

$$= \frac{16}{52} = 0.3077$$

Events favorable to A

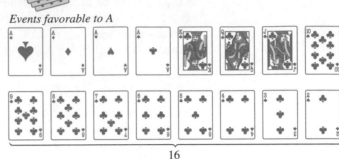

where A = drawing and ace or a club

16

(b) By the addition rule

Events favorable to A

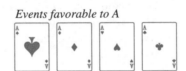

$$p(A \text{ or } B) = p(A) + p(B) - p(A \text{ and } B)$$

$$= \frac{4}{52} + \frac{13}{52} - \frac{1}{52}$$

$$= \frac{16}{52} = 0.3077$$

Events favorable to B

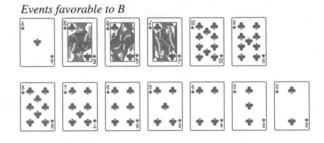

where A = drawing an ace
B = drawing a club

Events favorable to A and B

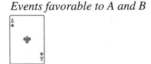

DEFINITION

- **Two events are mutually exclusive** *if both cannot occur together.* Another way of saying this is that **two events are mutually exclusive** *if the occurrence of one precludes the occurrence of the other.*

The events of rolling a one and of rolling a two in one roll of a die are mutually exclusive. If the roll ends with a one, it cannot also be a two. The events of picking an ace and a king in one draw from a deck of ordinary playing cards are mutually exclusive. If the card is an ace, it precludes the card also being a king. This can be contrasted with the events of picking an ace and a club in one draw from the deck. These events are not mutually exclusive because there is a card that is both an ace and a club (the ace of clubs).

When the events are mutually exclusive, the probability of both events occurring together is zero. Thus, $p(A$ and $B) = 0$ when A and B are mutually exclusive. Under these conditions, the addition rule simplifies to

$$p(A \text{ or } B) = p(A) + p(B) \qquad \textit{addition rule when A and B are mutually exclusive}$$

Let's practice solving some problems involving situations where A and B are mutually exclusive.

PRACTICE PROBLEM 8.1

What is the probability of randomly picking a 10 or a 4 in one draw from a deck of ordinary playing cards?

The solution is shown in Figure 8.3. Since we want either a 10 *or* a 4 and since these two events are mutually exclusive, the addition rule with mutually exclusive events is appropriate. Thus, $p(10 \text{ or } 4) = p(10) + p(4)$. Since there are four 10s, four 4s, and 52 cards, $p(10) = \frac{4}{52}$ and $p(4) = \frac{4}{52}$. Thus, $p(10 \text{ or } 4) = \frac{4}{52} + \frac{4}{52} = \frac{8}{52} = 0.1538$.

FIGURE 8.3   Determining the probability of randomly picking a 10 or a 4 in one draw from a deck of ordinary playing cards: Addition rule with mutually exclusive events

$$p(A \text{ or } B) = p(A) + p(B)$$

$$p(\text{a } 10 \text{ or a } 4) = p(10) + p(4)$$

$$= \frac{4}{52} + \frac{4}{52}$$

$$= \frac{8}{52} = 0.1538$$

where A = drawing a 10
B = drawing a 4

10 or 4

Events favorable to A

Events favorable to B

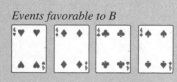

PRACTICE PROBLEM 8.2

In rolling a fair die once, what is the probability of rolling a one or an even number?

The solution is shown in Figure 8.4. Since the events are mutually exclusive and the problem asks for either a one *or* an even number, the addition rule with mutually exclusive events applies. Thus, p(one or an even number) = p(one) + p(an even number). There is one way to roll a one, three ways to roll an even number (2, 4, 6), and six possible outcomes. Thus, p(one) = $\frac{1}{6}$, p(an even number) = $\frac{3}{6}$, and p(one or an even number) = $\frac{1}{6} + \frac{3}{6} = \frac{4}{6} = 0.6667$.

FIGURE 8.4   Determining the probability of rolling a one or an even
....   number in rolling a fair die once: Addition rule with
mutually exclusive events

$$p(A \text{ or } B) = p(A) + p(B)$$

p(a one or an even number) = p(a one) + p(an even number)

$$= \frac{1}{6} + \frac{3}{6} = \frac{4}{6}$$

$$= 0.6667$$

Events favorable to A

where A = rolling a one
         B = rolling an even number

Events favorable to B

PRACTICE PROBLEM 8.3

Suppose you are going to randomly sample 1 individual from a population of 130 people. In the population there are 40 children under 12, 60 teenagers, and 30 adults. What is the probability the individual you select will be a teenager or an adult?

The solution is shown in Figure 8.5. Since the events are mutually exclusive, and we want a teenager or an adult, the addition rule with mutually exclusive events is appropriate. Thus, p(teenager or adult) = p(teenager) + p(adult). Since there are 60 teenagers, 30 adults, and 130 people in the population, p(teenager) = $\frac{60}{130}$ and p(adult) = $\frac{30}{130}$. Thus, p(teenager or adult) = $\frac{60}{130} + \frac{30}{130} = \frac{90}{130} = 0.6923$.

FIGURE 8.5   Determining the probability of randomly sampling a teenager or an adult in one draw from a population of 40 children under 12 years of age, 60 teenagers, and 30 adults; sampling is one at a time with replacement: Addition rule with mutually exclusive events

$$p(A \text{ or } B) = p(A) + p(B)$$

$$p(\text{a teenager or an adult}) = p(\text{a teenager}) + p(\text{an adult})$$

$$= \frac{60}{130} + \frac{30}{130}$$

$$= \frac{90}{130} = 0.6923$$

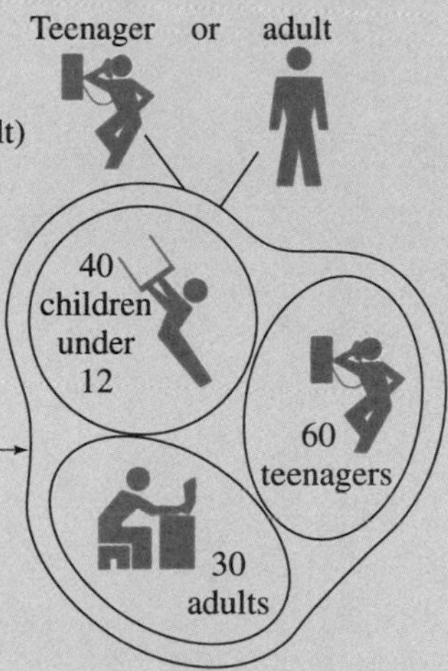

where   A = a teenager
        B = an adult

The addition rule may also be used in situations where there are more than two events. This is accomplished by a simple extension of the equation used for two events. Thus, when there are more than two events, and the events are mutually exclusive, the probability of occurrence of any one of the events is equal to the sum of the probability of each event. The equation follows:

$$p(A \text{ or } B \text{ or } C \text{ or } \dots \text{ or } Z) = p(A) + p(B) + p(C) + \cdots + p(Z)$$

addition rule with more than two mutually exclusive events

where   Z is the last event.

Very often we shall encounter situations where the events are not only mutually exclusive but also exhaustive. We have already defined mutually exclusive but not *exhaustive*.

DEFINITION

• A set of events is exhaustive *if the set includes all of the possible events.*

For example, in rolling a die once, the set of events of getting a one, two, three, four, five, or six is exhaustive because the set includes all of the possible events. When a set of events is both exhaustive and mutually exclusive a very useful relationship exists. Under these conditions, the sum of the individual probabilities of each event in the set must equal 1. Thus,

$$p(A) + p(B) + p(C) + \cdots + p(Z) = 1.00$$ *when events are exhaustive and mutually exclusive*

where   A, B, C, . . . , Z are the events.

To illustrate this relationship, let's consider the set of events of getting a one, two, three, four, five, or six in rolling a fair die once. Since the events are exhaustive and mutually exclusive, the sum of their probabilities must equal 1. We can see this is true since $p(\text{one}) = \frac{1}{6}$, $p(\text{two}) = \frac{1}{6}$, $p(\text{three}) = \frac{1}{6}$, $p(\text{four}) = \frac{1}{6}$, $p(\text{five}) = \frac{1}{6}$, and $p(\text{six}) = \frac{1}{6}$. Thus,

$$p(\text{one}) + p(\text{two}) + p(\text{three}) + p(\text{four}) + p(\text{five}) + p(\text{six}) = 1.00$$
$$\tfrac{1}{6} + \tfrac{1}{6} + \tfrac{1}{6} + \tfrac{1}{6} + \tfrac{1}{6} + \tfrac{1}{6} = 1.00$$

When there are only two events, and the events are mutually exclusive, it is customary to assign the symbol P to the probability of occurrence of one of the events and Q to the probability of occurrence of the other event. For example, if I were flipping a penny and only allowed it to come up heads or tails, this would be a situation where there are only two possible events (a head or a tail) with each flip, and the events are mutually exclusive (if it is a head, it can't be a tail). It is customary to let P equal the probability of occurrence of one of the events, say, a head, and Q equal the probability of occurrence of the other event, a tail. In this case, if the coin were a fair coin, $P = \frac{1}{2}$ and $Q = \frac{1}{2}$. Since the events are exhaustive and mutually exclusive, their probabilities must equal 1. Thus,

$$P + Q = 1.00$$ *when two events are exhaustive and mutually exclusive*

We shall be using the symbols P and Q extensively in Chapter 9 in conjunction with the binomial distribution.

The Multiplication Rule

Whereas the addition rule gives the probability of occurrence of any one of several events, the multiplication rule is concerned with the joint or successive occurrence of several events. Note that the multiplication rule often deals with what happens on more than one roll or draw, whereas the addition rule covers just one roll or one draw. If we are interested in the joint or successive occurrence of two events A and B, the multiplication rule states the following:

DEFINITION

• **The probability of occurrence of both A and B** *is equal to the probability of occurrence of A times the probability of occurrence of B given A has occurred.*

In equation form, the multiplication rule is

$$p(A \text{ and } B) = p(A)p(B|A)$$ *multiplication rule with two events—general equation*

Note that the symbol $p(B|A)$ is read "probability of occurrence of B given A has occurred." It does not mean B divided by A. Note also that the multiplication rule is concerned with the occurrence of A *and* B, whereas the addition rule applies to the occurrence of A *or* B.

In discussing the multiplication rule, it is useful to distinguish among three conditions: when the events are mutually exclusive, when the events are independent, and when the events are dependent.

Multiplication Rule: Mutually Exclusive Events

We have already discussed the joint occurrence of A and B when A and B are mutually exclusive. You will recall that if A and B are mutually exclusive,

$$p(A \text{ and } B) = 0 \quad \text{\textit{multiplication rule with mutually exclusive events}}$$

because when events are mutually exclusive, the occurrence of one precludes the occurrence of the other. The probability of their joint occurrence is zero.

Multiplication Rule: Independent Events

To understand how the multiplication rule applies in this situation we must first define *independent*.

DEFINITION

• **Two events are independent** *if the occurrence of one has no effect on the probability of occurrence of the other.*

Sampling with replacement illustrates this condition well. For example, suppose we are going to draw two cards, one at a time, with replacement, from a deck of ordinary playing cards. We can let A be the card drawn first and B be the card drawn second. Since A is replaced before drawing B, the occurrence of A on the first draw has no effect on the probability of occurrence of B. For instance, if A were an ace, since it is replaced in the deck before picking the second card, the occurrence of an ace on the first draw has no effect on the probability of occurrence of the card picked on the second draw. If A and B are independent, then the probability of B occurring is unaffected by A. Therefore, $p(B|A) = p(B)$. Under this condition, the multiplication rule becomes

$$p(A \text{ and } B) = p(A)p(B|A) = p(A)p(B) \quad \text{\textit{multiplication rule with independent events}}$$

Let's see how to use this equation. Suppose we are going to randomly draw two cards, one at a time, with replacement, from a deck of ordinary playing cards. What is the probability both cards will be aces?

The solution is shown in Figure 8.6. Since the problem requires an ace on the first draw *and* an ace on the second draw, the multiplication rule is appropriate. We can let A be an ace on the first draw and B be an ace on the second draw. Since sampling is with replacement, A and B are independent. Thus, p(an ace

FIGURE 8.6

Determining the probability of randomly sampling two aces in two draws from a deck of ordinary playing cards; sampling is one at a time with replacement: Multiplication rule with independent events

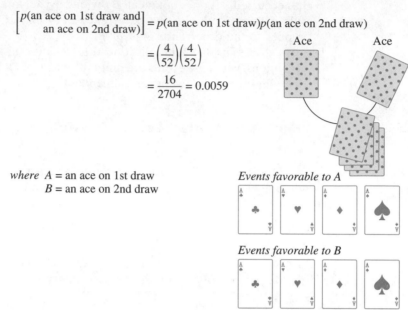

$$p(A \text{ and } B) = p(A)p(B)$$

$$\left[\begin{array}{l} p(\text{an ace on 1st draw and} \\ \text{an ace on 2nd draw)} \end{array} \right] = p(\text{an ace on 1st draw})p(\text{an ace on 2nd draw})$$

$$= \left(\frac{4}{52}\right)\left(\frac{4}{52}\right)$$

$$= \frac{16}{2704} = 0.0059$$

where A = an ace on 1st draw
       B = an ace on 2nd draw

Ace   Ace

Events favorable to A

Events favorable to B

on first draw and an ace on second draw) = p(an ace on first draw)p(an ace on second draw). Since there are four aces possible on the first draw, four aces possible on the second draw (sampling is *with* replacement), and 52 cards in the deck, p(an ace on first draw) = $\frac{4}{52}$ and p(an ace on second draw) = $\frac{4}{52}$. Thus, p(an ace on first draw and an ace on second draw) = $\frac{4}{52}(\frac{4}{52})$ = $\frac{16}{2704}$ = 0.0059.

Let's do a few more problems for practice.

PRACTICE PROBLEM 8.4

Suppose we roll a pair of fair dice once. What is the probability of obtaining a two on die 1 and a four on die 2?

The solution is shown in Figure 8.7. Since there is independence between the dice and since the problem asks for a two *and* a four, the multiplication rule with independent events applies. Thus, p(a two on die 1 and a four on die 2) = p(a two on die 1)p(a four on die 2). There is one way to get a two on die 1, one way to get a four on die 2, and six possible outcomes with each die. Therefore, p(a two on die 1) = $\frac{1}{6}$, p(a four on die 2) = $\frac{1}{6}$, and p(a two on die 1 and a four on die 2) = $\frac{1}{6}(\frac{1}{6})$ = $\frac{1}{36}$ = 0.0278.

FIGURE 8.7   Determining the probability of obtaining a two on die 1 and a four on die 2 in one throw of two fair dice: Multiplication rule with independent events

$$p(A \text{ and } B) = p(A)p(B)$$

p(a two on die 1 and a four on die 2) = p(a two on die 1)p(a four on die 2)

$$= \left(\frac{1}{6}\right)\left(\frac{1}{6}\right) = \frac{1}{36} = 0.0278$$

Events favorable to A
Die 1

where A = a two on die 1
B = a four on die 2

Events favorable to B
Die 2

PRACTICE PROBLEM 8.5

If two pennies are flipped once, what is the probability both pennies will turn up heads? Assume that the pennies are fair coins and that a head or tail is the only possible outcome with each coin.

The solution is shown in Figure 8.8. Since the outcome with the first coin has no effect on the outcome of the second coin, there is independence between events. Since the problem requires a head with the first coin *and* a head with the second coin, the multiplication rule with independent events is appropriate. Thus, p(a head with the first penny and a head with the second penny) = p(a head with first penny)p(a head with second penny). Since there is only one way to get a head with each coin and two possibilities with each coin (a head or a tail), p(a head with first penny) = $\frac{1}{2}$ and p(a head with second penny) = $\frac{1}{2}$. Thus, p(head with first penny and head with second penny) = $\frac{1}{2}(\frac{1}{2})$ = $\frac{1}{4}$ = 0.2500.

FIGURE 8.8   Determining the probability of obtaining two heads in one
....   flip of two fair pennies: Multiplication rule with
independent events

$$p(A \text{ and } B) = p(A)p(B)$$

$$\begin{bmatrix} p(\text{a head with 1st penny and} \\ \text{a head with 2nd penny}) \end{bmatrix} = p(\text{a head with 1st penny})p(\text{a head with 2nd penny})$$

$$= \left(\frac{1}{2}\right)\left(\frac{1}{2}\right) = 0.2500$$

where A = a head with 1st penny   *Events favorable to A*
B = a head with 2nd penny   *First penny*

Events favorable to B
Second penny

Suppose you are randomly sampling from a bag of fruit. The bag contains 4 apples, 6 oranges, and 5 peaches. If you sample two fruits, one at a time, with replacement, what is the probability you will get an orange and an apple in that order?

The solution is shown in Figure 8.9. Since there is independence between draws (sampling is with replacement) and since we want an orange *and* an apple, the multiplication rule with independent events applies. Thus, p(an orange on first draw and an apple on second draw) = p(an orange on first draw)p(an apple on second draw). Since there are 6 oranges and 15 pieces of fruit in the bag, p(an orange on first draw) = $\frac{6}{15}$. Because the fruit selected on the first draw is replaced before the second draw, it has no effect on the fruit picked on the second draw. Since there are 4 apples and 15 pieces of fruit, p(an apple on second draw) = $\frac{4}{15}$. Therefore, p(an orange on first draw and an apple on second draw) = $\frac{6}{15}(\frac{4}{15})$ = 0.1067.

FIGURE 8.9   Determining the probability of randomly sampling an orange and an apple in two draws from a bag containing 6 oranges, 4 apples, and 5 peaches; sampling is with replacement: Multiplication rule with independent events

$$p(A \text{ and } B) = p(A)p(B)$$

$$\begin{bmatrix} p(\text{an orange on 1st draw and} \\ \text{an apple on 2nd draw}) \end{bmatrix} = p(\text{an orange on 1st draw})p(\text{an apple on 2nd draw})$$

$$= \left(\frac{6}{15}\right)\left(\frac{4}{15}\right)$$

$$= \frac{24}{225} = 0.1067$$

Orange   Apple

where A = an orange on 1st draw
B = an apple on 2nd draw

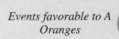

Events favorable to A
Oranges

Events favorable to B
Apples

PRACTICE PROBLEM 8.7

Suppose you are randomly sampling 2 individuals from a population of 110 men and women. There are 50 men and 60 women in the population. Sampling is one at a time, with replacement. What is the probability the sample will contain all women?

The solution is shown in Figure 8.10. Since the problem requires a woman on the first draw and a woman on the second draw and since there is independence between these two events (sampling is with replacement), the multiplication rule with independent events is appropriate. Thus, p(a woman on first draw and a woman on second draw) = p(a woman on first draw)p(a woman on second draw). Since there are 60 women and 110 people in the population, p(a woman on first draw) = $\frac{60}{110}$, and p(a woman on second draw) = $\frac{60}{110}$. Therefore, p(a woman on first draw and a woman on second draw) = $\frac{60}{110}(\frac{60}{110}) = \frac{3600}{12,100} = 0.2975$

FIGURE 8.10   Determining the probability of randomly sampling 2 women in two draws from a population of 50 men and 60 women; sampling is one at a time with replacement: Multiplication rule with independent events

$$p(A \text{ and } B) = p(A)p(B)$$

$$\left[\begin{array}{c} p(\text{a woman on 1st draw and} \\ \text{a woman on 2nd draw}) \end{array}\right] = p(\text{a woman on 1st draw})p(\text{a woman on 2nd draw})$$

$$= \left(\frac{60}{110}\right)\left(\frac{60}{110}\right)$$

$$= \frac{3600}{12,100} = 0.2975$$

where A = a woman on 1st draw
B = a woman on 2nd draw

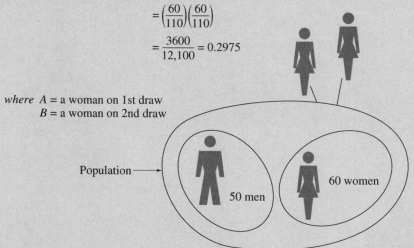

Population →   50 men   60 women

The multiplication rule with independent events also applies to situations where there are more than two events. In such cases, the probability of the joint occurrence of the events is equal to the product of the individual probabilities of each event. In equation form,

$$p(A \text{ and } B \text{ and } C \text{ and} \ldots \text{ and } Z) = p(A)p(B)p(C) \cdots p(Z)$$

multiplication rule with more than two independent events

To illustrate the use of this equation, let's suppose that instead of sampling two individuals from the population in Practice Problem 8.7, you are going to sample 4 persons. Otherwise the problem is the same. The population is composed of 50 men and 60 women. As before, sampling is one at a time, with replacement. What is the probability you will pick 3 women and 1 man in that order? The solution is shown in Figure 8.11. Since the problem requires a woman on the first *and* second *and* third draws *and* a man on the fourth draw and since sampling is with replacement, the multiplication rule with more than two independent events is appropriate. This rule is just like the multiplication rule with two independent events except there are more terms to multiply. Thus, p(a woman on first draw and a woman on second draw and a woman on third draw and a man on fourth draw) = p(a woman on first draw)p(a woman on second draw)p(a woman on third draw)p(a man on fourth draw). There are 60 women, 50 men, and 110 people in the population. Since sampling is with replacement, p(a woman on first draw) = $\frac{60}{110}$, p(a woman on second draw) = $\frac{60}{110}$, p(a woman on third draw) = $\frac{60}{110}$, and p(a man on fourth draw) = $\frac{50}{110}$. Thus, p(a woman on first draw

FIGURE 8.11

Determining the probability of randomly sampling 3 women and 1 man, in that order, in four draws from a population of 50 men and 60 women; sampling is one at a time with replacement: Multiplication rule with several independent events

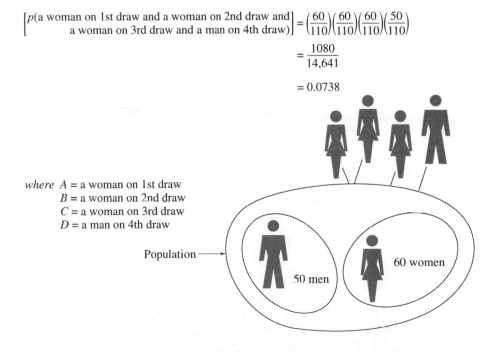

$$p(A \text{ and } B \text{ and } C \text{ and } D) = p(A)p(B)p(C)p(D)$$

$$\begin{bmatrix} p(\text{a woman on 1st draw and a woman on 2nd draw and} \\ \text{a woman on 3rd draw and a man on 4th draw}) \end{bmatrix} = \left(\frac{60}{110}\right)\left(\frac{60}{110}\right)\left(\frac{60}{110}\right)\left(\frac{50}{110}\right)$$

$$= \frac{1080}{14,641}$$

$$= 0.0738$$

where A = a woman on 1st draw
B = a woman on 2nd draw
C = a woman on 3rd draw
D = a man on 4th draw

Population →

50 men

60 women

and a woman on second draw and a woman on third draw and a man on fourth draw) $= \frac{60}{110}(\frac{60}{110})(\frac{60}{110})(\frac{50}{110}) = 1080/14{,}641 = 0.0738$.

Multiplication Rule: Dependent Events

When A and B are dependent, the probability of occurrence of B is affected by the occurrence of A. In this case, we cannot simplify the equation for the probability of A and B. We must use it in its original form. Thus, if A and B are dependent,

$$p(A \text{ and } B) = p(A)p(B|A) \quad \textit{multiplication rule with dependent events}$$

Sampling without replacement provides a good illustration for dependent events. Suppose you are going to draw two cards, one at a time, *without* replacement, from a deck of ordinary playing cards. What is the probability both cards will be aces?

The solution is shown in Figure 8.12. We can let A be an ace on the first draw and B be an ace on the second draw. Since sampling is without replacement (whatever card is picked the first time is kept out of the deck), the occurrence of A *does* affect the probability of B. A and B are dependent. Since the problem asks for an ace on the first *and* an ace on the second draw and since these events are dependent, the multiplication rule with dependent events is appropriate. Thus, p(an ace on first draw and an ace on second draw) = p(an ace on first draw)p(an ace on second draw, given an ace was obtained on first draw). For the first draw, there are 4 aces and 52 cards. Therefore, p(an ace on first draw) = $\frac{4}{52}$. Since sampling is without replacement, p(an ace on second draw given an ace on first draw) = $\frac{3}{51}$. Thus, p(an ace on first draw and an ace on second draw) = $\frac{4}{52}(\frac{3}{51}) = \frac{12}{2652} = 0.0045$.

FIGURE 8.12

Determining the probability of randomly picking two aces in two draws from a deck of ordinary playing cards; sampling is one at a time without replacement: Multiplication rule with dependent events

$$p(A \text{ and } B) = p(A)p(B|A)$$

$$\begin{bmatrix} p(\text{an ace on 1st draw and} \\ \text{an ace on 2nd draw)} \end{bmatrix} = \begin{bmatrix} p(\text{an ace on 1st draw})p(\text{an ace on 2nd draw given} \\ \text{an ace on 1st draw)} \end{bmatrix}$$

$$= \left(\frac{4}{52}\right)\left(\frac{3}{51}\right)$$

$$= \frac{12}{2652} = 0.0045$$

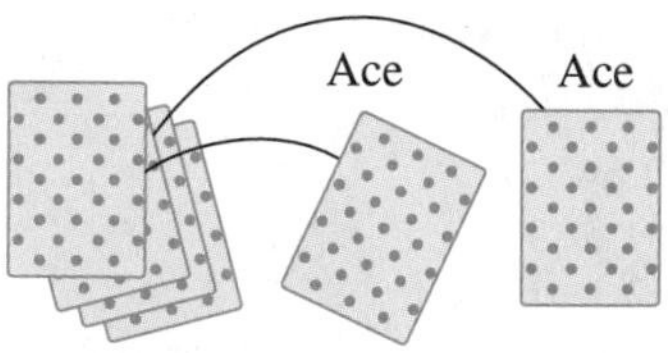

where A = drawing an ace on 1st draw
B = drawing an ace on 2nd draw

Events favorable to A

Events favorable to B

Suppose you are randomly sampling two fruits, one at a time, from the bag of fruit in Practice Problem 8.6. As before, the bag contains 4 apples, 6 oranges, and 5 peaches. However, this time you are sampling *without* replacement. What is the probability you will get an orange and an apple in that order?

The solution is shown in Figure 8.13. Since the problem requires an orange *and* an apple and since sampling is without replacement, the multiplication rule with dependent events applies. Thus, p(an orange on first draw and an apple on second draw) = p(an orange on first draw)p(an apple on second draw given an orange was obtained on first draw). On the first draw, there are 6 oranges and 15 fruits. Therefore p(an orange on first draw) = $\frac{6}{15}$. Since sampling is without replacement, p(an apple on second draw given an orange on first draw) = $\frac{4}{14}$. Therefore, p(an orange on first draw and an apple on second draw) = $\frac{6}{15}(\frac{4}{14})$ = $\frac{24}{210}$ = 0.1143.

FIGURE 8.13   Determining the probability of randomly sampling an orange and an apple in two draws from a bag containing 4 apples, 6 oranges, and 5 peaches; sampling is one at a time without replacement: Multiplication rule with dependent events

$$p(A \text{ and } B) = p(A)p(B|A)$$

$$\begin{bmatrix} p(\text{an orange on 1st draw and} \\ \text{an apple on 2nd draw}) \end{bmatrix} = \begin{bmatrix} p(\text{an orange on 1st draw})p(\text{an apple on 2nd draw} \\ \text{given an orange on 1st draw}) \end{bmatrix}$$

$$= \left(\frac{6}{15}\right)\left(\frac{4}{14}\right)$$

$$= \frac{24}{210} = 0.1143$$

Orange   Apple

where A = an orange on 1st draw
B = an apple on 2nd draw

Events favorable to A
Oranges

Events favorable to B
Apples

PRACTICE PROBLEM 8.9

In a particular college class, there are 15 music majors, 24 history majors, and 46 psychology majors. If you randomly sample 2 students from the class, what is the probability they will both be history majors? Sampling is one at a time, without replacement.

The solution is shown in Figure 8.14. Since the problem requires a history major on the first draw *and* a history major on the second draw and sampling is without replacement, the multiplication rule with dependent events is appropriate. Thus, p(a history major on first draw and a history major on second draw) = p(a history major on first draw)p(a history major on second draw given a history major was obtained on first draw). On the first draw, there were 24 history majors and 85 people in the population. Therefore, p(a history major on first draw) = $\frac{24}{85}$. Since sampling is without replacement, p(a history major on second draw given a history major on first draw) = $\frac{23}{84}$. Therefore, p(a history major on first draw and a history major on second draw) = $\frac{24}{85}(\frac{23}{84})$ = $\frac{552}{7140}$ = 0.0773.

FIGURE 8.14   Determining the probability of randomly sampling 2 history
....   majors on two draws from a population of 15 music majors,
24 history majors, and 46 psychology majors; sampling is
one at a time without replacement: Multiplication rule with
dependent events

$$p(A \text{ and } B) = p(A)p(B|A)$$

$$\left[\begin{matrix} p(\text{a history major on 1st draw and} \\ \text{a history major on 2nd draw)} \end{matrix}\right] = \left[\begin{matrix} p(\text{a history major on 1st draw})p(\text{a history} \\ \text{major on 2nd draw given a history major} \\ \text{on 1st draw)} \end{matrix}\right]$$

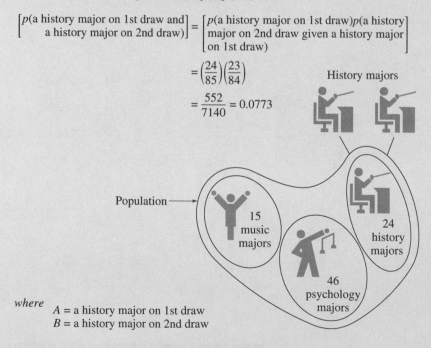

$$= \left(\frac{24}{85}\right)\left(\frac{23}{84}\right)$$

$$= \frac{552}{7140} = 0.0773$$

where A = a history major on 1st draw
B = a history major on 2nd draw

Like the multiplication rule with independent events, the multiplication rule with dependent events also applies to situations where there are more than two events. In such cases, the equation becomes

$p(A$ and B and C and ... and $Z)$

$$= p(A)p(B|A)p(C|AB) \cdots p(Z|ABC \ldots)$$

multiplication rule with more than two dependent events

where

$p(A)$ = probability of A

$p(B|A)$ = probability of B given A has occurred

$p(C|AB)$ = probability of C given A and B have occurred

$p(Z|ABC \cdots)$ = probability of Z given A, B, C, and all other events have occurred

To illustrate how to use this equation let's do a problem that involves more than two dependent events. Suppose you are going to sample 4 students from the college class given in Practice Problem 8.9. In that class there were 15 music majors, 24 history majors, and 46 psychology majors. If sampling is one at a time, without replacement, what is the probability you will obtain 4 history majors?

The solution is shown in Figure 8.15. Since the problem requires a history major on the first *and* second *and* third *and* fourth draws, and since sampling is without replacement, the multiplication rule with more than two dependent events is appropriate. This rule is very much like the multiplication rule with two dependent events, except more multiplying is required. Thus, for this problem, p(a history major on first draw and a history major on second draw and a

FIGURE 8.15

Determining the probability of randomly sampling 4 history majors on four draws from a population of 15 music majors, 24 history majors, and 46 psychology majors; sampling is one at a time without replacement: Multiplication rule with several dependent events

$p(A$ and B and C and $D) = p(A)p(B|A)p(C|AB)p(D|ABC)$

$$= \left(\frac{24}{85}\right)\left(\frac{23}{84}\right)\left(\frac{22}{83}\right)\left(\frac{21}{82}\right)$$

$$= \frac{255{,}024}{48{,}594{,}840} = 0.0052$$

where A = a history major on 1st draw
B = a history major on 2nd draw
C = a history major on 3rd draw
D = a history major on 4th draw

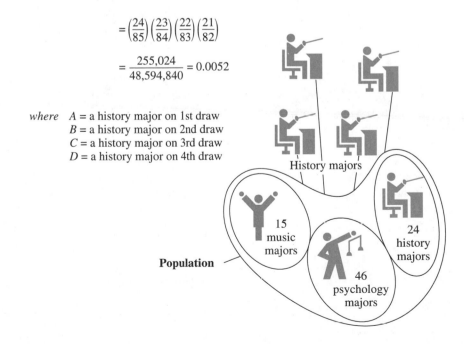

History majors

15 music majors

24 history majors

Population

46 psychology majors

history major on third draw and a history major on fourth draw) = p(a history major on first draw)p(a history major on second draw given a history major on first draw)p(a history major on third draw given a history major on first and second draws)p(a history major on fourth draw given a history major on first, second, and third draws). On the first draw, there are 24 history majors and 85 individuals in the population. Thus, p(a history major on first draw) = $\frac{24}{85}$. Since sampling is without replacement, p(a history major on second draw given a history major on first draw) = $\frac{23}{84}$, p(a history major on third draw given a history major on first and second draws) = $\frac{22}{83}$, and p(a history major on fourth draw given a history major on first, second, and third draws) = $\frac{21}{82}$. Therefore, p(a history major on first draw and a history major on second draw and a history major on third draw and a history major on fourth draw) = $\frac{24}{85}(\frac{23}{84})(\frac{22}{83})(\frac{21}{82})$ = 255,024/48,594,840 = 0.0052.

Multiplication and Addition Rules

Some situations require that we use both the multiplication and addition rules for their solution. For example, suppose that I am going to roll two fair dice once. What is the probability the sum of the numbers showing on the dice will equal 11? The solution is shown in Figure 8.16. There are two possible outcomes that yield a sum of 11 (die 1 = 5 and die 2 = 6, which we shall call outcome A; and die 1 = 6 and die 2 = 5, which we shall call outcome B). Since the dice are independent, we can use the multiplication rule with independent events to find the probability of each outcome. By using this rule, $p(A) = \frac{1}{6}(\frac{1}{6}) = \frac{1}{36}$, and $p(B) = \frac{1}{6}(\frac{1}{6}) = \frac{1}{36}$. Since either of the outcomes yields a sum of 11, p(sum of 11) = $p(A \text{ or } B)$. Since these outcomes are mutually exclusive, we can use the addition rule with mutually exclusive events to find $p(A \text{ or } B)$. Thus, p(sum of 11) = $p(A \text{ or } B) = p(A) + p(B) = \frac{1}{36} + \frac{1}{36} = \frac{2}{36} = 0.0556$.

FIGURE 8.16
....

Determining the probability of rolling a sum of 11 in one roll of two fair dice: Multiplication and addition rules

$p(A) = p$(5 on die 1 and 6 on die 2)

$\quad = p$(5 on die 1)p(6 on die 2)

$$= \left(\frac{1}{6}\right)\left(\frac{1}{6}\right) = \frac{1}{36}$$

$p(B) = p$(6 on die 1 and 5 on die 2)

$\quad = p$(6 on die 1)p(5 on die 2)

$$= \left(\frac{1}{6}\right)\left(\frac{1}{6}\right) = \frac{1}{36}$$

p(sum of 11) = $p(A \text{ or } B) = p(A) + p(B)$

$$= \frac{1}{36} + \frac{1}{36} = \frac{2}{36} = 0.0556$$

Possible outcomes yielding a sum of 11

Die 1   *Die 2*

A

B

Let's try one more problem that involves both the multiplication and addition rules.

PRACTICE PROBLEM 8.10

Suppose you have arrived in Las Vegas and you are going to try your "luck" on a one-armed bandit (slot machine). In case you are not familiar with slot machines, basically a slot machine has three wheels that rotate independently. Each wheel contains pictures of different objects. Let's assume the one you are playing has seven different fruits on wheel 1. There are a lemon, a plum, an apple, an orange, a pear, some cherries, and a banana. Wheels 2 and 3 have the same fruits as wheel 1. When the lever is pulled down, the three wheels rotate independently and then come to rest. On the slot machine, there is a window in front of each wheel. The pictures of the fruits pass under the window during rotation. When the wheel stops, one of the fruits from each wheel will be in view. We shall assume that each fruit on a wheel has an equal probability of appearing under the window at the end of rotation. You insert your quarter and pull down the lever. What is the probability that two lemons and a pear will appear? Order is not important: All you care about is getting two lemons and a pear, in any order.

The solution is shown in Figure 8.17. There are three possible orders of two lemons and a pear: lemon lemon pear, lemon pear lemon, and pear lemon lemon. Since the wheels rotate independently, we can use the multiplication rule with independent events to determine the probability of each order. Since each fruit is equally likely, p(lemon and lemon and pear) = p(lemon)p(lemon)p(pear) = $\frac{1}{7}(\frac{1}{7})(\frac{1}{7})$ = $\frac{1}{343}$. The same probability also applies to the other two orders. Since the three orders give two lemons and a pear, p(two lemons and a pear) = p(order 1, 2, or 3). By using the addition rule with independent events, p(order 1, 2, or 3) = $\frac{3}{343}$ = 0.0087. Thus, the probability of getting two lemons and a pear, without regard to order, equals 0.0087.

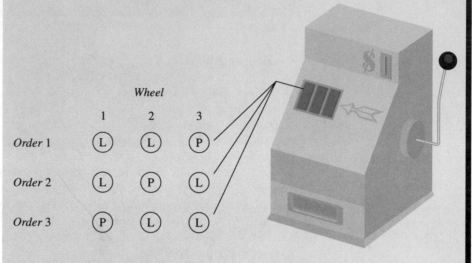

FIGURE 8.17   Determining the probability of obtaining two lemons and a pear without regard to order on a slot machine with three wheels and seven fruits on each wheel: Multiplication and addition rules

p(order 1) = p(lemon on wheel 1 and lemon on wheel 2 and pear on wheel 3)

= p(lemon on wheel 1)p(lemon on wheel 2)p(pear on wheel 3)

$$=\left(\frac{1}{7}\right)\left(\frac{1}{7}\right)\left(\frac{1}{7}\right)=\frac{1}{343}$$

p(order 2) = p(lemon on wheel 1 and pear on wheel 2 and lemon on wheel 3)

= p(lemon on wheel 1)p(pear on wheel 2)p(lemon on wheel 3)

$$=\left(\frac{1}{7}\right)\left(\frac{1}{7}\right)\left(\frac{1}{7}\right)=\frac{1}{343}$$

p(order 3) = p(pear on wheel 1 and lemon on wheel 2 and lemon on wheel 3)

= p(pear on wheel 1)p(lemon on wheel 2)p(lemon on wheel 3)

$$=\left(\frac{1}{7}\right)\left(\frac{1}{7}\right)\left(\frac{1}{7}\right)=\frac{1}{343}$$

$$\left(\begin{array}{c}p(2 \text{ lemons}\\ \text{and a pear})\end{array}\right) = p(\text{order 1 or 2 or 3}) = p(\text{order1}) + p(\text{order 2}) + p(\text{order 3})$$

$$=\frac{1}{343}+\frac{1}{343}+\frac{1}{343}=\frac{3}{343}=0.0087$$

Probability and Normally Distributed Continuous Variables

So far in our discussion of probability, we have considered variables that have been discrete, e.g., sampling from a deck of cards or rolling a pair of dice. However, many of the dependent variables that are evaluated in experiments are continuous, not discrete. When a variable is continuous

$$p(A) = \frac{\text{Area under the curve corresponding to } A}{\text{Total area under the curve}}$$ *probability of A with a continuous variable*

Since often (although not always) these variables are normally distributed, we shall concentrate our discussion on normally distributed continuous variables.

To illustrate the use of probability with continuous variables that are normally distributed, suppose we have measured the weights of all the sophomore women at your college. Let's assume this is a population set of scores that is normally distributed, with a mean of 120 pounds and a standard deviation of 8 pounds. If we randomly sampled one score from the population, what is the probability it would be equal to or greater than a score of 134?

The population is drawn in Figure 8.18. The mean of 120 and the score of 134 are located on the X axis. The shaded area represents all the scores that are equal to or greater than 134. Since sampling is random, each score has an equal chance of being selected. Thus, the probability of obtaining a score equal to or greater than 134 can be found by determining the proportion of the total scores that are contained in the shaded area. Since the scores are normally distributed, we can find this proportion by converting the raw score to its z-transformed value and then looking up the area in Table A in Appendix D. Thus,

$$z = \frac{X - \mu}{\sigma} = \frac{134 - 120}{8} = \frac{14}{8} = 1.75$$

From Table A, column C,

$$p(X \geq 134) = 0.0401$$

FIGURE 8.18

Probability of obtaining $X \geq 134$ if randomly sampling one score from a normal population with $\mu = 120$ and $\sigma = 8$

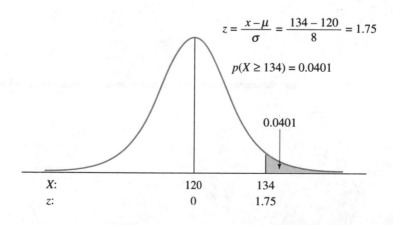

We are sure you will recognize that this type of problem is quite similar to those presented in Chapter 5 when dealing with standard scores. The main difference is that, in this chapter, the problem has been cast in terms of probability rather than asking for the proportion or percentage of scores as was done in Chapter 5. Since you are already familiar with this kind of problem, we don't think it necessary to give a lot of practice problems. However, let's try a couple just to be sure.

PRACTICE PROBLEM 8.11

Consider the same population of sophomore women. If one score is randomly sampled from the population, what is the probability it will be equal to or less than 110?

The solution is presented in Figure 8.19. The shaded area represents all the scores that are equal to or less than 110. Since sampling is random, each score has an equal chance of being selected. To find $p(X \leq 110)$, first we must transform the raw score of 110 to its z score. Then we can find the proportion of the total scores that are contained in the shaded area by using Table A. Thus,

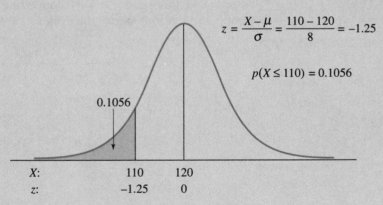

$$z = \frac{X - \mu}{\sigma} = \frac{110 - 120}{8} = -1.25$$

$$p(X \leq 110) = 0.1056$$

X:	110	120
z:	−1.25	0

FIGURE 8.19   Probability of obtaining $X \leq 110$ if randomly sampling one score from a normal population with $\mu = 120$ and $\sigma = 8$

Considering the same population again, what is the probability of randomly sampling a score that is as far or farther from the mean than a score of 138?

The solution is shown in Figure 8.20. The score of 138 is 18 units above the mean. Since the problem asks for scores as far or farther from the mean, we must also consider scores that are 18 units or more below the mean. The shaded areas contain all of the scores that are 18 units or more away from the mean. Since sampling is random, each score has an equal chance of being selected. To find $p(X \le 102 \text{ or } X \ge 138)$, first we must transform the raw scores of 102 and 138 to their z scores. Then we can find the proportion of the total scores that are contained in the shaded areas by using Table A. $p(X \le 102 \text{ or } X \ge 138)$ equals this proportion. Thus,

$$z = \frac{X - \mu}{\sigma} = \frac{102 - 120}{8} = -\frac{18}{8} \qquad z = \frac{X - \mu}{\sigma} = \frac{138 - 120}{8} = \frac{18}{8}$$

$$= -2.25 \qquad\qquad\qquad = 2.25$$

From Table A, column C,

$$p(X \le 102 \text{ or } X \ge 138) = 0.0122 + 0.0122 = 0.0244$$

FIGURE 8.20   Probability of obtaining a score as far or farther from the mean than a score of 138 from a population with $\mu = 120$ and $\sigma = 8$

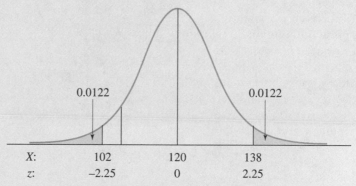

WHAT IS THE TRUTH?

"NOT GUILTY, I'M A VICTIM OF COINCIDENCE": GUTSY PLEA OR TRUTH?

Despite a tradition of qualitatively rather than quantitatively based decision making, the legal field is increasingly using statistics as a basis for decisions. The following case from Sweden is an example.

In a Swedish trial, the defendant was contesting a charge of overtime parking. An officer had marked the position of the valves of the front and rear tires of the accused car, according to a clock representation, e.g., front valve to one o'clock and rear valve to six o'clock, in both cases to the nearest hour (see diagram). After the allowed time had elapsed, the car was still there, with the two valves pointing to one and six o'clock as before. The accused was given a parking ticket. In court, however, he pleaded innocent, claiming that he had left the parking spot in time, but returned to it later, and the valves had just happened to come to rest in the same position as before. The judge, not having taken a basic course in statistics, called in a statistician to evaluate the defendant's claim of coincidence. Is the defendant's claim reasonable? Assume you are the statistician. What would you tell the judge? In formulating your answer, assume independence between the wheels, as did the statistician who advised the judge.

ANSWER As a statistician, your job is to determine how reasonable the plea of coincidence really is. If we assume the defendant's story is true about leaving and coming back to the parking spot, what is the probability of the two valves returning to their one and six o'clock positions? Since there are 12 possible positions for each valve, assuming independence between the wheels, using the multiplication rule,

$$p(\text{one and six}) = \left(\tfrac{1}{12}\right)\left(\tfrac{1}{12}\right) = \tfrac{1}{144}$$

$$= 0.0069$$

Thus, if coincidence (or chance alone) is at work, the probability of the valves returning to their original positions is about 7 times in one thousand. What do you think the judge did when given this information? Believe

it or not, the judge acquitted the defendant, saying that if all four wheels had been checked and found to point in the same directions as before ($p = \tfrac{1}{12} \times \tfrac{1}{12} \times \tfrac{1}{12} \times \tfrac{1}{12} = \tfrac{1}{20736} = 0.00005$), then the coincidence claim would have been rejected as too improbable and the defendant convicted. Thus, the judge considered the coincidence explanation as too probable to reject, even though the results would be obtained only 1 out of 144 times if coincidence was at work. Actually, because the wheels do not rotate independently, the formulas used most likely understate somewhat the probability of a chance return to the original position. (How did you do? Can we call on you in the future as a statistical expert to help mete out justice?)

WHAT IS THE TRUTH?

SPERM COUNT DECLINE—MALE OR RESEARCH INADEQUACY?

The headline of an article that appeared in 1995 in a leading metropolitan newspaper read, "20-year study shows sperm count decline among fertile men." Excerpts from the article are reproduced here.

A new study has found a marked decline in sperm counts among fertile men over the past 20 years. . . .

The paper, published today in The New England Journal of Medicine, was based on data collected over a 20-year period at a Paris sperm bank. Some experts in the United States took strong exception to the findings. . . .

The new study, by Dr. Pierre Jouannet of the Center for the

Study of the Conservation of Human Eggs and Sperm in Paris, examined semen collected by a sperm bank in Paris beginning in 1973. They report that sperm counts fell by an average of 2.1 percent a year, going from 89 million sperm per milliliter in 1973 to an average of 60 million per milliliter in 1992. At the same time they found the percentages of sperm that moved normally and were properly formed declined by 0.5 to 0.6 of 1 percent a year.

The paper is accompanied by an invited editorial by an expert on male infertility, Dr. Richard Sherins, director of the division of andrology at the Genetics and IVF institute in Fairfax, Va., who said the current studies and sev-

eral preceding it suffered from methodological flaws that made their data uninterpretable.

Sherins said that the studies did not look at sperm from randomly selected men and that sperm counts and sperm quality vary so much from week to week that it is hazardous to rely on single samples to measure sperm quality, as these studies did.

What do you think? Why might it be important to use samples from randomly selected men rather than from men who deposit their sperm at a sperm bank? Why might large week to week variability in sperm counts and sperm quality complicate interpretation of the data?

SUMMARY

In this chapter, we have discussed the topics of random sampling and probability. A random sample is defined as a sample that has been selected from a population by a process that assures that (1) each possible sample of a given size has an equal chance of being selected and (2) all members of the population have an equal chance of being selected into the sample. After defining and discussing the importance of random sampling, we described various methods for obtaining a random sample. In the last section on random sampling, we discussed sampling with and without replacement.

In presenting probability, we pointed out that probability may be approached from two viewpoints, *a priori* and *a posteriori*. According to the *a priori* view, $p(A)$ is defined as

$$p(A) = \frac{\text{Number of events classifiable as } A}{\text{Total number of possible events}}$$

From an *a posteriori* standpoint, $p(A)$ is defined as

$$p(A) = \frac{\text{Number of times } A \text{ has occurred}}{\text{Total number of occurrences}}$$

Since probability is fundamentally a proportion, it ranges from 0.00 to 1.00. Next, we presented two probability rules necessary for understanding inferential statistics, the addition rule and the multiplication rule. Assuming there are two events (A and B), the addition rule gives the probability of A or B, whereas the multiplication rule gives the probability of A and B. The addition rule states the following:

$$p(A \text{ or } B) = p(A) + p(B) - p(A \text{ and } B)$$

If the events are mutually exclusive,

$$p(A \text{ or } B) = p(A) + p(B)$$

If the events are mutually exclusive and exhaustive,

$$p(A) + p(B) = 1.00$$

The multiplication rule states the following:

$$p(A \text{ and } B) = p(A)p(B|A)$$

If the events are mutually exclusive,

$$p(A \text{ and } B) = 0$$

If the events are independent,

$$p(A \text{ and } B) = p(A)p(B)$$

If the events are dependent, we must use the general equation

$$p(A \text{ and } B) = p(A)p(B|A)$$

In addition, we discussed (1) the generalization of these equations to situations where there were more than two events and (2) situations that required both the addition and multiplication rules for their solution. Finally, we discussed the probability of A with continuous variables and described how to find $p(A)$ when the variable was both normally distributed and continuous. The equation for determining the probability of A when the variable is continuous is

$$p(A) = \frac{\text{Area under the curve corresponding to } A}{\text{Total area under the curve}}$$

IMPORTANT TERMS

Addition rule (p. 161)
A posteriori (p. 160)
A priori (p. 159)
Exhaustive (p. 165)

Independent (p. 167)
Multiplication rule (p. 166)
Mutually exclusive (p. 162)
Probability (p. 159)

Random sample (p. 156)
Sampling with replacement (p. 158)
Sampling without replacement
    (p. 159)

QUESTIONS AND PROBLEMS

1. Define or identify each term in the "Important Terms" section.
2. What two purposes does random sampling serve?
3. Assume you want to form a random sample of 20 subjects from a population of 400 individuals. Sampling will be without replacement, and you plan to use Table J in Appendix D to accomplish the randomization. Explain how you would use the table to select the sample.
4. What is the difference between *a priori* and *a posteriori* probability?
5. The addition rule gives the probability of occurrence of any one of several events, whereas the multiplication rule gives the probability of the joint or successive occurrence of several events. Is this statement correct? Explain, using examples to illustrate your explanation.
6. When solving problems involving the multiplication rule, is it useful to distinguish among three conditions? What are these conditions? Why is it useful to distinguish among them?
7. What is the definition of probability when the variable is continuous?
8. Which of the following are examples of independent events?
   a. Obtaining a three and a four in one roll of two fair dice
   b. Obtaining an ace and a king in that order by drawing twice without replacement from a deck of cards
   c. Obtaining an ace and a king in that order by drawing twice with replacement from a deck of playing cards
   d. A cloudy sky followed by rain
   e. A full moon and eating a hamburger
9. Which of the following are examples of mutually exclusive events?
   a. Obtaining a 4 and a 7 in one draw from a deck of ordinary playing cards
   b. Obtaining a three and a four in one roll of two fair dice
   c. Being male and becoming pregnant
   d. Obtaining a one and an even number in one roll of a fair die
   e. Getting married and remaining a bachelor
10. Which of the following are examples of exhaustive events?
   a. Flipping a coin and obtaining a head or a tail (edge not allowed)
   b. Rolling a die and obtaining a two
   c. Taking an exam and either passing or failing
   d. Going out on a date and having a good time
11. If you draw a single card once from a deck of ordinary playing cards, what is the probability that it will be
   a. The ace of diamonds?
   b. A 10?
   c. A queen or a heart?
   d. A 3 or a black card?
12. If you roll two fair dice once, what is the probability that you will obtain
   a. A two on die 1 and a five on die 2?
   b. A two and a five without regard to which die has the two or five?
   c. At least one two or one five?
   d. A sum of seven?
13. If you are randomly sampling one at a time with replacement from a bag that contains eight blue marbles, seven red marbles, and five green marbles, what is the probability of obtaining
   a. A blue marble in one draw from the bag?
   b. Three blue marbles in three draws from the bag?
   c. A red, a green, and a blue marble in that order in three draws from the bag?
   d. At least two red marbles in three draws from the bag?
14. Answer the same questions as in Problem 13, except sampling is one at a time without replacement.
15. You are playing the one-armed bandit (slot machine) described in Practice Problem 8.10, p. 177. There are three wheels, and on each wheel there is a picture of a lemon, a plum, an apple, an orange, a pear, cherries, and a banana (seven different pictures). You insert your quarter and pull down the lever. What is the probability that
   a. Three oranges will appear?
   b. Two oranges and a banana will appear, without regard to order?
   c. At least two oranges will appear?
16. You want to call a friend on the telephone. You remember the first three digits of her phone number, but you have forgotten the last four digits.

What is the probability that you will get the correct number merely by guessing?

17. You are planning to make a "killing" at the race track. In a particular race, there are seven horses entered. If the horses are all equally matched, what is the probability of your correctly picking the winner and runner-up?

18. A gumball dispenser has 38 orange gumballs, 30 purple ones, and 18 yellow ones. The dispenser operates such that one penny delivers 1 gumball.
    a. Using three pennies, what is the probability of obtaining 3 gumballs in the order orange, purple, orange?
    b. Using one penny, what is the probability of obtaining 1 gumball that is either purple or yellow?
    c. Using three pennies, what is the probability that of the 3 gumballs obtained, exactly 1 will be purple and 1 will be yellow?

19. If two cards are randomly drawn from a deck of ordinary playing cards, one at a time with replacement, what is the probability of obtaining at least one ace?

20. A state lottery is paying $1,000,000 to the holder of the ticket with the correct 8-digit number. Tickets cost a dollar apiece. If you buy one ticket, what is the probability you will win? Assume there is only one ticket for each possible 8-digit number and the winning number is chosen by a random process.

21. Given a population comprised of 30 bats, 15 gloves, and 60 balls, if sampling is random and one at a time without replacement,
    a. What is the probability of obtaining a glove if one object is sampled from the population?
    b. What is the probability of obtaining a bat and a ball if two objects are sampled from the population?
    c. What is the probability of obtaining a bat, a glove, and a bat in that order if three objects are sampled from the population?

22. A distribution of scores is normally distributed with a mean $\mu = 85$ and a standard deviation $\sigma = 4.6$. If one score is randomly sampled from the distribution, what is the probability that it will be
    a. Greater than 96?
    b. Between 90 and 97?
    c. Less than 88?

23. Assume the IQ scores of the students at your university are normally distributed, with $\mu = 115$ and $\sigma = 8$. If you randomly sample one score from this distribution, what is the probability it will be
    a. Higher than 130?
    b. Between 110 and 125?
    c. Lower than 100?

24. An ethologist is interested in how long it takes a certain species of water shrew to catch its prey. On 20 occasions each day, he lets a dragonfly loose inside the cage of a shrew and times how long it takes until the shrew catches the dragonfly. After months of research, the ethologist concludes that the mean prey-catching time was 30 seconds, the standard deviation was 5.5 seconds, and the scores were normally distributed. Based on the shrew's past record, what is the probability
    a. It will catch a dragonfly in less than 18 seconds?
    b. It will catch a dragonfly in between 22 and 45 seconds?
    c. It will take longer than 40 seconds to catch a dragonfly?

25. An instructor at the U.S. Navy's underwater demolition school believes he has developed a new technique for staying underwater longer. The school commandant gives him permission to try his technique with a student who has been randomly selected from the current class. As part of their qualifying exam, all students are tested to see how long they can stay underwater without an air tank. Past records show that the scores are normally distributed with a mean $\mu = 130$ seconds and a standard deviation $\sigma = 14$ seconds. If the new technique has no additional effect, what is the probability that the randomly selected student will stay underwater for
    a. Over 150 seconds?
    b. Between 115 and 135 seconds?
    c. Less than 90 seconds?

26. If you are randomly sampling two scores one at a time with replacement from a population comprised of the scores 2, 3, 4, 5, and 6, what is the probability that
    a. The mean of the sample ($\overline{X}$) will equal 6.0?
    b. $\overline{X} \geq 5.5$?
    c. $\overline{X} \leq 2.0$?
    Hint: All of the possible samples of size 2 are listed on p. 156.

NOTES

8.1 We realize that if the process assures that each possible sample of a given size has an equal chance of being selected, then it also assures that all the members of the population have an equal chance of being selected into the sample. We included the latter statement in the definition because we felt it is sufficiently important to deserve this special emphasis.

9 | BINOMIAL DISTRIBUTION

INTRODUCTION

In Chapter 10, we'll discuss the topic of hypothesis testing. This topic is a very important one. It forms the basis for most of the material taken up in the remainder of the textbook. For reasons explained in Chapter 10, we've chosen to introduce the concepts of hypothesis testing by using a simple inference test called the sign test. However, to understand and use the sign test, we must first discuss a probability distribution called the binomial distribution.

DEFINITION AND ILLUSTRATION OF THE BINOMIAL DISTRIBUTION

The binomial distribution may be defined as follows:

DEFINITION

- *The **binomial distribution** is a probability distribution that results when the following five conditions are met: (1) there is a series of N trials; (2) on each trial there are only two possible outcomes; (3) on each trial, the two possible outcomes are mutually exclusive; (4) there is independence between the outcomes of each trial; and (5) the probability of each possible outcome on any trial stays the same from trial to trial. When these requirements are met, the binomial distribution tells us each possible outcome of the N trials and the probability of getting each of these outcomes.*

Let's use coin flipping as an illustration for generating the binomial distribution. Suppose we flip a fair or unbiased penny once. Suppose further that we restrict the possible outcomes at the end of the flip to either a head or a tail. You will recall from Chapter 8 that a fair coin means the probability of a head with the coin equals the probability of a tail. Since there are only two possible outcomes in one flip,

$$p(\text{head}) = p(H) = \frac{\text{Number of outcomes classifiable as heads}}{\text{Total number of outcomes}}$$

$$= \frac{1}{2} = 0.5000$$

$$p(\text{tail}) = p(T) = \frac{\text{Number of outcomes classifiable as tails}}{\text{Total number of outcomes}}$$

$$= \frac{1}{2} = 0.5000$$

Now suppose we flip two pennies that are unbiased. The flip of each penny is considered a trial. Thus, with two pennies, there are two trials ($N = 2$). The possible outcomes of flipping two pennies are given in Table 9.1. There are four possible outcomes: one in which there are 2 heads (row 1), two in which there are 1 head and 1 tail (rows 2 and 3), and one in which there are 2 tails (row 4).

TABLE 9.1   All Possible Outcomes of Flipping Two Coins Once

Row No.	Penny 1	Penny 2	No. of Outcomes
1	H	H	1
2	H	T	2
3	T	H	
4	T	T	1
		Total outcomes	4

Next, let's determine the probability of getting each of these outcomes due to chance. If chance alone is operating, then each of the outcomes is equally likely. Thus,

$$p(2 \text{ heads}) = p(HH) = \frac{\text{Number of outcomes classifiable as 2 heads}}{\text{Total number of outcomes}}$$

$$= \frac{1}{4} = 0.2500$$

$$p(1 \text{ head}) = p(HT \text{ or } TH) = \frac{\text{Number of outcomes classifiable as 1 head}}{\text{Total number of outcomes}}$$

$$= \frac{2}{4} = 0.5000$$

$$p(0 \text{ head}) = p(TT) = \frac{\text{Number of outcomes classifiable as 0 heads}}{\text{Total number of outcomes}}$$

$$= \frac{1}{4} = 0.2500$$

You should note that we could have also found these probabilities from the multiplication and addition rules. For example, $p(1 \text{ head})$ could have been found from a combination of the addition and multiplication rules as follows:

$$p(1 \text{ head}) = p(HT \text{ or } TH)$$

Using the multiplication rule, we obtain

$$p(HT) = p(\text{head on coin 1 and tail on coin 2})$$
$$= p(\text{head on coin 1})p(\text{tail on coin 2})$$
$$= \tfrac{1}{2}(\tfrac{1}{2}) = \tfrac{1}{4}$$
$$p(TH) = p(\text{tail on coin 1 and head on coin 2})$$
$$= p(\text{tail on coin 1})p(\text{head on coin 2})$$
$$= \tfrac{1}{2}(\tfrac{1}{2}) = \tfrac{1}{4}$$

Using the addition rule, we obtain

$$p(1 \text{ head}) = p(HT \text{ or } TH)$$
$$= p(HT) + p(TH)$$
$$= \tfrac{1}{4} + \tfrac{1}{4} = \tfrac{2}{4} = 0.5000$$

Next, suppose we increase N from 2 to 3. The possible outcomes of flipping three unbiased pennies once are shown in Table 9.2. This time there are eight possible outcomes: one way to get 3 heads (row 1), three ways to get 2 heads and 1 tail (rows 2, 3, and 4), three ways to get 1 head and 2 tails (rows 5, 6, and

TABLE 9.2   All Possible Outcomes of Flipping Three Pennies Once

Row No.	Penny 1	Penny 2	Penny 3	No. of Outcomes
1	H	H	H	1
2	H	H	T	
3	H	T	H	3
4	T	H	H	
5	T	T	H	
6	T	H	T	3
7	H	T	T	
8	T	T	T	1
			Total outcomes	8

TABLE 9.3

Binomial Distribution for
Coin Flipping When the
Number of Coins Equals
1, 2, or 3

N	Possible Outcomes	Probability
1	1 H	0.5000
	0 H	0.5000
2	2 H	0.2500
	1 H	0.5000
	0 H	0.2500
3	3 H	0.1250
	2 H	0.3750
	1 H	0.3750
	0 H	0.1250

7), and one way to get 0 heads (row 8). Since each outcome is equally likely,

$$p(3 \text{ heads}) = \tfrac{1}{8} = 0.1250$$

$$p(2 \text{ heads}) = \tfrac{3}{8} = 0.3750$$

$$p(1 \text{ head}) = \tfrac{3}{8} = 0.3750$$

$$p(0 \text{ heads}) = \tfrac{1}{8} = 0.1250$$

The distributions resulting from flipping one, two, or three fair pennies are shown in Table 9.3. These are binomial distributions because they are probability distributions that have been generated by a situation where there is a series of trials ($N = 1, 2,$ or 3), where on each trial there are only two possible outcomes (head or tail), the outcomes are mutually exclusive (if it's a head, it cannot be a tail), there is independence between trials (there is independence between the outcomes of each coin), and the probability of a head or tail on any trial stays the same from trial to trial. In addition, each distribution gives two pieces of information: (1) all possible outcomes of the N trials and (2) the probability of getting each of the outcomes.

GENERATING THE BINOMIAL DISTRIBUTION FROM THE BINOMIAL EXPANSION

We could continue this enumeration process for larger values of N, but it becomes too laborious. It would indeed be a dismal prospect if we had to use enumeration for every value of N. Think about what happens when N gets to 15. With 15 pennies there are $(2)^{15} = 32,768$ different ways that the 15 coins could fall. Fortunately, there is a mathematical expression that allows us to generate in a simple way everything we've been considering. The expression is called the *binomial expansion*. The binomial expansion is given by

$$(P + Q)^N \qquad \textit{binomial expansion}$$

where P = probability of one of the two possible outcomes on a trial
Q = probability of the other possible outcome
N = number of trials

To generate the possible outcomes and associated probabilities we arrived at in the previous coin-flipping experiments, all we need to do is expand the expression $(P + Q)^N$ for the number of coins in the experiment and evaluate each term in the expansion. For example, if there are two coins, $N = 2$ and

2P events   *1P and 1Q event*   *2Q events*

$$(P + Q)^N = (P + Q)^2 = P^2 + 2P^1Q^1 + Q^2$$

The terms P^2, $2P^1Q^1$, and Q^2 represent all the possible outcomes of flipping two coins once.

*The letters of each term (**P** or **PQ** or **Q**) tell us the kinds of events that comprise the outcome; the exponent of each letter tells us how many of that kind of event* *there are in the outcome; and the coefficient of each term tells us how many ways there are of obtaining the outcome.*

Thus,

1. P^2 indicates that one possible outcome is comprised of two P events. The P alone tells us this outcome is comprised entirely of P events. The exponent 2 indicates there are two of this kind of event. If we associate P with heads, P^2 tells us one possible outcome is two heads.
2. $2P^1Q^1$ indicates that another possible outcome is one P and one Q event, or one head and one tail. The coefficient 2 tells us there are two ways to obtain one P and one Q event.
3. Q^2 represents an outcome of two Q events, or two tails (zero heads).

The probability of getting each of these possible outcomes is found by evaluating their respective terms using the numerical values of P and Q. If the coins are fair coins, then $P = Q = 0.50$. Thus,

$$p(2 \text{ heads}) = P^2 = (0.50)^2 = 0.2500$$

$$p(1 \text{ head}) = 2P^1Q^1 = 2(0.50)(0.50) = 0.5000$$

$$p(0 \text{ heads}) = Q^2 = (0.50)^2 = 0.2500$$

These results are the same as those obtained by enumeration. Note that in using the binomial expansion to find the probability of each possible outcome, we do not add the terms but use them separately. At this point, it probably seems much easier to use enumeration than the binomial expansion. However, the situation reverses itself quickly as N gets larger.

Let's do one more example, this time with $N = 3$. As before, we need to expand $(P + Q)^N$ and evaluate each term in the expansion using $P = Q = 0.50$.* Thus,

$$(P + Q)^N = (P + Q)^3 = P^3 + 3P^2Q + 3PQ^2 + Q^3$$

The terms P^3, $3P^2Q$, $3PQ^2$, and Q^3 represent all of the possible outcomes of flipping three pennies once. P^3 tells us there are three P events, or 3 heads. The term $3P^2Q$ indicates that this outcome has two P events and one Q event, or 2 heads and 1 tail. The term $3PQ^2$ represents one P event and two Q events, or 1 head and 2 tails. Finally, the term Q^3 designates three Q events and zero P events, or 3 tails and 0 heads. We can find the probability of each of these outcomes by evaluation of their respective terms. Since each coin is a fair coin, $P = Q = 0.50$. Thus,

$$p(3 \text{ heads}) = P^3 = (0.50)^3 = 0.1250$$

$$p(2 \text{ heads}) = 3P^2Q = 3(0.50)^2(0.50) = 0.3750$$

$$p(1 \text{ head}) = 3PQ^2 = 3(0.50)(0.50)^2 = 0.3750$$

$$p(0 \text{ heads}) = Q^3 = (0.50)^3 = 0.1250$$

These are the same results we derived previously by enumeration.

---

* See Note 9.1 for the general equation to expand $(P + Q)^N$.

The binomial distribution may be generated for any N, P, and Q by using the binomial expansion. We have graphed the binomial distributions for $N = 3$, 8, and 15 in Figure 9.1. $P = Q = 0.50$ for each of these distributions.

Note that (1) with $P = 0.50$, the binomial distribution is symmetrical; (2) it has two tails (i.e., it tails off as we go from the center toward either end); (3) it involves a discrete variable (e.g., we can't have $2\frac{1}{2}$ heads); and (4) as N increases, the binomial distribution gets closer to the shape of a normal curve.

FIGURE 9.1

Binomial distribution for $N = 3$, $N = 8$, and $N = 15$; $P = 0.50$

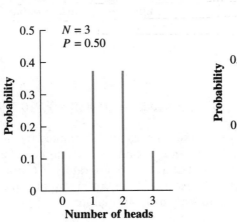

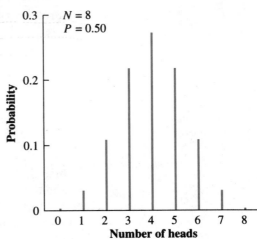

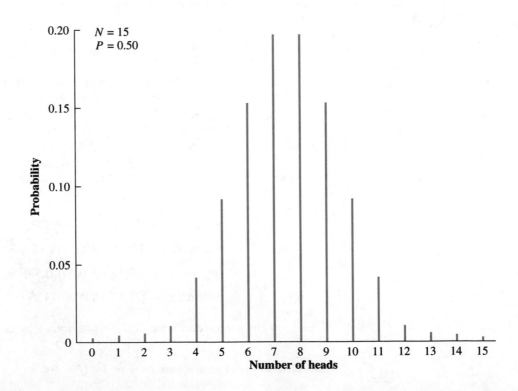

USING THE BINOMIAL TABLE

Although in principle any problem involving binomial data can be answered by directly substituting into the binomial expansion, mathematicians have saved us the work. They have solved the binomial expansion for many values of N and reported the results in tables. One such table is Table B in Appendix D. This table gives the binomial distribution for values of N up to 20. Glancing at Table B, you observe that N (the number of trials) is given in the first column and the possible outcomes are given in the second column, which is headed by "No. of P or Q Events." The rest of the columns contain probability entries for various values of P or Q. The values of P or Q are given at the top of each column. Thus, the second column contains probability values for P or $Q = 0.10$ and the last column the values for P or $Q = 0.50$. In practice, any problem involving binomial data can be solved by looking up the appropriate probability in this table. This, of course, applies only for $N \leq 20$ and the P or Q values given in the table.

The reader should note that Table B can be used to solve problems in terms of P or Q. Thus, with the exception of the first column, the column headings are given in terms of P or Q. To emphasize which we are using (P or Q) in a given problem, if we are entering the table under P and the number of P events, we shall refer to the second column as "number of P events" and the remaining column headings as "P" probability values. If we are entering Table B under Q and the number of Q events, we shall refer to the second column heading as "number of Q events" and the rest of the column headings as "Q" probability values. Let's now see how to use this table to solve problems involving binomial situations.

EXAMPLE

If I flip three unbiased coins once, what is the probability of getting 2 heads and 1 tail? Assume each coin can only be a head or tail.

SOLUTION

In this problem, N is the number of coins, which equals 3. We can let P equal the probability of a head in one flip of any coin. Since the coins are unbiased, $P = 0.50$. Since we want to determine the probability of getting 2 heads, the number of P events equals 2. Having determined the foregoing, all we need do is enter Table B under $N = 3$. Next, we locate the 2 in the number of P events column. The answer is found where the row containing the 2 intersects the column headed by $P = 0.50$. Thus,

p(2 heads and 1 tail) = 0.3750

	Table Entry	
N	No. of P Events	P .50
3	2	0.3750

Note that this is the same answer we arrived at before. In fact, if you look at the remaining entries in that column ($P = 0.50$) for $N = 2$ and 3, you will see that they are the same as we arrived at earlier using the binomial expansion—and they ought to be, since the table entries are taken from the binomial expansion. Let's try some practice problems using this table.

PRACTICE PROBLEM 9.1

If six unbiased coins are flipped once, what is the probability of getting

a. Exactly 6 heads?
b. 4, 5, or 6 heads?

SOLUTION

a. Given there are six coins, $N = 6$. Again, we can let P = the probability of a head in one flip of any coin. Since the coins are unbiased, $P = 0.50$. Since we want to know the probability of getting exactly 6 heads, the number of P events = 6. Entering Table B under $N = 6$, number of P events = 6, and $P = 0.50$, we find

p(exactly 6 heads) = 0.0156

	Table Entry	
N	No. of P Events	P .50
6	6	0.0156

b. Again, $N = 6$ and $P = 0.50$. We can find the probability of 4, 5, and 6 heads by entering Table B under number of P events = 4, 5, and 6, respectively. Thus,

p(4 heads) = 0.2344

p(5 heads) = 0.0938

p(6 heads) = 0.0156

	Table Entry	
N	No. of P Events	P .50
6	4	0.2344
	5	0.0938
	6	0.0156

From the addition rule with mutually exclusive events,

$$p(4, 5, \text{or } 6 \text{ heads}) = p(4) + p(5) + p(6)$$
$$= 0.2344 + 0.0938 + 0.0156$$
$$= 0.3438$$

PRACTICE PROBLEM 9.2

If 10 unbiased coins are flipped once, what is the probability of getting a result as extreme or more extreme than 9 heads?

SOLUTION

Since there are 10 coins, $N = 10$. As before, we shall let P = the probability of getting a head in one flip of any coin. Since the coins are unbiased, $P = Q = 0.50$. The phrase "as extreme or more extreme than" means "as far from the center of the distribution or farther from the center of the distribution than." Thus, "as extreme or more extreme than 9 heads" means results that are as far from the center of the distribution or farther from the center of the distribution than 9 heads. Thus, the number of P events = 0, 1, 9, or 10. In Table B under $N = 10$, number of P events = 0, 1, 9, or 10, and $P = 0.50$, we find

$$p\begin{pmatrix} \text{as extreme or} \\ \text{more extreme} \\ \text{than 9 heads} \end{pmatrix} = p(0, 1, 9, \text{ or } 10)$$

$$= p(0) + p(1) + p(9) + p(10)$$

$$= 0.0010 + 0.0098 + 0.0098 + 0.0010$$

$$= 0.0216$$

	Table Entry	
N	No. of P Events	P .50
10	0	0.0010
	1	0.0098
	9	0.0098
	10	0.0010

The binomial expansion is very general. It is not limited to values where $P = 0.50$. Thus, Table B also lists probabilities for values of P other than 0.50. Let's try some problems where P is not equal to 0.50.

PRACTICE PROBLEM 9.3

Assume you have eight biased coins. Each coin is weighted such that the probability of a head with it is 0.30. If the eight biased coins are flipped once, what is the probability of getting

a. 7 heads?
b. 7 or 8 heads?

c. The probability found in part **a** comes from evaluating which of the term(s) in the following binomial expansion?

$$P^8 + 8P^7Q^1 + 28P^6Q^2 + 56P^5Q^3 + 70P^4Q^4 + 56P^3Q^5 + 28P^2Q^6 + 8P^1Q^7 + Q^8$$

d. With your calculator, evaluate the term(s) selected in part **c** using $P = 0.30$. Compare your answer with the answer in part **a**. Explain.

S O L U T I O N

a. Given there are eight coins, $N = 8$. Let $P =$ the probability of getting a head in one flip of any coin. Since the coins are biased such that the probability of a head on any coin is 0.30, $P = 0.30$. Since we want to determine the probability of getting exactly 7 heads, the number of P events $= 7$. In Table B under $N = 8$, number of P events $= 7$, and $P = 0.30$, we find the following:

		Table Entry	
	N	No. of P Events	P .30
$p(\text{exactly 7 heads}) = 0.0012$	8	7	0.0012

b. Again, $N = 8$ and $P = 0.30$. We can find the probability of 7 and 8 heads in Table B under number of P events $= 7$ and 8, respectively. Thus,

		Table Entry	
	N	No. of P Events	P .30
$p(\text{7 heads}) = 0.0012$	8	7	0.0012
$p(\text{8 heads}) = 0.0001$		8	0.0001

From the addition rule with mutually exclusive events,

$$p(7 \text{ or } 8 \text{ heads}) = p(7) + p(8)$$

$$= 0.0012 + 0.0001$$

$$= 0.0013$$

c. $8P^7Q^1$

d. $8P^7Q^1 = 8(0.30)^7(0.7) = 0.0012$. As expected, the answers are the same. The table entry was computed using $8P^7Q^1$ with $P = 0.30$ and $Q = 0.70$.

Thus, using Table B in situations where P is less than 0.50 is very similar to situations where $P = 0.50$. We just look in the table under the new P value rather than under $P = 0.50$.

Table B can also be used in situations where $P > 0.50$. To illustrate, consider the following example.

EXAMPLE ($P > 0.50$)

If five biased coins are flipped once, what is the probability of getting (a) 5 heads and (b) 4 or 5 heads? Each coin is weighted such that the probability of a head on any coin is 0.75.

SOLUTION

(a) 5 heads. Since there are five coins, $N = 5$. Again, let $P =$ the probability of getting a head in one flip of any coin. Since the bias is such that the probability of a head on any coin is 0.75, $P = 0.75$. Since we want to determine the probability of getting 5 heads, the number of P events equals 5. Following our usual procedure, we would enter Table B under $N = 5$, number of P events $= 5$, and $P = 0.75$. However, Table B does not have a column headed by 0.75. All of the column headings are equal to or less than 0.50. Nevertheless, we can use Table B to solve this problem.

When $P > 0.50$, all we need do is solve the problem in terms of Q and the number of Q events, rather than P and the number of P events. Since the probability values given in Table B are for either P or Q, once the problem is put in terms of Q, we can refer to Table B using Q rather than P. Translating the problem into Q terms involves two steps: determining Q and determining the number of Q events. Let's follow these steps using the present example:

1. *Determining Q.*

$$Q = 1 - P = 1 - 0.75 = 0.25$$

2. *Determining the number of Q events.*

$$\textit{Number of Q events} = N - \textit{Number of P events} = 5 - 5 = 0$$

Thus, to solve this example, we refer to Table B under $N = 5$, number of Q events $= 0$, and $Q = 0.25$. Thus,

| | | Table Entry | |
	N	No. of Q Events	Q .25
$p(5 \text{ heads}) = p(0 \text{ tails}) = 0.2373$	5	0	0.2373

(b) 4 or 5 heads. Again, $N = 5$ and $Q = 0.25$. This time, the number of Q events $= 0$ or 1. Thus,

		Table Entry	
	N	No. of Q Events	Q .25
$p(4 \text{ or } 5 \text{ heads}) = p(0 \text{ or } 1 \text{ tail})$			
$= 0.2373 + 0.3955$	5	0	0.2373
$= 0.6328$		1	0.3955

We are now ready to try a practice problem.

PRACTICE PROBLEM 9.4

If 12 biased coins are flipped once, what is the probability of getting

a. Exactly 10 heads?
b. 10 or more heads?

The coins are biased such that the probability of a head with any coin equals 0.65.

SOLUTION

a. Given there are 12 coins, $N = 12$. Let $P =$ the probability of a head in one flip of any coin. Since the probability of a head with any coin equals 0.65, $P = 0.65$. Since $P > 0.50$, we shall enter Table B with Q rather than P. If there are 10 P events, there must be 2 Q events ($N = 12$). If $P = 0.65$, then $Q = .35$. Using Q in Table B, we obtain

		Table Entry	
	N	No. of Q Events	Q .35
$p(10 \text{ heads}) = 0.1088$			
	12	2	0.1088

b. Again, $N = 12$, and $P = 0.65$. This time, the number of P events equals 10, 11, or 12. Since $P > 0.50$, we must use Q in Table B rather than P.

With $N = 12$, the number of Q events equals 0, 1, or 2 and $Q = 0.35$. Using Q in Table B, we obtain

		Table Entry	
$p(10, 11, \text{or } 12 \text{ heads}) = 0.1088 + 0.0368$	N	No. of Q Events	Q .35
$+ 0.0057$	12	0	0.0057
$= 0.1513$		1	0.0368
		2	0.1088

So far, we have dealt exclusively with coin flipping. However, the binomial distribution is not just limited to coin flipping. It applies to all situations involving a series of trials where on each trial there are only two possible outcomes: the outcomes are mutually exclusive, there is independence between the outcomes on each trial, and the probability of each possible outcome on any trial stays the same from trial to trial. There are many situations that fit these requirements. To illustrate, let's do a couple of practice problems.

PRACTICE PROBLEM 9.5

A student is taking a multiple-choice exam with 15 questions. Each question has five choices. If the student guesses on each question, what is the probability of passing the test? The lowest passing score is 60% of the questions answered correctly. Assume that the choices for each question are equally likely.

SOLUTION

This problem fits the binomial requirements. There is a series of trials (questions). On each trial, there are only two possible outcomes. The student is either right or wrong. The outcomes are mutually exclusive. If she is right on a question, she can't be wrong. There is independence between the outcomes of each trial. If she is right on question 1, it has no effect on the outcome of question 2. Finally, if we assume the student guesses on each trial, then the probability of being right and the probability of being wrong on any trial stays the same from trial to trial. Thus, the binomial distribution and Table B apply.

We can consider each question a trial (no pun intended). Given there are 15 questions, $N = 15$. We can let $P =$ the probability that she will guess correctly on any question. Since there are five choices that are equally likely on each question, $P = 0.20$. A passing grade equals 60% correct answers or more. Therefore, a student will pass if she gets 9 or more answers correct (60% of 15 is 9). Thus, the number of P events equals 9, 10, 11, 12, 13, 14, and 15. Looking in Table B under $N = 15$, number of P events = 9, 10, 11, 12, 13, 14, and 15, and $P = 0.20$, we obtain

		Table Entry	
	N	No. of P Events	P .20
	15	9	0.0007
		10	0.0001
$p(9, 10, 11, 12, 13, 14$		11	0.0000
or 15 correct guesses$) = 0.0007 + 0.0001$		12	0.0000
		13	0.0000
$= 0.0008$		14	0.0000
		15	0.0000

PRACTICE PROBLEM 9.6

Your friend claims to be a bourbon whiskey connoisseur. He always orders MacNaughtons and claims no other bourbon even comes close to tasting so good. You suspect that he is being a little grandiose. In fact, you really wonder whether he can even taste the difference between MacNaughtons and the local "cheapie" bourbon, which we shall call Brand X. Your friend agrees to the following experiment. While blindfolded, he is given six opportunities to taste two glasses of bourbon and tell you which of the two glasses contains MacNaughtons. The glasses are identical, except that one contains MacNaughtons and the other Brand X. After each tasting, you of course refill the glasses, remove any tell-tale signs, and randomize which glass he is given first. Your friend correctly identifies MacNaughtons on all six trials. What do you conclude? Can you think of a way to increase your confidence in your conclusion?

SOLUTION

The logic of our analysis is as follows. We will assume that your friend really can't tell the difference between the two whiskeys. He must then be guessing on each trial. We will compute the probability of getting six out of six correct, assuming guessing on each trial. If this probability is very low, we will reject guessing as a reasonable explanation and conclude that your friend can really taste the difference.

This experiment fits the requirements for the binomial distribution. Each comparison of the two whiskeys can be considered a trial (agian, no pun intended). On each trial, there are only two possible outcomes. Your friend is either right or wrong. The outcomes are mutually exclusive. There is independence between trials. If your friend is correct on trial 1, it has no effect on the outcome of trial 2. (We are of course assuming that your friend sips only a small quantity of the whiskey each time. Otherwise, even though he might get happier over trials, his taste discrimination might be adversely affected, a condition that, as good experimenters, we cannot allow to happen!!) Finally, if we assume your friend guesses on any trial, then the probability of being correct and the probability of being wrong on any trial stays the same from trial to trial.

Given each whiskey comparison is a trial, $N = 6$. We can let $P =$ the probability your friend will guess correctly on any trial. Since there are only two whiskeys, $P = 0.50$. Your friend was correct on all six trials. Therefore, the number of P events = 6. Thus,

	Table Entry		
p(6 correct guesses) = 0.0156	N	No. of P Events	P .50
	6	6	0.0156

Assuming your friend is guessing, the probability of his getting six out of six correct is 0.0156. Since this is a fairly low value, you would probably reject guessing because it is an unreasonable explanation and conclude that your friend can really taste the difference. To increase your confidence in rejecting guessing, you could include more whiskeys on each trial, or you could increase the number of trials. For example, even with only two whiskeys, the probability of guessing correctly on twelve out of twelve trials is 0.0002.

SUMMARY

In this chapter, we have discussed the binomial distribution. The binomial distribution is a probability distribution that results when the following conditions are met: (1) There are a series of N trials; (2) on each trial, there are only two possible outcomes; (3) the outcomes are mutually exclusive; (4) there is independence between trials; and (5) the probability of each possible outcome on any trial stays the same from trial to trial. When these conditions are met, the binomial distribution tells us each possible outcome of the N trials and the probability of getting each of these outcomes. We illustrated the binomial distribution through coin-flipping experiments and then showed how the binomial distribution could be generated through the binomial expansion. The binomial expansion is given by $(P + Q)^N$, where P = the probability of occurrence of one of the events and Q = the probability of occurrence of the other event. Finally, we showed how to use the binomial table (Table B in Appendix D) to solve problems involving the binomial distribution. The binomial distribution is appropriate whenever the four conditions listed at the beginning of this summary are met.

IMPORTANT TERMS

Biased coins (p. 199)
Binomial distribution (p. 191)
Binomial expansion (p. 194)
Binomial table (p. 197)
Fair coins (p. 191)
Number of P events (p. 197)
Number of Q events (p. 197)

QUESTIONS AND PROBLEMS

1. Briefly define or explain each of the terms in the "Important Terms" section.
2. What are the five conditions necessary for the binomial distribution to be appropriate?
3. In a binomial situation, if $P = 0.10$, $Q =$ _____ .
4. Using Table B, if $N = 6$ and $P = 0.40$,
   a. The probability of getting exactly five P events = _____ .
   b. This probability comes from evaluating which of the terms in the following equation?

   $$P^6 + 6P^5Q + 15P^4Q^2 + 20P^3Q^3 + 15P^2Q^4 + 6PQ^5 + Q^6$$

   c. Evaluate the term(s) of your answer in part b using $P = 0.40$, and compare your answer with part a.
5. Using Table B, if $N = 12$ and $P = 0.50$,
   a. What is the probability of getting exactly 10 P events?
   b. What is the probability of getting 11 or 12 P events?

c. What is the probability of getting at least 10 P events?
   d. What is the probability of getting a result as extreme as or more extreme than 10 P events?
6. Using Table B, if $N = 14$ and $P = 0.70$,
   a. What is the probability of getting exactly 13 P events?
   b. What is the probability of getting at least 13 P events?
   c. What is the probability of getting a result as extreme as or more extreme than 13 P events?
7. Using Table B, if $N = 20$ and $P = 0.20$,
   a. What is the probability of getting exactly two P events?
   b. What is the probability of getting two or fewer P events?
   c. What is the probability of getting a result as extreme as or more extreme than two P events?
8. An individual flips nine fair coins. If she allows only a head or a tail with each coin,

a. What is the probability they all will fall heads?

b. What is the probability there will be seven or more heads?

c. What is the probability there will be a result as extreme as or more extreme than seven heads?

9. Someone flips 15 biased coins once. The coins are weighted such that the probability of a head with any coin is 0.85.

a. What is the probability of getting exactly 14 heads?

b. What is the probability of getting at least 14 heads?

c. What is the probability of getting exactly 3 tails?

10. A key shop advertises that the keys made there have a $P = 0.90$ of working effectively. If you bought 10 keys from the shop, what is the probability that all of the keys would work effectively?

11. A student is taking a true/false examination with 15 questions. If he guesses on each question, what is the probability he will get at least 13 questions correct?

12. A student is taking a multiple-choice exam with 16 questions. Each question has five alternatives. If the student guesses on 12 of the questions, what is the probability she will guess at least 8 correct? Assume all of the alternatives are equally likely for each question on which the student guesses.

13. You are interested in determining whether a particular child can discriminate the color green from blue. Therefore, you show the child five wooden blocks. The blocks are identical except that two are green and three are blue. You randomly arrange the blocks in a row and ask him to pick out a green block. After a block is picked, you replace it and randomize the order of the blocks once more. Then you again ask him to pick out a green block. This procedure is repeated until the child has made 14 selections. If he really can't discriminate green from blue, what is the probability he will pick a green block at least 11 times?

14. Let's assume you are an avid horse race fan. You are at the track and there are eight races. On this day, the horses and their riders are so evenly matched that chance alone determines the finishing order for each race. There are 10 horses in every race. If, on each race, you bet on one horse to show (to finish third, second, or first),

a. What is the probability that you will win your bet in all eight races?

b. What is the probability that you will win in at least six of the races?

15. A manufacturer of valves admits that its quality control has gone radically "downhill" such that currently the probability of producing a defective valve is 0.50. If it manufactures one million valves in a month and you randomly sample from these valves 10,000 samples each composed of 15 valves,

a. In how many samples would you expect to find exactly 13 good valves?

b. In how many sample would you expect to find at least 13 good valves?

16. Assume that 15% of the population is left-handed and the remainder is right-handed (there are no ambidextrous individuals). If you stop the next 5 persons you meet, what is the probability that

a. All will be left-handed?

b. All will be right-handed?

c. Exactly 2 will be left-handed?

d. At least 1 will be left-handed?

For the purposes of this problem, assume independence in the selection of the 5 individuals.

17. In your voting district, 25% of the voters are against a particular bill, and the rest favor it. If you randomly poll 4 voters from your district, what is the probability that

a. None will favor the bill?

b. All will favor the bill?

c. At least 1 will be against the bill?

18. At your university, 30% of the undergraduates are from out of state. If you randomly select 8 of the undergraduates, what is the probability that

a. All are from within the state?

b. All are from out-of-state?

c. Exactly 2 are from within the state?

d. At least 5 are from within the state?

19. A large bowl contains 1 million marbles. Half of the marbles have a plus ($+$) painted on them and other half a minus ($-$).

a. If you randomly sample 10 marbles, one at a time with replacement from the bowl, what is the probability you will select 9 marbles with pluses and 1 with a minus?

b. If you take 1000 random samples of 10 marbles, one at a time with replacement, how many of the samples would you expect to be all pluses?

NOTES

9.1 The equation for expanding $(P + Q)^N$ is

$$(P + Q)^N = P^N + \frac{N}{1}P^{N-1}Q + \frac{N(N-1)}{1(2)}P^{N-2}Q^2$$

$$+ \frac{N(N-1)(N-2)}{1(2)(3)}P^{N-3}Q^3 + \cdots + Q^N$$

10

INTRODUCTION TO HYPOTHESIS TESTING USING THE SIGN TEST

INTRODUCTION

We pointed out previously that inferential statistics has two main purposes: (1) hypothesis testing and (2) parameter estimation. By far, most of the applications of inferential statistics are in the area of hypothesis testing. As discussed in Chapter 1, scientific methodology depends on this application of inferential statistics. Without objective verification, science would cease to exist, and objective verification is often impossible without inferential statistics. You will recall that at the heart of scientific methodology is an experiment. Usually the experiment has been designed to test a hypothesis, and the resulting data must be analyzed. Occasionally, the results are so clear-cut that statistical inference is not necessary. However, such experiments are rare. Because of the variability that is inherent from subject to subject in the variable being measured, it is often quite difficult to detect the effect of the independent variable without the help of inferential statistics. In this chapter, we shall begin the fascinating journey into how experimental design, in conjunction with mathematical analysis, can be used to verify truth assertions or hypotheses, as we have been calling them. We urge you to apply yourself to this chapter with special rigor. The material it contains applies to all of the inference tests we shall take up (which constitutes most of the remaining text).

AN EXPERIMENT: MARIJUANA AND THE TREATMENT OF AIDS PATIENTS

Let's assume that you are a social scientist working in a metropolitan hospital that services a large population of AIDS patients. You are very concerned about

209

the pain and suffering that afflicts these patients. In particular, although you are not yet convinced, you think there may be an ethically proper place for using marijuana in the treatment of these patients, particularly in the more advanced stages of the illness. Of course, before seriously considering the other issues involved in advocating the usage of marijuana for this purpose, you must be convinced that it does have important positive effects. Thus far, although there have been many anecdotal reports from AIDS patients that using marijuana decreases their nausea, increases their appetite, and increases their desire to socialize, there have not been any scientific experiments to shore up these reports.

As a scientist, you realize that although personal reports are suggestive, they are not conclusive. Experiments must be done before one can properly assess cause and effect—in this case, the effects claimed for marijuana. Since this is very important to you, you decide to embark on a research program directed to this end. The first experiment you plan is to investigate the effect of marijuana on appetite in AIDS patients. Of course, if marijuana actually decreases appetite rather than increases it, you want to be able to detect this as well, since it has important practical consequences. Therefore, this will be a basic fact-finding experiment in which you attempt to determine whether marijuana has any effect at all, either to increase or to decrease appetite. The first experiment will be a modest one. You plan to randomly sample 10 individuals from the population of AIDS patients who are being treated at your hospital and who are in the advanced stages of that illness. We will refer to these individuals as "advanced-stage" AIDS patients. You realize that the generalization will be limited to this population, but for many reasons you are willing to accept this limitation for this initial experiment. After getting permission from the appropriate authorities, you conduct the following experiment.

A random sample of 10 AIDS patients who agree to participate in the experiment is selected from a rather large population of advanced-stage AIDS patients being treated on an outpatient basis at your hospital. None of the patients in this population are being treated with marijuana. Each patient is admitted to the hospital for a week to participate in the experiment. The first two days are used to allow each patient to get used to the hospital. On the third day, half of the patients receive a pill containing a synthetic form of marijuana's active ingredient, THC, prior to eating each meal, and on the sixth day, a placebo pill before each meal. The other half of the patients are treated the same as in the experimental condition, except that they receive the pills in the reverse order, that is, the placebo pills on the third day and the THC pills on the sixth day. The dependent variable is the amount of food eaten by each patient on day 3 and day 6.

In this experiment, each subject is tested under two conditions, an experimental condition and a control condition. We have labeled the condition in which the subject receives the THC pills as the experimental condition and the condition in which the subject receives the placebo pills as the control condition. Thus, there are two scores for each subject: the amount of food eaten (calories) in the experimental condition and the amount of food eaten in the control condition. If marijuana really does affect appetite, we would expect different scores for the two conditions. For example, if marijuana increases appetite, then more food should be eaten in the experimental condition. If the control score for each subject is subtracted from his experimental score, we would expect a predominance of positive difference scores. The results of the experiment are given in the table.

Patient No.	Experimental Condition THC Pill Food Eaten (calories)	Control Condition Placebo Pill Food Eaten (calories)	Difference Score (calories)
1	1325	1012	+313
2	1350	1275	+ 75
3	1248	950	+298
4	1087	840	+247
5	1047	942	+105
6	943	860	+ 83
7	1118	1154	− 36
8	908	763	+145
9	1084	920	+164
10	1088	876	+212

These data could be analyzed with several different statistical inference tests, e.g., sign test, Wilcoxon matched-pairs signed ranks test, and Student's *t* test for correlated groups. The choice of which test to use in an actual experiment is an important one. It depends on the sensitivity of the test and on whether the data of the experiment meet the assumptions of the test. We shall discuss each of these points in subsequent chapters. In this chapter, we shall analyze the data of your experiment with the sign test. We have chosen the sign test because (1) it is easy to understand and (2) all of the major concepts concerning hypothesis testing can be illustrated clearly and simply.

The sign test ignores the magnitude of the difference scores and considers only their direction or sign. This omits a lot of information that makes the test rather insensitive (but much easier to understand). If we consider only the signs of the difference scores, then your experiment produced 9 out of 10 pluses. The amount of food eaten in the experimental condition was greater after taking the THC pill in all but one of the patients. Are we therefore justified in concluding that marijuana produces an increase in appetite? Not necessarily.

Suppose that marijuana has absolutely no effect on appetite. Isn't it still possible to have obtained 9 out of 10 pluses in your experiment? Yes, it is. If marijuana has no effect on appetite, then each subject would have received two conditions that were identical except for chance factors. Perhaps when subject 1 was run in the THC condition, he had slept better the night before and his appetite was higher than when run in the control condition before any pills were taken. If so, we would expect him to eat more food in the THC condition even if THC has no effect on appetite. Perhaps subject 2 had a cold when run in the placebo condition, which blunted her appetite relative to when run in the experimental condition. Again we would expect more food to be eaten in the experimental condition even if THC has no effect. We could go on giving examples for the other subjects. The point is that these explanations of the greater amount eaten in the THC condition are chance factors. They are different factors, independent of one another, and they could just as easily have occurred on either of the two test days. It seems unlikely to get 9 out of 10 pluses simply as a result of chance factors. The crucial question really is, "How unlikely is it?" Suppose we know that if chance alone is responsible we shall get 9 out of 10 pluses only

1 time in 1 billion. Since this is such a rare occurrence, we would no doubt reject chance and, with it, the explanation that marijuana has no effect on appetite. We would then conclude by accepting the hypothesis that marijuana affects appetite, since it is the only other possible explanation. Since the sample was a random one, we can assume it was representative of the advanced-stage AIDS patients being treated at your hospital, and we therefore would generalize the results to that population.

Suppose, however, that the probability of getting 9 out of 10 pluses due to chance alone is really 1 in 3, not 1 in 1 billion. Can we reject chance as a cause of the data? The decision is not as clear-cut this time. What we need is a rule for determining when the obtained probability is small enough to reject chance as an underlying cause. We shall see that this involves setting a critical probability level (called the alpha level) against which to compare the results.

Let's formalize some of the concepts we've been presenting.

Repeated Measures Design   The experimental design that we have been using is called the repeated measures, replicated measures, or correlated groups design. *The essential features are that there are paired scores in the conditions and the differences between the paired scores are analyzed.* In the marijuana experiment, we used the same subjects in each condition. Thus, the subjects served as their own controls. Their scores were paired, and the differences between these pairs were analyzed. Instead of the same subjects, we could have used identical twins or subjects that were matched in some other way. In animal experimentation, littermates have often been used for pairing. The most basic form of this design employs just two conditions, an experimental and a control condition. The two conditions are kept as identical as possible except for values of the independent variable, which, of course, are intentionally made different. In our example, marijuana is the independent variable.

Alternative Hypothesis (H_1)   In any experiment, there are two hypotheses that compete for explaining the results: the *alternative hypothesis* and the *null hypothesis*. The alternative hypothesis is the one that claims the difference in results between conditions is due to the independent variable. In this case, it is the hypothesis that claims "marijuana affects appetite." The alternative hypothesis can be directional or nondirectional. The hypothesis "marijuana affects appetite" is nondirectional, because it does not specify the direction of the effect. If the hypothesis specifies the direction of the effect, it is a directional hypothesis. "Marijuana increases appetite" is an example of a directional alternative hypothesis.

Null Hypothesis (H_0)   The null hypothesis is set up to be the logical counterpart of the alternative hypothesis such that if the null hypothesis is false, the alternative hypothesis must be true. Therefore, these two hypotheses must be mutually exclusive and exhaustive. If the alternative hypothesis is nondirectional, it specifies that the independent variable has an effect on the dependent variable. For this nondirectional alternative hypothesis, the null hypothesis asserts that the independent variable has *no* effect on the dependent variable. In the present example, since the alternative hypothesis is nondirectional, the null hypothesis specifies that "marijuana has no effect on appetite." We pointed out previously that the alternative hypothesis specifies "marijuana affects appetite." You can see that these two hypotheses are mutually exclusive and exhaustive. If the null hypothesis is false, then the alternative hypothesis must be true. As you will see, we always first evaluate the null hypothesis and try to show that it is false. If we can show it to be false, then the alternative hypothesis must be true.*

If the alternative hypothesis is directional, the null hypothesis asserts that the independent variable does not have an effect in the direction specified by the alternative hypothesis.† For example, for the alternative hypothesis "marijuana increases appetite," the null hypothesis asserts that "marijuana does not increase appetite." Again, note that the two hypotheses are mutually exclusive and exhaustive. If the null hypothesis is false, then the alternative hypothesis must be true.

Decision Rule (α Level)   We always evaluate the results of an experiment by assessing the null hypothesis. The reason we directly assess the null hypothesis instead of the alternative hypothesis is that we can calculate the probability of chance events, but there is no way to calculate the probability of the alternative hypothesis. We evaluate the null hypothesis by assuming it is true and testing the reasonableness of this assumption by calculating the probability of getting the results *if chance alone is operating.* If the obtained probability turns out to be equal to or less than a critical probability level called the *alpha (α) level,* we reject the null hypothesis. Rejecting the null hypothesis allows us, then, to accept indirectly the alternative hypothesis since, if the experiment is done properly, it is the only other possible explanation. When we reject H_0, we say the results are *significant* or *reliable.* If the obtained probability is greater than the alpha level, we conclude by failing to reject H_0. Since the experiment does not allow rejection of H_0, we retain H_0 as a reasonable explanation of the data. Throughout the text, we shall use the expressions "failure to reject H_0" and "retain H_0" interchangeably. When we retain H_0, we say the results are not significant or reliable. Of course, when the results are not significant, we cannot accept the alternative hypothesis. Thus, the decision rule states:

If the obtained probability $\leq \alpha$, reject H_0.

If the obtained probability $> \alpha$, fail to reject H_0, retain H_0.

The alpha level is set at the beginning of the experiment. Commonly used alpha levels are $\alpha = 0.05$ and $\alpha = 0.01$. Later in this chapter, we shall discuss the rationale underlying the use of these levels.

For now let's assume $\alpha = 0.05$ for the marijuana data. Thus, to evaluate the results of the marijuana experiment, we need (1) to determine the probability of getting 9 out of 10 pluses if chance alone is responsible and (2) to compare this probability with alpha.

Evaluating the Marijuana Experiment Using the Binomial Distribution

The data of this experiment fit the requirements for the binomial distribution. The experiment consists of a series of trials (the exposure of each patient to the experimental and control conditions is a trial). On each trial there are only two possible outcomes, a plus and a minus. Note that this model does not allow ties.

---

* See Note 10.1.
† See Note 10.2.

If any ties occur, they must be discarded and the N reduced accordingly. The outcomes are mutually exclusive (a plus and a minus cannot occur simultaneously), there is independence between trials (the score of patient 1 in no way influences the score of patient 2, etc.), and the probability of a plus and the probability of a minus stay the same from trial to trial. Since the binomial distribution is appropriate, we can use Table B in Appendix D to determine the probability of getting 9 pluses out of 10 trials when chance alone is responsible. We solve this problem in the same way we did the coin-flipping problems.

Given there are 10 patients, $N = 10$. We can let P = the probability of getting a plus with any patient.* If chance alone is operating, the probability of a plus is equal to the probability of a minus. Since there are only two equally likely alternatives, $P = 0.50$. Since we want to determine the probability of 9 pluses, the number of P events = 9. In Table B under $N = 10$, number of P events = 9, and $P = 0.50$, we obtain

		Table Entry	
	N	No. of P Events	P .50
$p(9\text{ pluses}) = 0.0098$	10	9	0.0098

Alpha has been set at 0.05. The analysis shows that only 98 times in 10,000 would we get 9 pluses if chance alone is the cause. Since 0.0098 is lower than alpha, we reject the null hypothesis.† It does not seem to be a reasonable explanation of the data. Therefore, we conclude by accepting the alternative hypothesis that marijuana affects appetite. It appears to increase it. Since the sample was randomly selected, we assume the sample is representative of the population. Therefore, it is legitimate to assume that this conclusion applies to the *population* of advanced-stage AIDS patients being treated at your hospital. It is worth noting that very often in practice the results of an experiment are generalized to groups that were not part of the population from which the sample was taken. For instance, on the basis of this experiment, we might be tempted to claim that marijuana would increase the appetites of advanced-stage AIDS patients being treated at other hospitals. Strictly speaking, the results of an experiment apply only to the population from which the sample was randomly selected. Therefore, generalization to other groups should be made with caution. This caution is necessary because the other groups may differ from the subjects in the original population in some way that would cause a different result. Of course, as the experiment is replicated in different hospitals with different patients, the legitimate generalization becomes much broader.

---

* Throughout this chapter and the next, whenever using the sign test, we shall always let P = the probability of a *plus* with any subject. This is strictly arbitrary. Of course, we could have chosen Q. However, using the same letter (P or Q) to designate the probability of a plus for all problems does avoid unnecessary confusion.

† This is really a simplification made here for clarity. In actual practice, we evaluate the probability of getting the obtained result *or any more extreme*. The point is discussed in detail later in this chapter in the section entitled "Evaluating the Tail of the Distribution."

TYPE I AND TYPE II ERRORS

When making decisions regarding the null hypothesis, it is possible to make errors of two kinds. These are called *Type I* and *Type II errors.*

DEFINITION

- *A Type I error is defined as a decision to reject the null hypothesis when the null hypothesis is true. A Type II error is defined as a decision to retain the null hypothesis when the null hypothesis is false.*

To illustrate these concepts, let's return to the marijuana example. Recall the logic of the decision process. First, we assume H_0 is true and evaluate the probability of getting the obtained differences between conditions if chance alone is responsible. If the obtained probability $\leq \alpha$, we reject H_0. If the obtained probability $> \alpha$, we retain H_0. In the marijuana experiment, the obtained probability $[p(9 \text{ pluses})] = 0.0098$. Since this was lower than alpha, we rejected H_0 and concluded that marijuana was responsible for the results. Can we be certain that we made the correct decision? How do we know that chance wasn't really responsible? Perhaps the null hypothesis is really true. Isn't it possible that this was one of those 98 times in 10,000 we would get 9 pluses if chance alone was operating? *The answer is that we never know for sure that chance wasn't responsible.* It is possible that the 9 pluses were really due to chance. If so, then we made an error by rejecting H_0. This is a Type I error—a rejection of the null hypothesis when it is true.

A Type II error occurs when we retain H_0 and it is false. Suppose that in the marijuana experiment $p(9 \text{ pluses}) = 0.2300$ instead of 0.0098. In this case, since $0.2300 > \alpha$, we would retain H_0. If H_0 is false, we have made a Type II error, i.e., retaining H_0 when it is false.

To help clarify the relationship between the decision process and possible error, we've summarized the possibilities in Table 10.1. The column heading is "State of Reality." This means the correct state of affairs regarding the null hypothesis. There are only two possibilities. Either H_0 is true or it is false. The row heading is the decision made when analyzing the data. Again, there are only two possibilities. Either we reject H_0 or we retain H_0. If we retain H_0 and H_0 is true, we've made a correct decision (see the first cell in the table). If we reject H_0 and H_0 is true, we've made a Type I error. This is shown in cell 3. If we retain H_0 and H_0 is false, we've made a Type II error (cell 2). Finally, if we reject H_0 and H_0 is false, we've made a correct decision (cell 4). Note that when we reject H_0, the only possible error is a Type I error. If we retain H_0, the only error we may make is a Type II error.

TABLE 10.1   Possible Conclusions and the State of Reality

	State of Reality	
Decision	*H_0 is true*	*H_0 is false*
Retain H_0	[1]Correct decision	[2]Type II error
Reject H_0	[3]Type I error	[4]Correct decision

You may wonder why we've gone to the trouble of analyzing all the logical possibilities. We've done so because it is very important to know the possible errors we may be making when we draw conclusions from an experiment. From the above analysis, we know there are only two such possibilities, a Type I error or a Type II error. Knowing these are possible, we can design experiments before conducting them so as to minimize the probability of making a Type I error or a Type II error. By minimizing the probability of making these errors, we maximize the probability of concluding correctly, regardless of whether the null hypothesis is true or false. We shall see in the next section that alpha limits the probability of making a Type I error. Therefore, by controlling the alpha level we can minimize the probability of making a Type I error. Beta (read "bayta") is defined as the probability of making a Type II error. We shall discuss ways to minimize beta at the end of this chapter.

 ## ALPHA LEVEL AND THE DECISION PROCESS

It should be clear that whenever we are using sample data to evaluate a hypothesis, we are never certain of our conclusion. When we reject H_0, we don't know for sure that it is false. We take the risk that we may be making a Type I error. Of course, the less reasonable it is that chance is the cause of the results, the more confident we are that we haven't made an error by rejecting the null hypothesis. For example, when the probability of getting the results is 1 in 1 million ($p = 0.000001$) under the assumption of chance, we are more confident that the null hypothesis is false than when the probability is 1 in 10 ($p = 0.10$).

DEFINITION

- *The* alpha level *that the scientist sets at the beginning of the experiment is the level to which he or she wishes to limit the probability of making a Type I error.*

Thus, when a scientist sets $\alpha = 0.05$, he is in effect saying that when he collects the data he will reject the null hypothesis if, under the assumption that chance alone is responsible, the obtained probability is equal to or less than 5 times in 100. In so doing, he is saying that he is willing to limit the probability of rejecting the null hypothesis when it is true to 5 times in 100. Thus, he limits the probability of making a Type I error to 0.05.

There is no magical formula that tells us what the alpha level should be to arrive at truth in each experiment. To determine a reasonable alpha level for an experiment, we must consider the consequences of making an error. In science, the effects of rejecting the null hypothesis when it is true (Type I error) are costly. When a scientist publishes an experiment in which he rejects the null hypothesis, other scientists either attempt to replicate the results or accept the conclusion as valid and design experiments based on the scientist having made a correct decision. Since many work hours and dollars go into these follow-up experiments, scientists would like to minimize the possibility that they are pursuing a false path. Thus, they set rather conservative alpha levels: $\alpha = 0.05$ and $\alpha = 0.01$ are commonly used. You might ask, "Why not set even more stringent criteria, such as $\alpha = 0.001$?" Unfortunately, when alpha is made more stringent, the probability of making a Type II error increases.

We can see this by considering an example. This example is best understood in conjunction with Table 10.2. Suppose we do an experiment and set $\alpha = 0.05$

TABLE 10.2   Effect on Beta of Making Alpha More Stringent

| Alpha Level | Obtained Probability | Decision | State of Reality | |
			H_0 is true	H_0 is false
0.05	0.02	Reject H_0	[1]Type I error	[2]Correct decision
0.01	0.02	Retain H_0	[3]Correct decision	[4]Type II error

(top row of Table 10.2). We evaluate chance and get an obtained probability of 0.02. We reject H_0. If H_0 is true, we have made a Type I error (cell 1). Suppose, however, that alpha had been set at $\alpha = 0.01$ instead of 0.05 (bottom row of Table 10.2). In this case, we would retain H_0 and no longer would be making a Type I error (cell 3). Thus, the more stringent the alpha level, the lower the probability of making a Type I error.

On the other hand, what happens if H_0 is really false (last column of the table)? With $\alpha = 0.05$ and the obtained probability = 0.02, we would reject H_0 and thereby make a correct decision (cell 2). However, if we changed alpha to $\alpha = 0.01$, we would retain H_0, and we would make a Type II error (cell 4). Thus, making alpha more stringent decreases the probability of making a Type I error, but increases the probability of making a Type II error. Because of this interaction between alpha and beta, the alpha level employed for an experiment depends on the intended use of the experimental results. As mentioned previously, if the results are to communicate a new fact to the scientific community, the consequences of a Type I error are great, and therefore stringent alpha levels are used (0.05 and 0.01). If, however, the experiment is exploratory in nature and the results are to guide the researcher in deciding whether to do a full-fledged experiment, it would be foolish to use such stringent levels. In such cases, alpha levels as high as 0.10 or 0.20 are often used.

Let's consider one more example. Imagine you are the president of a drug company. One of your leading biochemists comes rushing into your office and tells you that she has discovered a drug that increases memory. You are of course elated, but you still ask to see the experimental results. Let's assume it will require a $3 million outlay to install the apparatus to manufacture the drug. This is quite an expense, but if the drug really does increase memory, the potential benefits and profits are well worth it. In this case, you would want to be very sure that the results are not due to chance. The consequences of a Type I error are great. You stand to lose $3 million. You will probably want to use an extremely stringent alpha level before deciding to reject H_0 and risk the $3 million.

We hasten to reassure you that *truth* is not dependent on the alpha level used in an experiment. Either marijuana affects appetite or it doesn't. Either the drug increases memory or it doesn't. Setting a stringent alpha level merely diminishes the possibility that we shall conclude for the alternative hypothesis when the null hypothesis is really true.

Since we never know for sure what the real truth is as a result of a single experiment, replication is a necessary and essential part of the scientific process. Before an "alleged fact" is accepted into the body of scientific knowledge, it must be demonstrated independently in several laboratories. The probability of making a Type I error decreases greatly with independent replication.

EVALUATING THE TAIL OF THE DISTRIBUTION

In the previous discussion, the obtained probability was found by using just the specific outcome of the experiment, i.e., 9 pluses. However, we did that to keep things simple for clarity when presenting the other major concepts. In fact, it is incorrect to use *just the specific outcome* when evaluating the results of an experiment. Instead, we must determine the probability of getting the obtained outcome or *any outcome even more extreme*. It is this probability that we compare with alpha to assess the reasonableness of the null hypothesis. In other words, we evaluate the *tail* of the distribution, beginning with the obtained result, rather than just the obtained result itself. If the alternative hypothesis is nondirectional, we evaluate the obtained result or any even more extreme in both directions (both tails). If the alternative hypothesis is directional, we evaluate only the tail of the distribution that is in the direction specified by H_1.

To illustrate, let's again evaluate the data in the present experiment, this time evaluating the tails rather than just the specific outcome. Figure 10.1 shows the binomial distribution for $N = 10$ and $P = 0.50$. The distribution has two tails, one containing few pluses and one containing many pluses. Since the alternative hypothesis is nondirectional, to calculate the obtained probability, we must determine the probability of getting the obtained result or a result even more extreme *in both directions*. Since the obtained result was 9 pluses, we must include outcomes as extreme as or more extreme than 9 pluses. From Figure 10.1, we can see that the outcome of 10 pluses is more extreme in one direction and the outcomes of 1 plus and 0 plus are as extreme or more extreme in the other direction. Thus, the obtained probability is as follows:

$$p(0, 1, 9, \text{or } 10 \text{ pluses}) = p(0) + p(1) + p(9) + p(10)$$

$$= 0.0010 + 0.0098 + 0.0098 + 0.0010$$

$$= 0.0216$$

It is this probability (0.0216, not 0.0098) that we compare with alpha to reject or retain the null hypothesis. This probability is called a *two-tailed probability* value because the outcomes we evaluate occur under both tails of the distribution.

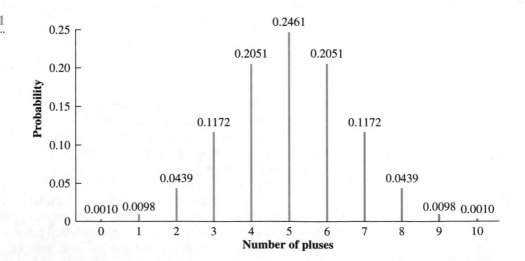

FIGURE 10.1

Binomial distribution for $N = 10$ and $P = 0.50$

Thus, alternative hypotheses that are nondirectional are evaluated with two-tailed probability values. If the alternative hypothesis is nondirectional, the alpha level must also be two-tailed. If $\alpha = 0.05_{2\,\text{tail}}$, this means that the two-tailed obtained probability value must be equal to or less than 0.05 to reject H_0. In this example, since 0.0216 is less than 0.05, we reject H_0 and conclude as we did before that marijuana affects appetite.

If the alternative hypothesis is directional, we evaluate the tail of the distribution that is in the direction predicted by H_1. To illustrate this point, suppose the alternative hypothesis was that "marijuana increases appetite" and the obtained result was 9 pluses and 1 minus. Since H_1 specifies that marijuana increases appetite, we evaluate the tail with the higher number of pluses. Remember that a plus means more food eaten in the marijuana condition. Thus, if marijuana increases appetite, we expect mostly pluses. The outcome of 10 pluses is the only possible result in this direction more extreme than 9 pluses. The obtained probability is

$$p(9 \text{ or } 10 \text{ pluses}) = 0.0098 + 0.0010$$

$$= 0.0108$$

This probability is called a *one-tailed probability* because all of the outcomes we are evaluating are under one tail of the distribution. Thus, alternative hypotheses that are directional are evaluated with one-tailed probabilities. If the alternative hypothesis is directional, the alpha level must be one-tailed. Thus, directional alternative hypotheses are evaluated against one-tailed alpha levels. In this example, if $\alpha = 0.05_{1\,\text{tail}}$, we would reject H_0 because 0.0108 is less than 0.05.

The reason we evaluate the tail has to do with the alpha level set at the beginning of the experiment. In the example we have been using, suppose the hypothesis is that "marijuana increases appetite." This is a directional hypothesis, so a one-tailed evaluation is appropriate. Assume N is 10 and $\alpha = 0.05_{1\,\text{tail}}$. By setting $\alpha = 0.05$ at the beginning of the experiment, the researcher desires to limit the probability of a Type I error to 5 in 100. Suppose the results of the experiment turn out to be 8 pluses and 2 minuses. Is this a result that allows rejection of H_0 consistent with the alpha level? Your first impulse is no doubt to answer "yes" because $p(8 \text{ pluses}) = 0.0439$. However, if we reject H_0 with 8 pluses, we must also reject it if the results are 9 or 10 pluses. Why? Because these outcomes are even more favorable to H_1 than 8 pluses and 2 minuses. Certainly, if marijuana really does increase appetite, obtaining 10 pluses and 0 minus is better evidence than 8 pluses and 2 minuses, and similarly for 9 pluses and 1 minus. Thus, if we reject with 8 pluses, we must also reject with 9 and 10 pluses. But what is the probability of getting 8, 9, or 10 pluses if chance alone is operating?

$$p(8, 9, \text{ or } 10 \text{ pluses}) = p(8) + p(9) + p(10)$$

$$= 0.0439 + 0.0098 + 0.0010$$

$$= 0.0547$$

The probability is greater than alpha. Therefore, we can't allow 8 pluses to be a result for which we could reject H_0; the probability of falsely rejecting H_0 would be greater than the alpha level. Note that this is true even though the probability of 8 pluses itself is less than alpha. Therefore, we don't evaluate the exact outcome, but rather we evaluate the tail so as to limit the probability of a Type I error to the alpha level set at the beginning of the experiment. The reason we use a two-tailed evaluation with a nondirectional alternative hypothesis is that results at both ends of the distribution are legitimate candidates for rejecting the null hypothesis.

ONE- AND TWO-TAILED PROBABILITY EVALUATIONS

When setting the alpha level, we must decide whether the probability evaluation should be one- or two-tailed. When making this decision, use the following rule:

The evaluation should always be two-tailed unless the experimenter will retain H_0 when results are extreme in the direction opposite to the predicted direction.

In following this rule, there are two situations commonly encountered that warrant directional hypotheses. First, when it makes no practical difference if the results turn out to be in the opposite direction, it is legitimate to use a directional hypothesis and a one-tailed evaluation. For example, if a manufacturer of automobile tires is testing a new type of tire supposed to last longer, a one-tailed evaluation is legitimate, because it doesn't make any practical difference if the experimental results turn out in the opposite direction. The conclusion will be to retain H_0, and the manufacturer will continue to use the old tires. Another situation in which it seems permissible to use a one-tailed evaluation is when there are good theoretical reasons, as well as strong supporting data, to justify the predicted direction. In this case, if the experimental results turn out to be in the opposite direction, the experimenter again will conclude by retaining H_0 (at least until the experiment is replicated) because the results fly in the face of previous data and theory.

In situations where the experimenter will reject H_0 if the results of the experiment are extreme in the direction opposite to the prediction direction, a two-tailed evaluation should be used. To understand why, let's assume the researcher goes ahead and uses a directional prediction, setting $\alpha = 0.05_{1\,\text{tail}}$, and the results turn out to be extreme in the opposite direction. If he is unwilling to conclude by retaining H_0, what he will probably do is shift, after seeing the data, to using a nondirectional hypothesis employing $\alpha = 0.05_{2\,\text{tail}}$ (0.025 under each tail) to be able to reject H_0. In the long run, following this procedure will result in a Type I error probability of 0.075 (0.05 under the tail in the predicted direction and 0.025 under the other tail). Thus, switching alternative hypotheses after seeing the data produces an inflated Type I error probability. It is of course even worse if, after seeing that the data are in the direction opposite to that predicted, the experimenter switches to $\alpha = 0.05_{1\,\text{tail}}$ in the direction of the outcome so as to reject H_0. In this case, the probability of a Type I error, in the long run, would be 0.10 (0.05 under each tail). Therefore, to maintain the Type I error probability at the desired level, it is important to decide at the beginning of the experiment whether H_1 should be directional or nondirectional and to set the alpha level accordingly. If a directional H_1 is used, the predicted direction must be adhered to, even if the results of the experiment turn out to be extreme in the opposite direction. Consequently, H_0 must be retained in such cases.

For solving the problems and examples contained in this textbook, we shall indicate whether a one- or two-tailed evaluation is appropriate; we would like you to practice both. Be careful when solving these problems. When a scientist conducts an experiment, she is often following a hunch that predicts a directional effect. The problems in this textbook are often stated in terms of the scientist's directional hunch. Nonetheless, *unless the scientist will conclude by retaining H_0 if the results turn out to be extreme in the opposite direction,* she should use a nondirectional H_1 and a two-tailed evaluation, even though her hunch is directional. Each textbook problem will tell you whether you should use a nondirectional or directional H_1 when it asks for the alternative hypothesis. If you are asked for a nondirectional H_1, you should assume that the appropriate criterion for a directional alternative hypothesis has not been met, regardless of whether the scientist's hunch in the problem is directional. If you are asked for a directional H_1, assume that the appropriate criterion has been met and it is proper to use a directional H_1.

We are now ready to do a complete problem in exactly the same way any scientist would if he or she were using the sign test to evaluate the data.

PRACTICE PROBLEM 10.1

Assume we have conducted an experiment to test the hypothesis that marijuana affects the appetites of advanced-stage AIDS patients. The procedure and population are the same as we described previously, except this time we have sampled 12 advanced-stage AIDS patients. The results are shown here (the scores are in calories):

							Patient					
Condition	1	2	3	4	5	6	7	8	9	10	11	12
THC	1051	1066	963	1179	1144	912	1093	1113	985	1271	978	951
Placebo	872	943	912	1213	1034	854	1125	1042	922	1136	886	902
	+	+	+	−	+	+	−	+	+	+	+	+

a. What is the nondirectional alternative hypothesis? $\frac{2}{12}$ $\frac{10}{12}$ +
b. What is the null hypothesis?
c. Using $\alpha = 0.05_{2\,\text{tail}}$, what do you conclude?
d. What error may you be making by your conclusion in part **c**?
e. To what population does your conclusion apply?

The solution is shown in Table 10.3.

SOLUTION

TABLE 10.3
....

a. Nondirectional alternative hypothesis: Marijuana affects appetites of advanced-stage AIDS patients who are being treated at your hospital.

b. Null hypothesis: Marijuana has no effect on appetites of advanced-stage AIDS patients who are being treated at your hospital.

c. Conclusion, using $\alpha = 0.05_{2\,tail}$:

Step 1: Calculate the number of pluses and minuses. The first step is to calculate the number of pluses and minuses in the sample. We have subtracted the "placebo" scores from the corresponding "THC" scores. The reverse could also have been done. There are 10 pluses and 2 minuses.

Step 2: Evaluate the number of pluses and minuses. Once we have calculated the obtained number of pluses and minuses, we must determine the probability of getting this outcome or any even more extreme in both directions since this is a two-tailed evaluation. The binomial distribution is appropriate for this determination. N = the number of difference scores (pluses and minuses) = 12. We can let P = the probability of a plus with any subject. If marijuana has no effect on appetite, chance alone accounts for whether any subject scores a plus or a minus. Therefore, $P = 0.50$. Since the obtained result was 10 pluses and 2 minuses, the number of P events = 10. The probability of getting an outcome as extreme as or more extreme than 10 pluses (two-tailed) equals the probability of 0, 1, 2, 10, 11, or 12 pluses. Since the distribution is symmetrical, $p(0, 1, 2, 10, 11,$ or 12 pluses) equals $p(10, 11,$ or 12 pluses) $\times$ 2. Thus, from Table B:

			Table Entry	
		N	No. of P Events	P .50
$p(0, 1, 2, 10,$				
11, or 12 pluses) $= p(10, 11,$ or 12 pluses) $\times$ 2		12	10	0.0161
$= [p(10) + p(11) + p(12)] \times 2$			11	0.0029
$= (0.0161 + 0.0029 + 0.0002) \times 2$			12	0.0002
$= 0.0384$				

The same value would have been obtained if we had added the six probabilities together rather than finding the one-tailed probability and multiplying by 2. Since $0.0384 < 0.05$, we reject the null hypothesis. It is not a reasonable explanation of the results. Therefore, we conclude that marijuana affects appetite. It appears to increase it.

d. Possible error: By rejecting the null hypothesis, you might be making a Type I error. In reality, the null hypothesis may be true and you have rejected it.

e. Population: These results apply to the population of advanced-stage AIDS patients from which the sample was taken.

$\mathbf{Y}$ou have good reason to believe a particular TV program is causing increased violence in teenagers. To test this hypothesis, you conduct an experiment in which 15 individuals are randomly sampled from the teenagers attending your neighborhood high school. Each subject is run in an experimental and a control condition. In the experimental condition, the teenagers watch the TV program for 3 months during which you record the number of violent acts committed. The control condition also lasts for 3 months, but the teenagers are not allowed to watch the program during this period. At the end of each 3-month period, you total the number of violent acts committed. The results are given here:

								Patient							
Condition	1	2	3	4	5	6	7	8	9	10	11	12	13	14	15
Viewing the program	25	35	10	8	24	40	44	18	16	25	32	27	33	28	26
Not viewing the program	18	22	7	11	13	35	28	12	20	18	38	24	27	21	22
	+	+	+	−	+	+	+	+	−	+	−	+	+	+	+

a. What is the directional alternative hypothesis? *Inc. Vio*
b. What is the null hypothesis? *no affect*
c. Using $\alpha = 0.01_{1\,tail}$, what do you conclude? *$\frac{3}{15}$ $\frac{12}{15}$*
d. What error may you be making by your conclusion in part c?
e. To what population does your conclusion apply?

The solution is shown in Table 10.4.

SOLUTION

TABLE 10.4

a. Directional alternative hypothesis: Watching the TV program causes increased violence in teenagers.
b. Null hypothesis: Watching the TV program does not cause increased violence in teenagers.
c. Conclusion, using $\alpha = 0.01_{1\,tail}$:

Step 1:   **Calculate the number of pluses and minuses.** The first step is to calculate the number of pluses and minuses in the sample from the data. We have subtracted the scores in the "not viewing" condition from the scores in the "viewing" condition. The obtained result is 12 pluses and 3 minuses.

Step 2: **Evaluate the number of pluses and minuses.** Next, we must determine the probability of getting this outcome or any even more extreme in the direction of the alternative hypothesis. This is a one-tailed evaluation because the alternative hypothesis is directional. The binomial distribution is appropriate. $N =$ the number of difference scores $= 15$. Let $P =$ the probability of a plus with any subject. We can evaluate the null hypothesis by assuming chance alone accounts for whether any subject scores a plus or minus. Therefore, $P = 0.50$. Since the obtained result was 12 pluses and 3 minuses, the number of P events $= 12$. The probability of 12 pluses or more equals the probability of 12, 13, 14, or 15 pluses. This can be found from Table B. Thus,

$p(12, 13, 14,$ or 15 pluses)

$= p(12) + p(13) + p(14) + p(15)$

$= 0.0139 + 0.0032 + 0.0005 + 0.0000$

$= 0.0176$

	Table Entry	
N	No. of P Events	P .50
15	12	0.0139
	13	0.0032
	14	0.0005
	15	0.0000

Since $0.0176 > 0.01$, we fail to reject the null hypothesis. Therefore, we retain H_0 and cannot conclude that the TV program causes increased violence in teenagers.

d. Possible error: By retaining the null hypothesis, you might be making a Type II error. The TV program may actually cause increased violence in teenagers.

e. Population: These results apply to the population of teenagers attending your neighborhood school.

PRACTICE PROBLEM 10.3

A corporation psychologist believes that exercise affects self-image. To investigate this possibility, 14 employees of the corporation are randomly selected to participate in a jogging program. Before beginning the program, they are given a questionnaire that measures self-image. Then they begin the jogging program. The program consists of jogging at a moderately taxing rate for 20 minutes a day, 4 days a week. Each employee's self-image is measured again after 2 months on the program. The results are shown here (the higher the score, the higher the self-image); a score of 20 is the highest score possible.

Subject	Before Jogging	After Jogging
1	14	20
2	13	16
3	8	15
4	14	12
5	12	15
6	7	13
7	10	12
8	16	13
9	10	16
10	14	18
11	6	14
12	15	17
13	12	18
14	9	15

a. What is the alternative hypothesis? Use a nondirectional hypothesis.
b. What is the null hypothesis?
c. Using $\alpha = 0.05_{2\,tail}$, what do you conclude?
d. What error may you be making by your conclusion in part **c**?
e. To what population does your conclusion apply?

The solution is shown in Table 10.5.

SOLUTION

TABLE 10.5
....

a. Nondirectional alternative hypothesis: Jogging affects self-image.
b. Null hypothesis: Jogging has no effect on self-image.
c. Conclusion, using $\alpha = 0.05_{2\,tail}$:

Step 1: **Calculate the number of pluses and minuses.** We have subtracted the "before jogging" from the "after jogging" scores. There are 12 pluses and 2 minuses.

Step 2: **Evaluate the number of pluses and minuses.** Because H_1 is nondirectional, we must determine the probability of getting a result as extreme as or more extreme than 12 pluses (two-tailed), assuming chance alone accounts for the differences. The binomial distribution is appropriate. $N = 14$, $P = 0.50$, and number of P events = 0, 1, 2, 12, 13, or 14. Thus, from Table B:

$$p(0, 1, 2, 12, 13, \text{ or } 14 \text{ pluses}) = p(0) + p(1) + p(2) + p(12)$$
$$+ p(13) + p(14)$$
$$= 0.0001 + 0.0009 + 0.0056$$
$$+ 0.0056 + 0.0009 + 0.0001$$
$$= 0.0132$$

Table Entry

N	No. of P Events	P .50
14	0	0.0001
	1	0.0009
	2	0.0056
	12	0.0056
	13	0.0009
	14	0.0001

The same value would have been obtained if we had found the one-tailed probability and multiplied by 2. Since $0.0132 < 0.05$, we reject the null hypothesis. It appears that jogging improves self-image.

d. Possible error: By rejecting the null hypothesis you might be making a Type I error. The null hypothesis may be true and it was rejected.

e. Population: These results apply to all the employees of the corporation who were employed at the time of the experiment.

SIGNIFICANT VERSUS IMPORTANT

The procedure we have been following in assessing the results of an experiment is first to evaluate directly the null hypothesis and then to conclude indirectly with regard to the alternative hypothesis. If we are able to reject the null hypothesis, we say the results are *significant*. What we really mean by "significant" is "statistically significant," i.e., that the results are probably not due to chance, the independent variable has had a real effect and that, if we repeat the experiment, we would again get results that would allow us to reject the null hypothesis. It might have been better to use the term *reliable* to convey this meaning, rather than significant. However, the usage of *significant* is well established, and so we will have to live with it. The point we wish to make is that we must not confuse statistically significant with practically or theoretically "important." A statistically significant effect says nothing about whether the effect is an important one. For example, suppose the real effect of marijuana is to increase appetite by only 10 calories. Using careful experimental design and a large enough sample, it is possible that we would be able to detect even this small an effect. If so, we would conclude that the result is significant (reliable), but then we still need to ask, How important is this real effect? For most purposes, except possibly theoretical ones, it would seem that this small an effect is not very important. For further discussion of this point, see "What Is the Truth? Much Ado About Almost Nothing," in Chapter 15.

 # POWER

Introduction

Thus far we have seen that there are two errors we might make when testing hypotheses. We have called them Type I and Type II errors. We have further pointed out that the alpha level limits the probability of making a Type I error. By setting alpha to 0.05 or 0.01, the experimenter can limit the probability that she will falsely reject the null hypothesis to these low levels. But what about Type II errors? We defined beta (β) as the probability of making a Type II error. We shall see a bit later that $\beta = 1 -$ power. By maximizing power, we minimize beta, which means that we minimize the probability of making a Type II error. Thus, power is a very important topic.

What Is Power?

Conceptually, the power of an experiment is a measure of the sensitivity of the experiment to detect a real effect of the independent variable. By "a real effect of the independent variable," we mean an effect that produces a change in the dependent variable. If the independent variable does not produce a change in the dependent variable, it has no effect and we say that the independent variable does not have a "real" effect.

We detect a real effect of the independent variable by rejecting the null hypothesis. Thus, power is defined mathematically in terms of rejecting H_0.

MATHEMATICAL DEFINITION

- *Mathematically,* **power** *is the probability that the results of an experiment will allow rejection of the null hypothesis if the independent variable has a real effect.*

For understanding the concept of power, it is helpful to refer to Table 10.6, which is a slightly modified version of Table 10.2. When we evaluate the results of an experiment, if the decision is to reject H_0 and H_0 is true, we have made a Type I error; the probability of making a Type I error is limited by α. On the other hand, if the decision is to reject H_0 and H_1 is true, then we have made a correct decision; the probability of making this correct decision is the power of the experiment. Finally, if the decision is to retain H_0 and H_1 is true, a Type II error has been made; the probability of making a type II error is β.

Since power is a probability, its value can vary from 0.00 to 1.00. The higher the power, the more sensitive the experiment is to detect a real effect of the

TABLE 10.6   Relationship Among State of Reality, Decision, and Power

		Reality	
		H_0 is true	H_1 is true (real effect)
Decision	Reject H_0	Type I error $\rightarrow \alpha$	Correct decision $\rightarrow$ POWER
	Retain H_0		Type II error $\rightarrow \beta$

independent variable. Experiments with power as high as 0.80 or higher are very desirable but rarely seen in the behavioral sciences. Values of 0.40 to 0.60 are much more common.

Variables That Affect Power

Alpha Level   We have already seen that if we decrease the alpha level, e.g., changing alpha from 0.05 to 0.01, with other variables held constant, beta increases. Since beta = 1 − power, if beta increases, power decreases. Thus decreasing alpha (making alpha more stringent) causes a decrease in power. By decreasing alpha, we have made it more difficult to falsely reject H_0, but in so doing, we have made the experiment less sensitive to detect a real effect of the independent variable. (There is no free ride.)

Number of Subjects (N)   There is a direct relationship between N and power. Consequently, increasing N causes an increase in power. This can be shown by considering the following example. Suppose we were to do the marijuana experiment discussed previously in this chapter, setting $\alpha = 0.05_{2\,\text{tail}}$, and running only 5 subjects in the experiment. The best possible result for rejecting H_0 would be 5 pluses or 5 minuses. However, even if we got this result, we would not be able to reject H_0; instead we would have to conclude by retaining H_0. This is because the obtained probability = p(5 pluses or 5 minuses, if $P = 0.50$) = $0.0312 + 0.0312 = 0.0624$. Since $0.0624 > 0.05$, we would be forced to retain H_0, even though the experiment had yielded the best possible result for rejecting H_0, all pluses or minuses. In this experiment, there is no possible way to reject H_0, no matter how strong the real effect actually is. For example, even if the real effect of marijuana is to increase eating by 10,000 calories in each subject, we would still be led by the experiment to retain H_0. This experiment has zero sensitivity to detect the real effect of the independent variable, if there is one. In this experiment, power = 0.00.

Next, consider what would happen if we were to run 10 subjects instead of 5. Again, assume we set $\alpha = 0.05_{2\,\text{tail}}$, and the results were all pluses or all minuses. This time the obtained probability = p(10 pluses or 10 minuses, if $P = 0.50$) = $0.0010 + 0.0010 = 0.0020$, and we would be able to reject H_0. Increasing N from 5 to 10 has caused the power of the experiment to go up. Now, there is at least one sample outcome that would allow rejection of H_0. (Actually, it turns out that there are four possible sample outcomes that will allow H_0 to be rejected.) If the magnitude of real effect were so strong that the only possible sample outcome was all pluses or all minuses, then this experiment would be sure to detect it. Thus, the power to detect this magnitude of real effect = 1.00. Power has gone from 0.00 to 1.00 by increasing N from 5 to 10. Increasing N is one of the most common ways of increasing the power of an experiment.

Magnitude of Real Effect   Intuitively, I think we would all agree that the greater the magnitude of real effect, the more likely the experiment would be to detect it, and hence the greater the power of the experiment. This intuition is correct. There is a direct relationship between the magnitude of real effect and power, which can be illustrated as follows. To simplify the illustration, let's deal only with real effects that are in a direction to produce pluses. P, the probability of a plus with any subject, is a measure of the real effect of the independent variable. If there is no real effect, then the probability of a plus with any subject equals the probability of a minus = 0.50. The greater the magnitude of the real effect, the more likely it will overcome opposite effects from other variables, and hence the more likely it will be to get a plus with any subject. If the effect were so strong that it overcame all variables with effects in the opposite direction, then $P = 1.00$ and each subject would show a plus.

In any experiment, if there were no effect, we would expect about 50% of the sample to show pluses. For example, if $N = 20$ and $P = 0.50$, the most likely result would be 10 pluses. If the real effect were small, we would expect a few more pluses than negatives, say, 14 pluses, just to name a number. If the effect were so strong as to swamp out all other variables acting in the opposite direction, then there would be 20 pluses. Obviously, when evaluating H_0, the obtained probability for 20 pluses will be lower than for 14 pluses and hence have a higher probability of allowing H_0 to be rejected. Thus the stronger the real effect, the higher the probability of rejecting H_0, hence the higher the power. In fact, it is often good strategy when conducting research to use levels of the independent variable that you suspect will maximize its real effect so that the experiment has the highest power to detect the effect. An obvious example is when doing a drug study. Instead of using random or arbitrarily determined drug concentrations, this strategy would suggest including some strong concentrations to maximize power. This assumes, of course, that there is a positive relationship between drug concentration and the drug's effect on the dependent variable.

Power and Beta

It is now time to show that beta = 1 − power. When we conclude an experiment, there are only two possible outcomes: Either we reject H_0, or we retain H_0. These outcomes are also mutually exclusive. Therefore, the sum of their probabilities must equal 1. Assuming H_1 is true,

$$p(\text{rejecting } H_0 \text{ if } H_1 \text{ is true}) + p(\text{retaining } H_0 \text{ if } H_1 \text{ is true}) = 1$$

but

$$\text{Power} = p(\text{rejecting } H_0 \text{ if } H_1 \text{ is true})$$

$$\text{Beta} = p(\text{retaining } H_0 \text{ if } H_1 \text{ is true})$$

Thus,   Power + Beta = 1

or   Beta = 1 − Power

Alpha, Power, and Reality

When we conduct an experiment, there are only two possibilities: Either H_0 is true or H_1 is true. By minimizing alpha and maximizing power, we maximize the likelihood that our conclusions will be correct regardless of which is the case. This can be seen as follows.

H_0 *is true.*   If H_0 is true, the probability of correctly concluding from the experiment is

$$p(\text{correctly concluding}) = p(\text{retaining } H_0) = 1 - \alpha$$

If alpha is set at a stringent level, say, 0.05, then

$$p(\text{correctly concluding}) = 1 - \alpha = 1 - 0.05 = 0.95$$

H_1 *is true.*   On the other hand, if H_1 is true

$$p(\text{correctly concluding}) = p(\text{rejecting } H_0) = \text{power}$$

If the power is high, say, equal to 0.90 to detect the minimum real effect of interest, then

$$p(\text{correctly concluding}) = \text{power} = 0.90$$

Thus, whichever is the true state of affairs (H_0 is true or H_1 is true), there is a high probability of correctly concluding when α is set at a stringent level and the power is high. One way of achieving a high power when α is set at a stringent level is to have a large N. Another way is to use the statistical inference test that is the most powerful for the data. A third way is to control the external conditions of the experiment such that the variability of the data is reduced. We shall discuss the latter two methods when we cover Student's t test in Chapters 13 and 14.

Interpreting Nonsignificant Results

Although considering power aids in designing an experiment, it is much more often used when interpreting the results of an experiment that has already been conducted and that has yielded nonsignificant results. As mentioned previously, nonsignificant results may occur because (1) H_0 is true or (2) H_1 is true, but the experiment had low power. It is because of the second possibility that we can never accept H_0 as being correct when an experiment fails to yield significance. We say the experiment has failed to allow the null hypothesis to be rejected. It is possible that H_1 is indeed true, but the experiment was insensitive; i.e., it didn't give H_0 much of a chance to be rejected. A case in point is the example we

[handwritten margin note:] (1) increase N (2) use test most powerful for data (3) control conditions so variability is reduced

presented previously with $N = 5$ [in the "Number of Subjects (N)" section on page 228]. In that experiment, whatever results we obtained, the results would not reach significance. We could not reject H_0 no matter how strong the real effect actually was. It would be a gross error to accept H_0 as a result of doing that experiment. The experiment did not give H_0 any chance of being rejected. *From this viewpoint, we can see that every experiment exists to give H_0 a chance to be rejected.* The higher the power, the more the experiment allows H_0 to be rejected if H_1 is true.

Perhaps an analogy will help in understanding this point. We can liken the power of an experiment to the use of a microscope. Neuroscientists have long been interested in what happens in the brain to allow the memory of an event to be recorded. One hypothesis states that a group of neurons fire together as a result of the stimulus presentation. With repeated firings (trials), there is growth across the synapses of the cells so that, after a while, they become activated together whenever the stimulus is presented. This "cell assembly" then becomes the physiological engram of the stimuli; that is, it is the memory trace.

To test this hypothesis, an experiment is done that involves visual recognition in rats. After some rats have practiced the task, the appropriate brain cells from each are prepared on slides to look for growth across the synapses. H_0 predicts no growth; H_1 predicts growth. First, the slides are examined with the naked eye; no growth is seen. Can we therefore accept H_0? No, because the eye is not powerful enough to see growth even if it were there. The same holds true for a low-power experiment. If the results are not significant, we cannot conclude by accepting H_0, because even if H_1 is true, the low power makes it unlikely that we would reject the null hypothesis. Next, a light microscope is used, and still there is no growth seen between synapses. Even though this is a more powerful experiment, can we conclude that H_0 is true? No, because a light microscope doesn't have enough power to see synapses clearly. Finally an electron microscope is used, producing a very powerful experiment in which all but the most minute structures at the synapse can be seen clearly. If H_1 is true, namely, if there is growth across the synapse, this powerful experiment has a higher probability of detecting it. Thus, the higher the power of an experiment, the more the experiment allows H_0 to be rejected if H_1 is true.

In light of the foregoing discussion, whenever an experiment fails to yield significant results, we must be careful in our interpretation. Certainly, we can't assert that the null hypothesis is correct. However, if the power of the experiment is high, we can say a little more than just that the experiment has failed to allow rejection of H_0. For example, if the power is 1.00 for an effect represented by $P = 1.00$ and we fail to reject H_0, we can at least conclude that the independent variable is not that strong. If the power is, say, 0.80 for a moderately strong effect, e.g., $P = 0.70$, we can be reasonably confident that the independent variable is not that effective. On the other hand, if the power is low, nonsignificant results tell us little about the true state of reality. Thus, a power analysis tells us how much confidence to place in experiments that fail to reject the null hypothesis. When we fail to reject the null hypothesis, the higher the power to detect a given real effect, the more confident we are that the independent variable is not that strong. However, as the real effect of the independent variable becomes very weak, the power of the experiment to detect it also becomes very low. *Thus, it is impossible to ever prove that the null hypothesis is true, because the power to detect very weak but real effects of the independent variable is always low.*

CHANCE OR REAL EFFECT?

An article appeared in *Time* magazine concerning the "Pepsi Challenge Taste Test." A Pepsi ad, which is shown here, appeared in the article. Taste Test participants were Coke drinkers from Michigan who were asked to drink from a glass of Pepsi and another glass of Coke and say which they preferred. To avoid obvious bias, the glasses were not labeled "Coke" or "Pepsi." Instead, to facilitate a "blind" administration of the drinks, the Coke glass was marked with a "Q" and the Pepsi glass with an "M." The results as stated in the ad are, "More than half the Coca-Cola drinkers tested in Michigan preferred Pepsi." Aside from a possible real preference for Pepsi in the population of Michigan Coke drinkers, can you think of any other possible explanation of these sample results?

ANSWER The most obvious alternative explanation of these results is that they are due to chance alone; that in the population the preference for Pepsi and Coke are equal ($P = 0.50$). You, of course, recognize this as the null hypothesis explanation. This explanation could and, in our opinion, should have been ruled out (within the limits of Type I error) by analyzing the sample data with the appropriate inference test. If the results really are

(continues)

WHAT IS THE TRUTH?

CHANCE OR REAL EFFECT?
(continued)

significant, it doesn't take much space in an ad to say so. This ad is like many that state sample results favoring their product, without evaluating chance as a reasonable explanation.

As an aside, Coke did not cry "chance alone," but instead claimed the study was invalid because people like the letter "M" better than "Q." Coke conducted a study to test its contention by putting Coke in both the "M" and "Q" glasses. Sure enough, more people preferred the drink in the "M" glass, even though it was Coke in both glasses. Pepsi responded by doing another Pepsi Challenge round, only this time revising

the letters to "S" and "L," with Pepsi always in the "L" glass. The sample results again favored Pepsi. Predictably, Coke executives again cried foul, claiming an "L" preference. A noted motivational authority was then consulted and he reported that he knew of no studies showing a

bias in favor of the letter "L." As a budding statistician, how might you design an experiment to determine whether there is a preference for Pepsi or Coke in the population and at the same time eliminate glass-preference as a possible explanation?

WHAT IS THE TRUTH?

"NO PRODUCT IS BETTER THAN OUR PRODUCT"

Often we see advertisements that present no data and make the assertion, "No product is better in doing X than our product." An ad regarding Excedrin, which was recently published in a national magazine, is an example of this kind of advertisement. The ad showed a large picture of a bottle of Excedrin tablets along with the statements,

"Nothing you can buy is stronger."
"Nothing you can buy works harder."
"Nothing gives you bigger relief."

The question is, How do we interpret these claims? Do we rush out and buy Excedrin because it is stronger, works harder, and gives bigger relief than any other headache remedy available? If there are experimental data that form the basis of this ad's claims, we wonder what the results really are. What is your guess?

ANSWER Of course we really don't know in every case, and therefore we don't intend our remarks to be directed at any specific ad. We have just chosen the Excedrin ad as an illustration of many such ads. However, we can't help but be suspicious that in most, if not all cases where sample data exist, the actual data show that there is no significant difference in doing X between the advertiser's product and the other products tested.

For the sake of discussion, let's call the advertiser's product "A." If the data had shown that "A" was better than the competing products, it seems reasonable that the advertiser would directly claim superiority for their product, rather than implying this indirectly through the weaker statement that no other product is better than theirs. Why, then, would the advertiser make this weaker statement? Probably because the actual data do not show product "A" to be superior at all. Most likely, the sample data show product "A" to be either equal to or inferior to the others, and the inference test shows no significant difference between the products.

Given such data, rather than saying that the research shows our product to be inferior or, at best, equal to the other products at doing X (which clearly would not sell a whole bunch of product "A"), the results are stated in this more positive, albeit, in our opinion, misleading way. Saying "No other product is better than ours in doing X" will obviously sell more products than "All products tested were equal in doing X." And, after all, if you read the weaker statement closely, it does not really say that product "A" is superior to the others.

Thus, in the absence of reported data to the contrary, we believe the most accurate interpretation of the claim "No other competitor's product is superior to ours at doing X" is that *the products are equal at doing X.*

WHAT IS THE TRUTH?

ASTROLOGY AND SCIENCE

A newspaper article appeared in a recent issue of the Pittsburgh *Post-Gazette,* with the headline, "When Clinical Studies Mislead." Excerpts from the article are reproduced here:

Shock waves rolled through the medical community two weeks ago when researchers announced that a frequently prescribed triad of drugs previously shown to be helpful after a heart attack had proved useless in new studies. . . .

"People are constantly dazzled by numbers, but they don't know what lies behind the numbers," said Alvan R. Feinstein, a professor of medicine and epidemiology at the Yale University School of Medicine. "Even scientists and physicians have been brainwashed into thinking that the magic phrase 'statistical significance' is the answer to everything."

The recent heart-drug studies belie that myth. Clinical trials involving thousands of patients over a period of several years had shown previously that nitrate-containing drugs such as nitroglycerine, the enzyme inhibitor captopril and magnesium all helped save lives when administered after a heart attack.

*Comparing the life spans of those who took the medicines with those who didn't, researchers found the difference to be statistically significant, and the drugs became part of the stan-*dard medical practice. In the United States, more than 80 percent of heart attack patients are given nitrate drugs.*

But in a new study involving more than 50,000 patients, researchers found no benefit from nitrates or magnesium and captopril's usefulness was marginal. Oxford epidemiologist Richard Peto, who oversaw the latest study, said the positive results from the previous trial must have been due to "the play of chance." . . . Faulty number crunching, Peto said, can be a matter of life and death.

He and his colleagues drove that point home in 1988 when they submitted a paper to the British medical journal the Lancet. Their landmark report showed that heart attack victims had a better chance of surviving if they were given aspirin within a few hours after their attacks. As Peto tells the story, the journal's editors wanted the researchers to break down the data into various subsets, to see whether certain kinds of patients who differed from each other by age or other characteristics were more or less likely to benefit from aspirin.*

Peto objected, arguing that a study's validity could be compromised by breaking it into too many pieces. If you compare enough subgroups, he said, you're bound to get some kind of correlation by chance alone. When the editors insisted, Peto

PRESCRIPTIONS AND ASTROLOGICAL ASPIRIN FORECASTS

SAGITTARIUS—NO ARIES—NO
CANCER—MAYBE TAURUS—NO
LEO—SURE GEMINI—NO
VIRGO—YES LIBRA
CAPRICORN—NOPE
AQUARIUS—YES
PISCES—NOT YET
SCORPIO—LATER

WHAT IS THE TRUTH?

ASTROLOGY AND SCIENCE
(continued)

capitulated, but among other things he divided his patients by zodiac birth signs and demanded that his findings be included in the published paper. Today, like a warning sign to the statistically uninitiated, the wacky numbers are there for all to see: Aspirin is useless for Gemini and Libra heart attack victims but is a lifesaver for people born under any other sign. . . .

Studies like these exemplify two of the more common statistical offenses committed by scientists—making too many comparisons and paying too little

attention to whether something makes sense—said James L. Mills, chief of the pediatric epidemiology section of the National Institute of Child Health and Human Development.

"People search through their results for the most exciting and positive things," he said. "But you also have to look at the biological plausibility. A lot of findings that don't withstand the test of time didn't really make any sense in the first place." . . .

In the past few years, many scientists have embraced larger and larger clinical trials to mini-

mize the chances of being deceived by a fluke.

What do you think? If you were a physician, would you continue to prescribe nitrates to heart attack patients? Is it really true that the early clinical trials are an example of Type I error, as suggested by Dr. Peto? Will larger and larger clinical trials minimize the chances of being deceived by a fluke? Finally, is aspirin really useless for Gemini and Libra heart attack victims but a lifesaver for people born under any other sign?

SUMMARY

In this chapter, we have discussed the topic of hypothesis testing, using the sign test as our vehicle. The sign test is used in conjunction with the repeated measures design. The essential features of the repeated measures design are that there are paired scores between conditions and difference scores are analyzed.

In any hypothesis-testing experiment, there are always two hypotheses that compete to explain the results: the alternative hypothesis and the null hypothesis. The alternative hypothesis specifies that the independent variable is responsible for the differences in score values between the conditions. The alternative hypothesis may be directional or nondirectional. It is legitimate to use a directional hypothesis when there is a good theoretical basis and good supporting evidence in the literature. If the experiment is a basic fact-finding experiment, ordinarily a nondirectional hypothesis should be used. A direc-

tional alternative hypothesis is evaluated with a one-tailed probability value and a nondirectional hypothesis with a two-tailed probability value.

The null hypothesis is the logical counterpart to the alternative hypothesis such that if the null hypothesis is false, the alternative hypothesis must be true. If the alternative hypothesis is nondirectional, the null hypothesis specifies that the independent variable has no effect on the dependent variable. If the alternative hypothesis is directional, the null hypothesis states that the independent variable has no effect in the direction specified.

In evaluating the data from an experiment, we never directly evaluate the alternative hypothesis. We always first evaluate the null hypothesis. The null hypothesis is evaluated by assuming chance alone is responsible for the differences in scores between conditions. In doing this evaluation, we calculate the probability of getting the obtained result or a result even more extreme if chance alone is responsible. If this obtained probability is equal to or lower than the alpha level, we consider the null hypothesis explanation unreasonable, and therefore we reject the null hypothesis. We conclude by accepting the alternative hypothesis, since it is the only other explanation. If the obtained probability is greater than the alpha level, we retain the null hypothesis. It is still considered a reasonable explanation of the data. Of course, if the null hypothesis is not rejected, the alternative hypothesis cannot be accepted. The conclusion applies legitimately only to the population from which the sample was randomly drawn. We must be careful to distinguish "statistically significant" from practically or theoretically "important."

The alpha level is usually set at 0.05 or 0.01 to minimize the probability of making a Type I error. A Type I error occurs when the null hypothesis is rejected and it is actually true. The alpha level limits the probability of making a Type I error. It is also possible to make a Type II error. This occurs when we retain the null hypothesis and it is false. Beta is defined as the probability of making a Type II error.

When alpha is made more stringent, beta increases. Power is the probability of rejecting H_0 when H_1 is true. Power varies directly with N, alpha, and magnitude of real effect of the independent variable. By minimizing α and maximizing power, it is possible to have a high probability of correctly concluding from an experiment regardless of whether H_0 or H_1 is true.

In analyzing the data of an experiment with the sign test, we ignore the magnitude of difference scores and just consider their direction. There are only two possible scores for each subject, a plus or a minus. We sum the pluses and minuses for all subjects, and the obtained result is the total number of pluses and minuses. To test the null hypothesis, we calculate the probability of getting the total number of pluses or a number of pluses even more extreme if chance alone is responsible. The binomial distribution with P(the probability of a plus) = 0.50 and N = the number of difference scores is appropriate for making this determination. An illustrative problem and several practice problems were given to show how to evaluate the null hypothesis using the binomial distribution.

IMPORTANT TERMS

Alpha level (α) (p. 213)
Alternative hypothesis (H_1) (p. 212)
Beta (β) (p. 216)
Binomial distribution (p. 213)
Correct decision (p. 215)
Correlated groups design (p. 212)
Directional hypothesis (p. 212)

Fail to reject null hypothesis (p. 213)
Important (p. 226)
Nondirectional hypothesis (p. 212)
Null hypothesis (H_0) (p. 212)
One-tailed probability (p. 219)
Power (p. 227)
Real effect (p. 227)
Reject null hypothesis (p. 213)
Repeated measures design (p. 212)

Replicated measures design (p. 212)
Retain null hypothesis (p. 213)
Sign test (p. 211)
Significant (p. 226)
State of reality (p. 215)
Two-tailed probability (p. 218)
Type I error (p. 215)
Type II error (p. 215)

QUESTIONS AND PROBLEMS

1. Briefly define or explain each of the terms in the "Important Terms" section.
2. Briefly describe the process involved in hypothesis testing. Be sure to include the alternative hypothesis, the null hypothesis, the decision rule, the possible type of error, and the population to which the results can be generalized.
3. Explain in your own words why it is important to know the possible errors we might make when rejecting or failing to reject the null hypothesis.
4. Does the null hypothesis for a nondirectional H_1

differ from the null hypothesis for a directional H_1? Explain.
5. Under what conditions is it legitimate to use a directional H_1? Why is it not legitimate to use a directional H_1 just because the experimenter has a "hunch" about the direction?
6. If the obtained probability in an experiment equals 0.0200, does this mean that the probability that H_0 is true equals 0.0200? Explain.
7. Discuss the difference between "significant" and "important."

8. What is power? How is it defined conceptually? How is it defined mathematically?

9. In hypothesis-testing experiments, why is the conclusion "We retain H_0" preferable to "We accept H_0 as true"?

10. In hypothesis-testing experiments, is it ever correct to conclude that the independent variable has had *no* effect? Explain.

11. Using α and power, explain how we can maximize the probability of correctly concluding from an experiment, regardless of whether H_0 is true or H_1 is true. As part of your explanation, choose values for α and power and determine the probability of correctly concluding when H_0 is true and when H_1 is true.

12. A primatologist believes that rhesus monkeys possess curiosity. She reasons that, if this is true, then they should prefer novel stimulation to repetitive stimulation. An experiment is conducted in which 12 rhesus monkeys are randomly selected from the university colony and taught to press two bars. Pressing bar 1 always produces the same sound, whereas bar 2 produces a novel sound each time it is pressed. After learning to press the bars, the monkeys are tested for 15 minutes during which they have free access to both bars. The number of presses on each bar during the 15 minutes is recorded. The resulting data are as follows:

Subject	Bar 1	Bar 2
1	20	40
2	18	25
3	24	38
4	14	27
5	5	31
6	26	21
7	15	32
8	29	38
9	15	25
10	9	18
11	25	32
12	31	28

a. What is the alternative hypothesis? In this case, assume a nondirectional hypothesis is appropriate because there is insufficient empirical basis to warrant a directional hypothesis.

b. What is the null hypothesis?

c. Using $\alpha = 0.05_{2\,\text{tail}}$, what is your conclusion?

d. What error may you be making by your conclusion in part c?

e. To what population does your conclusion apply?

13. A school principal is interested in a new method for teaching eighth-grade social studies, which he believes will increase the amount of material learned. To test this method, the principal conducts the following experiment. The eighth-grade students in the school district are grouped into pairs based on matching their IQs and past grades. Twenty matched pairs are randomly selected for the experiment. One member of each pair is randomly assigned to a group that receives the new method, and the other member of each pair to a group that receives the standard instruction. At the end of the course, all students take a common final exam. The following are the results:

Pair No.	New Method	Standard Instruction
1	95	83
2	75	68
3	73	80
4	85	82
5	78	84
6	86	78
7	93	85
8	88	82
9	75	84
10	84	68
11	72	81
12	84	91
13	75	72
14	87	81
15	94	83
16	82	87
17	70	65
18	84	76
19	72	63
20	83	80

a. What is the alternative hypothesis? Use a directional hypothesis.

b. What is the null hypothesis?

c. Using $\alpha = 0.05_{1\,\text{tail}}$, what is your conclusion?

d. What error may you be making by your conclusion in part c?

e. To what population does your conclusion apply?

14. A physiologist believes that the hormone angiotensin II is important in regulating thirst. To investigate this belief, she randomly samples 16 rats from the vivarium of the drug company where she works and places them in individual cages with free access to food and water. After they have grown acclimated to their new "homes," the experimenter measures the amount of water each rat drinks in a 20-minute period. Then she injects each animal intravenously with a known concentration (100 micrograms per kilogram) of angiotensin II. The rats are then put back into their home cages, and the amount each drinks for another 20-minute period is measured. The results are shown in the following table. Scores are in milliliters drunk per 20 minutes.

Subject	Before Injection	After Injection
1	1.2	11.3
2	0.8	10.7
3	0.5	10.3
4	1.3	11.5
5	0.6	9.6
6	3.5	3.3
7	0.7	10.5
8	0.4	11.4
9	1.1	12.0
10	0.3	12.8
11	0.6	11.4
12	0.3	9.8
13	0.5	10.6
14	4.1	3.2
15	0.4	12.1
16	1.0	11.2

a. What is the nondirectional alternative hypothesis?
b. What is the null hypothesis?
c. Using $\alpha = 0.05_{2\,tail}$, what is your conclusion? Assume the injection itself had no effect on drinking behavior.
d. What error may you be making by your conclusion in part **c**?
e. To what population does your conclusion apply?

15. A leading toothpaste manufacturer advertises that, in a recent medical study, 70% of the people tested had brighter teeth after using its toothpaste (called Very Bright) as compared to using the leading competitor's brand (called Brand X). The advertisement continues, "Therefore, use Very Bright and get brighter teeth." In point of fact, the data upon which the above statements were based were collected from a random sample of 10 employees from the Pasadena plant. In the experiment, each employee used both toothpastes. Half of the employees used Brand X for 3 weeks, followed by Very Bright for the same time period. The other half used Very Bright first, followed by Brand X. A brightness test was given at the end of each 3-week period. Thus, there were two scores for each employee, one from the brightness test following use of Brand X and one following the use of Very Bright. The following table shows the scores (the higher, the brighter):

Subject	Very Bright	Brand X
1	5	4
2	4	3
3	4	2
4	2	3
5	3	1
6	4	1
7	1	3
8	3	4
9	6	5
10	6	4

a. What is the alternative hypothesis? Use a directional hypothesis.
b. What is the null hypothesis?
c. Using $\alpha = 0.05_{1\,tail}$, what do you conclude?
d. What error may you be making by your conclusion in part **c**?
e. To what population does your conclusion apply?
f. Does the advertising seem misleading?

16. A researcher is interested in determining whether acupuncture affects pain tolerance. An experiment is performed in which 15 students are randomly chosen from a large pool of university undergraduate volunteers. Each subject serves in

two conditions. In both conditions, each subject receives a short-duration electric shock to the pulp of a tooth. The shock intensity is set to produce a moderate level of pain to the unanesthetized subject. After the shock is terminated, each subject rates the perceived level of pain on a scale of 0–10, with 10 being the highest level. In the experimental condition, each subject receives the appropriate acupuncture treatment prior to receiving the shock. The control condition is made as similar to the experimental condition as possible, except a placebo treatment is given instead of acupuncture. The two conditions are run on separate days at the same time of day. The pain ratings in the accompanying table are obtained.

a. What is the alternative hypothesis? Assume a nondirectional hypothesis is appropriate.
b. What is the null hypothesis?
c. Using $\alpha = 0.05_{2\,tail}$, what is your conclusion?
d. What error may you be making by your conclusion in part **c**?

e. To what population does your conclusion apply?

Subject	Acupuncture	Placebo
1	4	6
2	2	5
3	1	5
4	5	3
5	3	6
6	2	4
7	3	7
8	2	6
9	1	8
10	4	3
11	3	7
12	4	8
13	5	3
14	2	5
15	1	4

NOTES

10.1 If the null hypothesis is false, then chance does not account for the results. Strictly speaking, this means that something systematic differs between the two groups. Ideally, the only systematic difference is due to the independent variable. Thus, we say that if the null hypothesis is false, the alternative hypothesis must be true. Practically speaking, however, the reader should be aware that it is hard to do the perfect experiment. Consequently, in addition to the alternative hypothesis, there are often additional possible explanations of the systematic difference. Therefore, when we say "we accept H_1" you should be aware that there may be additional explanations of the systematic difference.

10.2 If the alternative hypothesis is directional, the null hypothesis asserts that the independent variable does not have an effect in the direction specified by the alternative hypothesis. This is true in the overwhelming number of experiments conducted. Occasionally, an experiment is conducted where the alternative hypothesis specifies not only the direction but also the magnitude of the effect. For example, in connection with the marijuana experiment, an alternative hypothesis of this type might be "Marijuana increases appetite so as to increase average daily eating by more than 200 calories." The null hypothesis for this alternative hypothesis is "Marijuana increases appetite so as to increase daily eating by 200 or less calories."

11 Mann–Whitney U Test

INTRODUCTION

In this chapter, we shall consider hypothesis testing using an independent groups design. We have chosen the Mann–Whitney U test as the inference test to use when first considering this design because it has a probability distribution that is easy to understand. Unlike the sign test, however, the Mann–Whitney U test is a powerful test and therefore has practical utility. Once you understand how the Mann–Whitney U test works, you will have covered two probability distributions (the binomial distribution and the distribution of U). With this background, you will be ready to grasp the importance of probability distributions to hypothesis testing. We shall treat this topic formally in Chapter 12.

INDEPENDENT GROUPS DESIGN

There are essentiallly two basic experimental designs used in studying behavior. We met the first when discussing the sign test. This design is called the repeated or replicated measures design. The simplest form of the design uses two conditions: an experimental and a control condition. The essential feature of the design is that there are paired scores between conditions, and *difference* scores are analyzed to determine whether chance alone can reasonably explain them.

The other type of design is called the *independent groups design*. In this design, subjects are randomly selected from the population and then randomly divided into two or more groups. The most basic experiment has only two groups. There is no basis for pairing of subjects, and each subject is tested only once. All of the subjects in one of the groups (it doesn't matter which group) are run in the experimental condition. These subjects are referred to as the experimental group. All of the subjects in the other group receive the control condition. Subjects in this condition constitute the control group. In analyzing the data, there is no basis for pairing scores between the conditions. Rather, a comparison is made between the scores of each group to determine whether chance alone is a reasonable explanation of the differences between the group scores.

EXAMPLE

Effect of High-Protein Diet on Intellectual Development

To illustrate this design, let's consider an example. Suppose you are a developmental psychologist with special competence in nutrition. Based on previous research and theory, you believe that a high-protein diet eaten during early childhood is important for intellectual development. The diet in the geographical area where you live is low in protein. You believe the low-protein diet eaten during the first few years of childhood is detrimental to intellectual development. If you are correct, a high-protein diet should result in higher intelligence. You decide to investigate this directional alternative hypothesis, and you choose the independent groups design for your experiment.

Alpha is set at the beginning of the experiment to $0.05_{1\,\text{tail}}$. Six children are randomly chosen from the 1-year-old children living in your city. Note that in an actual experiment the sample size would be much larger. We have limited it to six in this example for clarity in probability determinations. The six children are then randomly divided into two groups of three each. One group is fed the usual low-protein diet for 3 years, whereas the other group receives a diet high in protein for the same duration. We shall call the low-protein group the control group and the high-protein group the experimental group. At the end of the 3 years, each child is given an IQ test. The following are the scores:

Control Group, Low Protein C 1	Experimental Group, High Protein E 2
84	94
88	101
98	105
$\overline{X}_C = 90$	$\overline{X}_E = 100$

What can we conclude from these data? Most of the scores in the experimental group are higher than the scores in the control group. The mean IQ of the experimental group ($\overline{X}_E = 100$) is higher than the mean IQ of the control group ($\overline{X}_C = 90$). These results are consistent with the hypothesis that a high-protein diet increases IQ. Can we therefore conclude that the high-protein diet was responsible for the higher IQ scores? Not necessarily.

Suppose that, instead of giving the experimental group a high-protein diet, we gave them the same diet as the control group. Isn't it possible that, just due to chance alone, we would get scores as high as or even higher than those obtained in the original experiment? The answer is yes. Thus, as with the repeated measures design, we need to evaluate chance before we can conclude for H_1. The null hypothesis for the independent groups design is the same as for the repeated measures design. Since the alternative hypothesis is directional for this experiment, the null hypothesis specifies that a high-protein diet during infancy does not increase intellectual development compared to a low-protein diet. As with the repeated measures design, we always evaluate H_0 first and then indirectly decide about H_1. Of course, we evaluate H_0 by determining whether chance is a reasonable explanation of the results.

In the independent groups design, we do not analyze *difference* scores as in the repeated measures design. Instead, we analyze the two groups of scores as *separate samples*. The control group scores can be considered a random sample taken from a theoretical population of IQ scores that would have resulted after giving *all* the 1-year-old children living in the city the low-protein diet for 3 years. The scores in the experimental group can be considered a random sample from a population of IQ scores that would have resulted had the same children been fed the high-protein rather than the low-protein diet. If chance is the correct explanation of the results, then the high- and low-protein diets have the same effect on IQ, and the two theoretical populations would have identical distributions.

To evaluate chance, we must determine the probability of getting the obtained results or results even more extreme if the two groups of sample scores are random samples from the two populations having identical scores.

As with the replicated measures design, in assessing H_0, we assume chance to be true and calculate the previously mentioned probability. If the obtained probability is equal to or lower than the alpha level, we reject H_0 and conclude in favor of H_1. If the obtained probability is higher than alpha, we retain H_0. Whichever way we conclude, we run the same risks as discussed previously.

ANALYSIS USING THE MANN–WHITNEY *U* TEST

The Mann–Whitney *U* test analyzes the *separation* between the two sets of sample scores and allows us to determine the probability of getting the obtained separation or even greater separation if both sets of sample scores are random samples from identical populations. Although separation between the two samples is not a quantity you are used to dealing with, it should be intuitively clear that the greater the separation between the two sets of scores, the more reasonable it is that they are not random samples from the same or identical populations. Conversely, the more overlap between the two sets of scores, the more reasonable chance becomes.

To illustrate, let's assume that if we gave the low-protein diet to all of the 1-year-old children in your geographical area, the resulting set of IQ scores would be normally distributed with $\mu = 90$ and $\sigma = 10$. Next, suppose the high-protein diet has a very large effect on IQ, such that if we gave it to the same children we would find $\mu = 170$ and $\sigma = 10$. The two IQ populations are shown in Figure 11.1. Since there is negligible overlap between them, random sampling from these populations would produce very little overlap of scores in the two samples. For instance, if we were randomly sampling three scores from each population, it is quite probable that all of the scores in the high-protein sample would be higher than the three scores in the low-protein sample. The following example

FIGURE 11.1

Population scores resulting from eating a low-protein or a high-protein diet

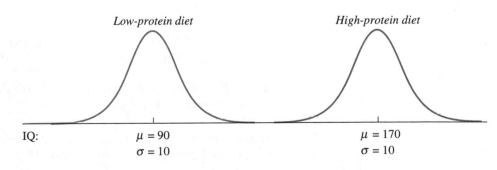

IQ: $\mu = 90$ $\mu = 170$
        $\sigma = 10$ $\sigma = 10$

uses the original three scores for the low-protein sample and, for the high-protein sample, three randomly selected scores from the high-protein population. The scores have been rank-ordered to show more clearly the separation between the groups. In this example, there is complete separation. All of the control group scores are lower than the experimental group scores.

Control Group Low Protein C			Experimental Group High Protein E		
84			163		
88			172		
98			175		
C_1	C_2	C_3	E_1	E_2	E_3
84	88	98	163	172	175

From Figure 11.1, we can readily see that, as the effect of high protein on IQ decreases, the high-protein population slides to the left, and its overlap with the low-protein population increases. Finally, when there is no difference in effect on IQ between the two diets, the populations superimpose on one another. Overlap is complete. In this case, random sampling from identical populations of scores accounts for any separation between the scores in the sample experimental and control groups. It should be apparent that, when the two populations have identical distributions, the overlap between *sample* group scores is likely to be greater than when there is a great separation between populations. Thus, the degree of separation in sample group scores is a measure of how reasonable chance is as an explanation. The more separated the sample group scores, the less reasonable it is to conclude that chance is responsible for the separation. Conversely, the less separated the sample group scores, the more reasonable it is to conclude in favor of chance.

Calculation of Separation (U_{obt} and U'_{obt})

The degree of separation between the two samples can be calculated in two ways: (1) by counting the total number of C scores that are lower than E scores or by counting the total number of E scores that are lower than C scores and (2) by using two equations. Whichever method is used, there are always two numbers that result. The smaller of the two numbers is arbitrarily called U_{obt} and the larger U'_{obt}. Thus,

> U_{obt} = Smaller of the two numbers indicating the degree of separation between the two samples

> U'_{obt} = Larger of the two numbers indicating the degree of separation between the two samples

The subscript "obt" stands for obtained, and U_{obt} means the U value calculated from the experimental data. Note that U_{obt} and U'_{obt} indicate the same degree of separation. Let's first see how to calculate U_{obt} and U'_{obt} by counting Es and Cs, and then by using the equations.

Determining U_{obt} and U'_{obt} by Counting Es and Cs

We'll begin with the example where there was complete separation between the samples. The scores were as follows:

Control Group C	Experimental Group E
84	163
88	172
98	175

There are two steps involved in calculating U and U' by this method.

Step 1: **Combine the scores from both groups and rank-order them:**

84	88	98	163	172	175
C_1	C_2	C_3	E_1	E_2	E_3

Step 2: **Count the number of E scores that are lower than C scores or the number of C scores that are lower than E scores:**

Es Lower than Cs	Cs Lower than Es
Es $< C_1 = 0$	Cs $< E_1 = 3$
Es $< C_2 = 0$	Cs $< E_2 = 3$
Es $< C_3 = \underline{0}$	Cs $< E_3 = \underline{3}$
Total Es $< C$s $= 0$	Total Cs $< E$s $= 9$

To count the total number of E scores that are lower than C scores, count the number of E scores that are lower than each C score and sum these values. This determination is shown in the table. There are no E scores lower than the C_1 score, no E scores lower than C_2, and no E scores lower than C_3. Their sum equals zero. Thus, there are zero E scores that are lower than C scores. The same procedure is followed in counting the number of C scores that are lower than E scores. In all, there are nine C scores that are lower than E scores: three lower than E_1, three lower than E_2, and three lower than E_3. Since U is always the lower of these two totals and U' the highest,

$$U_{obt} = 0$$

$$U'_{obt} = 9$$

This is the greatest degree of separation possible for three C scores and three E scores.

Now let's calculate the U value for the data of the original experiment. You will recall that the obtained IQ scores were as follows:

Control Group C	Experimental Group E
84	94
88	101
98	105

Step 1: **Combine the scores and rank-order them:**

C_1	C_2	E_1	C_3	E_2	E_3
84	88	94	98	101	105

We can see that these scores are not as separate as in the previous example. There is some overlap (E_1 is lower than C_3).

Step 2: **Count the total number of E scores that are lower than C scores or the number of C scores that are lower than E scores:**

Es Lower than Cs	Cs Lower than Es
Es $< C_1 = 0$	Cs $< E_1 = 2$
Es $< C_2 = 0$	Cs $< E_2 = 3$
Es $< C_3 = \underline{1}$	Cs $< E_3 = \underline{3}$
Total Es $< C$s $= 1$	Total Cs $< E$s $= 8$

Thus, for the original data,

$$U_{\text{obt}} = 1$$

$$U'_{\text{obt}} = 8$$

Calculating U_{obt} and U'_{obt} from Equations

Next, let's illustrate how to calculate U and U' from equations. The equations are as follows:

$$U_{\text{obt}} = n_1 n_2 + \frac{n_1(n_1 + 1)}{2} - R_1 \qquad \textit{general equation for finding } U_{\text{obt}} \textit{ or } U'_{\text{obt}}$$

$$U_{\text{obt}} = n_1 n_2 + \frac{n_2(n_2 + 1)}{2} - R_2 \qquad \textit{general equation for finding } U_{\text{obt}} \textit{ or } U'_{\text{obt}}$$

where n_1 = Number of scores in group 1
n_2 = Number of scores in group 2
R_1 = Sum of ranks for scores in group 1
R_2 = Sum of ranks for scores in group 2

According to this method, we identify one of the samples as group 1 and the other as group 2. Then we just go ahead and solve the equations. It doesn't matter which sample is labeled 1 and which is labeled 2. We still obtain the same two numbers from the equations. What does change with labeling is which equation yields the high number and which yields the low number. Since this depends on which group is labeled 1 and which 2, these equations are both written in terms of U. In an actual analysis, the equation that yields the lower number is the U equation; the one that yields the higher number is the U' equation.

Let's try this method on the previous problem. The scores are repeated here for your convenience:

Control Group 1	Experimental Group 2
84	94
88	101
98	105

We have labeled the control group as 1 and the experimental group as 2.
This method of calculating U and U' involves three steps:

Step 1: Combine the scores, rank-order them, and assign each a rank score using 1 for the lowest score:

Original score	84	88	94	98	101	105
Rank	1	2	3	4	5	6

Step 2: Sum the ranks for each group; i.e., determine R_1 and R_2. To find R_1, we add the ranked scores for group 1, and to find R_2, we add the ranked scores for group 2, as shown here:

Control Group 1		Experimental Group 2	
Original Score	*Rank*	*Original Score*	*Rank*
84	1	94	3
88	2	101	5
98	4	105	6
	$R_1 = \overline{7}$		$R_2 = \overline{14}$
	$n_1 = 3$		$n_2 = 3$

Step 3: Solve the equations for U and U':

$$U_{obt} = n_1 n_2 + \frac{n_1(n_1 + 1)}{2} - R_1 \qquad\qquad U_{obt} = n_1 n_2 + \frac{n_2(n_2 + 1)}{2} - R_2$$

$$= 3(3) + \frac{3(4)}{2} - 7 \qquad\qquad\qquad = 3(3) + \frac{3(4)}{2} - 14$$

$$= 9 + 6 - 7 = 8 \qquad\qquad\qquad\quad = 9 + 6 - 14 = 1$$

Thus,

$$U_{obt} = 1$$

$$U'_{obt} = 8$$

Again, we point out that both equations are written in terms of U rather than U'. In this case, the equation with R_2 yielded U, and the one with R_1 gave U', because the group we labeled 1 had the lower sum of ranks ($R_1 < R_2$). If we had labeled the experimental group as 1 and the control group as 2, the equation with R_2 would have yielded U', and the one with R_1 would have given U. Therefore, we can't label the equations generally as U and U'. We assign U and U' after doing the calculations. Note that the answers we obtained for U and U' are the same as those we calculated by counting Es and Cs. The equation solution can be used under all circumstances. The counting Es and Cs solution cannot be used when there are tied scores between samples. After all, which score would we put first when doing the rank ordering, the tied E score or the C score? Since the equation solution is more general, we shall use it for the remainder of the chapter.

Generalizations

Some generalizations are now in order:

1. The Mann–Whitney U test analyzes the separation between the control and experimental group scores. For every experiment there will be two numbers that indicate the degree of separation between the groups. They are U and U'. Both numbers indicate the same degree of separation for that set of data. Thus, in the first example, $U = 0$ and $U' = 9$. Both 0 and 9 indicate the same degree of separation for the data in that example.
2. A U value of 0 represents the greatest degree of separation possible. It means there is no overlap between the groups. This is true regardless of the number of scores in each group. In the second example, there was not complete separation between groups. The E_1 score intruded into the C scores, and the C_3 score intruded into the E scores. In this case, the U_{obt} value was 1. Thus, as U gets larger, the scores become more mixed.
3. The sum of U_{obt} and U'_{obt} must equal $n_1 n_2$. Thus,

$$U_{obt} + U'_{obt} = n_1 n_2$$

In the two examples,

$$0 + 9 = n_1 n_2$$
$$9 = 3(3)$$
$$9 = 9$$

and

$$1 + 8 = n_1 n_2$$
$$9 = 3(3)$$
$$9 = 9$$

Since both U_{obt} and U'_{obt} indicate the same degree of separation, it is only necessary to evaluate one of them. We shall always evaluate U_{obt} because we know its smallest possible value is always 0 irrespective of n_1 and n_2. However, as a computation check, it is a good idea to calculate both of them and see that the calculated values satisfy the relationship

$$U_{obt} + U'_{obt} = n_1 n_2$$

4. When using the Mann–Whitney U test to evaluate H_0, we must determine the following:
   a. The degree of separation between the control and experimental group scores. We do this by calculating U_{obt} or U'_{obt}. For the data of our protein-IQ experiment, $U_{obt} = 1$ and $U'_{obt} = 8$. Since both U_{obt} and U'_{obt} indicate the same degree of separation, it is only necessary to calculate one of them. We shall always work with U_{obt}.
   b. The probability of getting a U value equal to or less than U_{obt} assuming the sample scores are random samples from populations with identical distributions. This value is one-tailed or two-tailed depending on H_1. In this experiment, we need to determine $p(U_{obt} \leq 1)_{1\,\text{tail}}$.

DETERMINING THE PROBABILITY OF U IF CHANCE ALONE IS RESPONSIBLE

If chance alone is responsible, then the separation between the groups is due to random sampling from populations with identical distributions. Suppose that we have two populations with identical distributions and that we randomly sample three scores from each population. Let's arbitrarily assign the three scores from one population to the experimental group and the other three to the control group. If we rank-order the scores, it turns out there are 20 different orders possible. These are listed in column 2 of Table 11.1. In order 1 ($CCCEEE$), there is complete separation between the E and C scores, with the C scores being lower than the E scores. Order 2 ($CCECEE$) is one step less separated. As we proceed from order 1 to order 10, the scores become more and more mixed. Beginning with order 11, the group scores begin to separate again. However, now the E scores are generally lower than the C scores. Finally, in order 20, the scores are again completely separated, with all of the E scores being lower than all of the C scores.

TABLE 11.1   Generation of the Probability Distribution of U and U' for $n_1 = n_2 = 3$

Order No. (1)	Order (2)	No. of Es Lower than Cs (3)	No. of Cs Lower than Es (4)	U (5)	U' (6)	$p(U)$ (7)
1	CCCEEE	0	9	0	9	0.05
2	CCECEE	1	8	1	8	0.05
3	CCEECE	2	7	2	7	0.10
4	CECCEE	2	7	2	7	
5	CCEEEC	3	6	3	6	
6	CECECE	3	6	3	6	0.15
7	ECCCEE	3	6	3	6	
8	CECEEC	4	5	4	5	
9	CEECCE	4	5	4	5	0.15
10	ECCECE	4	5	4	5	
11	CEECEC	5	4	4	5	
12	ECCEEC	5	4	4	5	0.15
13	ECECCE	5	4	4	5	
14	CEEECC	6	3	3	6	
15	ECECEC	6	3	3	6	0.15
16	EECCCE	6	3	3	6	
17	ECEECC	7	2	2	7	0.10
18	EECCEC	7	2	2	7	
19	EECECC	8	1	1	8	0.05
20	EEECCC	9	0	0	9	0.05

In columns 5 and 6 of Table 11.1, we have listed the U and U' values that correspond to each order. Note that U can take on a value of 0, 1, 2, 3, or 4, whereas U' can be 9, 8, 7, 6, or 5. Thus, no matter what the absolute magnitudes of the three scores in the control group and the three scores in the experimental group, these are the only possible values for U and U'.

Now, let's determine the probability of getting each U value. In calculating these values, we shall consider each half of the distribution separately. For the half where C scores are lower than E scores,

$$p(U = 0) = p(\text{order } 1) = p(CCCEEE)$$

$$= \frac{\text{Number of favorable events}}{\text{Total number of possible events}} = \frac{1}{20} = 0.05$$

$$p(U = 1) = p(\text{order } 2) = p(CCECEE) = \frac{1}{20} = 0.05$$

$$p(U = 2) = p(\text{order } 3 \text{ or order } 4) = 0.05 + 0.05 = 0.10$$

$$p(U = 3) = p(\text{order } 5, 6, \text{ or } 7) = 0.05 + 0.05 + 0.05 = 0.15$$

$$p(U = 4) = p(\text{order } 8, 9, \text{ or } 10) = 0.05 + 0.05 + 0.05 = 0.15$$

FIGURE 11.2   Two probability distributions for *U*

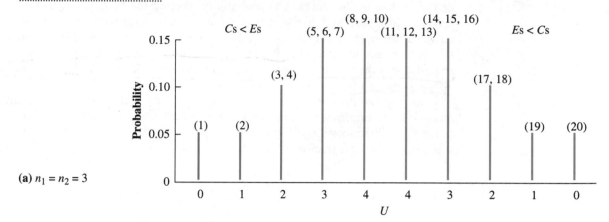

(a) $n_1 = n_2 = 3$

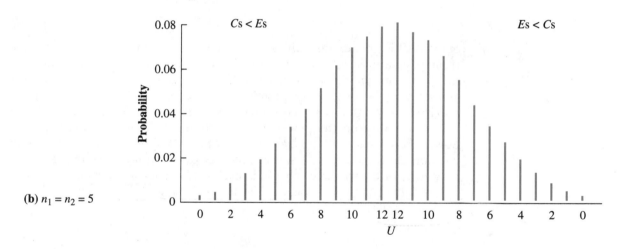

(b) $n_1 = n_2 = 5$

For the half of the distribution where *E* scores are lower than *C* scores,

$$p(U = 4) = p(\text{order } 11, 12, \text{ or } 13) = 0.05 + 0.05 + 0.05 = 0.15$$

$$p(U = 3) = p(\text{order } 14, 15, \text{ or } 16) = 0.05 + 0.05 + 0.05 = 0.15$$

$$p(U = 2) = p(\text{order } 17 \text{ or } 18) = 0.05 + 0.05 = 0.10$$

$$p(U = 1) = p(\text{order } 19) = 0.05$$

$$p(U = 0) = p(\text{order } 20) = 0.05$$

These values have been entered in column 7 of Table 11.1.*

The probability distribution of *U* for n_1 and $n_2 = 3$ has been graphed in Figure 11.2(a). The distribution is symmetrical and tails off from the center in two

---

* See Note 11.1 for an alternate method of calculating these probabilities.

directions. The direction of course depends on whether the *C* scores are higher or lower than the *E* scores. The probability distribution of *U* for n_1 and $n_2 = 5$ is shown in Figure 11.2(b). Note that it too is symmetrical and has a shape very close to the normal curve. From these two examples, we would like to make the generalizations that (1) all probability distributions of *U* are symmetrical regardless of the size of n_1 and n_2, and (2) as n_1 and n_2 increase, the distribution approaches normality.

Knowing the probability distribution of *U* for n_1 and $n_2 = 3$ when chance alone is operating, we are now in a position to evaluate H_0 for our experiment. Since H_1 is directional, i.e., it predicts the *E*s to be higher than the *C*s, we shall use a one-tailed evaluation. Our obtained *U* value was $U_{obt} = 1$.

$$p(U \le 1)_{1\,\text{tail}} = p(U = 0)_{1\,\text{tail}} + p(U = 1)_{1\,\text{tail}}$$

$$= p(\text{order } 1) + p(\text{order } 2)$$

$$= 0.05 + 0.05 = 0.10$$

Since 0.10 is greater than alpha ($\alpha = 0.05$), we retain H_0. Random sampling from populations with identical distributions is a reasonable explanation of the data. The experiment therefore fails to establish that a high-protein diet increases IQ relative to a low-protein diet. In concluding this way, we may be making a Type II error. The probability of making a Type II error (β) depends of course on the power of the experiment. Everything we said about power in Chapter 10 applies to the Mann–Whitney *U* test. Actual calculation of power for this test is complicated and beyond the scope of this text. However, with only three subjects in each group, you can guess that the power is low for all but the strongest effects. Increasing n_1 and n_2 will, of course, increase power.

 ## USING TABLES C.1–C.4

In evaluating the protein experiment data, we developed the probability distribution of *U* for $n_1 = n_2 = 3$. That distribution was the basis for our decision to retain H_0. Fortunately, we do not have to derive a probability distribution of *U* each time we do an experiment. The probability distributions of *U* with different n_1 and n_2 have already been worked out. Since it would take too many pages to give the $p(U)$ for each n_1 and n_2, summary tables have been derived. These are presented in Tables C.1–C.4 in Appendix D. For each cell there are two entries. The upper entry is the highest value of *U* for various n_1 and n_2 combinations that will allow rejection of H_0. The lower entry is the lowest value of *U′* that will allow rejection of H_0. Since both *U* and *U′* measure the same degree of separation, we shall only consider the values of *U*. Each table is for a different alpha level. Let's see how to use these tables.

PRACTICE PROBLEM 11.1

Suppose $n_1 = 5$ and $n_2 = 7$, $\alpha = 0.01_{1\,\text{tail}}$, and $U_{\text{obt}} = 2$. Can we reject H_0?

SOLUTION

Step 1: **Find the appropriate U table.** Since $\alpha = 0.01_{1\,\text{tail}}$, Table C.2 applies.

Step 2: **Locate the U value at the intersection of n_1 and n_2.** Since $n_1 = 5$ and $n_2 = 7$,

$$U = 3$$

This value is the highest value of U that will allow rejection of H_0 at the alpha level heading the table. Thus, a U value of 3 or less will allow rejection of H_0 for an alpha level of $0.01_{1\,\text{tail}}$. This value has been derived by determining the probability distribution of U for $n_1 = 5$ and $n_2 = 7$. A U value of 3 or less occurs due to chance less than 1 time in 100.

Step 3: **Draw conclusion.** Since $U_{\text{obt}} < 3$, we reject H_0.

Let's do another problem.

PRACTICE PROBLEM 11.2

Suppose $n_1 = 8$ and $n_2 = 10$. Alpha has been set at $0.05_{2\,\text{tail}}$. $U_{\text{obt}} = 15$. Can we reject H_0?

SOLUTION

Step 1: **Find the appropriate U table.** Since $\alpha = 0.05_{2\,\text{tail}}$, Table C.3 applies.

Step 2: **Locate the U value at the intersection of n_1 and n_2.** Since $n_1 = 8$ and $n_2 = 10$,

$$U = 17$$

Step 3: **Conclude:** Reject H_0 because $U_{\text{obt}} < 17$.

Now let's evaluate the protein and IQ data. In this experiment, $n_1 = n_2 = 3$. Alpha has been set at $0.05_{1 \text{ tail}}$, and $U_{obt} = 1$. Can we reject H_0?

SOLUTION

Step 1: **Find the appropriate U table.** Since $\alpha = 0.05_{1 \text{ tail}}$, Table C.4 applies.

Step 2: **Locate the U value at the intersection of n_1 and n_2.** Since $n_1 = n_2 = 3$,

$$U = 0$$

Step 3: **Conclude:** Retain H_0 because $U_{obt} > 0$.

SUMMARY OF PROTEIN–IQ DATA ANALYSIS

In explaining the use of an independent groups design and the analysis of the data with the Mann–Whitney U test, we necessarily digressed at various stages in the analysis. We would now like to present the total analysis without digressions.

EXAMPLE

The Protein–IQ Experiment Revisited

You believe a high-protein diet during infancy will increase intellectual functioning in children in your geographical area. You decide to investigate this hypothesis using the independent groups design. Alpha is set at the beginning of the experiment to $0.05_{1 \text{ tail}}$. Six children are randomly sampled from all children 1 year old living in your geographical area, and then further randomly divided into two groups of three children each. The control group is fed a low-protein diet for 3 years, and the experimental group gets a high-protein diet for the same time period. At the end of the 3 years, each child is given an IQ test. The following are the results:

Control Group 1	Experimental Group 2
84	94
88	101
98	105

1. What is the directional alternative hypothesis?
2. What is the null hypothesis?
3. What can we conclude? Use $\alpha = 0.05_{1 \text{ tail}}$.
4. What type error may be involved?
5. To what population does this conclusion apply?

SOLUTION

1. Directional alternative hypothesis: The directional alternative hypothesis states that a high-protein diet eaten during infancy will increase intellectual functioning relative to a low-protein diet.

2. Null hypothesis: The null hypothesis states that a high-protein as compared to a low-protein diet eaten during infancy does not increase intellectual functioning.

3. Conclusion, using $\alpha = 0.05_{1\ \text{tail}}$:

Step 1: Calculate U_{obt} for the data:

a. Combine the scores, rank-order them, and assign them each a rank score, using 1 for the lowest score:

Original score	84	88	94	98	101	105
Rank	1	2	3	4	5	6

b. Sum the ranks for each group; i.e., determine R_1 and R_2:

Control Group 1		Experimental Group 2	
Original Score	*Rank*	*Original Score*	*Rank*
84	1	94	3
88	2	101	5
98	4	105	6
	$R_1 = 7$		$R_2 = 14$
	$n_1 = 3$		$n_2 = 3$

c. Solve the equations for U_{obt} and U'_{obt}:

$$U_{\text{obt}} = n_1 n_2 + \frac{n_1(n_1 + 1)}{2} - R_1 \qquad\qquad U_{\text{obt}} = n_1 n_2 + \frac{n_2(n_2 + 1)}{2} - R_2$$

$$= 3(3) + \frac{3(4)}{2} - 7 \qquad\qquad\qquad = 3(3) + \frac{3(4)}{2} - 14$$

$$= 9 + 6 - 7 = 8 \qquad\qquad\qquad\qquad = 9 + 6 - 14 = 1$$

Thus,

$$U_{\text{obt}} = 1$$

$$U'_{\text{obt}} = 8$$

Step 2: Evaluate U_{obt}. With $\alpha = 0.05_{1\ \text{tail}}$, Table C.4 is appropriate. With $n_1 = n_2 = 3$, the U value needed to reject H_0 is $U = 0$. Since $U_{\text{obt}} > 0$, we retain H_0. The null hypothesis is a reasonable explanation of the data. Since we have not rejected the null hypothesis, we cannot accept H_1.

4. Type error: We may have made a Type II error. This high-protein diet may be responsible for the generally higher IQ scores in the experimental group, but due to relatively low power, we have not been able to reject H_0.

5. Population: This conclusion applies to the 1-year-old children living in your geographical area at the time of the experiment.

PRACTICE PROBLEM 11.3

Let's increase the N in the previous experiment to increase power and analyze the results. The problem is the same as before except $n_1 = 9$ and $n_2 = 8$. The following results are obtained:

Control Group 1	Experimental Group 2
102	110
104	115
105	117
107	122
108	125
111	130
113	135
118	140
120	

a. What is the directional alternative hypothesis?
b. What is the null hypothesis?
c. What can we conclude? Use $\alpha = 0.05_{1\,tail}$.
d. What type error may be involved?
e. To what population does this conclusion apply?

SOLUTION

a. Directional alternative hypothesis: same as before.
b. Null hypothesis: same as before.
c. Conclusion, using $\alpha = 0.05_{1\,tail}$:

Step 1: **Calculate U_{obt} for the data:**
   a. Combine the scores, rank-order them, and assign each a rank score, using 1 for the lowest score:

Original score	102	104	105	107	108	110	111	113	115
Rank	1	2	3	4	5	6	7	8	9

Original score	117	118	120	122	125	130	135	140
Rank	10	11	12	13	14	15	16	17

b. Sum the ranks for each group; i.e., determine R_1 and R_2:

Control Group 1		Experimental Group 2	
Original Score	*Rank*	*Original Score*	*Rank*
102	1	110	6
104	2	115	9
105	3	117	10
107	4	122	13
108	5	125	14
111	7	130	15
113	8	135	16
118	11	140	17
120	12		$R_2 = 100$
	$R_1 = 53$		$n_2 = 8$
	$n_1 = 9$		

c. Solve the equations for U_{obt} and U'_{obt}:

$$U_{obt} = n_1 n_2 + \frac{n_1(n_1 + 1)}{2} - R_1 \qquad U_{obt} = n_1 n_2 + \frac{n_2(n_2 + 1)}{2} - R_2$$

$$= 9(8) + \frac{9(10)}{2} - 53 \qquad\qquad = 9(8) + \frac{8(9)}{2} - 100$$

$$= 72 + 45 - 53 \qquad\qquad\qquad = 72 + 36 - 100$$

$$= 64 \qquad\qquad\qquad\qquad\qquad = 8$$

Therefore,

$$U_{obt} = 8$$

$$U'_{obt} = 64$$

Step 2: **Evaluate U_{obt}.** With $\alpha = 0.05_{1\,tail}$, Table C.4 is appropriate. With $n_1 = 9$ and $n_2 = 8$, the U value needed to reject H_0 is $U = 18$. Since $U_{obt} < 18$, we reject H_0. Increasing the power of the experiment allowed H_0 to be rejected. We can now accept H_1; that is, a high-protein diet eaten during infancy increases intellectual functioning relative to a low-protein diet.

d. Type error: We may have made a Type I error, rejecting H_0 when it is true.

e. Population: same as before.

TIED RANKS

Thus far, none of the problems involved two or more scores of the same value. When this occurs, U_{obt} and U'_{obt} are still computed with the equations we've just been using. However, ranking the scores is a little more complicated, because of the ties. We've already shown how to do the ranking when tied scores are involved when we discussed the Spearman rho correlation coefficient (p. 117). To review, tied scores are handled by assigning them the average of the tied ranks. For example, consider the following two sets of scores:

Group 1	Group 2
12	11
14	12
15	16
17	17
18	17
	20

To rank-order the combined scores, we proceed as follows. First, the scores are arranged in ascending order. Thus,

Raw score	11	12	12	14	15	16	17	17	17	18	20
Rank	1	2.5	2.5	4	5	6	8	8	8	10	11

Next, we assign each raw score its rank beginning with 1 for the lowest score. This has been shown previously. Note that the two raw scores of 12 are tied at the ranks of 2 and 3. They are assigned the average of these tied ranks. Thus, they each get a rank of 2.5 [(2 + 3)/2 = 2.5]. Since we have used up the ranks of 2 and 3, the next score gets a rank of 4. The raw scores of 17 are tied at the ranks of 7, 8, and 9. Therefore, they receive the rank of 8, which is the average of 7, 8, and 9 [(7 + 8 + 9)/3 = 8]. Note that the next rank is 10 (not 9) because we've already used ranks 7, 8, and 9 in computing the average. If the ranking is done correctly, unless there are tied ranks at the end, the last raw score should have a rank equal to N. In this case, $N = 11$ and so does the rank of the last score. Once the ranks have been assigned, U_{obt} and U'_{obt} are calculated in the usual way.

PRACTICAL CONSIDERATIONS IN USING THE MANN–WHITNEY U TEST

The Mann–Whitney U test is used with an independent groups design. It does not depend on the shape of the population of scores. It is therefore appropriate for testing the hypothesis that both sets of scores are random samples from identical populations regardless of the shape of the populations. It does, however, require that the data be at least ordinal. You must be able to rank-order the scores. This is a fairly powerful, practical test that is often used when the data are only ordinal. It is also used as a substitute for Student's t test when the assumptions of that test are not met. However, since it only uses the ordinal property of the scores, it is not as powerful as the t test, which uses the interval property of the scores.

SUMMARY

In this chapter, we have discussed the Mann–Whitney U test and the independent groups design. In the independent groups design, subjects are randomly selected from the population and randomly assigned to conditions. In the most basic form of the design, there are only two conditions: the experimental and control conditions. Each subject serves in only one condition, and there is no basis for pairing of scores. The scores of each sample are analyzed separately. If chance alone is responsible for differences between sample scores, both sets of sample scores can be considered random samples from the same population of scores.

The Mann–Whitney U test analyzes the degree of separation between the samples. The less the separation, the more reasonable chance is as the underlying explanation. For any analysis, there are two values indicating the degree of separation. They both indicate the same degree of separation. The lower value is arbitrarily called U_{obt}, and the higher value is called U'_{obt}. The lower the U_{obt} value, the greater the separation. The higher the U'_{obt} value, the greater the separation. Tables C.1–C.4 give all the values of U_{obt} and U'_{obt} that allow rejection of the null hypothesis for various combinations of n_1 and n_2. As with the repeated measures design, the null hypothesis is evaluated directly. If it is rejected, H_1 is accepted. If not, we can't accept H_1. With the Mann–Whitney U test, we calculate the U_{obt} or U'_{obt} value of the data. Since they both indicate the same degree of separation, we just used the U_{obt} value. If U_{obt} is equal to or less than the tabled U value for rejecting the null hypothesis, we reject H_0. If not, we retain H_0. The Mann–Whitney U test is appropriate for an independent groups design where the data are at least ordinal in scaling. It is a powerful test, often used in place of Student's t test when the data do not meet the assumptions of the t test.

IMPORTANT TERMS

Control group (p. 242)
Degree of separation (p. 244)
Experimental group (p. 242)
Independent groups design (p. 241)

Mann–Whitney U test (p. 243)
Population of scores (p. 242)
Rank order (p. 247)

Tied ranks (p. 258)
U_{obt} (p. 244)
U'_{obt} (p. 244)

QUESTIONS AND PROBLEMS

1. Briefly define or explain each of the terms in the "Important Terms" section.
2. Briefly contrast the repeated measures and independent groups design.
3. What are the practical considerations in using the Mann–Whitney U test?
4. Briefly describe the process of hypothesis testing using the Mann–Whitney U test.
5. In an independent groups design, if the alternative hypothesis is nondirectional, what does the null hypothesis say with respect to the two samples?
6. An independent groups experiment is conducted to see whether treatment A differs from treatment B. Eight subjects are randomly assigned to treatment A and 7 to treatment B. The following data are collected:

Treatment A	Treatment B
30	14
35	8
34	25
40	16
19	26
32	28
21	9
23	

a. What is the alternative hypothesis? Use a non-directional hypothesis.
b. What is the null hypothesis?
c. What is your conclusion? Use $\alpha = 0.05_{2\,tail}$.

7. A social scientist believes that university theology professors are more conservative in political orientation than their colleagues in psychology. A random sample of 8 professors from the theology department and 12 professors from the psychology department at a local university are given a 50-point questionnaire that measures the degree of political conservatism. The following scores were obtained. Higher scores indicate greater conservatism.
   a. What is the alternative hypothesis? In this case, assume a nondirectional hypothesis is appropriate because there are insufficient theoretical and empirical bases to warrant a directional hypothesis.
   b. What is the null hypothesis?
   c. What is your conclusion? Use $\alpha = 0.05_{2\,tail}$.
   d. What error may you be making by concluding as you did in part **c**?
   e. To what population do the results apply?

Theology Professors	Psychology Professors
36	13
42	25
22	40
48	29
31	10
35	26
47	43
38	17
	12
	32
	27
	32

8. Someone has told you that men are better in abstract reasoning than women. You are skeptical, so you decide to test this idea using a nondirectional hypothesis. You randomly select eight adult men and eight adult women living in your hometown and administer an abstract reasoning test. A higher score reflects better abstract reasoning abilities. You obtain the following scores:

Men	Women
70	81
86	80
60	50
92	95
82	93
65	85
74	90
94	75

a. What is the alternative hypothesis? Assume a nondirectional hypothesis is appropriate.
b. What is the null hypothesis?
c. Using $\alpha = 0.05_{2\,tail}$, what do you conclude?
d. What error may you have made by concluding as you did in part **c**?
e. To what population do your results apply?

9. An ornithologist thinks that injections of follicle-stimulating hormone (FSH) increase the singing rate of his captive male cotingas (birds). To test this hypothesis, he randomly selects 20 singing cotingas and divides them into two groups of 10 birds each. The first group receives injections of FSH and the second injections of saline solution, as a control for the trauma of receiving an injection. He then records the singing rate (in songs per hour) for both groups. The results are given in the following table. Note that two of the FSH birds escaped during injection and were not replaced.

Saline	FSH
17	10
31	29
14	37
12	41
29	16
23	45
7	34
19	57
28	
3	

a. What is the alternative hypothesis? Use a directional alternative hypothesis.
b. What is the null hypothesis?
c. Using $\alpha = 0.05_{1\ tail}$, what is your conclusion?

10. A psychologist is interested in determining whether left-handed and right-handed people differ in spatial ability. She randomly selects 10 left-handers and 10 right-handers from the students enrolled in the university where she works and administers a test that measures spatial ability. The following are the scores (a higher score indicates better spatial ability). Note that one of the subjects did not show up for the testing.

Left-handers	Right-handers
87	47
94	68
56	92
74	73
98	71
83	82
92	55
84	61
76	75
	85

a. What is the alternative hypothesis? Use a nondirectional hypothesis.
b. What is the null hypothesis?
c. Using $\alpha = 0.05_{2\ tail}$, what do you conclude?

11. A university counselor believes that hypnosis is more effective than the standard treatment given to students who have high test anxiety. To test his belief, he randomly divides 22 students with high test anxiety into two groups. One of the groups receives the hypnosis treatment and the other group the standard treatment. When the treatments are concluded, each student is given a test anxiety questionnaire. High scores on the questionnaire indicate high anxiety. The following are the results:

Hypnosis Treatment	Standard Treatment
20	42
21	35
33	30
40	53
24	57
43	26
48	37
31	30
22	51
44	62
30	59

a. What is the alternative hypothesis? Assume there is sufficient basis for a directional hypothesis.
b. What is the null hypothesis?
c. Using $\alpha = 0.05_{1\ tail}$, what do you conclude?

NOTES

11.1 These probabilities could also have been calculated from multiplication and addition rules. For example, to determine $p(U = 0)$ such that all the C scores are lower than E scores, we need to determine $p(CCCEEE)$. Assume we have a bag containing three C scores and three E scores and we are randomly sampling one score at a time without replacement. What is the probability we shall get the order $CCCEEE$? The solution is as follows. By using the multiplication rule,

$$p(U = 0) = p(CCCEEE)$$

$$= \tfrac{3}{6} \cdot \tfrac{2}{5} \cdot \tfrac{1}{4} \cdot \tfrac{3}{3} \cdot \tfrac{2}{2} \cdot \tfrac{1}{1} = 0.05$$

Note that each order yields the same probability value (0.05). Thus, to find the other probabilities, all we need do is multiply 0.05 by the number of orders yielding the particular U value. For example, find $p(U = 2)$ with Cs lower than Es, since there are two orders that yield $U = 2$ in this direction:

$$p(U = 2) = 2(0.05) = 0.10$$

12 | Sampling Distributions, Sampling Distribution of the Mean, the Normal Deviate (z) Test

INTRODUCTION

In Chapters 10 and 11, we have seen how to use the scientific method to investigate hypotheses. We have introduced the replicated measures and the independent groups designs and discussed how to analyze the resulting data. At the heart of the analysis is the ability to answer the question, *What is the probability of getting the obtained result or results even more extreme if chance alone is responsible for the differences between the experimental and control scores?*

Although it hasn't been emphasized, the answer to this question involves two steps: (1) calculating the appropriate statistic and (2) evaluating the statistic based on its sampling distribution. In this chapter, we shall more formally discuss the topic of a statistic and its sampling distribution. Then we shall begin our analysis of single sample experiments, using the mean of the sample as a statistic. This involves the sampling distribution of the mean and the normal deviate (z) test.

SAMPLING DISTRIBUTIONS

What is a sampling distribution?

 DEFINITION

- *The sampling distribution of a statistic gives* (1) *all the values that the statistic can take and* (2) *the probability of getting each value under the assumption that it resulted from chance alone.*

In the replicated measures design, we used the sign test to analyze the data. The statistic calculated was the *number of pluses* in the sample of N difference scores. In one version of the "marijuana and appetite" experiment, we obtained nine pluses and one minus. This result was evaluated by using the binomial distribution. The binomial distribution with $P = 0.50$ lists all the possible values of the statistic, the number of pluses, along with the probability of getting each value under the assumption that chance alone produced it. *The binomial distribution with $P = 0.50$ is the sampling distribution of the statistic used in the sign test.* Note that there is a different sampling distribution for each sample size (N).

With the independent groups design, we used the Mann–Whitney U test to analyze the data. The statistic calculated was U or U'. This statistic measures the degree of separation between the two groups of scores. In the "protein and IQ" experiment, we wanted to know the probability of getting $U \leq 1$ with $n_1 = n_2 = 3$, if chance alone was responsible. We answered this question by determining all the possible values that U and U' could take, along with the probability of getting each value if chance alone was operating (see Table 11.1, p. 250). *In fact, we constructed the sampling distribution of U and U' for $n_1 = n_2 = 3$.* The critical values of U and U' reported in Tables C.1–C.4 in Appendix D are based on the sampling distribution of U and U' for the n_1 and n_2 given. Again, note that there is a different sampling distribution for each n_1 and n_2 combination.

By generalizing from these two examples, it can be seen that data analysis basically involves two steps:

1. Calculating the appropriate statistic—for example, number of pluses and minuses for the sign test or U or U' for the Mann–Whitney U test
2. Evaluating the statistic based on its sampling distribution

If the probability of getting the obtained value of the statistic or any value more extreme is equal to or less than the alpha level, we reject H_0 and accept H_1. If not, we retain H_0. If we reject H_0 and it is true, we've made a Type I error. If we retain H_0 and it's false, we've made a Type II error. This process applies to all experiments involving hypothesis testing. *What changes from experiment to experiment is the statistic used and its accompanying sampling distribution.* Once you understand this concept, you can appreciate that a large part of teaching inferential statistics is devoted to presenting the most often used statistics, their sampling distributions, and the conditions under which each statistic is appropriately used.

Generating Sampling Distributions   We have defined a sampling distribution as a probability distribution of all the possible values of a statistic under the assumption that chance alone is operating. One way of deriving sampling distributions is from basic probability considerations. We used this approach in generating the binomial distribution and the sampling distribution of U. Sampling distributions can also be derived from an empirical sampling approach. In this approach, we have an actual or theoretical set of population scores that exists if the independent variable has no effect. We derive the sampling distribution of the statistic by

1. Determining all the possible different samples of size N that can be formed from the population
2. Calculating the statistic for each of the samples
3. Calculating the probability of getting each value of the statistic if chance alone is operating

To illustrate the sampling approach, let's suppose we are conducting an experiment with a sample size $N = 2$, using the sign test for analysis. We can imagine a theoretical set of scores that would result if the experiment were done on the entire population and the independent variable had no effect. This population set of scores is called the *null-hypothesis population.*

DEFINITION

• *The* **null-hypothesis population** *is an actual or theoretical set of population scores that would result if the experiment were done on the entire population and the independent variable had no effect. It is called the null-hypothesis population because it is used to test the validity of the null hypothesis.*

In the case of the sign test, if the independent variable had no effect, the null-hypothesis population would have an equal number of pluses and minuses ($P = Q = 0.50$).

For computational ease in generating the sampling distribution, let's assume there are only six scores in the population: three pluses and three minuses. To derive the sampling distribution of "the number of pluses" with $N = 2$, we must first determine all the different samples of size N that can be formed from the population. Sampling is one at a time with replacement. Figure 12.1 shows the

FIGURE 12.1

All of the possible samples of size 2 that can be drawn from a population of 3 pluses and 3 minuses; sampling is one at a time with replacement

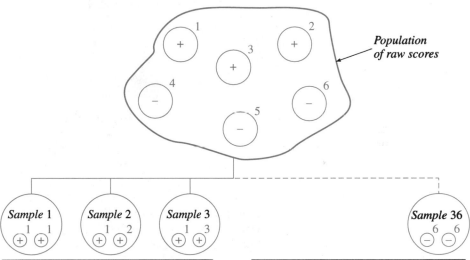

Sample composition				Sample composition			
Sample number (1)	*Element numbers* (2)	*Actual scores* (3)	*Statistic, no. of pluses* (4)	*Sample number* (1)	*Element numbers* (2)	*Actual scores* (3)	*Statistic, no. of pluses* (4)
1	1, 1	++	2+	19	4, 1	−+	1+
2	1, 2	++	2+	20	4, 2	−+	1+
3	1, 3	++	2+	21	4, 3	−+	1+
4	1, 4	+−	1+	22	4, 4	−−	0+
5	1, 5	+−	1+	23	4, 5	−−	0+
6	1, 6	+−	1+	24	4, 6	−−	0+
7	2, 1	++	2+	25	5, 1	−+	1+
8	2, 2	++	2+	26	5, 2	−+	1+
9	2, 3	++	2+	27	5, 3	−+	1+
10	2, 4	+−	1+	28	5, 4	−−	0+
11	2, 5	+−	1+	29	5, 5	−−	0+
12	2, 6	+−	1+	30	5, 6	−−	0+
13	3, 1	++	2+	31	6, 1	−+	1+
14	3, 2	++	2+	32	6, 2	−+	1+
15	3, 3	++	2+	33	6, 3	−+	1+
16	3, 4	+−	1+	34	6, 4	−−	0+
17	3, 5	+−	1+	35	6, 5	−−	0+
18	3, 6	+−	1+	36	6, 6	−−	0+

FIGURE 12.2

Sampling distribution of
"number of pluses" with
$N = 2$ and $P = 0.50$

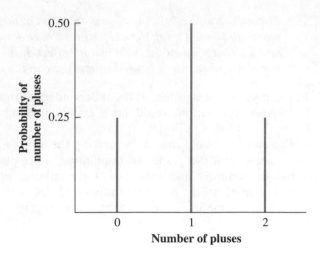

population and, schematically, the different samples of size 2 that can be drawn from it. It turns out that there are 36 different samples of size 2 possible. These are listed in the table of Figure 12.1, column 2. Next, we must calculate the value of the statistic for each sample. This information is presented in the table of Figure 12.1, columns 3 and 4. Note that of the 36 different samples possible, 9 have two pluses, 18 have one plus, and 9 have no pluses. The last step is to calculate the probability of getting each value of the statistic. If chance alone is operating, each sample is equally likely. Thus,

$$p(2 \text{ pluses}) = \tfrac{9}{36} = 0.2500$$

$$p(1 \text{ plus}) = \tfrac{18}{36} = 0.5000$$

$$p(0 \text{ pluses}) = \tfrac{9}{36} = 0.2500$$

We have now derived the sampling distribution for $N = 2$ of the statistic "number of pluses." The distribution is plotted in Figure 12.2. In this example, we used a population in which there were only six scores. The identical sampling distribution would have resulted (even though there would be many more "different" samples) had we used a larger population so long as the number of pluses equaled the number of minuses and the sample size equaled 2. Note that this is the same sampling distribution we arrived at through basic probability considerations when we were discussing the binomial distribution with $N = 2$ (see Figure 12.3 for a comparison). This time, however, we generated it by sampling from the null hypothesis population. The sampling distribution of a statistic is often defined in terms of this process. Viewed in this manner, we obtain the following definition.

DEFINITION

- *A **sampling distribution** gives all the values a statistic can take, along with the probability of getting each value if sampling is random from the null-hypothesis population.*

FIGURE 12.3

Comparison of empirical sampling approach and *a priori* approach for generating sampling distributions

Empirical Sampling Approach				A Priori Approach		
Draw 1	Draw 2	Number of Ways		Coin 1	Coin 2	Number of Ways
+	+	9		H	H	1
+ −	− + }	18		H T	T H }	2
−	−	9		T	T	1
		— 36				— 4

$$p(2+) = \frac{9}{36} = 0.2500 \qquad p(2\text{H}) = \frac{1}{4} = 0.2500$$

$$p(1+) = \frac{18}{36} = 0.5000 \qquad p(1\text{H}) = \frac{2}{4} = 0.5000$$

$$p(0+) = \frac{9}{36} = 0.2500 \qquad p(0\text{H}) = \frac{1}{4} = 0.2500$$

THE NORMAL DEVIATE (z) TEST

Although much of the foregoing has been abstract and seemingly impractical, it is necessary to understand the sampling distributions underlying many of the statistical tests that follow. One such test, the normal deviate (z) test, is a test that is used when we know the parameters of the null-hypothesis population. The z test uses the mean of the sample as a basic statistic. Let's consider an experiment where the z test is appropriate.

AN EXPERIMENT

Evaluating the Reading Program at the Local Public Schools

Assume you are superintendent of public schools for the city in which you live. Recently, local citizens have been concerned that the reading program in the public schools may be an inferior one. Since this is a serious issue, you decide to conduct an experiment to investigate the matter. You set $\alpha = 0.05_{1\,\text{tail}}$ for making your decision. You begin by comparing the reading level of current high school seniors with established norms. The norms are based on scores from a reading proficiency test administered nationally to a large number of high school seniors. The scores of this population are normally distributed with $\mu = 75$ and $\sigma = 16$. For your experiment, you administer the reading test to 100 randomly selected high school seniors in your city. The obtained mean of the sample ($\overline{X}_{\text{obt}}$) = 72. What is your conclusion?

There is no doubt that the sample mean of 72 is lower than the national population mean of 75. Is it significantly lower, however? If chance alone is at work, then we can consider the 100 sample scores to be a random sample from a population with $\mu = 75$. The null hypothesis for this experiment asserts that such is the case. What is the probability of getting a mean score as low as or even lower than 72 if the 100 scores are a random sample from a normally distributed population having a mean of 75 and standard deviation of 16? If the probability is equal to or lower than alpha, we reject H_0 and accept H_1. If not, we retain H_0. It is clear that the statistic we are using is the *mean* of the sample. To determine the appropriate probability, we must know the *sampling distribution of the mean*.

In the following section we shall discuss the sampling distribution of the mean. For the time being, set aside the "Super" and his problem. We shall return to him soon enough. For now, it is sufficient to realize that we are going to use the mean of a sample to evaluate H_0, and to do that, we must know the sampling distribution of the mean.

Sampling Distribution of the Mean

Applying the definition of the sampling distribution of a statistic to the mean, we obtain the following:

DEFINITION

- *The sampling distribution of the mean gives all the values the mean can take, along with the probability of getting each value if sampling is random from the null-hypothesis population.*

The sampling distribution of the mean can be determined empirically and theoretically, the latter through use of a theorem called the Central Limit Theorem. The theoretical derivation is complex and beyond the level of this textbook. Therefore, for pedagogical reasons, we prefer to present the empirical approach. When we follow this approach, we can determine the sampling distribution of the mean by actually taking a specific population of raw scores having a mean μ and standard deviation σ and (1) drawing all possible different samples of a fixed size N, (2) calculating the mean of each sample, and (3) calculating the probability of getting each mean value if chance alone were operating. This process is shown in Figure 12.4. After performing these three steps, we would have derived the sampling distribution of the mean for samples of size N taken from a specific population with mean μ and standard deviation σ. This sampling distribution of the mean would give all the values that the mean could take for samples of size N, along with the probability of getting each value if sampling is random from the specified population. By repeating the three-step process for populations of different score values and by systematically varying N, it can be determined that the sampling distribution of the mean has the following general characteristics. For samples of any size N, the sampling distribution of the mean

1. Is a distribution of scores, each score of which is a sample mean. This distribution has a mean and a standard deviation. The distribution is shown in the bottom part of Figure 12.4. You should note that this is a *population* set of scores even though the scores are based on samples, because the distribution contains the *complete* set of sample means. We shall symbolize the mean of the distribution as $\mu_{\bar{X}}$ and the standard deviation as $\sigma_{\bar{X}}$. Thus,

$\mu_{\bar{X}}$ = Mean of the sampling distribution of the mean

$\sigma_{\bar{X}}$ = Standard deviation of the sampling distribution of the mean

FIGURE 12.4

Generating the sampling distribution of the mean for samples of size N taken from a population of raw scores

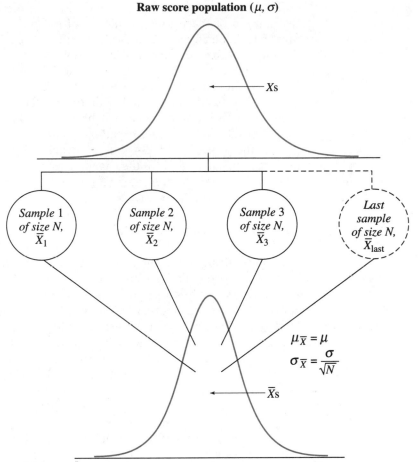

Raw score population (μ, σ)

Xs

Sample 1 of size N, $\bar{X}_1$

Sample 2 of size N, $\bar{X}_2$

Sample 3 of size N, $\bar{X}_3$

Last sample of size N, $\bar{X}_{last}$

$\mu_{\bar{X}} = \mu$

$\sigma_{\bar{X}} = \dfrac{\sigma}{\sqrt{N}}$

$\bar{X}s$

Sampling distribution of the mean for samples of size N

$\sigma_{\bar{X}}$ is also called the *standard error of the mean* because each sample mean can be considered an estimate of the mean of the raw-score population. Variability between sample means then occurs due to errors in estimation—hence the phrase standard *error* of the mean for $\sigma_{\bar{X}}$.

2. Has a mean equal to the mean of the raw-score population. In equation form,

$$\mu_{\bar{X}} = \mu$$

3. Has a standard deviation equal to the standard deviation of the raw-score population divided by $\sqrt{N}$. In equation form,

$$\sigma_{\bar{X}} = \frac{\sigma}{\sqrt{N}}$$

4. Is normally shaped, depending on the shape of the raw-score population and on sample size.

The first characteristic is rather obvious. It merely states that the sampling distribution of the mean is made up of sample mean scores. As such, it, too, must have a mean and a standard deviation. The second characteristic says that the mean of the sampling distribution of the mean is equal to the mean of the raw scores ($\mu_{\bar{X}} = \mu$). We can gain some insight into this relationship by recognizing that each sample mean is an estimate of the mean of the raw-score population. Each will differ from the mean of the raw-score population due to chance. Sometimes the sample mean will be greater than the population mean, and sometimes it will be smaller because of chance factors. As we take more sample means, the average of these sample means will get closer to the mean of the raw-score population, since the chance factors will cancel. Finally, when we have all of the possible different sample means, their average will equal the mean of the raw-score population ($\mu_{\bar{X}} = \mu$).

The third characteristic says that the standard deviation of the sampling distribution of the mean is equal to the standard deviation of the raw-score population divided by $\sqrt{N}$ ($\sigma_{\bar{X}} = \sigma/\sqrt{N}$). This says that the standard deviation of the sampling distribution of the mean varies directly with the standard deviation of the raw-score population and inversely with $\sqrt{N}$. It is fairly obvious why $\sigma_{\bar{X}}$ should vary directly with σ. If the scores in the population are more variable, σ goes up, and so will the variability between the means based on these scores. Understanding why $\sigma_{\bar{X}}$ varies inversely with $\sqrt{N}$ is a little more difficult. Recognizing that each sample mean is an estimate of the mean of the raw-score population is the key. As N (the number of scores in each sample) goes up, each sample mean becomes a more accurate estimate of μ. Since the sample means are more accurate, they will vary less from sample to sample, causing the variance ($\sigma_{\bar{X}}^2$) of the sample means to decrease. Thus, $\sigma_{\bar{X}}^2$ varies inversely with N. Since $\sigma_{\bar{X}} = \sqrt{\sigma_{\bar{X}}^2}$, then $\sigma_{\bar{X}}$ varies inversely with $\sqrt{N}$. We would like to further point out that, since the standard deviation of the sampling distribution of the mean ($\sigma_{\bar{X}}$) changes with sample size, there is a different sampling distribution of the mean for each different sample size. This seems reasonable, because if the sample size changes, then the scores in each sample change and, consequently, so do the sample means. Thus, the sampling distribution of the mean for samples of size 10 should be different from the sampling distribution of the mean for samples of size 20 and so forth.

Regarding the fourth point, there are two factors that determine the shape of the sampling distribution of the mean: (1) the shape of the population raw scores and (2) the sample size (N). Concerning the first factor, if the population of raw scores is normally distributed, the sampling distribution of the mean will also be normally distributed, regardless of sample size. However, if the population of raw scores is not normally distributed, the shape of the sampling distribution depends on the sample size. *The Central Limit Theorem tells us that, regardless of the shape of the population of raw scores, the sampling distribution of the mean approaches a normal distribution as sample size N increases.* If N is sufficiently large, the sampling distribution of the mean is approximately normal. How large must N be for the sampling distribution of the mean to be considered normal? This depends on the shape of the raw-score population. The further the raw scores deviate from normality, the larger the sample size must be for the sampling distribution of the mean to be normally shaped. If $N \geq 300$, the shape of the population of raw scores is no longer important. With this size N, regardless of the shape of the raw-score population, the sampling distribution of the mean will deviate so little from normality that, for statistical calculations, we can consider it normally distributed. Since most populations encountered in the behavioral sciences do not differ greatly from normality, if $N \geq 30$, it is usually assumed that the sampling distribution of the mean will be normally shaped.*

Although it is beyond the scope of this text to prove these characteristics, we can demonstrate them, as well as gain more understanding about the sampling distribution of the mean, by considering a population and deriving the sampling distribution of the mean for samples taken from it. To simplify computation, let's use a population with a small number of scores. For the purposes of this illustration, assume the population raw scores are 2, 3, 4, 5, and 6. The mean of the population (μ) equals 4.00, and the standard deviation (σ) equals 1.41. We want to derive the sampling distribution of the mean for samples of size 2 taken from this population. Again, assume sampling is one score at a time with replacement. The first step is to draw all possible different samples of size 2 from the population. Figure 12.5 shows the population raw scores and, schematically, the different samples of size 2 that can be drawn from it. There are 25 different samples of size 2 possible. These are listed in the table of Figure 12.5, column 2. Next, we must calculate the mean of each sample. The results are shown in column 3 of this table. It is now a simple matter to calculate the probability of getting each mean value. Thus,

$$p(\overline{X} = 2.0) = \frac{\text{Number of possible } \overline{X}\text{s of 2.0}}{\text{Total number of } \overline{X}\text{s}} = \frac{1}{25} = 0.04$$

$$p(\overline{X} = 2.5) = \tfrac{2}{25} = 0.08$$

$$p(\overline{X} = 3.0) = \tfrac{3}{25} = 0.12$$

$$p(\overline{X} = 3.5) = \tfrac{4}{25} = 0.16$$

$$p(\overline{X} = 4.0) = \tfrac{5}{25} = 0.20$$

$$p(\overline{X} = 4.5) = \tfrac{4}{25} = 0.16$$

$$p(\overline{X} = 5.0) = \tfrac{3}{25} = 0.12$$

$$p(\overline{X} = 5.5) = \tfrac{2}{25} = 0.08$$

$$p(\overline{X} = 6.0) = \tfrac{1}{25} = 0.04$$

We have now derived the sampling distribution of the mean for samples of $N = 2$ taken from a population comprising the raw scores 2, 3, 4, 5, and 6. We have determined all the mean values possible from sampling two scores from the given population, along with the probability of obtaining each mean value if sampling is random from the population. The complete sampling distribution is shown in Table 12.1.

Suppose, for some reason, we wanted to know the probability of obtaining an $\overline{X} \geq 5.5$ due to randomly sampling two scores, one at a time, with replacement, from the raw-score population. We can determine the answer by consulting the sampling distribution of the mean for $N = 2$. Why? Because this distribution contains all of the possible mean values and their probability under the assumption of random sampling. Thus,

$$p(\overline{X} \geq 5.5) = 0.08 + 0.04 = 0.12$$

TABLE 12.1

Sampling Distribution of the Mean with $N = 2$ and Population Scores of 2, 3, 4, 5, and 6

$\overline{X}$	$p(\overline{X})$
2.0	0.04
2.5	0.08
3.0	0.12
3.5	0.16
4.0	0.20
4.5	0.16
5.0	0.12
5.5	0.08
6.0	0.04

* There are some notable exceptions to this rule, e.g., reaction-time scores.

FIGURE 12.5

All of the possible samples of size 2 that can be drawn from a population comprising the raw scores 2, 3, 4, 5, and 6; sampling is one at a time with replacement

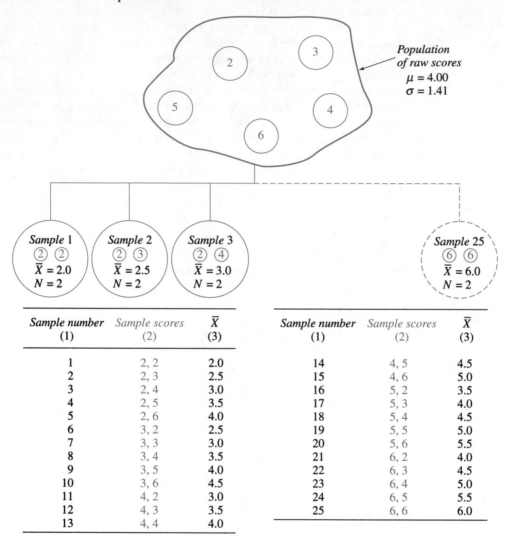

Sample number (1)	Sample scores (2)	$\overline{X}$ (3)	Sample number (1)	Sample scores (2)	$\overline{X}$ (3)
1	2, 2	2.0	14	4, 5	4.5
2	2, 3	2.5	15	4, 6	5.0
3	2, 4	3.0	16	5, 2	3.5
4	2, 5	3.5	17	5, 3	4.0
5	2, 6	4.0	18	5, 4	4.5
6	3, 2	2.5	19	5, 5	5.0
7	3, 3	3.0	20	5, 6	5.5
8	3, 4	3.5	21	6, 2	4.0
9	3, 5	4.0	22	6, 3	4.5
10	3, 6	4.5	23	6, 4	5.0
11	4, 2	3.0	24	6, 5	5.5
12	4, 3	3.5	25	6, 6	6.0
13	4, 4	4.0			

Now, let's consider the characteristics of this distribution: first, its shape. The original population of raw scores and the sampling distribution have been plotted in Figure 12.6(a) and (b). In part (c), we have plotted the sampling distribution of the mean with $N = 3$. Note that the shape of the two sampling distributions differs greatly from the population of raw scores. *Even with an N as small as 3 and a very nonnormal population of raw scores, the sampling distribution of the mean has a shape that approaches normality.* This is an illustration of what the Central Limit Theorem is telling us—namely, that as N increases, the shape of the sampling distribution of the mean approaches that of a normal distribution. Of course, if the shape of the raw-score population were normal, the shape of the sampling distribution of the mean would be too.

FIGURE 12.6
....

Population scores and the
sampling distribution of
the mean for samples of
size $N = 2$ and $N = 3$

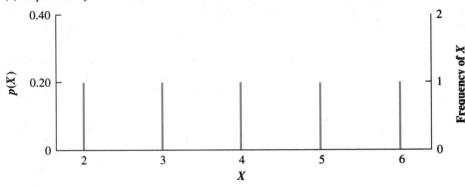

(a) *Population of raw scores*

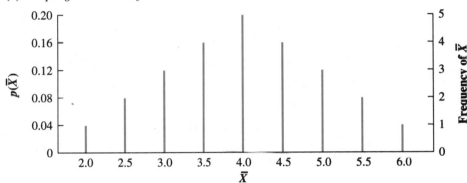

(b) *Sampling distribution of $\overline{X}$ with $N = 2$*

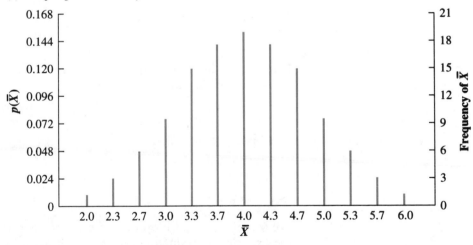

(c) *Sampling distribution of $\overline{X}$ with $N = 3$*

Next, let's demonstrate that $\mu_{\overline{X}} = \mu$:

$$\mu = \frac{\Sigma X}{\text{Number of raw scores}} \qquad\qquad \mu_{\overline{X}} = \frac{\Sigma \overline{X}}{\text{Number of mean scores}}$$

$$= \frac{20}{5} = 4.00 \qquad\qquad\qquad\qquad = \frac{100}{25} = 4.00$$

Thus,

$$\mu_{\bar{X}} = \mu$$

The mean of the raw scores is found by dividing the sum of the raw scores by the number of raw scores: $\mu = 4.00$. The mean of the sampling distribution of the mean is found by dividing the sum of the sample mean scores by the number of mean scores: $\mu_{\bar{X}} = 4.00$. Thus, $\mu_{\bar{X}} = \mu$.

Using $\sigma_{\bar{X}} = \sigma/\sqrt{N}$	Using the Sample Mean Scores
$\sigma_{\bar{X}} = \dfrac{\sigma}{\sqrt{N}}$	$\sigma_{\bar{X}} = \sqrt{\dfrac{\Sigma\,(\bar{X} - \mu_{\bar{X}})^2}{\text{Number of mean scores}}}$
$= \dfrac{1.41}{\sqrt{2}}$	$= \sqrt{\dfrac{(2.0 - 4.0)^2 + (2.5 - 4.0)^2 + \cdots + (6.0 - 4.0)^2}{25}}$
$= 1.00$	$= \sqrt{\dfrac{25}{25}} = 1.00$

Finally, we need to show that $\sigma_{\bar{X}} = \sigma/\sqrt{N}$. $\sigma_{\bar{X}}$ can be calculated in two ways: (1) from the equation $\sigma_{\bar{X}} = \sigma/\sqrt{N}$ and (2) directly from the sample mean scores themselves. Our demonstration will involve calculating $\sigma_{\bar{X}}$ in both ways, showing that they lead to the same value. The calculations are shown in the above table. Since both methods yield the same value ($\sigma_{\bar{X}} = 1.00$), we have demonstrated that

$$\sigma_{\bar{X}} = \frac{\sigma}{\sqrt{N}}$$

Note that N in the previous equation is the number of scores in each sample. Thus, we have demonstrated that

1. $\mu_{\bar{X}} = \mu$
2. $\sigma_{\bar{X}} = \sigma/\sqrt{N}$
3. The sampling distribution of the mean takes on a shape similar to normal even if the raw scores are nonnormal.

The Reading Proficiency Experiment Revisited

We are now in a position to return to the "Super" and evaluate the data from the experiment evaluating reading proficiency. Let's restate the experiment.

You are now superintendent of public schools and have conducted an experiment to investigate whether the reading proficiency of high school seniors living in your city is deficient. A random sample of 100 high school seniors from this population had a mean reading score of 72 ($\bar{X}_{\text{obt}} = 72$). National norms of reading proficiency for high school seniors show a normal distribution of scores with a mean of 75 ($\mu = 75$) and a standard deviation of 16 ($\sigma = 16$). Is it reasonable to consider the 100 scores a random sample from a normally distributed population of reading scores where $\mu = 75$ and $\sigma = 16$? Use $\alpha = 0.05_{1\,\text{tail}}$.

If we take all possible samples of size 100 from the population of normally distributed reading scores, we can determine the sampling distribution of the mean samples with $N = 100$. From what has been said before, this distribution (1) is normally shaped, (2) has a mean $\mu_{\bar{X}} = \mu = 75$, and (3) has a standard deviation $\sigma_{\bar{X}} = \sigma/\sqrt{N} = 16/\sqrt{100} = 1.6$. The two distributions are shown in Figure 12.7. Note that the sampling distribution of the mean contains all the possible mean scores from samples of size 100 drawn from the null-hypothesis population ($\mu = 75$, $\sigma = 16$). For the sake of clarity in the following exposition, we have redrawn the sampling distribution of the mean alone in Figure 12.8.

The shaded area of Figure 12.8 contains all the mean values of samples with $N = 100$ that are as low as or lower than $\bar{X}_{obt} = 72$. The proportion of shaded area to total area will tell us the probability of obtaining a sample mean equal to or less than 72, if chance alone is at work (another way of saying this is, "if the sample is a random sample from the null-hypothesis population"). Since the sampling distribution of the mean is normally shaped, we can find the proportion of the shaded area by (1) calculating the z transform (z_{obt}) for $\bar{X}_{obt} = 72$ and (2) determining the appropriate area from Table A, Appendix D, using z_{obt}.

FIGURE 12.7

Sampling distribution of the mean for samples of size $N = 100$ drawn from a population of raw scores with $\mu = 75$ and $\sigma = 16$

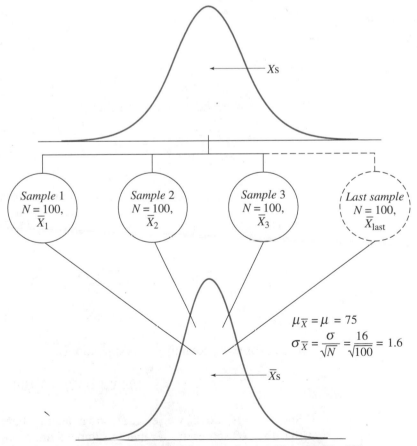

Sampling distribution of the mean for samples with $N = 100$

FIGURE 12.8

Evaluation of reading proficiency data comparing the obtained probability with the alpha level

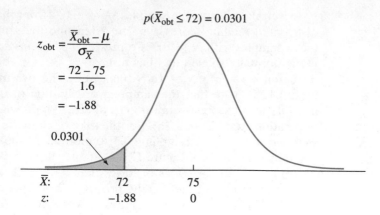

The equation for z_{obt} is very similar to the z equation in Chapter 5, except, instead of dealing with raw scores, we are dealing with mean values. The two equations are shown here:

Raw Scores	Mean Scores
$z = \dfrac{X - \mu}{\sigma}$	$z_{obt} = \dfrac{\overline{X}_{obt} - \mu_{\overline{X}}}{\sigma_{\overline{X}}}$

Since $\mu_{\overline{X}} = \mu$, the z_{obt} equation simplifies to

$$z_{obt} = \frac{\overline{X}_{obt} - \mu}{\sigma_{\overline{X}}} \qquad \textit{z transformation for } \overline{X}_{obt}$$

Calculating z_{obt} for the present experiment, we obtain

$$z_{obt} = \frac{\overline{X}_{obt} - \mu}{\sigma_{\overline{X}}}$$

$$= \frac{72 - 75}{1.6}$$

$$= -1.88 \qquad \textit{for } \overline{X}_{obt} = 72$$

From Table A, column C, in Appendix D,

$$p(\overline{X}_{obt} \leq 72) = 0.0301$$

Since $0.0301 < 0.05$, we reject H_0 and conclude that it is unreasonable to assume that the 100 scores are a random sample from a population where $\mu = 75$. The reading proficiency of high school seniors in your city appears to be deficient.

Alternate Solution Using z_{obt} and the Critical Region for Rejection of H_0

The results of this experiment can be analyzed in another way. This method is actually the preferred method, because it is simpler and it sets the pattern for the inference tests to follow. However, it builds upon the previous method and therefore couldn't be presented until now. To use this method, we must first define some terms.

DEFINITION

....

- *The critical region for rejection of the null hypothesis is the area under the curve that contains all the values of the statistic that allow rejection of the null hypothesis.*

- *The critical value of a statistic is the value of the statistic that bounds the critical region.*

To analyze the data using the alternate method, all we need do is calculate z_{obt}, determine the critical value of z (z_{crit}), and assess whether z_{obt} falls within the critical region for rejection of H_0. We already know how to calculate z_{obt}.

The critical region for rejection of H_0 is determined by the alpha level. For example, if $\alpha = 0.05_{1\,tail}$ in the direction predicting a negative z_{obt} value, as in the previous example, then the critical region for rejection of H_0 is the area under the left tail of the curve that equals 0.0500. We find z_{crit} for this area by using Table A in a reverse manner. Referring to Table A and skimming column C until we locate 0.0500, we can determine the z value that corresponds to 0.0500. It turns out that 0.0500 falls midway between the z scores of 1.64 and 1.65. Therefore, the z value corresponding to 0.0500 is 1.645. Since we are dealing with the left tail of the distribution,

$$z_{crit} = -1.645$$

This score defines the critical region for rejection of H_0 and, hence, is called z_{crit}. If z_{obt} falls in the critical region for rejection, we will reject H_0. These relationships are shown in Figure 12.9(a). If $\alpha = 0.05_{1\,tail}$ in the direction predicting a positive z_{obt} value, then

$$z_{crit} = 1.645$$

This is shown in Figure 12.9(b). If $\alpha = 0.05_{2\,tail}$, then the combined area under the two tails of the curve must equal 0.0500. Thus, the area under each tail must equal 0.0250 (see Figure 13.9(c)). For this area,

$$z_{crit} = \pm 1.96$$

To reject H_0, the obtained sample mean ($\overline{X}_{obt}$) must have a z-transformed value (z_{obt}) that falls within the critical region for rejection.

Let's now use these concepts to analyze the reading data. First, we calculate z_{obt}:

$$z_{obt} = \frac{\overline{X}_{obt} - \mu}{\sigma_{\overline{X}}}$$

$$= \frac{72 - 75}{1.6} = \frac{-3}{1.6}$$

$$= -1.88$$

FIGURE 12.9

Critical region of rejection for H_0 for (a) $\alpha = 0.05_{1\,tail}$, z_{obt} negative; (b) $\alpha = 0.05_{1\,tail}$, z_{obt} positive; and (c) $\alpha = 0.05_{2\,tail}$

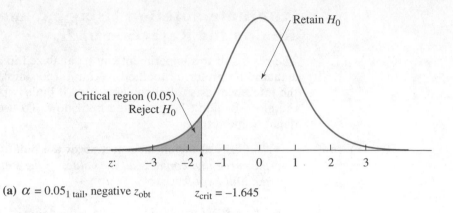

(a) $\alpha = 0.05_{1\,tail}$, negative z_{obt}

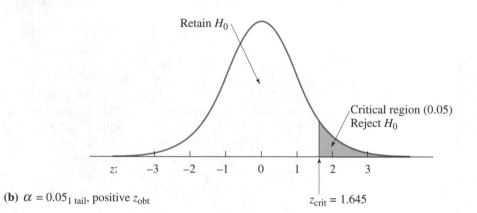

(b) $\alpha = 0.05_{1\,tail}$, positive z_{obt}

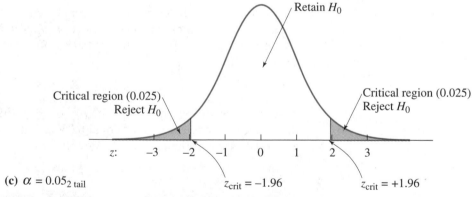

(c) $\alpha = 0.05_{2\,tail}$

(Adapted from *Fundamental Statistics for Psychology,* Second Edition by Robert B. McCall, © 1975 by Harcourt Brace Jovanovich, Inc.)

FIGURE 12.10

Solution to reading
proficiency experiment
using z_{obt} and the critical
region

Step 1: Calculate the appropriate statistic:

$$z_{obt} = \frac{\overline{X}_{obt} - \mu}{\sigma_{\overline{X}}} = \frac{72 - 75}{1.6} = -1.88$$

Step 2: Evaluate the statistic based on its sampling distribution. The decision rule is as
follows: If $|z_{obt}| \geq |z_{crit}|$, reject H_0. Since $\alpha = 0.05_{1tail}$, from Table A,

$$z_{crit} = -1.65$$

Since $|z_{obt}| > 1.645$, it falls within the critical region for rejection of H_0.
Therefore, we reject H_0.

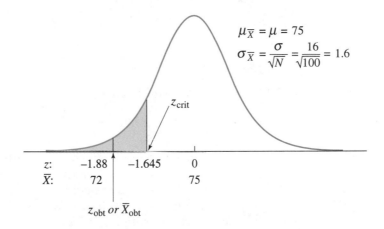

$$\mu_{\overline{X}} = \mu = 75$$
$$\sigma_{\overline{X}} = \frac{\sigma}{\sqrt{N}} = \frac{16}{\sqrt{100}} = 1.6$$

The next step is to determine z_{crit}. Since $\alpha = 0.05_{1\ tail}$, the area under the left
tail equals 0.0500. For this area, from Table A we obtain

$$z_{crit} = -1.645$$

Finally, we must determine whether z_{obt} falls within the critical region. If it does,
we reject the null hypothesis. If it doesn't, we retain the null hypothesis. The
decision rule states the following:

If $|z_{obt}| \geq |z_{crit}|$, rejects the null hypothesis. If not, retain the null hypothesis.

Note that this equation is just a shorthand way of specifying that, if z_{obt} is positive,
it must be equal to or greater than $+z_{crit}$ to fall within the critical region. If z_{obt}
is negative, it must be equal to or less than $-z_{crit}$ to fall within the critical region.
In the present example, since $|z_{obt}| > 1.645$, we reject the null hypothesis. The
complete solution using this method is shown in Figure 12.10. We would like to
point out that, in using this method, we are following the two-step procedure
outlined previously in this chapter for analyzing data: (1) calculating the appropriate statistic and (2) evaluating the statistic based on its sampling distribution.
Actually, the experimenter calculates two statistics, $\overline{X}_{obt}$ and z_{obt}. The final one
evaluated is z_{obt}. If the sampling distribution of $\overline{X}$ is normally shaped, then the
z distribution will also be normal and the appropriate probabilities given by
Table A. Of course, the z distribution has a mean of 0 and a standard deviation
of 1, as discussed in Chapter 5.
Let's try another problem using this approach.

A university president believes that over the past few years the average age of students attending his university has changed. To test this hypothesis, an experiment is conducted in which the age of 150 students who have been randomly sampled from the student body is measured. The mean age is 23.5 years. A complete census taken at the university a few years prior to the experiment showed a mean age of 22.4 years, with a standard deviation of 7.6.

a. What is the nondirectional alternative hypothesis?
b. What is the null hypothesis?
c. Using $\alpha = 0.05_{2\,\text{tail}}$, what is the conclusion?

SOLUTION

The solution is shown in Table 12.2.

> **TABLE 12.2**
> ....
>
> a. Nondirectional alternative hypothesis: Over the past few years, the average age of students at the university has changed. Therefore, the sample with $\overline{X}_{\text{obt}} = 23.5$ is a random sample from a population where $\mu \neq 22.4$.
> b. Null hypothesis: The null hypothesis asserts that it is reasonable to consider the sample with $\overline{X}_{\text{obt}} = 23.5$ a random sample from a population with $\mu = 22.4$.
> c. Conclusion, using $\alpha = 0.05_{2\,\text{tail}}$:
>
> Step 1: **Calculate the appropriate statistic.** The data are given in the problem.
>
> $$z_{\text{obt}} = \frac{\overline{X}_{\text{obt}} - \mu}{\sigma_{\overline{X}}}$$
>
> $$= \frac{\overline{X}_{\text{obt}} - \mu}{\sigma/\sqrt{N}} = \frac{23.5 - 22.4}{7.6/\sqrt{150}}$$
>
> $$= \frac{1.1}{0.6205} = 1.77$$
>
> Step 2: **Evaluate the statistic based on its sampling distribution.** The decision rule is as follows: If $|z_{\text{obt}}| \geq |z_{\text{crit}}|$, reject H_0. If not, retain H_0. Since $\alpha = 0.05_{2\,\text{tail}}$, from Table A,
>
> $$z_{\text{crit}} = \pm 1.96$$

Since $|z_{\text{obt}}| < 1.96$, it does not fall within the critical region for rejection of H_0. Therefore, we retain H_0. We cannot conclude that the average age of students attending the president's university has changed.

Let's try one more problem.

PRACTICE PROBLEM 12.2

A gasoline manufacturer believes a new additive will result in more miles per gallon. A large number of mileage measurements on the gasoline without the additive have been made by the company under rigorously controlled conditions. The results show a mean of 24.7 miles per gallon and a standard deviation of 4.8. Tests are conducted on a sample of 75 cars using the gasoline plus the additive. The sample mean equals 26.5 miles per gallon.

a. What is the directional alternative hypothesis?
b. What is the null hypothesis?
c. What is the conclusion? Let's assume there is adequate basis for a one-tailed test. Use $\alpha = 0.05_{1\,\text{tail}}$.

SOLUTION

The solution is shown in Table 12.3.

TABLE 12.3

a. Directional alternative hypothesis: The new additive increases the number of miles per gallon. Therefore, the sample with $\overline{X}_{\text{obt}} = 26.5$ is a random sample from a population where $\mu > 24.7$.
b. Null hypothesis: H_0: The sample with $\overline{X}_{\text{obt}} = 26.5$ is a random sample from a population with $\mu \leq 24.7$.
c. Conclusion, using $\alpha = 0.05_{1\,\text{tail}}$:

Step 1: **Calculate the appropriate statistic.** The data are given in the problem.

$$z_{\text{obt}} = \frac{\overline{X}_{\text{obt}} - \mu}{\sigma/\sqrt{N}}$$

$$= \frac{26.5 - 24.7}{4.8/\sqrt{75}} = \frac{1.8}{0.5543}$$

$$= 3.25$$

Step 2: **Evaluate the statistic based on its sampling distribution.** The decision rule states that if $|z_{\text{obt}}| \geq |z_{\text{crit}}|$, reject H_0. If not, retain H_0. Given $\alpha = 0.05_{1\,\text{tail}}$, from Table A,

$$z_{\text{crit}} = 1.645$$

Since $|z_{\text{obt}}| > 1.645$, it falls within the critical region for rejection of H_0. Therefore, we reject the null hypothesis and conclude that the gasoline additive does increase miles per gallon.

Conditions Under Which the z Test Is Appropriate

The z test is appropriate when the experiment involves a single sample mean ($\overline{X}_{obt}$) and the parameters of the null-hypothesis population are known, i.e., when μ and σ are known. In addition, to use this test, the sampling distribution of the mean should be normally distributed. This, of course, requires that $N \geq 30$ or that the null-hypothesis population itself is normally distributed.* This normality requirement is spoken of as "the mathematical assumption underlying the z test."

Power and the z Test

In Chapter 10, we presented an introduction to power. Let's review some of the main points made in that chapter.

1. Conceptually, power is the sensitivity of the experiment to detect a real effect of the independent variable, if there is one.
2. Power is defined mathematically as the probability that the experiment will result in rejecting the null hypothesis if the independent variable has a real effect.
3. Power + Beta = 1. Thus, power varies inversely with beta.
4. Power varies directly with N. Increasing N increases power.
5. Power varies directly with the magnitude of the real effect of the independent variable. The power of an experiment is greater for large effects than for small effects.
6. Power varies directly with alpha level. If alpha is made more stringent, power decreases.

In this section, we will again illustrate these conclusions, only this time in conjunction with the normal deviate test. We will begin with a discussion of power and sample size.

EXAMPLE

Power and Sample Size (N)

Let's return to the illustrative experiment at the beginning of this chapter. We'll assume you are again wearing the hat of superintendent of public schools. This time, however, you are just designing the experiment. It has not yet been conducted. You want to determine whether the reading program for high school seniors in your city is deficient. As described previously, the national norms of reading proficiency of high school seniors is a normal distribution of population scores with $\mu = 75$, and $\sigma = 16$. You plan to test a random sample of high school seniors from your city and you are trying to determine how large the sample size should be. You will use $\alpha = 0.05_{1\,tail}$ in evaluating the data when collected. You want to be able to detect proficiency deficiencies in your program of 3 or more mean points from the national norms. That is to say, if the mean reading proficiency of the population of high school seniors in your city is lower than the national norms by 3 or more points, you want your experiment to have a high probability to detect it.

a. If you decide to use a sample size of 25 ($N = 25$), what is the power of your experiment to detect a population deficiency in reading proficiency of 3 mean points from the national norms?

---

* Many authors would limit the use of the z test to data that are of interval or ratio scaling. Please see the footnote in Chapter 2, p. 27, for references discussing this point.

b. If you increase the sample size to $N = 100$, what is the power now to detect a population deficiency in reading proficiency of 3 mean points?

c. What size N should you use for the power to be approximately 0.9000 to detect a population deficiency in reading proficiency of 3 mean points?

SOLUTION

a. Power with $N = 25$.

As discussed in Chapter 10, power is the probability of rejecting H_0 if the independent variable has a real effect. In computing the power to detect a hypothesized real effect, we must first determine the sample outcomes that will allow rejection of H_0. Then, we must determine the probability of getting any of these sample outcomes if the independent variable has the hypothesized real effect. The resulting probability is the power to detect the hypothesized real effect. Thus, there are two steps in computing power:

Step 1: Determine the possible sample mean outcomes in the experiment that would allow H_0 to be rejected. With the z test, this means determining the critical region for rejection of H_0, using $\overline{X}$ as the statistic.

Step 2: Assuming the hypothesized real effect of the independent variable is the true state of affairs, determine the probability of getting a sample mean in the critical region for rejection of H_0.

Let's now compute the power to detect a population deficiency in reading proficiency of 3 mean points from the national norms, using $N = 25$.

Step 1: Determine the possible sample mean outcomes in the experiment that would allow H_0 to be rejected. With the z test, this means determining the critical region for rejection of H_0, using $\overline{X}$ as the statistic.

When evaluating H_0 with the z test, we assume the sample is a random sample from the null-hypothesis population. We will symbolize the mean of the null-hypothesis population as μ_{null}. In the present example, the null-hypothesis population is the set of scores established by national testing, that is, a normal population of scores with $\mu_{\text{null}} = 75$. With $\alpha = 0.05_{1\,\text{tail}}$, $z_{\text{crit}} = -1.645$. To determine the critical value of $\overline{X}$, we can use the z equation solved for $\overline{X}_{\text{crit}}$:

$$z_{\text{crit}} = \frac{\overline{X}_{\text{crit}} - \mu_{\text{null}}}{\sigma_{\overline{X}}}$$

$$\overline{X}_{\text{crit}} = \mu_{\text{null}} + \sigma_{\overline{X}}(z_{\text{crit}})$$

Substituting the data with $N = 25$,

$$\overline{X}_{\text{crit}} = 75 + 3.2(-1.645) \qquad \sigma_{\overline{X}} = \frac{\sigma}{\sqrt{N}} = \frac{16}{\sqrt{25}} = 3.2$$

$$= 75 - 5.264$$

$$= 69.74$$

Thus, with $N = 25$, we will reject H_0 if, when we conduct the experiment, the mean of the sample ($\overline{X}_{\text{obt}}$) ≤ 69.74. See Figure 12.11 for a pictorial representation of these relationships.

Step 2: Assuming the hypothesized real effect of the independent variable is the true state of affairs, determine the probability of getting a sample mean in the critical region for rejection of H_0.

FIGURE 12.11

Power for $N = 25$

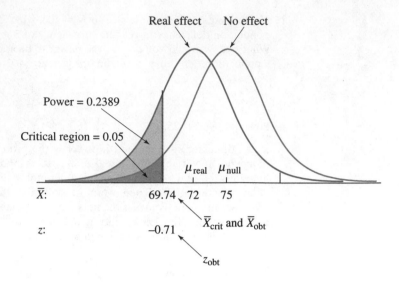

If the independent variable has the hypothesized real effect, then the sample scores in your experiment are not a random sample from the null-hypothesis population. Instead, they are a random sample from a population having a mean as specified by the hypothesized real effect. We shall symbolize this mean as μ_{real}. Thus, if the reading proficiency of the population of seniors in your city is 3 mean points lower than the national norms, then the sample in your experiment is a random sample from a population where $\mu_{real} = 72$. The probability of your sample mean falling in the critical region if the sample is actually a random sample from a population where $\mu_{real} = 72$ is found by obtaining the z transform for $\overline{X}_{obt} = 69.74$ and looking up its corresponding area in Table A. Thus,

$$z_{obt} = \frac{\overline{X}_{obt} - \mu_{real}}{\sigma_{\overline{X}}} \qquad \sigma_{\overline{X}} = \frac{\sigma}{\sqrt{N}} = \frac{16}{\sqrt{25}} = 3.2$$

$$= \frac{69.74 - 72}{3.2}$$

$$= -0.71$$

From Table A,

$$p(\overline{X}_{obt} \le 69.74) = 0.2389$$

Thus,

$$\text{Power} = 0.2389$$

$$\text{Beta} = 1 - \text{Power} = 1 - 0.2389 = 0.7611$$

Thus, the power to detect a deficiency of 3 mean points with $N = 25$ is 0.2389 and beta = 0.7611.

Since the probability of a Type II error is too high, you decide not to go ahead and run the experiment with $N = 25$. Let's now see what happens to power and beta if N is increased to 100.

b. If $N = 100$, what is the power to detect a population difference in reading proficiency of 3 mean points?

Step 1: Determine the possible sample mean outcomes in the experiment that would allow H_0 to be rejected. With the z test, this means determining the critical region for rejection of H_0, using $\overline{X}$ as the statistic:

$$\overline{X}_{\text{crit}} = \mu_{\text{null}} + \sigma_{\overline{X}}(z_{\text{crit}}) \qquad \sigma_{\overline{X}} = \frac{\sigma}{\sqrt{N}} = \frac{16}{\sqrt{100}} = 1.6$$

$$= 75 + 1.6(-1.645)$$

$$= 75 - 2.632$$

$$= 72.37$$

Step 2: Assuming the hypothesized real effect of the independent variable is the true state of affairs, determine the probability of getting a sample mean in the critical region for rejection of H_0.

$$z_{\text{obt}} = \frac{\overline{X}_{\text{obt}} - \mu_{\text{real}}}{\sigma_{\overline{X}}} \qquad \sigma_{\overline{X}} = \frac{\sigma}{\sqrt{N}} = \frac{16}{\sqrt{100}} = 1.6$$

$$= \frac{72.37 - 72}{1.6}$$

$$= 0.23$$

From Table A,

$$p(\overline{X}_{\text{obt}} \leq 72.37) = 0.5000 + 0.0910 = 0.5910$$

Thus,

$$\text{Power} = 0.5910$$

$$\text{Beta} = 1 - \text{Power} = 1 - 0.5910 = 0.4090$$

Thus, by increasing N from 25 to 100, the power to detect a deficiency of 3 mean points has increased from 0.2389 to 0.5910. Beta has decreased from 0.7611 to 0.4090. This is a demonstration that power varies directly with N and beta varies inversely with N. Thus, increasing N causes an increase in power and a decrease in beta. Figure 12.12 summarizes the relationships for this problem.

FIGURE 12.12

Power for $N = 100$

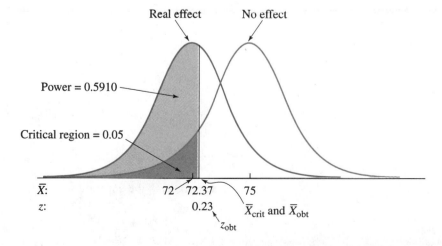

FIGURE 12.13
....

Determining N for
power = 0.9000

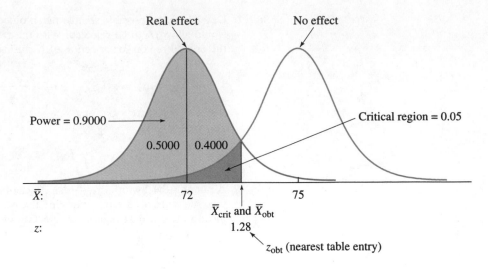

c. What size N should you use for the power to be approximately 0.9000?

For the power to be 0.9000 to detect a population deficiency of 3 mean points, the probability that $\overline{X}_{obt}$ will fall in the critical region must be equal to 0.9000. As shown in Figure 12.13, this dictates that the area between z_{obt} and μ_{real} = 0.4000. From Table A, $z_{obt} = 1.28$. (Note that we have taken the closest table reading, rather than interpolating. This will result in a power close to 0.9000, but not exactly equal to 0.9000.) By solving the z_{obt} equation for $\overline{X}_{obt}$ and setting $\overline{X}_{obt}$ equal to $\overline{X}_{crit}$, we can determine N. Thus,

$$\overline{X}_{obt} = \mu_{real} + \sigma_{\overline{X}}(z_{obt})$$

$$\overline{X}_{crit} = \mu_{null} + \sigma_{\overline{X}}(z_{crit})$$

Setting $\overline{X}_{obt} = \overline{X}_{crit}$, we have,

$$\mu_{real} + \sigma_{\overline{X}}(z_{obt}) = \mu_{null} + \sigma_{\overline{X}}(z_{crit})$$

Solving for N,

$$\mu_{real} - \mu_{null} = \sigma_{\overline{X}}(z_{crit}) - \sigma_{\overline{X}}(z_{obt})$$

$$\mu_{real} - \mu_{null} = \sigma_{\overline{X}}(z_{crit} - z_{obt})$$

$$\mu_{real} - \mu_{null} = \frac{\sigma}{\sqrt{N}}(z_{crit} - z_{obt})$$

$$\sqrt{N}(\mu_{real} - \mu_{null}) = \sigma(z_{crit} - z_{obt})$$

$$N = \left[\frac{\sigma(z_{crit} - z_{obt})}{\mu_{real} - \mu_{null}} \right]^2$$

Thus, to determine N, the equation we use is

$$N = \left[\frac{\sigma(z_{crit} - z_{obt})}{\mu_{real} - \mu_{null}} \right]^2 \qquad \textit{equation for determining } N$$

Applying this equation to the problem we have been considering, we get

$$N = \left[\frac{\sigma(z_{crit} - z_{obt})}{\mu_{real} - \mu_{null}} \right]^2$$

$$= \left[\frac{16(-1.645 - 1.28)}{72 - 75} \right]^2$$

$$= 243$$

Thus, if you increase N to 243 subjects, the power will be approximately 0.9000 (power = 0.8997) to detect a population deficiency in reading proficiency of 3 mean points. I suggest you confirm this power calculation yourself using $N = 243$ as a practice exercise.

Power and Alpha Level   Next, let's take a look at the relationship between power and alpha. Suppose you had set $\alpha = 0.01_{1\,tail}$ instead of $0.05_{1\,tail}$. What happens to the resulting power? (We'll assume $N = 100$ in this question.)

SOLUTION

Step 1: Determine the possible sample mean outcomes in the experiment that would allow H_0 to be rejected. With the z test, this means determining the critical region for rejection of H_0, using $\overline{X}$ as the statistic:

$$\overline{X}_{crit} = \mu_{null} + \sigma_{\overline{X}}(z_{crit}) \qquad \sigma_{\overline{X}} = \frac{\sigma}{\sqrt{N}} = \frac{16}{100} = 1.6$$

$$= 75 + 1.6(-2.33) \qquad z_{crit} = -2.33$$

$$= 71.27$$

Step 2: Assuming the hypothesized real effect of the independent variable is the true state of affairs, determine the probability of getting a sample mean in the critical region for rejection of H_0.

$$z_{obt} = \frac{\overline{X}_{obt} - \mu_{real}}{\sigma_{\overline{X}}} \qquad \sigma_{\overline{X}} = \frac{\sigma}{\sqrt{N}} = \frac{16}{100} = 1.6$$

$$= \frac{71.27 - 72}{1.6}$$

$$= -0.46$$

From Table A,

$$p(\overline{X}_{obt} \le 71.27) = 0.3228$$

Thus,

$$\text{Power} = 0.3228$$

$$\text{Beta} = 1 - \text{Power} = 0.6772$$

FIGURE 12.14

Power for $N = 100$ and
$\alpha = 0.01_{1\,\text{tail}}$

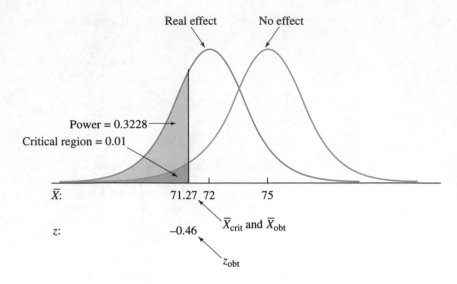

Thus, by making alpha more stringent (changing it from $0.05_{1\,\text{tail}}$ to $0.01_{1\,\text{tail}}$), power has decreased from 0.5910 to 0.3288. Beta has increased from 0.4090 to 0.6772. This demonstrates that there is a direct relationship between alpha and power and an inverse relationship between alpha and beta. Figure 12.14 shows the relationships for this problem.

Relationship Between Magnitude of Real Effect and Power   Next, let's investigate the relationship between the size of the real effect and power. To do this, let's calculate the power to detect a population deficiency in reading proficiency of 5 mean points from the national norms. We'll assume $N = 100$ and $\alpha = 0.05_{1\,\text{tail}}$. Figure 12.15 shows the relationships for this problem.

FIGURE 12.15   Power for $N = 100$ and $\mu_{\text{real}} = 70$

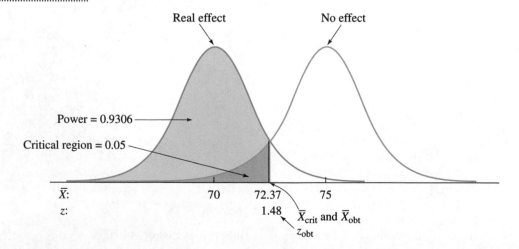

SOLUTION

Step 1: Determine the possible sample mean outcomes in the experiment that would allow H_0 to be rejected. With the z test, this means determining the critical region for rejection of H_0, using $\overline{X}$ as the statistic:

$$\overline{X}_{\text{crit}} = \mu_{\text{null}} + \sigma_{\overline{X}}(z_{\text{crit}}) \qquad \sigma_{\overline{X}} = \frac{\sigma}{\sqrt{N}} = \frac{16}{100} = 1.6$$

$$= 75 + 1.6(-1.645)$$

$$= 72.37$$

Step 2: Assuming the hypothesized real effect of the independent variable is the true state of affairs, determine the probability of getting a sample mean in the critical region for rejection of H_0:

$$z_{\text{obt}} = \frac{\overline{X}_{\text{obt}} - \mu_{\text{real}}}{\sigma_{\overline{X}}} \qquad \sigma_{\overline{X}} = \frac{\sigma}{\sqrt{N}} = \frac{16}{100} = 1.6$$

$$= \frac{72.37 - 70}{1.6}$$

$$= 1.48$$

From Table A,

$$p(\overline{X}_{\text{obt}} \leq 72.37) = 0.5000 + 0.4306 = 0.9306$$

Thus,

$$\text{Power} = 0.9306$$

$$\text{Beta} = 1 - \text{Power} = 1 - 0.9306 = 0.0694$$

Thus, by increasing the magnitude of the real effect from 3 to 5 mean points, power has increased from 0.5910 to 0.9306. Beta has decreased from 0.4090 to 0.0694. This demonstrates that there is a direct relationship between the magnitude of the real effect and the power to detect it.

SUMMARY

In this chapter, we have discussed the topics of the sampling distribution of a statistic, how to generate sampling distributions from an empirical sampling approach, the sampling distribution of the mean, and analyzing single-sample experiments with the z test. We pointed out that the procedure for analyzing data in most hypothesis-testing experiments is to calculate the appropriate statistic and then evaluate the statistic based on its sampling distribution. The sampling distribution of a statistic gives all the values that the statistic can take, along with the probability of getting each value if sampling is random from the null-hypothesis population. The sampling distribution can be generated theoretically by the Central Limit Theorem or empirically by (1) determining all the possible different samples of size N that can be formed from the raw-score population, (2) calculating the statistic for each of the samples, and (3) calculating the probability of getting each value of the statistic if sampling is random from the null-hypothesis population.

The sampling distribution of the mean is a distribution of sample mean values having a mean ($\mu_{\bar{X}}$) equal to μ and a standard deviation ($\sigma_{\bar{X}}$) equal to $\sigma/\sqrt{N}$. It is normally distributed if the raw-score population is normally distributed or if $N \geq 30$, assuming the raw-score population is not radically different from normality. The z test is appropriate for analyzing single sample experiments, where μ and σ are known, and the sample mean is used as the basic statistic. In using this test, z_{obt} is calculated and then evaluated to determine whether it falls in the critical region for rejecting the null hypothesis. To use the z test, the sampling distribution of the mean must be normally distributed. This in turn requires that the null-hypothesis population be normally distributed or that $N \geq 30$.

Finally, we discussed power in conjunction with the z test. Power is the probability of rejecting H_0 if the independent variable has a real effect. To calculate power, we followed a two-step procedure: determining the possible sample means that allowed rejection of H_0 and finding the probability of getting any of these sample means assuming the hypothesized real effect of the independent variable is true. Power varies directly with N, alpha, and the magnitude of the real effect of the independent variable. Power varies inversely with beta.

IMPORTANT TERMS

Critical region (p. 277)
Critical value(s) of a statistic
  (p. 277)
Critical value(s) of $\bar{X}$ (p. 283)
Critical value(s) of z (p. 277)
Mean of the sampling distribution
  of the mean (p. 268)

Null-hypothesis population (p. 265)
Power (p. 282)
Sampling distribution (p. 263)
Sampling distribution of the
  mean (p. 268)

Standard error of the mean
  (p. 269)
μ_{null} (p. 283)
μ_{real} (p. 284)

QUESTIONS AND PROBLEMS

1. Define each of the terms in the "Important Terms" section.
2. Why is the sampling distribution of a statistic important to be able to use the statistic in hypothesis testing? Explain in a short paragraph.
3. How are sampling distributions generated using the empirical sampling approach?
4. What are the two basic steps used when analyzing data?
5. What are the assumptions underlying the use of the z test?
6. What are the characteristics of the sampling distribution of the mean?
7. Explain why the standard deviation of the sampling distribution of the mean is sometimes referred to as the "standard error of the mean."
8. How do each of the following differ?
   a. s and $s_{\bar{X}}$ b. s^2 and σ^2
   c. μ and $\mu_{\bar{X}}$ d. σ and $\sigma_{\bar{X}}$

9. Explain why $\sigma_{\bar{X}}$ should vary directly with σ and inversely with N.
10. Why should $\mu_{\bar{X}} = \mu$?
11. Is the shape of the sampling distribution of the mean always the same as the shape of the null-hypothesis population? Explain.
12. In using the z test, why is it important that the sampling distribution of the mean be normally distributed?
13. If the assumptions underlying the z test are met, what are the characteristics of the sampling distribution z?
14. Define power, both conceptually and mathematically.
15. Explain what happens to the power of the z test when each of the following variables increases.
   a. N b. Alpha level
   c. Magnitude of real effect of the independent variable
   d. σ

16. Given the population set of scores 3, 4, 5, 6, 7,
    a. Determine the sampling distribution of the mean for sample sizes of 2. Assume sampling is one at a time with replacement.
    b. Demonstrate that $\mu_{\bar{X}} = \mu$.
    c. Demonstrate that $\sigma_{\bar{X}} = \sigma/\sqrt{N}$.

17. If a population of raw scores is normally distributed and has a mean $\mu = 80$ and a standard deviation $\sigma = 8$, determine the parameters ($\mu_{\bar{X}}$ and $\sigma_{\bar{X}}$) of the sampling distribution of the mean for the following sample sizes.
    a. $N = 16$
    b. $N = 35$
    c. $N = 50$
    d. Explain what happens as N gets larger.

18. Is it reasonable to consider a sample of 40 scores with $\bar{X}_{obt} = 65$ to be a random sample from a population of scores that is normally distributed, with $\mu = 60$ and $\sigma = 10$? Use $\alpha = 0.05_{2\,tail}$ in making your decision.

19. A set of sample scores from an experiment has an $N = 30$ and an $\bar{X}_{obt} = 19$.
    a. Can we reject the null hypothesis that the sample is a random sample from a normal population with $\mu = 22$ and $\sigma = 8$? Use $\alpha = 0.01_{1\,tail}$. Assume the sample mean is in the correct direction.
    b. What is the power of the experiment to detect a real effect such that $\mu_{real} = 20$?
    c. What is the power to detect a $\mu_{real} = 20$ if N is increased to 100?
    d. What value does N have to equal to achieve a power of 0.8000 to detect a $\mu_{real} = 20$? Use the nearest table value for z_{obt}.

20. On the basis of her newly developed technique, a student believes she can reduce the amount of time schizophrenics spend in an institution. As director of training at a nearby institution, you agree to let her try her method on 20 schizophrenics, randomly sampled from your institution. The mean duration that schizophrenics stay at your institution is 85 weeks with a standard deviation of 15 weeks. The scores are normally distributed. The results of the experiment show that the patients treated by the student stay a mean duration of 78 weeks with a standard deviation of 20 weeks.
    a. What is the alternative hypothesis? In this case, assume a nondirectional hypothesis is appropriate because there are insufficient theoretical and empirical bases to warrant a directional hypothesis.
    b. What is the null hypothesis?
    c. What do you conclude about the student's technique? Use $\alpha = 0.05_{2\,tail}$.

21. A professor has been teaching statistics for many years. His records show that the overall mean for final exam scores is 82 with a standard deviation of 10. The professor believes that this year's class is superior to his previous ones. The mean for final exam scores for this year's class of 65 students is 87. What do you conclude? Use $\alpha = 0.05_{1\,tail}$.

22. An engineer believes that her newly designed engine will be a great gas saver. A large number of tests on engines of the old design yielded a mean gasoline consumption of 19.5 miles per gallon with a standard deviation of 5.2. Fifteen new engines are tested. The mean gasoline consumption is 21.6 miles per gallon. What is your conclusion? Use $\alpha = 0.05_{1\,tail}$.

23. Refer to Practice Problem 12.2, p. 281, in answering this question. Suppose that before doing the experiment, the manufacturer wants to determine the probability he will be able to detect a real mean increase of 2.0 miles per gallon with the additive, if the additive is at least that effective.
    a. If he tests 20 cars, what is the power to detect a mean increase of 2.0 miles per gallon?
    b. If he increases the N to 75 cars, what is the power to detect a mean increase of 2.0 miles per gallon?
    c. How many cars should he use if he wants to have a 99% chance of detecting a mean increase of 2.0 miles per gallon?

24. A physical education professor believes that exercise can slow down the aging process. For the past 10 years, he has been conducting an exercise class for 14 individuals who are currently 50 years old. Normally, as one ages, maximum oxygen consumption decreases. The national norm for maximum oxygen consumption in 50-year-old individuals is 30 milliliters per kilogram per minute with a standard deviation of 8.6. The mean of the 14 individuals is 40 milliliters per kilogram per minute. What do you conclude? Use $\alpha = 0.05_{1\,tail}$.

13 STUDENT'S *t* TEST FOR SINGLE SAMPLES

INTRODUCTION

In Chapter 12, we discussed the *z* test and determined that it was appropriate in situations where both the mean and the standard deviation of the null-hypothesis population were known. However, these situations are relatively rare. It is more common to encounter situations in which the mean of the null-hypothesis population can be specified and the standard deviation is *unknown*. In these cases, the *z* test cannot be used. Instead, another test, called *Student's t test,* is employed. The *t* test is very similar to the *z* test. It was developed by W. S. Gosset, writing under the pen name of "Student." Student's *t* test is a practical, quite powerful test widely used in the behavioral sciences. In this chapter, we shall discuss the *t* test in conjunction with experiments involving a single sample. In Chapter 14, we shall discuss the *t* test as it applies to experiments using two samples or conditions.

COMPARISON OF THE *z* AND *t* TESTS

Since the *z* and *t* tests for single sample experiments are quite alike, let's first compare their equations before considering an example:

z Test	*t* Test
$z_{\text{obt}} = \dfrac{\bar{X} - \mu}{\sigma/\sqrt{N}}$	$t_{\text{obt}} = \dfrac{\bar{X} - \mu}{s/\sqrt{N}}$
$= \dfrac{\bar{X} - \mu}{\sigma_{\bar{X}}}$	$= \dfrac{\bar{X} - \mu}{s_{\bar{X}}}$

where s = Estimate of σ

$s_{\bar{X}}$ = Estimate of $\sigma_{\bar{X}}$

In comparing these equations, we can see that the only difference is that the z test uses the standard deviation of the null-hypothesis population (σ), whereas the t test uses the standard deviation of the sample (s). When σ is unknown, we estimate it using the estimate given by s, and the resulting statistic is called t. Thus, the denominator of the t test is $s/\sqrt{N}$ rather than $\sigma/\sqrt{N}$. The symbol $s_{\bar{X}}$ replaces $\sigma_{\bar{X}}$ where

$$s_{\bar{X}} = \frac{s}{\sqrt{N}} \qquad \textit{estimated standard error of the mean}$$

We are ready now to consider an experiment using the t test to analyze the data.

AN EXPERIMENT

Technique for Increasing Early Speaking in Children

Suppose you have a technique that you believe will affect the age at which children begin speaking. In your locale, the average age of first word utterances is 13.0 months. The standard deviation is unknown. You apply your technique to a random sample of 15 children. The results show that the sample mean age of first word utterances is 11.0 months with a standard deviation of 3.34.

1. What is the nondirectional alternative hypothesis?
2. What is the null hypothesis?
3. Did the technique work? Use $\alpha = 0.05_{2\,\text{tail}}$.

SOLUTION

1. Alternative hypothesis: The technique affects the age at which children begin speaking. Therefore, the sample with $\bar{X}_{\text{obt}} = 11.0$ is a random sample from a population where $\mu \neq 13.0$.
2. Null hypothesis: H_0: The sample with $\bar{X}_{\text{obt}} = 11.0$ is a random sample from a population with $\mu = 13.0$.
3. Conclusion using $\alpha = 0.05_{2\,\text{tail}}$:

Step 1: Calculate the appropriate statistic. Since σ is unknown, it is impossible to determine z_{obt}. However, since s is known, we can calculate t_{obt}. Thus,

$$t_{\text{obt}} = \frac{\bar{X}_{\text{obt}} - \mu}{s/\sqrt{N}}$$

$$= \frac{11.0 - 13.0}{3.34/\sqrt{15}}$$

$$= \frac{-2}{0.862}$$

$$= -2.32$$

The next step ordinarily would be to evaluate t_{obt} using the sampling distribution of t. However, because this distribution is not yet familiar, we need to discuss it before we can proceed with the evaluation.

THE SAMPLING DISTRIBUTION OF *t*

Using the definition of sampling distribution developed in Chapter 12, we note the following.

• *The sampling distribution of t is a probability distribution of the t values that would occur if all possible different samples of a fixed size N were drawn from the null-hypothesis population. It gives (1) all the possible different t values for samples of size N and (2) the probability of getting each value if sampling is random from the null-hypothesis population.*

As with the sampling distribution of the mean, the sampling distribution of *t* can be determined theoretically or empirically. Again, for pedagogical reasons, we prefer the empirical approach. The sampling distribution of *t* can be derived empirically by taking a specific population of raw scores, drawing all possible different samples of a fixed size *N*, and then calculating the *t* value for each sample. Once all the possible *t* values are obtained, it is a simple matter to calculate the probability of getting each different *t* value under the assumption of random sampling from the population. By varying *N* and the population scores, one can derive sampling distributions for various populations and sample sizes. Empirically or theoretically, it turns out that, if the null hypothesis population is normally shaped, or if $N \geq 30$, the *t* distribution looks very much like the *z* distribution except that there is a *family* of *t* curves that vary with sample size. You will recall that the *z* distribution has only one curve for all sample sizes (the values represented in Table A in Appendix D). On the other hand, the *t* distribution, like the sampling distribution of the mean, has many curves, depending on sample size. Since we are estimating σ by using *s* in the *t* equation, and since the size of the sample influences the accuracy of the estimate, it makes sense that there should be a different sampling distribution of *t* for different sample sizes.

Degrees of Freedom

Although the *t* distribution varies with sample size, Gosset found that it varies *uniquely* with the *degrees of freedom* associated with *t*, rather than simply with sample size. Why this is so will not be apparent until Chapter 14. For now, let's just pursue the concept of degrees of freedom.

• *The degrees of freedom (df) for any statistic is the number of scores that are free to vary in calculating that statistic.*

For example, there are *N* degrees of freedom associated with the mean. How do we know this? For any set of scores, *N* is given. If there are three scores and we know the first two scores, the last score can take on any value. It has no restrictions. There is no way to tell what it must be by knowing the other two scores. The same is true for the first two scores. Thus, all three scores are free to vary when calculating the mean. Thus, there are *N* degrees of freedom.

Contrast this with calculating the standard deviation:

$$s = \sqrt{\frac{\Sigma (X - \bar{X})^2}{N - 1}}$$

Since the sum of deviations about the mean must equal zero, only $N - 1$ of the deviation scores are free to take on any value. Thus, there are $N - 1$ degrees of freedom associated with *s*. Why is this so? Consider the raw scores 4, 8, and 12. The mean is 8. Note what happens when calculating *s*:

X	$\bar{X}$	$X - \bar{X}$
4	8	-4
8	8	0
12	8	?

Since the mean is 8, the deviation score for the raw score of 4 is -4 and for the raw score of 8 is 0. Since $\Sigma (X - \bar{X}) = 0$, the last deviation is fixed by the other deviations. It must be $+4$ (see the table). Therefore only two of the three deviation scores are free to vary. Whatever value these take, the third is fixed. In calculating *s*, only $N - 1$ deviation scores are free to vary. Thus, there are $N - 1$ degrees of freedom associated with the standard deviation.

In calculating *t* for single samples, we must first calculate *s*. Since we lose 1 degree of freedom in calculating *s*, there are $N - 1$ degrees of freedom associated with *t*. Thus, for the *t* test,

$$\text{df} = N - 1 \qquad \textit{degrees of freedom for t test (single sample)}$$

t AND *z* DISTRIBUTIONS COMPARED

Figure 13.1 shows the *t* distribution for various degrees of freedom. The *t* distribution is symmetrical about zero and becomes closer to the normally distributed *z* distribution with increasing df. Notice how quickly it approaches the normal curve. Even with df as small as 20, the *t* distribution rather closely approximates the normal curve. Theoretically when df $= \infty$,* the *t* distribution is identical to the *z* distribution. This makes sense because as the df increase, sample size increases, and the estimate *s* gets closer to σ. At any df other than ∞, the *t* distribution has more extreme *t* values than the *z* distribution, since there is more variability in *t* because we used *s* to estimate σ. Another way of saying this is that the tails of the *t* distribution are elevated relative to the *z* distribution. Thus, for a given alpha level, the critical value of *t* is higher than for *z*, making the *t* test less sensitive than the *z* test. That is, for any alpha level, t_{obt} must be higher than z_{obt} to reject the null hypothesis. Table 13.1 shows the critical values of *z* and *t* at the 0.05 and 0.01 alpha levels. As the df increase, the critical value of *t* approaches that of *z*. The critical *z* value, of course, doesn't change with

---

* As df approaches infinity, the *t* distribution approaches the normal curve.

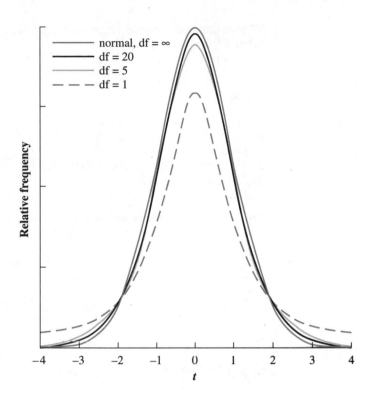

FIGURE 13.1

t distribution for various degrees of freedom

TABLE 13.1   Critical Values of *z* and *t* at the 0.05 and 0.01 Alpha Levels, One-Tailed

df	$z_{0.05}$	$t_{0.05}$	$z_{0.01}$	$t_{0.01}$
5	1.645	2.015	2.326	3.365
30	1.645	1.697	2.326	2.457
60	1.645	1.671	2.326	2.390
∞	1.645	1.645	2.326	2.326

sample size. Critical values of *t* for various alpha levels and df are contained in Table D of Appendix D. These values have been obtained from the sampling distribution of *t* for each df. The table may be used for evaluating t_{obt} for any experiment. We are now ready to return to the illustrative example.

FIRST WORD UTTERANCES EXPERIMENT REVISITED

You are investigating a technique purported to affect the age at which children begin speaking. $\mu = 13.0$ months; σ is unknown; the sample of 15 children using your technique has a mean for first word utterances of 11.0 months and a standard deviation of 3.34.

sample $\bar{X}_{obt} = 11.0$
$\mu \neq 13$.

1. What is the nondirectional alternative hypothesis?
2. What is the null hypothesis?
3. Did the technique work? Use $\alpha = 0.05_{2\,tail}$.

SOLUTION

1. Alternative hypothesis: The technique affects the age at which children begin speaking. Therefore, the sample with $\bar{X}_{obt} = 11.0$ is a random sample from a population where $\mu \neq 13.0$.
2. Null hypothesis: H_0: The sample with $\bar{X}_{obt} = 11.0$ is a random sample from a population with $\mu = 13.0$.
3. Conclusion using $\alpha = 0.05_{2\,tail}$:

> **Step 1: Calculate the appropriate statistic.** Since this is a single sample experiment with unknown σ, t_{obt} is appropriate:
>
> $$t_{obt} = \frac{\bar{X}_{obt} - \mu}{s/\sqrt{N}}$$
> $$= \frac{11.0 - 13.0}{3.34/\sqrt{15}}$$
> $$= -2.32$$

> **Step 2: Evaluate the statistic based on its sampling distribution.** Just as with the z test, if
>
> $$|t_{obt}| \geq |t_{crit}|$$

FIGURE 13.2

Solution to the first word utterance experiment using Student's *t* test

Step 1: Calculate the appropriate statistic. Since σ is unknown, t_{obt} is appropriate.

$$t_{obt} = \frac{\bar{X}_{obt} - \mu}{s/\sqrt{N}} = \frac{11.0 - 13.00}{3.34/\sqrt{15}} = -2.32$$

Step 2: Evaluate the statistic. If $|t_{obt}| \geq |t_{crit}|$, reject H_0. Since $\alpha = 0.05_{2\,tail}$ and $df = N - 1 = 15 - 1 = 14$, from Table D,

$$t_{crit} = \pm 2.145$$

Since $|t_{obt}| > 2.145$, it falls within the critical region. Therefore, we reject H_0.

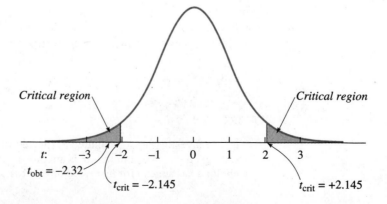

then it falls within the critical region for rejection of the null hypothesis. t_{crit} is found in Table D under the appropriate alpha level and df. For this example, with $\alpha = 0.05_{2\,tail}$ and df $= N - 1 = 15 - 1 = 14$, from Table D,

$$t_{crit} = \pm 2.145$$

Since $|t_{obt}| > 2.145$, we reject H_0 and conclude that the technique does affect the age at which children in your locale first begin speaking. The solution is shown in Figure 13.2.

 ## CALCULATING t_{obt} FROM ORIGINAL SCORES

If in a given situation the original scores are available, t can be calculated directly without first having to calculate s. The appropriate equation is given here:*

$$t_{obt} = \frac{\bar{X}_{obt} - \mu}{\sqrt{\dfrac{SS}{N(N-1)}}} \qquad \textit{equation for computing } t_{obt} \textit{ from raw scores}$$

where $\quad SS = \Sigma\, (X - \bar{X})^2 = \Sigma\, X^2 - \dfrac{(\Sigma\, X)^2}{N}$

EXAMPLE

Suppose the original data in the previous problem were as follows:

Age (months) X	Computation of X^2
8	64
9	81
10	100
15	225
18	324
17	289
12	144
11	121
7	49
8	64
10	100
11	121
8	64
9	81
12	144
165	1971

$$N = 15 \qquad \bar{X}_{obt} = \tfrac{165}{15} = 11.0$$

Let's calculate t_{obt} directly from these raw scores.

---

* The derivation is presented in Note 13.1.

SOLUTION

$$t_{obt} = \frac{\bar{X}_{obt} - \mu}{\sqrt{\dfrac{SS}{N(N-1)}}} \qquad\qquad SS = \Sigma X^2 - \frac{(\Sigma X)^2}{N}$$

$$= \frac{11.0 - 13.0}{\sqrt{\dfrac{156}{15(14)}}} \qquad\qquad = 1971 - \frac{(165)^2}{15}$$

$$= 156$$

$$= \frac{-2}{0.862} = -2.32$$

This is the same value arrived at previously. Note that it is all right to first calculate *s* and then use the original t_{obt} equation. However, the answer is more subject to rounding error.

Let's try another problem.

PRACTICE   PROBLEM   13.1

A researcher believes that in recent years women have been getting taller. She knows that 10 years ago the average height of young adult women living in her city was 63 inches. The standard deviation is unknown. She randomly samples eight young adult women currently residing in her city and measures their heights. The following data are obtained:

Height (in.) X	Calculation of X^2
64	4,096
66	4,356
68	4,624
60	3,600
62	3,844
65	4,225
66	4,356
63	3,969
514	33,070
$N = 8$	$\bar{X}_{obt} = \frac{514}{8} = 64.25$

a. What is the alternative hypothesis? In evaluating this experiment, assume a nondirectional hypothesis is appropriate because there are insufficient theoretical and empirical bases to warrant a directional hypothesis.
b. What is the null hypothesis?
c. What is your conclusion? Use $\alpha = 0.01_{2\,tail}$.

SOLUTION

The solution is shown in Table 13.2.

TABLE 13.2
....

a. Nondirectional alternative hypothesis: In recent years the height of women has been changing. Therefore, the sample with $\bar{X}_{obt} = 64.25$ is a random sample from a population where $\mu \neq 63$.

b. Null hypothesis: The null hypothesis asserts that it is reasonable to consider the sample with $\bar{X}_{obt} = 64.25$ a random sample from a population with $\mu = 63$.

c. Conclusion, using $\alpha = 0.01_{2\,tail}$:

Step 1: Calculate the appropriate statistic. The data were given previously. Since σ is unknown, t_{obt} is appropriate. There are two ways to find t_{obt}: (1) by calculating s first and then t_{obt} and (2) by calculating t_{obt} directly from the raw scores. Both methods are shown here:

s first and then t_{obt}:

$$s = \sqrt{\frac{SS}{N-1}} = \sqrt{\frac{45.5}{7}} \qquad SS = \sum X^2 - \frac{(\sum X)^2}{N} = 33{,}070 - \frac{(514)^2}{8}$$

$$= \sqrt{6.5} = 2.550 \qquad\qquad = 33{,}070 - 33{,}024.5 = 45.5$$

$$t_{obt} = \frac{\bar{X}_{obt} - \mu}{s/\sqrt{N}} = \frac{64.25 - 63}{2.550/\sqrt{8}}$$

$$= \frac{1.25}{0.902} = 1.39$$

directly from the raw scores:

$$t_{obt} = \frac{\bar{X}_{obt} - \mu}{\sqrt{\dfrac{SS}{N(N-1)}}} = \frac{64.25 - 63}{\sqrt{\dfrac{45.5}{8(7)}}} \qquad SS = \sum X^2 - \frac{(\sum X)^2}{N}$$

$$= \frac{1.25}{\sqrt{0.812}} = 1.39 \qquad\qquad = 33{,}070 - \frac{(514)^2}{8}$$

$$= 33{,}070 - 33{,}024.5 = 45.5$$

Step 2: Evaluate the statistic. If $|t_{obt}| \geq |t_{crit}|$, reject H_0. If not, retain H_0. With $\alpha = 0.01_{2\,tail}$ and df $= N - 1 = 8 - 1 = 7$, from Table D,

$$t_{crit} = \pm 3.499$$

Since $|t_{obt}| < 3.499$, it doesn't fall in the critical region. Therefore, we retain H_0. We cannot conclude that young adult women in the researcher's city have been changing in height in recent years.

PRACTICE PROBLEM 13.2

A friend of yours has been "playing" the stock market. He claims he has spent years doing research in this area and has devised an empirically successful method for investing. Since you are not averse to becoming a little richer, you are considering giving him some money to invest for you. However, before you do, you decide to evaluate his method. He agrees to a "dry run" during which he will use his method, but instead of actually buying and selling, you will just monitor the stocks he recommends to see whether his method really works. During the trial time period, the recommended stocks showed the following price changes (a plus score means an increase in price, and a minus indicates the stock went down):

Stock	Price Change ($) X	Calculation of X^2
A	+4.52	20.430
B	+5.15	26.522
C	+3.28	10.758
D	+4.75	22.562
E	+6.03	36.361
F	+4.09	16.728
G	+3.82	14.592
	31.64	147.953

$$N = 7 \qquad \overline{X}_{obt} = 4.52$$

During the same time period, the average price change of the stock market as a whole was +$3.25. Since you want to know whether the method does better or worse than chance, you decide to use a two-tailed evaluation.

a. What is the nondirectional alternative hypothesis?
b. What is the null hypothesis?
c. What is your conclusion? Use $\alpha = 0.05_{2\,tail}$.

SOLUTION

The solution is shown in Table 13.3.

TABLE 13.3
....

a. Nondirectional alternative hypothesis: Your friend's method results in a choice of stocks whose change in price differs from that expected due to random sampling from the stock market in general. Thus, the sample with $\overline{X}_{obt} = \$4.52$ cannot be considered a random sample from a population where $\mu = \$3.25$.

b. Null hypothesis: Your friend's method results in a choice of stocks whose change in price doesn't differ from that expected due to random sampling from the stock market in general. Therefore, the sample with $\overline{X}_{obt} =$ \$4.52 can be considered a random sample from a population where $\mu =$ \$3.25.

c. Conclusion, using $\alpha = 0.05_{2\,tail}$:

Step 1: Calculate the appropriate statistic. The data are given in the previous table. Since σ is unknown, t_{obt} is appropriate.

$$t_{obt} = \frac{\overline{X}_{obt} - \mu}{\sqrt{\dfrac{SS}{N(N-1)}}} \qquad SS = \sum X^2 - \frac{(\sum X)^2}{N}$$

$$= \frac{4.52 - 3.25}{\sqrt{\dfrac{4.940}{7(6)}}} \qquad\qquad = 147.953 - \frac{(31.64)^2}{7}$$

$$\qquad\qquad\qquad\qquad = 4.940$$

$$= \frac{1.27}{0.343} = 3.70$$

Step 2: Evaluate the statistic. With $\alpha = 0.05_{2\,tail}$ and df $= N - 1 = 7 - 1 = 6$, from Table D,

$$t_{crit} = \pm 2.447$$

Since $|t_{obt}| > 2.447$, we reject H_0. Your friend appears to be a winner. His method does seem to work! However, before investing heavily, we suggest you run the experiment at least one more time to guard against Type I error. Remember that replication is essential before accepting a result as factual. Better to be safe than poor.

CONDITIONS UNDER WHICH THE *t* TEST IS APPROPRIATE

The *t* test (single sample) is appropriate when the experiment has only one sample, μ is specified, σ is unknown, and the mean of the sample is used as the basic statistic. Like the *z* test, the *t* test requires that the sampling distribution of $\overline{X}$ be normal. For the sampling distribution of $\overline{X}$ to be normal, *N* must be ≥ 30 or the population of raw scores must be normal.*

---

* Many authors would limit the use of the *t* test to data that are of interval or ratio scaling. Please see the footnote in Chapter 2, p. 27, for references discussing this point.

t TEST: CONFIDENCE INTERVALS FOR THE POPULATION MEAN

Sometimes it is desirable to know the value of a population mean. Since it is very uneconomical to measure everyone in the population, a random sample is taken, and the sample mean is used as an estimate of the population mean. To illustrate, suppose a university administrator is interested in the average IQ of professors at her university. A random sample is taken, and $\overline{X} = 135$. The estimate, then, would be 135. The value 135 is called a *point estimate* because it uses only one value for the estimate. However, if we asked the administrator whether she thought the population mean was exactly 135, her answer would almost certainly be no. Well, then, how close is 135 to the population mean?

The usual way to answer this question is to give a range of values for which one is reasonably confident that the range includes the population mean. This is called *interval* estimation. For example, the administrator might have some confidence that the population mean lies within the range 130–140. Certainly she would have more confidence in the range of 130–140 than in the single value of 135. How about the range 110–160? Clearly there would be more confidence in this range than in the range 130–140. Thus, the wider the range, the greater the confidence that it contains the population mean.

DEFINITIONS

- *A* **confidence interval** *is a range of values that probably contains the population value.*

- **Confidence limits** *are the values that state the boundaries of the confidence interval.*

It is possible to be more quantitative about the degree of confidence we have that the interval contains the population mean. In fact, we can construct confidence intervals about which there are specified degrees of confidence. For example, we could construct the 95% confidence interval:

- *The* 95% *confidence interval is an interval such that the probability is* 0.95 *that the interval contains the population value.*

Although there are many different intervals we could construct, in practice the 95% and 99% confidence intervals are most often used. Let's consider how to construct these intervals.

Construction of 95% Confidence Interval

Suppose we have randomly sampled a set of N scores from a population of raw scores having a mean $= \mu$ and have calculated t_{obt}. Assuming the assumptions of t are met, we see that the probability is 0.95 that the following inequality is true:

$$-t_{0.025} \leq t_{obt} \leq +t_{0.025}$$

$t_{0.025}$ is the critical value of t for $\alpha = 0.025_{1\,tail}$ and df $= N - 1$. All this inequality says is that if we randomly sample N scores from a population of raw scores

FIGURE 13.3

Percentage of *t* scores between $\pm t_{crit}$ for $\alpha = 0.05_{2\,tail}$ and df $= N - 1$

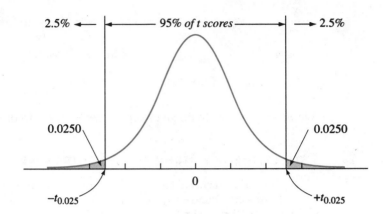

FIGURE 13.3

Percentage of *t* scores between $\pm t_{crit}$ for $\alpha = 0.05_{2\,tail}$ and df $= N - 1$

having a mean of μ and calculate t_{obt}, the probability is 0.95 that t_{obt} will lie between $-t_{0.025}$ and $+t_{0.025}$. The truth of this statement can be understood best by referring to Figure 13.3. This figure shows the *t* distribution for $N - 1$ degrees of freedom. We've located $+t_{0.025}$ and $-t_{0.025}$ on the distribution. Remember that these values are the critical values of *t* for $\alpha = 0.025_{1\,tail}$. By definition, 2.5% of the *t* values must lie under each tail, and 95% of the values must lie between $-t_{0.025}$ and $+t_{0.025}$. It follows, then, that the probability is 0.95 that t_{obt} will lie between $-t_{0.025}$ and $+t_{0.025}$.

We can use the previously given inequality to derive an equation for estimating the value of an unknown μ. Thus,

$$-t_{0.025} \leq t_{obt} \leq t_{0.025}$$

but

$$t_{obt} = \frac{\overline{X}_{obt} - \mu}{s_{\overline{X}}}$$

Therefore,

$$-t_{0.025} \leq \frac{\overline{X}_{obt} - \mu}{s_{\overline{X}}} \leq t_{0.025}$$

Solving this inequality for μ, we obtain*

$$\overline{X}_{obt} - s_{\overline{X}} t_{0.025} \leq \mu \leq \overline{X}_{obt} + s_{\overline{X}} t_{0.025}$$

This states that the chances are 95 in 100 that the interval $\overline{X}_{obt} \pm s_{\overline{X}} t_{0.025}$ contains the population mean. Thus, the interval $\overline{X}_{obt} + s_{\overline{X}} t_{0.025}$ is the 95% confidence

---

* See Note 13.2 for the intermediate steps in this derivation.

interval. The lower and upper confidence limits are given by

$$\mu_{\text{lower}} = \bar{X}_{\text{obt}} - s_{\bar{X}}t_{0.025} \qquad \textit{lower limit for 95\% confidence interval}$$

$$\mu_{\text{upper}} = \bar{X}_{\text{obt}} + s_{\bar{X}}t_{0.025} \qquad \textit{upper limit for 95\% confidence interval}$$

We are now ready to do an example. Let's return to the university administrator.

EXAMPLE

Estimating the Mean IQ of Professors at a University

Suppose a university administrator is interested in determining the average IQ of professors at her university. Since it is too costly to test all of the professors, a random sample of 20 is drawn from the population. Each professor is given an IQ test, and the results show a sample mean of 135 and a sample standard deviation of 8. Construct the 95% confidence interval for the population mean.

SOLUTION

The 95% confidence interval for the population mean can be found by solving the equations for the upper and lower confidence limits. Thus,

$$\mu_{\text{lower}} = \bar{X}_{\text{obt}} - s_{\bar{X}}t_{0.025} \quad \text{and} \quad \mu_{\text{upper}} = \bar{X}_{\text{obt}} + s_{\bar{X}}t_{0.025}$$

Solving for $s_{\bar{X}}$,

$$s_{\bar{X}} = \frac{s}{\sqrt{N}} = \frac{8}{\sqrt{20}} = 1.789$$

From Table D, with $\alpha = 0.025_{1\,\text{tail}}$ and df $= N - 1 = 20 - 1 = 19$,

$$t_{0.025} = 2.093$$

Substituting the values for $s_{\bar{X}}$ and $t_{0.025}$ in the confidence limit equations, we obtain

$$\mu_{\text{lower}} = \bar{X}_{\text{obt}} - s_{\bar{X}}t_{0.025} \qquad \text{and} \qquad \mu_{\text{upper}} = \bar{X}_{\text{obt}} + s_{\bar{X}}t_{0.025}$$

$$= 135 - 1.789(2.093) \qquad\qquad\qquad = 135 + 1.789(2.093)$$

$$= 135 - 3.744 \qquad\qquad\qquad\qquad = 135 + 3.744$$

$$= 131.26 \quad \textit{lower limit} \qquad\qquad = 138.74 \quad \textit{upper limit}$$

Thus, the 95% confidence interval = 131.26–138.74.

What precisely does it mean to say that the 95% confidence interval equals a certain range? In the case of the previous sample, the range is 131.26–138.74. A second sample would yield a different $\bar{X}_{\text{obt}}$ and a different range, perhaps $\bar{X}_{\text{obt}} = 138$ and a range of 133.80–142.20. If we took all of the different possible samples from the population, we would have derived the sampling distribution of the 95% confidence interval. *The important point here is that 95% of these intervals will contain the population mean; 5% of the intervals will not.* Thus, when we say "the 95% confidence interval is 131.26–138.74," what we mean is that the probability is 0.95 that the interval contains the population mean. Note that the probability value applies to the interval and not to the population mean. The population mean is constant. What varies from sample to sample is the interval. Thus, it is not technically proper to state "the probability is 0.95 that the population mean lies within the interval." Rather, the proper statement is "the probability is 0.95 that the interval contains the population mean."

General Equations for Any Confidence Interval

The equations we have presented thus far deal only with the 95% confidence interval. However, they are easily extended to form general equations for any confidence interval. Thus,

$$\mu_{lower} = \overline{X}_{obt} - s_{\overline{X}} t_{crit} \qquad \textit{general equation for lower confidence limit}$$

$$\mu_{upper} = \overline{X}_{obt} + s_{\overline{X}} t_{crit} \qquad \textit{general equation for upper confidence limit}$$

where t_{crit} = the critical one-tailed value of *t* corresponding to the desired confidence interval.

Thus, if we were interested in the 99% confidence interval, $t_{crit} = t_{0.005}$ = the critical value of *t* for $\alpha = 0.005_{1\,tail}$. To illustrate, let's solve the previous problem for the 99% confidence interval.

SOLUTION

From Table D, with df = 19 and $\alpha = 0.005_{1\,tail}$,

$$t_{0.005} = 2.861$$

From the previous solution, $s_{\overline{X}} = 1.789$. Substituting these values into the equations for confidence limits, we have

$\mu_{lower} = \overline{X}_{obt} - s_{\overline{X}} t_{crit}$	and	$\mu_{upper} = \overline{X}_{obt} + s_{\overline{X}} t_{crit}$
$= \overline{X}_{obt} - s_{\overline{X}} t_{0.005}$		$= \overline{X}_{obt} + s_{\overline{X}} t_{0.05}$
$= 135 - 1.789(2.861)$		$= 135 + 1.789(2.861)$
$= 135 - 5.118$		$= 135 + 5.118$
$= 129.88$ *lower limit*		$= 140.12$ *upper limit*

Thus, the 99% confidence interval = 129.88–140.12.

Note that this interval is larger than the 95% confidence interval (131.26–138.74). As discussed previously, the larger the interval, the more confidence we have that it contains the population mean.

Let's try a practice problem.

PRACTICE PROBLEM 13.3

An ethologist is interested in determining the average weight of adult Olympic marmots (found only on the Olympic Peninsula in Washington). Since it would be expensive and impractical to trap and measure the whole population, a random sample of 15 adults is trapped and weighed. The sample has a mean of 7.2 kilograms and a standard deviation of 0.48. Construct the 95% confidence interval for the population mean.

SOLUTION

The solution is shown in Table 13.4.

TABLE 13.4

The data are given in the problem. The 95% confidence interval for the population mean is found by determining the upper and lower confidence limits. Thus,

$$\mu_{lower} = \overline{X}_{obt} - s_{\overline{X}}t_{0.025} \quad \text{and} \quad \mu_{upper} = \overline{X}_{obt} + s_{\overline{X}}t_{0.025}$$

Solving for $s_{\overline{X}}$,

$$s_{\overline{X}} = \frac{s}{\sqrt{N}} = \frac{0.48}{\sqrt{15}} = 0.124$$

From Table D, with $\alpha = 0.025_{1\,tail}$ and df $= N - 1 = 15 - 1 = 14$,

$$t_{0.025} = 2.145$$

Substituting the values for $s_{\overline{X}}$ and $t_{0.025}$ in the confidence limit equations, we obtain

$$\mu_{lower} = \overline{X}_{obt} - s_{\overline{X}}t_{0.025} \quad\quad \text{and} \quad\quad \mu_{upper} = \overline{X}_{obt} + s_{\overline{X}}t_{0.025}$$
$$= 7.2 - 0.124(2.145) \quad\quad\quad\quad = 7.2 + 0.124(2.145)$$
$$= 7.2 - 0.266 \quad\quad\quad\quad\quad = 7.2 + 0.266$$
$$= 6.93 \quad \textit{lower limit} \quad\quad\quad = 7.47 \quad \textit{upper limit}$$

Thus, the 95% confidence interval = 6.93–7.47 kilograms.

Let's do another practice problem.

PRACTICE PROBLEM 13.4

To estimate the average life of their 100-watt light bulbs, the manufacturer randomly samples 200 light bulbs and keeps them lit until they burn out. The sample has a mean life of 215 hours and a standard deviation of 8 hours. Construct the 99% confidence limits for the population mean. In solving this problem, use the closest table value for degrees of freedom.

SOLUTION

The solution is shown in Table 13.5.

TABLE 13.5
....

The data are given in the problem.

$$\mu_{\text{lower}} = \overline{X}_{\text{obt}} - s_{\overline{X}} t_{0.005} \quad \text{and} \quad \mu_{\text{upper}} = \overline{X}_{\text{obt}} + s_{\overline{X}} t_{0.005}$$

$$s_{\overline{X}} = \frac{s}{\sqrt{N}} = \frac{8}{\sqrt{200}} = 0.567$$

From Table D, with $\alpha = 0.005_{1\,\text{tail}}$ and df $= N - 1 = 200 - 1 = 199$,

$$t_{0.005} = 2.617$$

Note that this is the closest table value available from Table D. Substituting the values for $s_{\overline{X}}$ and $t_{0.005}$ in the confidence limit equations, we obtain

$$\mu_{\text{lower}} = \overline{X}_{\text{obt}} - s_{\overline{X}} t_{0.005} \qquad \text{and} \qquad \mu_{\text{upper}} = \overline{X}_{\text{obt}} + s_{\overline{X}} t_{0.005}$$

$$= 215 - 0.567(2.617) \qquad\qquad\qquad = 215 + 0.567(2.617)$$

$$= 213.52 \quad \textit{lower limit} \qquad\qquad = 216.48 \quad \textit{upper limit}$$

Thus, the 99% confidence interval = 213.52–216.48 hours.

TESTING THE SIGNIFICANCE OF PEARSON *r*

When a correlational study is conducted, it is quite rare for the whole population to be involved. Rather, the usual procedure is to randomly sample from the population and calculate the correlation coefficient on the sample data. To determine whether a correlation exists in the population, we must test the significance of the obtained *r* (r_{obt}). Of course, this is the same procedure we have used all along for testing hypotheses. The population correlation coefficient is symbolized by the Greek letter ρ (rho). A nondirectional alternative hypothesis asserts that $\rho \neq 0$. A directional alternative hypothesis asserts that ρ is positive or negative depending on the predicted direction of the relationship. The null hypothesis is tested by assuming that the sample set of *X* and *Y* scores having a correlation equal to r_{obt} is a random sample from a population where $\rho = 0$. The sampling distribution of *r* can be generated empirically by taking all samples of size *N* from a population in which $\rho = 0$ and calculating *r* for each sample. By systematically varying the population scores and *N*, the sampling distribution of *r* is generated.

The significance of *r* can be evaluated using the *t* test. Thus,

$$t_{obt} = \frac{r_{obt} - \rho}{s_r} \qquad \textit{t test for testing the significance of r}$$

where r_{obt} = Correlation obtained on a sample of *N* subjects
    ρ = Population correlation coefficient
    s_r = Estimate of the standard deviation of the sampling distribution
       of *r*

Note that this is very similar to the *t* equation used when dealing with the mean of a single sample. The only difference is that the statistic we are dealing with is *r* rather than $\overline{X}$.

$$t_{obt} = \frac{r_{obt} - \rho}{s_r} = \frac{r_{obt}}{\sqrt{\dfrac{1 - r_{obt}^2}{N - 2}}}$$

where $\rho = 0$
    $s_r = \sqrt{(1 - r_{obt}^2)/(N - 2)}$
    $df = N - 2$

Let's use this equation to test the significance of the correlation obtained in the "IQ and grade point average" problem presented in Chapter 6, p. 111. Assume that the 12 students were a random sample from a population of university undergraduates and that we want to determine whether there is a correlation in the population. We'll use $\alpha = 0.05_{2\,tail}$ in making our decision.

Ordinarily, the first step in a problem of this sort is to calculate r_{obt}. However, we have already done this and found that $r_{obt} = 0.856$. Substituting this value into the t equation, we obtain

$$t_{obt} = \frac{r_{obt}}{\sqrt{\dfrac{1 - r_{obt}^2}{N-2}}} = \frac{0.856}{\sqrt{\dfrac{1 - (0.856)^2}{10}}} = \frac{0.856}{0.163} = 5.252 = 5.25$$

From Table D, with df $= N - 2 = 10$ and $\alpha = 0.05_{2\,tail}$,

$$t_{crit} = \pm 2.228$$

Since $|t_{obt}| > 2.228$, we reject H_0 and conclude that there is a significant positive correlation in the population.

Although the foregoing method works, there is an even easier way to solve this problem. By substituting t_{crit} into the t equation, r_{crit} can be determined for any df and any α level. Once r_{crit} is known, all we need do is compare r_{obt} with r_{crit}. The decision rule is

$$\text{If } |r_{obt}| \geq |r_{crit}|, \text{ reject } H_0.$$

Statisticians have already calculated r_{crit} for various df and α levels. These are shown in Table E in Appendix D. This table is used in the same way as the t table (Table D) except the entries list r_{crit} rather than t_{crit}.

Applying the r_{crit} method to the present problem, we would first calculate r_{obt} and then determine r_{crit} from Table E. Finally, we would compare r_{obt} with r_{crit} using the decision rule. In the present example, we have already determined that $r_{obt} = 0.856$. From Table E, with df $= 10$ and $\alpha = 0.05_{2\,tail}$,

$$r_{crit} = \pm 0.5760$$

Since $|r_{obt}| > 0.5760$, we reject H_0, as before. This solution is preferred because it is shorter and easier than the solution that involves comparing t_{obt} with t_{crit}.

Let's try some problems for practice.

PRACTICE PROBLEM 13.5

Folklore has it that there is an inverse correlation between mathematical and artistic ability. A psychologist decides to determine whether there is anything to this notion. She randomly samples 15 undergraduates and gives them tests measuring these two abilities. The resulting data are shown in Table 13.6. Is there a correlation in the population between mathematical ability and artistic ability? Use $\alpha = 0.01_{2\,tail}$.

TABLE 13.6   Mathematical and Artistic Ability

Subject No.	Math Ability, X	Artistic Ability, Y	X^2	Y^2	XY
1	15	19	225	361	285
2	30	22	900	484	660
3	35	17	1,225	289	595
4	10	25	100	625	250
5	28	23	784	529	644
6	40	21	1,600	441	840
7	45	14	2,025	196	630
8	24	10	576	100	240
9	21	18	441	324	378
10	25	19	625	361	475
11	18	30	324	900	540
12	13	32	169	1,024	416
13	9	16	81	256	144
14	30	28	900	784	840
15	23	24	529	576	552
Total	366	318	10,504	7,250	7,489

SOLUTION

Step 1: Calculate the appropriate statistic:

$$r_{obt} = \frac{\sum XY - \dfrac{(\sum X)(\sum Y)}{N}}{\sqrt{\left[\sum X^2 - \dfrac{(\sum X)^2}{N}\right]\left[\sum Y^2 - \dfrac{(\sum Y)^2}{N}\right]}}$$

$$= \frac{7489 - \dfrac{366(318)}{15}}{\sqrt{\left[10,504 - \dfrac{(366)^2}{15}\right]\left[7250 - \dfrac{(318)^2}{15}\right]}}$$

$$= \frac{-270.2}{894.437}$$

$$= -0.302$$

$$= -0.30$$

Step 2: Evaluate the statistic. From Table E, with df $= N - 2 = 15 - 2 = 13$ and $\alpha = 0.01_{2\,tail}$,

$$r_{crit} = \pm 0.6411$$

Since $|r_{obt}| < 0.6411$, we conclude by retaining H_0.

PRACTICE PROBLEM 13.6

In Chapter 6, Practice Problem 6.2, we calculated the Pearson r for the relationship between similarity of attitudes and attraction in a sample of 15 college students. In that example, $r_{obt} = 0.94$. Using $\alpha = 0.05_{2\,tail}$, let's now determine whether this is a significant value for r_{obt}.

SOLUTION

From Table E, with df $= N - 2 = 13$ and $\alpha = 0.05_{2\,tail}$,

$$r_{crit} = \pm 0.5139$$

Since $|r_{obt}| > 0.5139$, we reject H_0 and conclude there is a significant correlation in the population.

SUMMARY

In this chapter, we have discussed the use of Student's t test for (1) testing hypotheses involving single sample experiments, (2) estimating the population mean by constructing confidence intervals, and (3) testing the significance of Pearson r.

In testing hypotheses involving single sample experiments, the t test is appropriate when the mean of the null-hypothesis population is known and the standard deviation is unknown. In this situation, we estimate σ by using the sample standard deviation. The equation for calculating t_{obt} is very similar to z_{obt} except we use s instead of σ. The sampling distribution of t is a family of curves that vary with the degrees of freedom associated with calculating t. There are $N - 1$ degrees of freedom associated with the t test for single samples. The sampling distribution curves are symmetrical, bell-shaped curves having a mean equal to 0. However, these are elevated at the tails relative to the normal distribution. In using the t test, t_{obt} is computed and then evaluated to determine whether it falls within the critical region. The t test

is appropriate when the sampling distribution of $\overline{X}$ is normal. For the sampling distribution of $\overline{X}$ to be normal, the population of raw scores should be normally distributed, or $N \geq 30$.

Next we discussed constructing confidence intervals for the population mean. A confidence interval was defined as a range of values that probably contains the population value. Confidence limits are the values that bound the confidence interval. In discussing this topic, we showed how to construct confidence intervals about which we have a specified degree of confidence that the interval contains the population mean. Illustrative and practice problems were given for constructing the 95% and 99% confidence intervals.

The last topic involved testing the significance of Pearson r. We pointed out that, since most correlative data are collected on samples, we must evaluate the sample r value (r_{obt}) to see whether there is a correlation in the population. The evaluation involves the t test. However, by substituting t_{crit} into the t equation, we can determine r_{crit} for any df and any alpha level. The value of r_{obt} is evaluated by comparing it with r_{crit} for the given df and alpha level. Several problems were given for practice in evaluating r_{obt}.

IMPORTANT TERMS

Confidence interval (p. 304)   Critical value(s) of r (p. 311)   Degrees of freedom (p. 295)
Confidence limits (p. 304)   Critical value(s) of t (p. 298)

QUESTIONS AND PROBLEMS

1. Define each of the terms in the "Important Terms" section.
2. Assuming the assumptions underlying the t test are met, what are the characteristics of the sampling distribution of t?
3. Explain what is meant by *degrees of freedom*. Use an example.
4. What are the assumptions underlying the proper use of the t test?
5. Discuss the similarities and differences between the z and t tests.
6. Explain in a short paragraph why the z test is more powerful than the t test.
7. Which of the following two statements is technically more correct? (1) We are 95% confident that the population mean lies in the interval 80–90, or (2) We are 95% confident that the interval 80–90 contains the population mean. Explain.
8. Explain why df $= N - 1$ when the t test is used with single samples.
9. If the sample correlation coefficient has a value different from zero, e.g., $r = 0.45$, this automatically means that the correlation in the population is also different from zero. Is this statement correct? Explain.
10. A sample set of 30 scores has a mean equal to 82 and a standard deviation of 12. Can we reject the hypothesis that this sample is a random sample from a normal population with $\mu = 85$? Use $\alpha = 0.01_{2\,tail}$ in making your decision.
11. A sample set of 29 scores has a mean of 76 and a standard deviation of 7. Can we accept the hypothesis that the sample is a random sample from a population with a mean greater than 72? Use $\alpha = 0.01_{1\,tail}$ in making your decision.

12. Is it reasonable to consider a sample with $N = 22$, $\bar{X}_{obt} = 42$, and $s = 9$ to be a random sample from a normal population with $\mu = 38$? Use $\alpha = 0.05_{1\,tail}$ in making your decision. Assume $\bar{X}_{obt}$ is in the right direction.
13. Using each of the following random samples, determine the 95% and 99% confidence intervals for the population mean:
    a. $\bar{X}_{obt} = 25$, $s = 6$, $N = 15$
    b. $\bar{X}_{obt} = 120$, $s = 8$, $N = 30$
    c. $\bar{X}_{obt} = 30.6$, $s = 5.5$, $N = 24$
    d. Redo part **a** with $N = 30$. What happens to the confidence interval as N increases?
14. Assume in Problem 20 of Chapter 12, p. 291, that the standard deviation of the population is unknown. Again using $\alpha = 0.05_{2\,tail}$, what do you conclude about the student's technique? Explain the difference in conclusion between Problem 20 and this one.
15. As the principal of a private high school, you are interested in finding out how the training in mathematics at your school compares with that of the public schools in your area. For the last 5 years, the public schools have given all graduating seniors a mathematics proficiency test. The distribution has a mean of 78. You give all the graduating seniors in your school the same mathematics proficiency test. The results show a distribution of 41 scores, with a mean of 83 and a standard deviation of 12.2.
    a. What is the alternative hypothesis? Use a nondirectional hypothesis.
    b. What is the null hypothesis?
    c. Using $\alpha = 0.05_{2\,tail}$, what do you conclude?
16. A college counselor wants to determine the aver-

age amount of time first-year students spend studying. He randomly samples 61 students from the freshman class and asks them how many hours a week they study. The mean of the resulting scores is 20 hours, and the standard deviation is 6.5 hours.

a. Construct the 95% confidence interval for the population mean.

b. Construct the 99% confidence interval for the population mean.

17. A professor in the women's studies program believes that the amount of smoking by women has increased in recent years. A complete census taken 2 years ago of women living in a neighboring city showed that the mean number of cigarettes smoked daily by the women was 5.4 with a standard deviation of 2.5. To assess her belief, the professor determined the daily smoking rate of a random sample of 200 women currently living in that city. The data show that the number of cigarettes smoked daily by the 200 women has a mean of 6.1 and a standard deviation of 2.7.

a. Is the professor's belief correct? Assume a directional H_1 is appropriate and use $\alpha = 0.05_{1\,tail}$ in making your decision. Be sure that the most sensitive test is used to analyze the data.

b. Assume the population mean is unknown and reanalyze the data using the same alpha level. What is your conclusion this time? Explain any differences between part **a** and part **b**.

18. A physician employed by a large corporation believes that due to an increase in sedentary life in the past decade, middle-age men have become fatter. In 1970, the corporation measured the percentage of fat in their employees. For the middle-age men, the scores were normally distributed with a mean of 22%. To test her hypothesis, the physician measures the fat percentage in a random sample of 12 middle-age men currently employed by the corporation. The fat percentages found were as follows: 24, 40, 29, 32, 33, 25, 15, 22, 18, 25, 16, 27. On the basis of these data, can we conclude that middle-age men employed by the corporation have become fatter? Assume a directional H_1 is legitimate and use $\alpha = 0.05_{1\,tail}$ in making your decision.

19. A local business school claims that their graduating seniors get higher-paying jobs than the national average. Last year's figures for salaries paid to all business school graduates on their first job showed a mean of $6.20 per hour. A random sample of 10 graduates from last year's class of the local business school showed the following hourly salaries for their first job: $5.40, $6.30, $7.20, $6.80, $6.40, $5.70, $5.80, $6.60, $6.70, $6.90. Since you are skeptical of the business school claim, evaluate the salary of the business school graduates using $\alpha = 0.05_{2\,tail}$.

20. You wanted to estimate the mean number of vehicles crossing a busy bridge in your neighborhood each morning during rush hour for the past year. To accomplish this, you stationed yourself and a few assistants at one end of the bridge on 18 randomly selected mornings during the year and counted the number of vehicles crossing the bridge in a 10-minute period during rush hour. You found the mean to be 125 vehicles per minute with a standard deviation of 32.

a. Construct the 95% confidence limits for the population mean (vehicles per minute).

b. Construct the 99% confidence limits for the population mean (vehicles per minute).

21. Refer to the data given in Problem 15 on p. 124 in answering the following questions.

a. Construct a scatter plot for these data.

b. Calculate the value of Pearson r.

c. Is the correlation between cigarettes smoked and days absent significant? Use $\alpha = 0.05_{2\,tail}$.

22. Refer to the scores given in Problem 16 on p. 125 in answering the following questions.

a. Calculate the value of Pearson r for the two administrations of the mechanical aptitude test.

b. Is the correlation significant? Use $\alpha = 0.05_{2\,tail}$.

23. Refer to the data given in Problem 14 on p. 124 in answering the following questions.

a. Calculate the value of Pearson r for the two tests.

b. Using $\alpha = 0.05_{2\,tail}$, determine whether the correlation is significant. If not, does this mean that $\rho = 0$? Explain.

c. Assume you increased the number of students to 20, and now $r = 0.653$. Using the same alpha level as in part **b**, what do you conclude this time? Explain.

24. A developmental psychologist is interested in whether tense parents tend to have tense children. A study is done involving one parent for each of 15 families and the oldest child in each family, measuring tension in each pair. Pearson $r = 0.582$. Using $\alpha - 0.05_{2\,tail}$, is the relationship significant?

NOTES

13.1 $t_{obt} = \dfrac{\overline{X}_{obt} - \mu}{s/\sqrt{N}}$

Given

$= \dfrac{\overline{X}_{obt} - \mu}{\sqrt{\dfrac{SS}{N-1}} \Big/ \sqrt{N}}$

Substituting $\sqrt{\dfrac{SS}{N-1}}$ for s

$(t_{obt})^2 = \dfrac{(\overline{X}_{obt} - \mu)^2}{\left(\dfrac{SS}{N-1}\right)\left(\dfrac{1}{N}\right)} = \dfrac{(\overline{X}_{obt} - \mu)^2}{\dfrac{SS}{N(N-1)}}$

Squaring both sides of the equation and rearranging terms

$t_{obt} = \dfrac{\overline{X}_{obt} - \mu}{\sqrt{\dfrac{SS}{N(N-1)}}}$

Taking the square root of both sides of the equation

13.2 $-t_{0.025} \le \dfrac{\overline{X}_{obt} - \mu}{s_{\overline{X}}} \le t_{0.025}$

Given

$-s_{\overline{X}} t_{0.025} \le \overline{X}_{obt} - \mu \le s_{\overline{X}} t_{0.025}$

Multiplying by $s_{\overline{X}}$

$-\overline{X}_{obt} - s_{\overline{X}} t_{0.025} \le -\mu \le -\overline{X}_{obt} + s_{\overline{X}} t_{0.025}$

Subtracting $\overline{X}_{obt}$

$\overline{X}_{obt} + s_{\overline{X}} t_{0.025} \ge \mu \ge \overline{X}_{obt} - s_{\overline{X}} t_{0.025}$

Multiplying by -1

$\overline{X}_{obt} - s_{\overline{X}} t_{0.025} \le \mu \le \overline{X}_{obt} + s_{\overline{X}} t_{0.025}$

Rearranging terms

14

STUDENT'S *t* TEST FOR CORRELATED AND INDEPENDENT GROUPS

INTRODUCTION

In Chapters 12 and 13, we have seen that hypothesis testing basically involves two steps: (1) calculating the appropriate statistic and (2) evaluating the statistic using its sampling distribution. We further discussed how to use the z and t tests to evaluate hypotheses that have been investigated with single sample experiments. In this chapter, we shall present the t test in conjunction with experiments involving two conditions or two samples.

We have already encountered the two-condition experiment when using the sign test and the Mann–Whitney U test. The two-condition experiment, whether of the correlated groups or independent groups design, has great advantages over the single sample experiment previously discussed. A major limitation of the single sample experiment is the requirement that at least one population parameter (μ) must be specified. In the great majority of cases, this information is not available. As will be shown later in this chapter, the two-treatment experiment completely eliminates the need to measure population parameters when testing hypotheses. This has obvious widespread practical utility.

A second major advantage of the two-condition experiment has to do with interpreting the results of the study. Correct scientific methodology does not often allow an investigator to use previously acquired population data when conducting an experiment. For example, in the illustrative problem involving first word utterances, we used a population mean value of 13.0 months. How do we really know the mean is 13.0 months? Suppose the figures were collected 3 to 5 years prior to performing the experiment. How do we know that infants haven't changed over those years? And what about the conditions under which the population data were collected? Were they the same as in the experiment? Isn't it possible that the people collecting the population data were not as motivated as the experimenter and, hence, were not as careful in collecting the data? Just how were the data collected? By being on hand at the moment that the child spoke the first word? Quite unlikely. The data probably were collected by asking parents when their children first spoke. How accurate, then, is the population mean? Even if the foregoing problems didn't exist, there are others having to do with the conduct of the experiment itself. For example, assuming 13.0 months is accurate and applies properly to the sample of 15 infants, how can we be sure it was the experimenter's technique that produced the early utterances? Couldn't it have been due to the extra attention or handling or stimulation given the children in conjunction with the method rather than the method itself?

Many of these problems can be overcome by the use of the two-condition experiment. By using two groups of infants, arrived at by matched pairs or random assignment, giving each group the same treatment except for the experimenter's particular technique (same in attention, handling, etc.), running both groups concurrently, using the same people to collect the data from both groups, and so forth, most alternative explanations of results can be ruled out. In the discussion that follows, we shall first consider the *t* test for the correlated groups design and then for the independent groups design.

STUDENT'S *t* TEST FOR CORRELATED GROUPS*

You will recall that, in the repeated measures or correlated groups design, each subject gets two or more treatments: A difference score is calculated for each subject, and the resulting difference scores are analyzed. The simplest experiment of this type uses two conditions, often called *control* and *experimental,* or *before* and *after*. In a variant of this design, instead of the same subject being used in both conditions, pairs of subjects that are matched on one or more characteristics serve in the two conditions. Thus, pairs might be matched on IQ, age, gender, and so forth. The difference scores between the matched pairs are then analyzed in the same manner as when the same subject serves in both conditions. This design is also referred to as a *correlated groups design* because the subjects in the groups are not independently assigned; i.e., the pairs share specifically matched common characteristics. In the independent groups design, which is discussed later in this chapter, there is no pairing.

We first encountered the correlated groups design when using the sign test. However, the sign test had low power because it ignored the magnitude of the difference scores. We used the sign test because of its simplicity. In the analysis of actual experiments another test, such as the *t* test, would probably be used. The *t* test for correlated groups allows utilization of *both* the magnitude and direction of the difference scores. Essentially, it treats the difference scores as though they were raw scores and tests the assumption that the difference scores are a random sample from a population of difference scores having a mean of zero. This can best be seen through an example.

AN EXPERIMENT:

Brain Stimulation and Eating

To illustrate, suppose a physiological psychologist believes that a brain region called the lateral hypothalamus is involved in eating behavior. One way to test this belief is to use a group of animals (e.g., rats) and electrically stimulate the lateral hypothalamus

---

* See Note 14.1.

through a chronically indwelling electrode. If the lateral hypothalamus is involved in eating behavior, electrical stimulation of the lateral hypothalamus might alter the amount of food eaten. To control for the effect of brain stimulation per se, another electrode would be implanted in each animal in a neutral brain area. Each area would be stimulated for a fixed period of time, and the amount of food eaten would be recorded. A difference score for each animal would then be calculated.

Let's assume there is insufficient supporting evidence to warrant a directional alternative hypothesis. Therefore, a two-tailed evaluation is planned. The alternative hypothesis states that electrical stimulation of the lateral hypothalamus affects the amount of food eaten. The null hypothesis specifies that electrical stimulation of the lateral hypothalamus does not affect the amount of food eaten. If H_0 is true, the difference score for each rat would be due to chance factors. Sometimes it would be positive, and sometimes it would be negative; sometimes it would be large in magnitude, and sometimes small. If the experiment were done on a large number of rats, say, the entire population, the mean of the difference scores would equal zero.* Figure 14.1 shows such a distribution. Note, carefully, that the mean of this population is known ($\mu_D = 0$) and that the standard deviation is unknown ($\sigma_D = ?$). The chance explanation assumes that the difference scores of the sample in the experiment are a random sample from this population of difference scores. Thus, we have a situation in which there is one set of scores (e.g., the sample difference scores), and we are interested in determining whether it is reasonable to consider these scores a random sample from a population of difference scores having a known mean ($\mu_D = 0$) and unknown standard deviation. This situation is almost identical to those we have previously considered regarding the *t* test with single samples. The only change is that in the correlated groups experiment we are dealing with *difference* scores rather than *raw* scores. It follows, then, that the equations for each should be quite similar. These equations are presented here:

t Test for Single Samples	*t* Test for Correlated Groups
$t_{obt} = \dfrac{\bar{X}_{obt} - \mu}{s/\sqrt{N}}$	$t_{obt} = \dfrac{\bar{D}_{obt} - \mu_D}{s_D/\sqrt{N}}$
$t_{obt} = \dfrac{\bar{X}_{obt} - \mu}{\sqrt{\dfrac{SS}{N(N-1)}}}$	$t_{obt} = \dfrac{\bar{D}_{obt} - \mu_D}{\sqrt{\dfrac{SS_D}{N(N-1)}}}$
$SS = \sum X^2 - \dfrac{(\sum X)^2}{N}$	$SS_D = \sum D^2 - \dfrac{(\sum D)^2}{N}$

where

D = Difference score
$\bar{D}_{obt}$ = Mean of the sample difference scores
μ_D = Mean of the population of difference scores
s_D = Standard deviation of the sample difference scores
N = Number of difference scores
$SS_D = \sum (D - \bar{D})^2$ = Sum of squares of sample difference scores

---

* See Note 14.2.

FIGURE 14.1

Null-hypothesis population
of difference scores

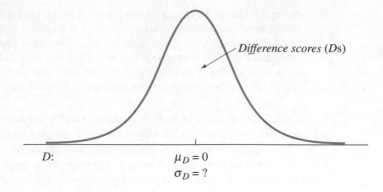

$$\mu_D = 0$$
$$\sigma_D = ?$$

It is obvious that the two sets of equations are identical except that, in the single sample case, we are dealing with raw scores, whereas in the correlated groups experiment we are analyzing difference scores. Let's now add some numbers to the brain stimulation experiment and see how to use the *t* test for correlated groups.

AN EXPERIMENT

Brain Stimulation Experiment Revisited

A psychologist believes that the lateral hypothalamus is involved in eating behavior. If so, then electrical stimulation of that area might affect the amount eaten. To test this possibility, 10 rats are implanted with chronically indwelling electrodes. Each rat has two electrodes, one implanted in the lateral hypothalamus and the other in an area where electrical stimulation is known to have no effect. After the animals have recovered from surgery, they each receive 30 minutes of electrical stimulation to each brain area, and the amount of food eaten during the stimulation is measured. The following amounts of food in grams were eaten during the stimulation to each area:

	Food Eaten		Difference	
Subject	*Lateral Hypothalamus* (g)	*Neutral Area* (g)	*D*	*D²*
1	10	6	4	16
2	18	8	10	100
3	16	11	5	25
4	22	14	8	64
5	14	10	4	16
6	25	20	5	25
7	17	10	7	49
8	22	18	4	16
9	12	14	−2	4
10	21	13	8	64
			53	379

$$N = 10 \qquad \bar{D}_{obt} = \frac{\Sigma D}{N} = \frac{53}{10} = 5.3$$

1. What is the alternative hypothesis? Assume a nondirectional hypothesis is appropriate.
2. What is the null hypothesis?
3. What is the conclusion? Use $\alpha = 0.05_{2\,\text{tail}}$.

SOLUTION

1. Alternative hypothesis: The alternative hypothesis specifies that electrical stimulation of the lateral hypothalamus affects the amount of food eaten. The sample difference scores having a mean $\bar{D}_{\text{obt}} = 5.3$ are a random sample from a population of difference scores having a mean $\mu_D \neq 0$.

2. Null hypothesis: The null hypothesis states that electrical stimulation of the lateral hypothalamus has no effect on the amount of food eaten. The sample difference scores having a mean $\bar{D}_{\text{obt}} = 5.3$ are a random sample from a population of difference scores having a mean $\mu_D = 0$.

3. Conclusion, using $\alpha = 0.05_{2\,\text{tail}}$:

   Step 1: **Calculate the appropriate statistic.** Since this is a correlated groups design, we are interested in the difference between the paired scores rather than the scores in each condition per se. The difference scores are shown in the table. Of the tests covered so far, both the sign test and the t test are possible choices. Since we want to use the test that is most sensitive, the t test has been chosen. From the data table, $N = 10$ and $\bar{D}_{\text{obt}} = 5.3$. The calculation of t_{obt} is as follows:

$$t_{\text{obt}} = \frac{\bar{D}_{\text{obt}} - \mu_D}{\sqrt{\dfrac{SS_D}{N(N-1)}}} \qquad SS_D = \Sigma D^2 - \frac{(\Sigma D)^2}{N}$$

$$= \frac{5.3 - 0}{\sqrt{\dfrac{98.1}{10(9)}}} \qquad\qquad = 379 - \frac{(53)^2}{10}$$

$$= \frac{5.3}{\sqrt{1.09}} \qquad\qquad\quad = 98.1$$

$$= 5.08$$

   Step 2: **Evaluate the statistic.** As with the t test for single samples, if t_{obt} falls within the critical region for rejection of H_0, the conclusion is to reject H_0. Thus, the same decision rule applies, namely,

$$\text{If } |t_{\text{obt}}| \geq |t_{\text{crit}}|, \text{reject } H_0.$$

   The degrees of freedom are equal to the number of difference scores minus one. Thus, df $= N - 1 = 10 - 1 = 9$. From Table D in Appendix D, with $\alpha = 0.05_{2\,\text{tail}}$ and df $= 9$,

$$t_{\text{crit}} = \pm 2.262$$

   Since $|t_{\text{obt}}| > 2.262$, we reject H_0 and conclude that electrical stimulation of the lateral hypothalamus affects eating behavior. It appears to increase the amount eaten.

PRACTICE PROBLEM 14.1

To motivate citizens to conserve gasoline, the government is considering mounting a nationwide conservation campaign. However, before doing so on a national level, it decides to conduct an experiment to evaluate the effectiveness of the campaign. For the experiment, the conservation campaign is conducted in a small but representative geographical area. Twelve families are randomly selected from the area, and the amount of gasoline they use is monitored for 1 month prior to the advertising campaign and for 1 month following the campaign. The following data are collected:

Family	Before the Campaign (gal/mo.)	After the Campaign (gal/mo.)	Difference D	Difference D^2
A	55	48	7	49
B	43	38	5	25
C	51	53	−2	4
D	62	58	4	16
E	35	36	−1	1
F	48	42	6	36
G	58	55	3	9
H	45	40	5	25
I	48	49	−1	1
J	54	50	4	16
K	56	58	−2	4
L	32	25	7	49
			35	235

$$N = 12 \qquad \bar{D}_{obt} = \frac{\Sigma D}{N} = \frac{35}{12} = 2.917$$

a. What is the alternative hypothesis? Use a nondirectional hypothesis.
b. What is the null hypothesis?
c. What is the conclusion? Use $\alpha = 0.05_{2\,tail}$.

SOLUTION

The solution is shown in Table 14.1.

TABLE 14.1

a. Alternative hypothesis: The conservation campaign affects the amount of gasoline used. The sample with $\overline{D}_{obt} = 2.917$ is a random sample from a population of difference scores where $\mu_D \neq 0$.
b. Null hypothesis: The conservation campaign has no effect on the amount of gasoline used. The sample with $\overline{D}_{obt} = 2.917$ is a random sample from a population of difference scores where $\mu_D = 0$.
c. Conclusion, using $\alpha = 0.05_{2\,tail}$:

Step 1: **Calculate the appropriate statistic.** The difference scores are included in the previous table. We have subtracted the "after" scores from the "before" scores. Assuming the assumptions of *t* are met, the appropriate statistic is t_{obt}. From the data table, $N = 12$ and $\overline{D}_{obt} = 2.917$.

$$t_{obt} = \frac{\overline{D}_{obt} - \mu_D}{\sqrt{\dfrac{SS_D}{N(N-1)}}} \qquad SS_D = \Sigma D^2 - \frac{(\Sigma D)^2}{N}$$

$$= \frac{2.917 - 0}{\sqrt{\dfrac{132.917}{12(11)}}} \qquad = 235 - \frac{(35)^2}{12}$$

$$= 2.91 \qquad\qquad = 132.917$$

Step 2: **Evaluate the statistic.** Degrees of freedom $= N - 1 = 12 - 1 = 11$. From Table D, with $\alpha = 0.05_{2\,tail}$ and 11 df,

$$t_{crit} = \pm 2.201$$

Since $|t_{obt}| > 2.201$, we reject H_0. The conservation campaign affects the amount of gasoline used. It appears to decrease gasoline consumption.

Let's do one more practice problem.

PRACTICE PROBLEM 14.2

Recently, an endogenous brain neurotransmitter called Galanin has been discovered that appears to specifically affect one's desire to eat foods with high fat content. The more of this naturally occurring neurotransmitter that an individual possesses, the higher is his or her craving for high-fat foods. Recently, a drug company developed an experimental drug that blocks Galanin without affecting the appetite for healthier (less fat) foods. A neuroscientist at the drug company believes that the experimental drug will be very useful for controlling obesity. More specifically, he believes that a daily regime of this drug will result in eating foods with less fat and thereby promote weight loss. An experiment is conducted in which 15 obese female volunteers are randomly selected and given the experimental drug for 6 months. Baseline and ending (6-month) weights are recorded for each subject. These weights are shown in the table.

Subject No.	Baseline Weight (lbs)	Ending Weight (lbs)	*D*	*D²*
1	165	145	20	400
2	143	137	6	36
3	175	170	5	25
4	135	136	−1	1
5	148	141	7	49
6	155	138	17	289
7	158	137	21	441
8	140	125	15	225
9	172	161	11	121
10	164	156	8	64
11	178	165	13	169
12	182	170	12	144
13	190	176	14	196
14	169	154	15	225
15	157	143	14	196
			177	2581

$$N = 15 \qquad \bar{D}_{obt} = \frac{\Sigma D}{N} = \frac{177}{15} = 11.80$$

a. What is the directional alternative hypothesis?
b. What is the null hypothesis?
c. What is the conclusion? Use $\alpha = 0.05_{1\,tail}$.

SOLUTION

a. Alternative hypothesis: Use of the experimental drug results in weight loss. The sample with $\bar{D}_{\text{obt}} = 11.80$ is a random sample from a population of difference scores where $\mu_D > 0$.

b. Null hypothesis: Use of the experimental drug either has no effect on weight loss or produces weight gain. The sample with $\bar{D}_{\text{obt}} = 11.80$ is a random sample from a population of difference scores where $\mu_D \leq 0$.

c. Conclusion, using $\alpha = 0.05_{1\,\text{tail}}$.

Step 1: Calculate the appropriate statistic. The difference scores are shown in the table. We have subtracted the Ending Weight scores from the Baseline Weight scores. Assuming the assumptions of t are met, the appropriate statistic is t_{obt}. From the data table, $N = 15$, $\bar{D}_{\text{obt}} = 11.80$.

$$t_{\text{obt}} = \frac{\bar{D}_{\text{obt}} - \mu_D}{\sqrt{\dfrac{SS_D}{N(N-1)}}} \qquad SS_D = \sum D^2 - \frac{(\sum D)^2}{N}$$

$$= \frac{11.80 - 0}{\sqrt{\dfrac{492.40}{15(14)}}} \qquad\qquad = 2581 - \frac{(177)^2}{15}$$

$$= 492.40$$

$$= 7.71$$

Step 2: Evaluate the statistic. Degrees of freedom $= N - 1 = 14$. From Table D, with $\alpha = 0.05_{1\,\text{tail}}$ and 14 df,

$$t_{\text{crit}} = 1.761$$

Since $|t_{\text{obt}}| > 1.761$, we reject H_0. Use of the experimental drug appears to produce weight loss.

t TEST AND SIGN TEST COMPARED

It would have been possible to solve either of the previous two problems using the sign test. We chose the t test because it is more sensitive. To illustrate this point, let's use the sign test to solve the problem dealing with gasoline conservation.

SOLUTION

Step 1: Calculate the statistic. There are 8 pluses in the sample.

Step 2: Evaluate the statistic. With $N = 12$, $P = 0.50$, and $\alpha = 0.05_{2\,\text{tail}}$,

$$p(8 \text{ or more pluses}) = p(8) + p(9) + p(10) + p(11) + p(12)$$

From Table B in Appendix D,

$$p(8 \text{ or more pluses}) = 0.1208 + 0.0537 + 0.0161 + 0.0029 + 0.0002$$
$$= 0.1937$$

Since the alpha level is two-tailed,

$$p(\text{outcome as extreme as 8 pluses}) = 2[p(8 \text{ or more pluses})]$$
$$= 2(0.1937)$$
$$= 0.3874$$

Since $0.3874 > 0.05$, we conclude by retaining H_0.

We are unable to reject H_0 with the sign test, but were able to reject H_0 with the t test. Does this mean the campaign is effective if we analyze the data with the t test and ineffective if we use the sign test? Obviously not. With the low power of the sign test, there is a high chance of making a Type II error; i.e., retaining H_0 when it is false. The t test is usually more powerful than the sign test. The additional power gives H_0 a better chance to be rejected if it is false. In this case, the additional power resulted in rejection of H_0. When several tests are appropriate for analyzing the data, *it is a general rule of statistical analysis to use the most powerful one,* since this gives the highest probability of rejecting H_0 when it is false.

Assumptions Underlying the t Test for Correlated Groups

The assumptions are very similar to those underlying the t test for single samples. The t test for correlated groups requires that the sampling distribution of $\bar{D}$ be normally distributed. This means that N should be ≥ 30, assuming the population shape doesn't differ greatly from normality, or the population scores themselves should be normally distributed.*

z AND t TESTS FOR INDEPENDENT GROUPS

The independent groups design was introduced previously in conjunction with the Mann–Whitney U test. This design involves random sampling of subjects from the population and then random assignment of the subjects to each condition. There can be many conditions. In this chapter, we shall consider the most simple experiment, which involves only two conditions. Each condition uses a different level of the independent variable. More complicated experiments will be considered in Chapter 15.

This design differs from the replicated measures or correlated groups design in that there is no basis for pairing scores and each subject gets tested only once. Analysis is performed separately on the raw scores of each sample, not on the

---

* Many authors limit the use of the t test to data that are of interval or ratio scaling. Please see the footnote in Chapter 2, p. 27, for references discussing this point.

difference scores. With the Mann–Whitney *U* test, we analyzed the separation between the sample scores. With the *t* test, we calculate the mean of each sample and analyze the difference between the sample means.

The sample scores in one of the conditions (say, condition 1) can be considered a random sample from a normally distributed population of scores that would result if all the individuals in the population received that condition (condition 1). Let's call the mean of this hypothetical population μ_1 and the standard deviation σ_1. Similarly, the sample scores in condition 2 can be considered a random sample from a normally distributed population of scores that would result if all the individuals were given condition 2. We can call the mean of this second population μ_2 and the standard deviation σ_2. Thus,

μ_1 = Mean of the population that receives condition 1

σ_1 = Standard deviation of the population that receives condition 1

μ_2 = Mean of the population that receives condition 2

σ_2 = Standard deviation of the population that receives condition 2

Changing the level of the independent variable is assumed to affect the mean of the distribution (μ_2), but not the standard deviation (σ_2) or variance (σ_2^2). Thus, under this assumption, if the independent variable has a real effect, the means of the populations will differ, but their variances will stay the same. Thus, σ_1^2 is assumed equal to σ_2^2. One way in which this assumption would be met is if the independent variable has an equal effect on each individual. A directional alternative hypothesis would predict that the samples are random samples from populations where $\mu_1 > \mu_2$ or $\mu_1 < \mu_2$, depending on the direction of the effect. A nondirectional alternative hypothesis would predict $\mu_1 \neq \mu_2$. If the independent variable has no effect, the samples would be random samples from populations where $\mu_1 = \mu_2$,* and chance alone would account for the differences between the sample means.

Before discussing the *t* test for independent groups, we shall present the *z* test. In the two-group situation, the *z* test is almost never used because it requires that σ_1^2 or σ_2^2 be known. However, it provides an important conceptual foundation for understanding the *t* test. After presenting the *z* test, we shall move to the *t* test.

z TEST FOR INDEPENDENT GROUPS

Let's begin with an experiment.

AN EXPERIMENT:

Hormone X and Sexual Behavior

A physiologist has the hypothesis that hormone X is important in producing sexual behavior. To investigate this hypothesis, 20 male rats were randomly sampled and

---

* In this case, there would be two null-hypothesis populations: one with a mean μ_1 and a standard deviation of σ_1 and the other with a mean μ_2 and standard deviation σ_2. However, since $\mu_1 = \mu_2$ and $\sigma_1 = \sigma_2$, the populations would be identical.

then randomly assigned to two groups. The animals in group 1 were injected with
hormone X and then were placed in individual housing with a sexually receptive
female. The animals in group 2 were given similar treatment except they were injected
with a placebo solution. The number of matings were counted over a 20-minute period.
The following results were obtained:

Hormone X, Group 1	Placebo, Group 2
8	5
10	6
12	3
6	4
6	7
7	8
9	6
8	5
7	4
11	8
84	56
$n_1 = 10$	$n_2 = 10$
$\overline{X}_1 = 8.4$	$\overline{X}_2 = 5.6$

$$\overline{X}_1 - \overline{X}_2 = 2.8$$

As shown in the table, the mean of group 1 is higher than the mean of group 2.
The difference between the means of the two samples is 2.8 and is in the direction
that indicates a positive effect. Is it legitimate to conclude that hormone X was responsible for the difference in means? The answer of course, is no. Before drawing this
conclusion, we must evaluate the null-hypothesis explanation. The statistic we are
using for this evaluation is the difference between the means of the two samples. As
with all other statistics, we must know its sampling distribution before we can evaluate
the null hypothesis.

The Sampling Distribution of the Difference Between Sample Means ($\overline{X}_1 - \overline{X}_2$)

Like the sampling distribution of the mean, this sampling distribution can be
determined theoretically or empirically. Again, for pedagogical purposes, we
prefer the empirical approach. To empirically derive the sampling distribution
of $\overline{X}_1 - \overline{X}_2$, all possible different samples of size n_1 would be drawn from a
population with a mean of μ_1 and variance of σ_1^2. Likewise, all possible samples
of size n_2 would be drawn from another population with a mean μ_2 and variance
σ_2^2. The values of $\overline{X}_1$ and $\overline{X}_2$ would then be calculated for each sample. Next,
$\overline{X}_1 - \overline{X}_2$ would be calculated for all possible pairings of samples of size n_1 and
n_2. The resulting distribution would contain all the possible different $\overline{X}_1 - \overline{X}_2$
scores that could be obtained from the populations when sample sizes are n_1
and n_2. Once this distribution has been obtained, it is a simple matter to calculate
the probability of obtaining each mean difference score ($\overline{X}_1 - \overline{X}_2$), assuming
sampling is random. This process would be repeated for different sample sizes

and population scores. Whether determined theoretically or empirically, the sampling distribution of the difference between sample means has the following characteristics:

1. If the populations from which the samples are taken are normal, then the distribution of the difference between sample means is also normal.
2. $\mu_{\bar{X}_1 - \bar{X}_2} = \mu_1 - \mu_2$, where $\mu_{\bar{X}_1 - \bar{X}_2}$ = the mean of the distribution of the difference between sample means.
3. $\sigma_{\bar{X}_1 - \bar{X}_2} = \sqrt{\sigma_{\bar{X}_1}^2 + \sigma_{\bar{X}_2}^2}$

where $\sigma_{\bar{X}_1 - \bar{X}_2}$ = Standard deviation of the sampling distribution of the difference between sample means; alternatively, standard error of the difference between sample means

   $\sigma_{\bar{X}_1}^2$ = Variance of the sampling distribution of the mean for samples of size n_1 taken from the first population

   $\sigma_{\bar{X}_2}^2$ = Variance of the sampling distribution of the mean for samples of n_2 taken from the second population

If, as mentioned previously, we assume that the variances of the two populations are equal ($\sigma_1^2 = \sigma_2^2$), then the equation for $\sigma_{\bar{X}_1 - \bar{X}_2}$ can be simplified as follows:

$$\sigma_{\bar{X}_1 - \bar{X}_2} = \sqrt{\sigma_{\bar{X}_1}^2 + \sigma_{\bar{X}_2}^2}$$
$$= \sqrt{\frac{\sigma_1^2}{n_1} + \frac{\sigma_2^2}{n_2}}$$
$$= \sqrt{\sigma^2 \left(\frac{1}{n_1} + \frac{1}{n_2} \right)}$$

where $\sigma^2 = \sigma_1^2 = \sigma_2^2$ = the variance of each population

The distribution is shown in Figure 14.2. Now let's return to the illustrative example.

FIGURE 14.2

Sampling distribution of the difference between sample mean scores

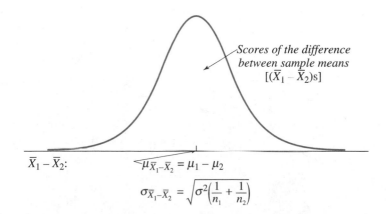

$\bar{X}_1 - \bar{X}_2$:

$\mu_{\bar{X}_1 - \bar{X}_2} = \mu_1 - \mu_2$

$\sigma_{\bar{X}_1 - \bar{X}_2} = \sqrt{\sigma^2 \left(\frac{1}{n_1} + \frac{1}{n_2} \right)}$

Scores of the difference between sample means $[(\bar{X}_1 - \bar{X}_2)s]$

AN EXPERIMENT: ## Hormone X Experiment Revisited

The results of the experiment showed that 10 rats injected with hormone X had a mean of 8.4 matings, whereas the mean of the 10 rats injected with a placebo was 5.6. Is the mean difference ($\overline{X}_1 - \overline{X}_2 = 2.8$) significant? Use $\alpha = 0.05_{2\,\text{tail}}$.

SOLUTION

The sampling distribution of $\overline{X}_1 - \overline{X}_2$ is shown in Figure 14.3. The shaded area contains all the mean difference scores of +2.8 or more. Assuming the sampling distribution of $\overline{X}_1 - \overline{X}_2$ is normal, if the sample mean difference (2.8) can be converted to its z value, we can use the z test to solve the problem. The equation for z_{obt} is similar to the other z equations we have already considered. However, here the value we are converting is $\overline{X}_1 - \overline{X}_2$. Thus,

$$z_{\text{obt}} = \frac{(\overline{X}_1 - \overline{X}_2) - \mu_{\overline{X}_1 - \overline{X}_2}}{\sigma_{\overline{X}_1 - \overline{X}_2}} \qquad \textit{equation for } z_{\text{obt}}, \textit{ independent groups design}$$

If hormone X had no effect on mating behavior, then both samples are random samples from populations where $\mu_1 = \mu_2$ and $\mu_{\overline{X}_1 - \overline{X}_2} = \mu_1 - \mu_2 = 0$. Thus,

$$z_{\text{obt}} = \frac{(\overline{X}_1 - \overline{X}_2) - 0}{\sqrt{\sigma^2\left(\dfrac{1}{n_1} + \dfrac{1}{n_2}\right)}} = \frac{2.8}{\sqrt{\sigma^2\left(\dfrac{1}{10} + \dfrac{1}{10}\right)}}$$

Note that, to calculate z_{obt}, the variance of the populations (σ^2) must be known before z_{obt} can be calculated. Since σ^2 is almost never known, this limitation severely restricts the practical use of the z test in this design. However, as you might guess, σ^2 can be

FIGURE 14.3

Sampling distribution of the difference between sample mean scores for the hormone problem

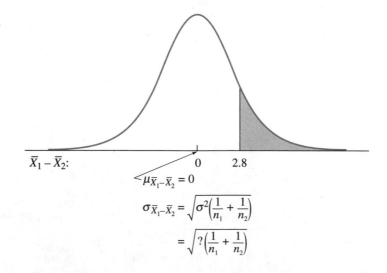

estimated from the sample data. When this is done, we have the *t* test for independent groups.

 ## STUDENT'S *t* TEST FOR INDEPENDENT GROUPS

Comparing the Equations for z_{obt} and t_{obt}

The equations for the *z* and *t* test are shown here:

z Test	*t* Test

$$z_{obt} = \frac{(\overline{X}_1 - \overline{X}_2) - \mu_{\overline{X}_1 - \overline{X}_2}}{\sigma_{\overline{X}_1 - \overline{X}_2}} \qquad t_{obt} = \frac{(\overline{X}_1 - \overline{X}_2) - \mu_{\overline{X}_1 - \overline{X}_2}}{s_{\overline{X}_1 - \overline{X}_2}}$$

$$= \frac{(\overline{X}_1 - \overline{X}_2) - \mu_{\overline{X}_1 - \overline{X}_2}}{\sqrt{\sigma^2 \left(\frac{1}{n_1} + \frac{1}{n_2} \right)}} \qquad = \frac{(\overline{X}_1 - \overline{X}_2) - \mu_{\overline{X}_1 - \overline{X}_2}}{\sqrt{s_W^2 \left(\frac{1}{n_1} + \frac{1}{n_2} \right)}}$$

where s_W^2 = Weighted estimate of σ^2
$s_{\overline{X}_1 - \overline{X}_2}$ = Estimate of $\sigma_{\overline{X}_1 - \overline{X}_2}$ = estimated standard error
of the difference between sample means

The *z* and *t* equations are identical except that the *t* equation uses s_W^2 to estimate the population variance (σ^2). This situation is analogous to the *t* test for single samples. You will recall in that situation we used the sample standard deviation(s) to estimate σ. However, in the *t* test for independent groups, there are two samples, and we wish to estimate σ^2. Since *s* is an accurate estimate of σ, s^2 is an accurate estimate of σ^2. Since there are two samples, either could be used to estimate σ^2, but we can get a more precise estimate by using both. It turns out the most precise estimate of σ^2 is obtained by using a weighted average of the sample variances s_1^2 and s_2^2. Weighting is done using degrees of freedom as the weights. Thus,

$$s_W^2 = \frac{df_1 s_1^2 + df_2 s_2^2}{df_1 + df_2} = \frac{(n_1 - 1)\left(\frac{SS_1}{n_1 - 1}\right) + (n_2 - 1)\left(\frac{SS_2}{n_2 - 1}\right)}{(n_1 - 1) + (n_2 - 1)} = \frac{SS_1 + SS_2}{n_1 + n_2 - 2}$$

where s_W^2 = Weighted estimate of σ^2
s_1^2 = Variance of the first sample
s_2^2 = Variance of the second sample
SS_1 = Sum of squares of the first sample
SS_2 = Sum of squares of the second sample

Substituting for s_W^2 in the *t* equation, we arrive at the computational equation for t_{obt}. Thus,

$$t_{obt} = \frac{(\overline{X}_1 - \overline{X}_2) - \mu_{\overline{X}_1 - \overline{X}_2}}{\sqrt{s_W^2 \left(\dfrac{1}{n_1} + \dfrac{1}{n_2}\right)}} = \frac{(\overline{X}_1 - \overline{X}_2) - \mu_{\overline{X}_1 - \overline{X}_2}}{\sqrt{\left(\dfrac{SS_1 + SS_2}{n_1 + n_2 - 2}\right)\left(\dfrac{1}{n_1} + \dfrac{1}{n_2}\right)}}$$

computational equation for t_{obt}, independent groups design

To evaluate the null hypothesis, we assume both samples are random samples from populations having the same mean value ($\mu_1 = \mu_2$). Therefore, $\mu_{\overline{X}_1 - \overline{X}_2} = 0$.* The previous equation reduces to

$$t_{obt} = \frac{\overline{X}_1 - \overline{X}_2}{\sqrt{\left(\dfrac{SS_1 + SS_2}{n_1 + n_2 - 2}\right)\left(\dfrac{1}{n_1} + \dfrac{1}{n_2}\right)}}$$

computational equation for t_{obt}, assuming $\mu_{\overline{X}_1 - \overline{X}_2} = 0$

We could go ahead and calculate t_{obt} for the hormone X example, but to evaluate t_{obt} we must know the sampling distribution of *t*. It turns out that, when one derives the sampling distribution of *t* for independent groups, the same family of curves are obtained as with sampling distribution of *t* for single samples, except that there is a different number of degrees of freedom. You will recall that 1 degree of freedom is lost each time a standard deviation is calculated. Since we calculate s_1^2 and s_2^2 for the two-sample case, we lose 2 degrees of freedom, one from each sample. Thus,

$$df = (n_1 - 1) + (n_2 - 1) = n_1 + n_2 - 2 = N - 2$$

where $N = n_1 + n_2$

Table D, then, can be used in the same manner as with the *t* test for single samples, except in the two-sample case we enter the table with $N - 2$ df. Thus, the *t* distribution varies both with *N* and degrees of freedom, but it varies uniquely only with degrees of freedom. That is, the *t* distribution corresponding to 13 df is the same whether it is derived from the single sample situation with $N = 14$ or the two-sample situation with $N = 15$.

At long last we are ready to evaluate the hormone data. The problem and data are restated for convenience.

AN EXPERIMENT

Hormone X Experiment One More Time

A physiologist has conducted an experiment to evaluate the effect of hormone X on sexual behavior. Ten rats were injected with hormone X, and 10 other rats received a placebo injection. The number of matings were counted over a 20-minute period.

---

* See Note 14.3.

The following results were obtained:

Group 1, Hormone X		Group 2, Placebo	
X_1	X_1^2	X_2	X_2^2
8	64	5	25
10	100	6	36
12	144	3	9
6	36	4	16
6	36	7	49
7	49	8	64
9	81	6	36
8	64	5	25
7	49	4	16
11	121	8	64
84	744	56	340

$$n_1 = 10 \qquad n_2 = 10$$
$$\overline{X}_1 = 8.4 \qquad \overline{X}_2 = 5.6$$
$$\overline{X}_1 - \overline{X}_2 = 2.8$$

1. What is the alternative hypothesis? Use a nondirectional hypothesis.
2. What is the null hypothesis?
3. What do you conclude? Use $\alpha = 0.05_{2\,\text{tail}}$.

SOLUTION

1. Alternative hypothesis: The alternative hypothesis specifies that hormone X affects sexual behavior. The sample mean difference of 2.8 is due to random sampling from populations where $\mu_1 \neq \mu_2$.
2. Null hypothesis: The null hypothesis states that hormone X is not related to sexual behavior. The sample mean difference of 2.8 is due to random sampling from populations where $\mu_1 = \mu_2$.
3. Conclusion, using $\alpha = 0.05_{2\,\text{tail}}$:

   Step 1: Calculate the appropriate statistic. Either the Mann–Whitney U test or the t test can be used. Since t is somewhat more powerful than U, if the data allow, t should be used. We shall discuss the assumptions of t in a later section. For now, assume t is appropriate. From the data table, $n_1 = 10$, $n_2 = 10$, $\overline{X}_1 = 8.4$, and $\overline{X}_2 = 5.6$. Solving for SS_1 and SS_2,

$$SS_1 = \sum X_1^2 - \frac{(\sum X_1)^2}{n_1} \qquad\qquad SS_2 = \sum X_2^2 - \frac{(\sum X_2)^2}{n_2}$$

$$= 744 - \frac{(84)^2}{10} \qquad\qquad\qquad = 340 - \frac{(56)^2}{10}$$

$$= 38.4 \qquad\qquad\qquad\qquad\qquad = 26.4$$

Substituting these values in the equation for t_{obt}, we have

$$t_{obt} = \frac{\overline{X}_1 - \overline{X}_2}{\sqrt{\left(\dfrac{SS_1 + SS_2}{n_1 + n_2 - 2}\right)\left(\dfrac{1}{n_1} + \dfrac{1}{n_2}\right)}} = \frac{8.4 - 5.6}{\sqrt{\left(\dfrac{38.4 + 26.4}{10 + 10 - 2}\right)\left(\dfrac{1}{10} + \dfrac{1}{10}\right)}} = 3.30$$

Step 2: **Evaluate the statistic.** As with the previous *t* tests, if t_{obt} falls in the critical region for rejection of H_0, we reject H_0. Thus,

If $|t_{obt}| \geq |t_{crit}|$, reject H_0.

If not, retain H_0.

The number of degrees of freedom is df = $N - 2 = 20 - 2 = 18$. From Table D, with $\alpha = 0.05_{2\,tail}$ and df = 18,

$$t_{crit} = \pm 2.101$$

Since $|t_{obt}| > 2.101$, we conclude by rejecting H_0.

Calculating t_{obt} When $n_1 = n_2$

When the samples sizes are equal, the equation for t_{obt} can be simplified. Thus,

$$t_{obt} = \frac{\overline{X}_1 - \overline{X}_2}{\sqrt{\left(\dfrac{SS_1 + SS_2}{n_1 + n_2 - 2}\right)\left(\dfrac{1}{n_1} + \dfrac{1}{n_2}\right)}}$$

but $n_1 = n_2 = n$. Substituting n for n_1 and n_2 in the equation for t_{obt},

$$t_{obt} = \frac{\overline{X}_1 - \overline{X}_2}{\sqrt{\left(\dfrac{SS_1 + SS_2}{n + n - 2}\right)\left(\dfrac{1}{n} + \dfrac{1}{n}\right)}} = \frac{\overline{X}_1 - \overline{X}_2}{\sqrt{\left(\dfrac{SS_1 + SS_2}{2(n - 1)}\right)\left(\dfrac{2}{n}\right)}} = \frac{\overline{X}_1 - \overline{X}_2}{\sqrt{\left(\dfrac{SS_1 + SS_2}{n(n - 1)}\right)}}$$

Thus,

$$t_{obt} = \frac{\overline{X}_1 - \overline{X}_2}{\sqrt{\dfrac{SS_1 + SS_2}{n(n - 1)}}} \qquad \textit{equation for calculating } t_{obt} \textit{ when } n_1 = n_2$$

Since $n_1 = n_2$ in the previous problem, we can use the simplified equation to calculate t_{obt}. Thus,

$$t_{obt} = \frac{\overline{X}_1 - \overline{X}_2}{\sqrt{\dfrac{SS_1 + SS_2}{n(n - 1)}}} = \frac{8.4 - 5.6}{\sqrt{\dfrac{38.4 + 26.4}{10(9)}}} = 3.30$$

This is the same value for t_{obt} that we obtained when using the more complicated equation. Whenever $n_1 = n_2$, it's easier to use the simplified equation. When $n_1 \neq n_2$, the more complicated equation *must* be used.

Let's do another problem involving equal *n*s.

A psychologist is interested in determining whether immediate memory capacity is affected by sleep loss. Immediate memory is defined as the amount of material that can be remembered immediately after it has been presented. Twelve students are randomly selected from Introductory Psychology and randomly assigned to two groups of 6 each. One of the groups is sleep-deprived for 24 hours before the material is presented. All subjects in the other group receive the normal amount of sleep (7–8 hours). The material consists of a series of slides, with each slide containing nine numbers. Each slide is presented for a short time interval (50 milliseconds), after which the subject must recall as many numbers as possible. The following are the results. The scores represent the percentage correctly recalled.

Normal Sleep Group 1		Sleep-Deprived Group 2	
X_1	X_1^2	X_2	X_2^2
68	4,624	70	4,900
73	5,329	62	3,844
72	5,184	68	4,624
65	4,225	63	3,969
70	4,900	69	4,761
73	5,329	60	3,600
421	29,591	392	25,698

$$n_1 = 6 \qquad n_2 = 6$$
$$\overline{X}_1 = 70.167 \qquad \overline{X}_2 = 65.333$$
$$\overline{X}_1 - \overline{X}_2 = 4.834$$

a. What is the alternative hypothesis? Use a nondirectional hypothesis.
b. What is the null hypothesis?
c. What do you conclude? Use $\alpha = 0.01_{2\,tail}$.

SOLUTION

The solution is shown in Table 14.2.

TABLE 14.2

a. Alternative hypothesis: The alternative hypothesis states that sleep deprivation has an effect on immediate memory. The difference between sample means of 4.834 is due to random sampling from populations where $\mu_1 \neq \mu_2$.

b. Null hypothesis: The null hypothesis states that sleep deprivation has no effect on immediate memory. The difference between sample means of 4.834 is due to random sampling from populations where $\mu_1 = \mu_2$.

c. Conclusion, using $\alpha = 0.01_{2\,\text{tail}}$:

Step 1: Calculate the appropriate statistic. Assuming the assumptions of t are met, t_{obt} is the appropriate statistic. From the data table, $n_1 = n_2 = n = 6$, $\overline{X}_1 = 70.167$, and $\overline{X}_2 = 65.333$. Solving for SS_1 and SS_2, we obtain

$$SS_1 = \sum X_1^2 - \frac{(\sum X_1)^2}{n_1} \qquad SS_2 = \sum X_2^2 - \frac{(\sum X_2)^2}{n_2}$$

$$= 29{,}591 - \frac{(421)^2}{6} \qquad\qquad = 25{,}698 - \frac{(392)^2}{6}$$

$$= 50.833 \qquad\qquad\qquad = 87.333$$

Substituting these values in the equation for t_{obt} with equal ns, we have

$$t_{\text{obt}} = \frac{\overline{X}_1 - \overline{X}_2}{\sqrt{\dfrac{SS_1 + SS_2}{n(n-1)}}} = \frac{70.167 - 65.333}{\sqrt{\dfrac{50.833 + 87.333}{6(5)}}} = 2.25$$

Step 2: Evaluate the statistic. Degrees of freedom $= N - 2 = 12 - 2 = 10$. From Table D, with $\alpha = 0.01_{2\,\text{tail}}$ and df $= 10$,

$$t_{\text{crit}} = \pm 3.169$$

Since $|t_{\text{obt}}| \leq 3.169$, we retain H_0. We cannot conclude that sleep deprivation has an effect on immediate memory.

Let's do one more problem, this time with unequal ns.

A neurosurgeon believes that lesions in a particular area of the brain, called the thalamus, will decrease pain perception. If so, this could be important in the treatment of terminal illness that is accompanied by intense pain. As a first attempt to test this hypothesis, he conducts an experiment in which 16 rats are randomly divided into two groups of 8 each. Animals in the experimental group receive a small lesion in the part of the thalamus thought to be involved with pain perception. Animals in the control group receive a comparable lesion in a brain area believed to be unrelated to pain. Two weeks after surgery each animal is given a brief electrical shock to the paws. The shock is administered in an ascending series, beginning with a very low intensity level and increasing until the animal first flinches. In this manner, the pain threshold to electric shock is determined for each rat. The following data are obtained. Each score represents the current level (milliamperes) at which flinching is first observed. The higher the current level, the higher is the pain threshold. Note that one animal died during surgery and was not replaced.

Neutral Area Lesions Control Group Group 1		Thalamic Lesions Experimental Group Group 2	
X_1	X_1^2	X_2	X_2^2
0.8	0.64	1.9	3.61
0.7	0.49	1.8	3.24
1.2	1.44	1.6	2.56
0.5	0.25	1.2	1.44
0.4	0.16	1.0	1.00
0.9	0.81	0.9	0.81
1.4	1.96	1.7	2.89
1.1	1.21	10.1	15.55
7.0	6.96		

$$n_1 = 8 \qquad n_2 = 7$$
$$\overline{X}_1 = 0.875 \qquad \overline{X}_2 = 1.443$$
$$\overline{X}_1 - \overline{X}_2 = -0.568$$

a. What is the alternative hypothesis? In this problem, assume there is sufficient theoretical and experimental basis to use a directional hypothesis.
b. What is the null hypothesis?
c. What do you conclude? Use $\alpha = 0.05_{1\,tail}$.

SOLUTION

The solution is shown in Table 14.3.

TABLE 14.3

a. Alternative hypothesis: The alternative hypothesis states that lesions of the thalamus decrease pain perception. The difference between sample means of -0.568 is due to random sampling from populations where $\mu_1 < \mu_2$.

b. Null hypothesis: The null hypothesis states that lesions of the thalamus either have no effect or they increase pain perception. The difference between sample means of -0.568 is due to random sampling from populations where $\mu_1 \geq \mu_2$.

c. Conclusion, using $\alpha = 0.05_{1\,\text{tail}}$:

Step 1: **Calculate the appropriate statistic.** Assuming the assumptions of *t* are met, t_{obt} is the appropriate statistic. From the data table, $n_1 = 8$, $n_2 = 7$, $\overline{X}_1 = 0.875$, and $\overline{X}_2 = 1.443$. Solving for SS_1 and SS_2, we obtain

$$SS_1 = \sum X_1^2 - \frac{(\sum X_1)^2}{n_1} \qquad SS_2 = \sum X_2^2 - \frac{(\sum X_2)^2}{n_2}$$

$$= 6.960 - \frac{(7)^2}{8} \qquad\qquad = 15.550 - \frac{(10.1)^2}{7}$$

$$= 0.835 \qquad\qquad\qquad = 0.977$$

Substituting these values into the general equation for t_{obt}, we have

$$t_{\text{obt}} = \frac{\overline{X}_1 - \overline{X}_2}{\sqrt{\left(\dfrac{SS_1 + SS_2}{n_1 + n_2 - 2}\right)\left(\dfrac{1}{n_1} + \dfrac{1}{n_2}\right)}}$$

$$= \frac{0.875 - 1.443}{\sqrt{\left(\dfrac{0.835 + 0.977}{8 + 7 - 2}\right)\left(\dfrac{1}{8} + \dfrac{1}{7}\right)}} = -2.94$$

Step 2: **Evaluate the statistic.** Degrees of freedom $= N - 2 = 15 - 2 = 13$. From Table D, with $\alpha = 0.05_{1\,\text{tail}}$ and df $= 13$,

$$t_{\text{crit}} = -1.771$$

Since $|t_{\text{obt}}| > 1.771$, we reject H_0 and conclude that lesions of the thalamus decrease pain perception.

Assumptions Underlying the *t* Test for Independent Groups

The assumptions underlying the *t* test for independent groups are as follows:

1. The sampling distribution of $\overline{X}_1 - \overline{X}_2$ is normally distributed. This means the populations from which the samples were taken should be normally distributed.

2. There is *homogeneity of variance.* You will recall that, at the beginning of our discussion concerning the *t* test for independent groups, we pointed out that the *t* test assumes that the independent variable affects the means of the populations but not their standard deviations ($\sigma_1 = \sigma_2$). Since the variance is just the square of the standard deviation, the *t* test for independent groups also assumes that the variances of the two populations are equal; i.e., $\sigma_1^2 = \sigma_2^2$. This is spoken of as the *homogeneity of variance assumption.* If the variances of the samples in the experiment (s_1^2 and s_2^2) are very different, e.g., if one variance is more than 4 times larger than the other, the two samples probably are not random samples from populations where $\sigma_1^2 = \sigma_2^2$. If this is true, the homogeneity of variance assumption is violated ($\sigma_1^2 \neq \sigma_2^2$).*

Violation of the Assumptions of the *t* Test for Independent Groups

Experiments have been conducted to determine the effect on the *t* test for independent groups of violating the assumptions of normality of the raw-score populations and homogeneity of variance. Fortunately, it turns out that the *t* test is a *robust* test. *A test is said to be robust if it is relatively insensitive to violations of its underlying mathematical assumptions.* The *t* test is relatively insensitive to violations of normality and homogeneity of variance, depending on sample size and the type and magnitude of the violation.† If $n_1 = n_2$ and if the size of each sample is equal to or greater than 30, then the *t* test for independent groups may be used without appreciable error despite moderate violation of the normality and/or the homogeneity of variance assumptions. If there are extreme violations of these assumptions, then an alternate test such as the Mann–Whitney *U* test should be used.

Before leaving this topic, it is worth noting that, when the two samples show large differences in their variances, it may indicate that the independent variable is not having an equal effect on all the subjects within a condition. This can be an important finding in its own right, leading to further experimentation into how the independent variable varies in its effects on different types of subjects.

---

* There are many inference tests to determine whether the data meet homogeneity of variance assumptions. However, this topic is beyond the scope of this textbook. See R. E. Kirk, *Experimental Design,* 3rd ed., Brooks/Cole, 1995, pp. 100–103.

    Some statisticians also require that the data be of interval or ratio scaling to use the *z* test, Student's *t* test, and the analysis of variance (covered in Chapter 15). For a discussion of this point, see the references contained in the Chapter 2 footnote on p. 27.

† For a review of this topic, see C. A. Boneau, "The Effects of Violations of Assumptions Underlying the *t* Test," *Psychological Bulletin,* 57 (1960), 49–64.

Power of the *t* Test

The three equations for t_{obt} are as follows:

Single Sample
$$t_{obt} = \frac{\bar{X}_{obt} - \mu}{\sqrt{\dfrac{SS}{N(N-1)}}}$$

Correlated Groups
$$t_{obt} = \frac{\bar{D}_{obt} - 0}{\sqrt{\dfrac{SS_D}{N(N-1)}}}$$

Independent Groups
$$t_{obt} = \frac{(\bar{X}_1 - \bar{X}_2) - 0}{\sqrt{\dfrac{SS_1 + SS_2}{n(n-1)}}}$$

[Handwritten margin notes: More powerful — ① large real effect of IV, ② increasing sample size, ③ decrease sample variability]

It seems fairly obvious that the larger t_{obt} is, the more likely H_0 will be rejected. Hence, anything that increases the likelihood of obtaining high values of t_{obt} will result in a more powerful *t* test. This can occur in several ways. First, the larger the real effect of the independent variable, the more likely $\bar{X}_{obt} - \mu$, $\bar{D}_{obt} - 0$, or $(\bar{X}_1 - \bar{X}_2) - 0$ will be large. Since these difference scores are in the numerator of the *t* equation, it follows that *the greater the effect of the independent variable, the higher the power of the t test* (other factors held constant). Of course, we don't know before doing the experiment what the actual effect of the independent variable is. If we did, then why do the experiment? Nevertheless, this analysis is useful because it suggests that, when designing an experiment, it is desirable to use the level of independent variable that the experimenter believes is the most effective to maximize the chances of detecting its effect. This analysis further suggests that, given meager resources for conducting an experiment, the experiment may still be powerful enough to detect the effect if the independent variable has a large effect.

The denominator of the *t* equation varies as a function of sample size and sample variability. As sample size increases, the denominator decreases. Therefore,

$$\sqrt{\frac{SS}{N(N-1)}}, \quad \sqrt{\frac{SS_D}{N(N-1)}}, \quad \text{and} \quad \sqrt{\frac{SS_1 + SS_2}{n(n-1)}}$$

decrease, causing t_{obt} to increase. *Thus, increasing sample size increases the power of the t test.*

The denominator also varies as a function of sample variability. In the single sample case, *SS* is the measure of variability. SS_D in the correlated groups experiments and $SS_1 + SS_2$ in the independent groups experiments reflect the variability. As the variability increases, the denominator in each case also increases, causing t_{obt} to decrease. *Thus, high sample variability decreases power.* Therefore, it is desirable to *decrease* variability as much as possible. One way to decrease variability is to carefully control the experimental conditions. For example, in a reaction-time experiment, the experimenter might use a warning signal that directly precedes the stimulus to which the subject must respond. In this way, variability due to attention lapses could be eliminated. Another way is to use the appropriate experimental design. For example, in certain situations, using a correlated groups design rather than an independent groups design will decrease variability.

Correlated Groups and Independent Groups Designs Compared

You are probably aware that many of the hypotheses presented in illustrative examples could have been investigated with either the correlated groups or the independent groups design. For instance, in Practice Problem 14.1, we presented an experiment that was conducted to evaluate the effect of a conservation campaign on gasoline consumption. The experiment used a correlated groups design, and the data were analyzed with the *t* test for correlated groups. For convenience, the data and analysis are provided again.

Family	Before the Campaign (gal) (1)	After the Campaign (gal) (2)	Difference	
			D	D^2
A	55	48	7	49
B	43	38	5	25
C	51	53	−2	4
D	62	58	4	16
E	35	36	−1	1
F	48	42	6	36
G	58	55	3	9
H	45	40	5	25
I	48	49	−1	1
J	54	50	4	16
K	56	58	−2	4
L	32	25	7	49
			35	235

$$N = 12 \qquad \bar{D}_{obt} = \frac{\Sigma D}{N} = \frac{35}{12} = 2.917$$

$$t_{obt} = \frac{\bar{D}_{obt} - \mu_D}{\sqrt{\dfrac{SS_D}{N(N-1)}}} \qquad SS_D = \Sigma D^2 - \frac{(\Sigma D)^2}{N}$$

$$= \frac{2.917 - 0}{\sqrt{\dfrac{132.917}{12(11)}}} \qquad \qquad = 235 - \frac{(35)^2}{12}$$

$$= 2.907 \qquad \qquad \qquad = 132.917$$

$$= 2.91$$

From Table D, with df = 11 and $\alpha = 0.05_{2\,tail}$,

$$t_{crit} = \pm 2.201$$

Since $|t_{obt}| > 2.201$, we rejected H_0 and concluded that the conservation campaign does indeed affect gasoline consumption. It significantly lowered the amount of gasoline used.

The conservation campaign could also have been evaluated using the independent groups design. Instead of using the same subjects in each condition, there would be two groups of subjects. One group would be monitored before the campaign and the other group monitored after the campaign. To evaluate the null hypothesis, each sample would be treated as an independent sample randomly selected from populations where $\mu_1 = \mu_2$. The basic statistic calculated would be $\overline{X}_1 - \overline{X}_2$. For the sake of comparison, let's analyze the conservation campaign data as though it were collected by using an independent groups design. Assume there are two different groups. The families in group 1 (before) are monitored for 1 month with respect to the amount of gasoline used before the conservation campaign is conducted, whereas the families in group 2 are monitored for 1 month after the campaign has been conducted.

Since $n_1 = n_2$, we can use the t_{obt} equation for equal n. In this experiment, $n_1 = n_2 = 12$. Solving for $\overline{X}_1$, $\overline{X}_2$, SS_1, and SS_2, we obtain

$$\overline{X}_1 = \frac{\Sigma X_1}{n_1} = \frac{587}{12} = 48.917 \qquad \overline{X}_2 = \frac{\Sigma X_2}{n_2} = \frac{552}{12} = 46.000$$

$$SS_1 = \Sigma X_1^2 - \frac{(\Sigma X_1)^2}{n_1} \qquad SS_2 = \Sigma X_2^2 - \frac{(\Sigma X_2)^2}{n_2}$$

$$= 29{,}617 - \frac{(587)^2}{12} \qquad = 26{,}496 - \frac{(552)^2}{12}$$

$$= 902.917 \qquad\qquad = 1104$$

Substituting these values into the equation for t_{obt} with equal n, we obtain

$$t_{obt} = \frac{\overline{X}_1 - \overline{X}_2}{\sqrt{\dfrac{SS_1 + SS_2}{n(n-1)}}} = \frac{48.917 - 46.000}{\sqrt{\dfrac{902.917 + 1104}{12(11)}}} = 0.748 = 0.75$$

From Table D, with df $= N - 2 = 24 - 2 = 22$ and $\alpha = 0.05_{2\,tail}$,

$$t_{crit} = \pm 2.074$$

Since $|t_{obt}| < 2.074$, we retain H_0.

Something seems strange here. When the data were collected with a correlated groups design, we were able to reject H_0. However, with an independent groups design, we were unable to reject H_0, even though the data were identical. Why? The correlated groups design allows us to use the subjects as their own control. This maximizes the possibility that there will be a high correlation between the scores in the two conditions. In the present illustration, Pearson r for the correlation between the paired before and after scores equals 0.938. When the correlation is high,* the difference scores will be much less variable than the original

---

* See Note 14.4 for a direct comparison between the two t equations that involves Pearson r.

scores. For example, consider the scores of families A and L. Family A uses quite a lot of gasoline (55 gallons), whereas family L uses much less (32 gallons). As a result of the conservation campaign, the scores of both families decrease by 7 gallons. Their difference scores are identical (7). There is no variability between the difference scores for these families, whereas there is a great variability between their raw scores. It is the potential for a high correlation and, hence, decreased variability that causes the correlated groups design to be potentially more powerful than the independent groups design. The decreased variability in the present illustration can be seen most clearly by viewing the two solutions side by side:

Correlated Groups	Independent Groups
$t_{obt} = \dfrac{\bar{D}}{\sqrt{\dfrac{SS_D}{N(N-1)}}}$	$t_{obt} = \dfrac{\bar{X}_1 - \bar{X}_2}{\sqrt{\dfrac{SS_1 + SS_2}{n(n-1)}}}$
$= \dfrac{2.917}{\sqrt{\dfrac{132.917}{12(11)}}}$	$= \dfrac{2.917}{\sqrt{\dfrac{902.917 + 1104}{12(11)}}}$
$= 2.91$	$= 0.75$

The two equations yield the same values except for SS_D in the correlated groups design and $SS_1 + SS_2$ in the independent groups design. SS_D is a measure of the variability of the difference scores. $SS_1 + SS_2$ are measures of the variability of the raw scores. $SS_D = 132.917$, whereas $SS_1 + SS_2 = 2006.917$; SS_D is much smaller than $SS_1 + SS_2$. It is this decreased variability that causes t_{obt} to be greater in the correlated groups analysis.

If the correlated groups design is potentially more powerful, why not always use a correlated groups design? First, the independent groups design is much more efficient from a df per measurement analysis. The degrees of freedom are important because the higher the df, the lower t_{crit}. In the present illustration, for the correlated groups design, there were 24 measurements taken, but only 11 df resulted. For the independent groups design, there were 24 measurements and 22 df. Thus, the independent groups design results in twice the df for the same number of measurements.

Second, many experiments preclude using the same subject in both conditions. For example, suppose we are interested in investigating whether men and women differ in aggressiveness. Obviously, the same subject could not be used in both conditions. Sometimes the effect of the first condition persists too long over time. If the experiment calls for the two conditions to be administered closely in time, it may not be possible to run the same subject in both conditions without the first condition affecting performance in the second condition. Often, when the subject is run in the first condition, he or she is "used up" or can't be run in the second condition. This is particularly true in learning experiments. For example, if we are interested in the effects of exercise on learning how to ski, we know that once the subjects have learned to ski, they can't be used in the second condition because they already know how to ski. When the same subject can't be used in the two conditions, then it is still possible to match subjects. However, matching is time-consuming and costly. Further, it is often true that the experimenter doesn't know which are the important variables for matching so as to produce a higher correlation. For all these reasons, the independent groups design is more often used than the correlated groups design.

SUMMARY

In this chapter, we have discussed the t test for correlated and independent groups. We pointed out that the t test for correlated groups was really just a special case of the t test for single samples. In the correlated groups design, the differences between paired scores are analyzed. If the independent variable has no effect and chance alone is responsible for the difference scores, then they can be considered a random sample from a population of difference scores where $\mu_D = 0$ and σ_D is unknown. But these are the exact conditions in which the t test for single samples applies. The only change is that, in the correlated groups design, we are analyzing difference scores, whereas in the single sample design, we analyze raw scores. After presenting some illustrative and practice problems, we concluded our discussion of the t test for correlated groups by comparing it with the sign test. We showed that although both are appropriate for the correlated groups design, as long as its assumptions are met, the t test should be used because it is more sensitive.

The t test for independent groups is employed when there are two independent groups in the experiment. The statistic that is analyzed is the difference between the means of the two samples ($\overline{X}_1 - \overline{X}_2$). The scores of sample 1 can be considered a random sample from a population having a mean μ_1 and a standard deviation σ_1. The scores of sample 2 are a random sample from a population having a mean μ_2 and a standard deviation σ_2.

If the independent variable has a real effect, then the difference between sample means is due to random sampling from populations where $\mu_1 \neq \mu_2$. Changing the level of the independent variable is assumed to affect the means of the populations but not their standard deviations or variances. If the independent variable has no effect and chance alone is responsible for the differences between the two samples, then the difference between sample means is due to random sampling from populations where $\mu_1 = \mu_2$. Under these conditions, the sampling distribution of $\overline{X}_1 - \overline{X}_2$ has a mean of zero and a standard deviation whose value depends on knowing the variance of the populations from which the samples were taken. Since this value is never known, the z test cannot be used. However, we can estimate the variance using a weighted estimate taken from both samples. When this is done, the resulting statistic is t_{obt}.

The t statistic, then, is also used for analyzing the data from the two-sample, independent groups experiment. The sampling distribution of t for this design is the same as for the single sample design, except the degrees of freedom are different. In the independent groups design, df $= N - 2$. After presenting some illustrative and practice problems, we discussed the assumptions underlying the t test for independent groups. We pointed out that this test requires that (1) the raw-score populations be normally distributed and (2) there be homogeneity of variance. We also pointed out that the t test is robust with regard to violations of the population normality and homogeneity of variance assumptions.

Next, we discussed the power of the t test. We showed that its power varies directly with the magnitude of the real effect of the independent variable and the N of the experiment, but varies inversely with the variability of the sample scores.

Finally, we compared the correlated groups and independent groups designs. When the correlation between paired scores is high, the correlated groups design is more sensitive than the independent groups design. However, it is easier and more efficient regarding degrees of freedom to conduct an independent groups experiment. In addition, there are many situations where the correlated groups design is inappropriate.

IMPORTANT TERMS

Correlated groups design (p. 318)
Degrees of freedom (p. 321)
Difference between sample
  means (p. 328)
Difference scores (p. 318)
Homogeneity of variance (p. 339)

Independent groups design (p. 326)
Mean of the difference scores
  (p. 319)
Sampling distribution of the
  difference between sample
  means (p. 328)

t test for correlated groups (p. 318)
t test for independent groups
  (p. 331)
Variability (p. 340)

QUESTIONS AND PROBLEMS

1. Identify or define the terms in the "Important Terms" section.
2. Discuss the advantages of the two-condition experiment as compared to the advantages of the single sample experiment.
3. The t test for correlated groups can be thought of as a special case of the t test for single samples, discussed in the previous chapter. Explain.
4. What are the characteristics of the sampling distribution of the difference between sample means?
5. Why is the z test for independent groups never used?
6. What is estimated in the t test for independent groups? How is the estimate obtained?
7. How does the t test for independent groups differ from the Mann–Whitney U test?
8. It is said that the variance of the sample data has an important bearing on the power of the t test. Is this statement true? Explain.
9. What are the advantages and disadvantages of using a correlated groups design as compared to an independent groups design?
10. What are the assumptions underlying the t test for independent groups?
11. Having just made what you feel to be a Type II error, using an independent groups design and a t test analysis, name all the things you might do in the next experiment to reduce the probability of a Type II error.

For each of the following problems, unless otherwise told, assume normality in the population.

12. You are interested in determining whether an experimental birth control pill has the side effect of changing blood pressure. You randomly sample 10 women from the city in which you live. You give 5 of them a placebo for a month and then measure their diastolic blood pressure. Then you switch them to the birth control pill for a month and again measure their blood pressure. The other 5 women receive the same treatment except they are given the birth control pill first for a month, followed by the placebo for a month. The blood pressure readings are shown here. Note that to safeguard the women from unwanted pregnancy, another means of birth control that does not interact with the pill was used for the duration of the experiment.

	Diastolic Blood Pressure	
Subject No.	Birth Control Pill	Placebo
1	108	102
2	76	68
3	69	66
4	78	71
5	74	76
6	85	80
7	79	82
8	78	79
9	80	78
10	81	85

a. What is the alternative hypothesis? Assume a nondirectional hypothesis is appropriate.
b. What is the null hypothesis?
c. What do you conclude? Use $\alpha = 0.01_{2\,\text{tail}}$.

13. Based on previous research and sound theoretical considerations, an experimental psychologist believes that memory for pictures is superior to memory for words. To test this hypothesis, the psychologist performs an experiment in which students from an introductory psychology class are used as subjects. Eight randomly selected students view 30 slides with nouns printed on them, and another group of eight randomly selected students view 30 slides with actual pictures of the same nouns. Each slide contains either one noun or one picture, and is viewed for 4 seconds. After viewing the slides, subjects are given a recall test, and the number of correctly remembered items is measured. The data are shown here:

No. of Pictures Recalled	No. of Nouns Recalled
18	12
21	9
14	21
25	17
23	16
19	10
26	19
15	22

a. What is the alternative hypothesis? Assume that a directional hypothesis is warranted.
b. What is the null hypothesis?
c. Using $\alpha = 0.05_{1\,\text{tail}}$, what is your conclusion?

14. A nurse was hired by a governmental ecology agency to investigate the impact of a lead smelter on the level of lead in the blood of children living near the smelter. Ten children were chosen at random from those living near the smelter. A comparison group of 7 children was randomly selected from those living in an area relatively free from possible lead pollution. Blood samples were taken from the children and lead levels determined. The following are the results (scores are in micrograms of lead per 100 milliliters of blood):

Lead Levels

Children Living Near Smelter	Children Living in Unpolluted Area
18	9
16	13
21	8
14	15
17	17
19	12
22	11
24	
15	
18	

Using $\alpha = 0.01_{1\,\text{tail}}$, what do you conclude?

15. The manager of the cosmetics section of a large department store wants to determine whether newspaper advertising really does affect sales. For her experiment, she randomly selects 15 items currently in stock and proceeds to establish a baseline. The 15 items are priced at their usual competitive values, and the quantity of each item sold for a 1-week period is recorded. Then, without changing their price, she places a large ad in the newspaper, advertising the 15 items. Again, she records the quantity sold for a 1-week period. The results follow.

Item	No. Sold Before Ad	No. Sold After Ad
1	25	32
2	18	24
3	3	7
4	42	40
5	16	19
6	20	25
7	23	23
8	32	35
9	60	65
10	40	43
11	27	28
12	7	11
13	13	12
14	23	32
15	16	28

Using $\alpha = 0.05_{2\,\text{tail}}$, what do you conclude?

16. Since muscle tension in the head region has been associated with tension headaches, you reason that if the muscle tension could be reduced, perhaps the headaches would decrease or go away altogether. You design an experiment in which nine subjects with tension headaches participate. The subjects keep daily logs of the number of headaches they experience during a 2-week baseline period. Then you train them to lower their muscle tension in the head region, using a biofeedback device. For this experiment, the biofeedback device is connected to frontalis muscle, a muscle in the forehead region. The device tells the subject the amount of tension in the muscle to which it is attached (in this case, frontalis) and helps them achieve low tension levels. After 6 weeks of training, during which the subjects have become successful at maintaining frontalis muscle tension low, they again keep a 2-week log of the number of headaches experienced. The following are the number of headaches recorded during each 2-week period.

	No. of Headaches	
Subject No.	Baseline	After Training
1	17	3
2	13	7
3	6	2
4	5	3
5	5	6
6	10	2
7	8	1
8	6	0
9	7	2

a. Using $\alpha = 0.05_{2\,tail}$, what do you conclude? Assume the sampling distribution of the mean of the difference scores $(\overline{D})$ is normally distributed. Assume a nondirectional hypothesis is appropriate because there is insufficient empirical basis to warrant a directional hypothesis.

b. If the sampling distribution of $\overline{D}$ is not normally distributed, what other test could you use to analyze the data? What would your conclusion be?

17. There is an interpretation difficulty with Problem 16. It is clear that the headaches decreased significantly. However, it is possible that the decrease was not due to the biofeedback training but rather to some other aspect of the situation, such as the attention shown the subjects. What is really needed is a group to control for this possibility. Assume another group of nine headache patients was run at the same time as the group in Problem 16. This group was treated in the same way except the subjects did not receive any training involving biofeedback. They just talked with you about their headaches each week for 6 weeks, and you showed them lots of warm, loving care and attention. The number of headaches for the baseline and 2-week follow-up period for the control group were as follows:

	No. of Headaches	
Subject No.	Baseline	Follow-up
1	5	4
2	8	9
3	14	12
4	16	15
5	6	4
6	5	3
7	8	7
8	10	6
9	9	7

Evaluate the effect of these other factors, such as attention, on the incidence of headaches. Use $\alpha = 0.05_{2\,tail}$.

18. Since the control group in Problem 17 also showed significant reductions in headaches, the interpretation of the results in Problem 16 is in doubt. Did relaxation training contribute to the headache decrease, or was the decrease due solely to other factors such as attention? To answer this question, we can compare the change scores between the two groups. These scores are shown here:

Headache Change Scores

Relaxation Training Group	Control Group
14	1
6	−1
4	2
2	1
−1	2
8	2
7	1
6	4
5	2

What is your conclusion? Use $\alpha = 0.05_{2\,\text{tail}}$.

19. The owner of a large factory is considering hiring part-time employees to fill jobs previously staffed with full-time workers. However, he wonders if doing so will affect productivity. Therefore, he conducts an experiment to evaluate the idea before implementing it factory-wide. Six full-time job openings, from the parts manufacturing division of the company, are each filled with two employees hired to work half-time. The output of these six half-time pairs is compared with the output of a randomly selected sample of six full-time employees from the same division. Note that all employees in the experiment are engaged in manufacturing the same parts. The average number of parts produced per day by the half-time pairs and full-time workers is shown here:

Parts Produced per Day

Half-Time Pairs	Full-Time Workers
24	20
26	28
46	40
32	36
30	24
36	30

Does the hiring of part-time workers affect productivity? Use $\alpha = 0.05_{2\,\text{tail}}$ in making your decision.

20. On the basis of her experience with clients, a clinical psychologist thinks that depression may affect sleep. She decides to test this idea. The sleep of nine depressed patients and eight normal controls is monitored for three successive nights. The average number of hours slept by each subject during the last two nights is shown in the following table:

Hours of Sleep

Depressed Patients	Normal Controls
7.1	8.2
6.8	7.5
6.7	7.7
7.3	7.8
7.5	8.0
6.2	7.4
6.9	7.3
6.5	6.5
7.2	

a. Is the clinician correct? Use $\alpha = 0.05_{2\,\text{tail}}$ in making your decision.

b. If the populations from which the samples were taken are very nonnormal, what other test could you use? What is your conclusion, using this test with $\alpha = 0.05_{2\,\text{tail}}$?

21. An educator wants to determine whether early exposure to school will affect IQ. He enlists the aid of the parents of 12 pairs of preschool-age identical twins who agree to let their twins participate in his experiment. One member of each twin pair is enrolled in preschool for 2 years, while the other member of each pair remains at home. At the end of the 2 years, the IQs of all the children are measured. The results follow.

	IQ	
Pair	**Twins at Preschool**	**Twins at Home**
1	110	114
2	121	118
3	107	103
4	117	112
5	115	117
6	112	106
7	130	125
8	116	113
9	111	109
10	120	122
11	117	116
12	106	104

Does early exposure to school affect IQ? Use $\alpha = 0.05_{2\,\text{tail}}$.

NOTES

14.1 Most textbooks present two methods for finding t_{obt} for the correlated groups design: (1) the direct-difference method and (2) a method that requires calculations of the degree of relationship (the correlation coefficient) existing between the two sets of raw scores. We have omitted the latter method, because it is rarely used in practice and, in our opinion, confuses many students. The direct-difference method flows naturally and logically from the discussion on the t test for single samples. It is much easier to use and much more frequently employed in practice.

14.2 Occasionally, in a repeated measures experiment where the alternative hypothesis is directional, the researcher may want to test whether the independent variable has an effect greater than some specified value other than 0. For example, assuming a directional hypothesis was justified in the present experiment, the researcher might want to test whether the average reward value of area A was greater than five bar presses per minute more than area B. In this case, the null hypothesis would be that the reward value of area A is not greater than five bar presses per minute more than area B. In this case, $\mu_D = 5$ rather than 0.

14.3 Occasionally, in an experiment involving independent groups, the alternative hypothesis is directional and specifies that the independent variable has an effect greater than some specified value other than 0. For example, in the "hormone X" experiment, assuming a directional hypothesis was legitimate, the physiologist might want to test whether hormone X has an average effect of over three matings more than the placebo. In this case, the null hypothesis would be that the average effect of the hormone X is ≤ three matings more than the placebo. We would test this hypothesis by assuming that the sample which received hormone X was a random sample from a population having a mean 3 units more than the population from which the placebo sample was taken. In this case, $\mu_{\bar{X}_1 - \bar{X}_2} = 3$, and

$$t_{\text{obt}} = \frac{(\bar{X}_1 - \bar{X}_2) - 3}{\sqrt{\left(\dfrac{SS_1 + SS_2}{n_1 + n_2 - 2}\right)\left(\dfrac{1}{n_1} + \dfrac{1}{n_2}\right)}}$$

14.4 Although we haven't previously presented the t equations in this form, the following can be shown:

t Test for Independent Groups	t Test for Correlated Groups
$t_{obt} = \dfrac{\overline{X}_1 - \overline{X}_2}{s_{\overline{x}_1 - \overline{x}_2}}$	$t_{obt} = \dfrac{\overline{D}}{s_{\overline{D}}}$
$= \dfrac{\overline{X}_1 - \overline{X}_2}{\sqrt{s_{\overline{x}_1}^2 + s_{\overline{x}_2}^2}}$	$= \dfrac{\overline{D}}{\sqrt{s_{\overline{x}_1}^2 + s_{\overline{x}_2}^2 - 2rs_{\overline{x}_1}s_{\overline{x}_2}}}$

Since $\overline{X}_1 - \overline{X}_2$ is equal to $\overline{D}$, the t_{obt} equations for independent groups and correlated groups are identical except for the term $-2rs_{\overline{x}_1}s_{\overline{x}_2}$ in the denominator of the correlated groups equation. Thus, the higher sensitivity of the correlated groups design depends on the magnitude of r. The higher the value of r, the more sensitive the correlated groups design will be relative to the independent groups design. Using the data of the conservation film experiment to illustrate the use of these equations, we obtain the following:

Independent Groups	Correlated Groups

Independent Groups:

$$t_{obt} = \frac{\overline{X}_1 - \overline{X}_2}{\sqrt{s_{\overline{x}_1}^2 + s_{\overline{x}_2}^2}}$$

where $s_{\overline{x}_1}^2 = \dfrac{s_1^2}{n_1} = \dfrac{82.083}{12} = 6.840$

and $s_{\overline{x}_2}^2 = \dfrac{s_2^2}{n_2} = \dfrac{100.364}{12} = 8.364$

Thus,

$$t_{obt} = \frac{48.917 - 46.000}{\sqrt{6.840 + 8.364}} = \frac{2.917}{\sqrt{15.204}} = 0.748$$

$$= 0.75$$

Correlated Groups:

$$t_{obt} = \frac{\overline{D}}{\sqrt{s_{\overline{x}_1}^2 + s_{\overline{x}_2}^2 - 2rs_{\overline{x}_1}s_{\overline{x}_2}}}$$

$$= \frac{2.917}{\sqrt{6.840 + 8.364 - 2(0.938)(2.615)(2.892)}}$$

$$= \frac{2.917}{\sqrt{1.017}}$$

$$= 2.91$$

Note that these are the same values obtained previously.

15

INTRODUCTION TO THE ANALYSIS OF VARIANCE

INTRODUCTION: THE *F* DISTRIBUTION

In Chapters 12, 13, and 14, we have been using the *mean* as the basic statistic for evaluating the null hypothesis. It's also possible to use the *variance* of the data for hypothesis testing. One of the most important tests that does this is called the *F* test, after R. A. Fisher, the statistician who developed it. In using this test, we calculate the statistic F_{obt}, which fundamentally is the ratio of two independent variance estimates of the same population variance σ^2. In equation form,

$$F_{obt} = \frac{\text{Variance estimate 1 of } \sigma^2}{\text{Variance estimate 2 of } \sigma^2}$$

The sampling distribution of *F* can be generated empirically by (1) taking all possible samples of size n_1 and n_2 from the same population, (2) estimating the population variance σ^2 from each of the samples using s_1^2 and s_2^2, (3) calculating F_{obt} for all possible combinations of s_1^2 and s_2^2, and then (4) calculating $p(F)$ for each different value of F_{obt}. The resulting distribution is the *sampling distribution of F*. Thus, as with all sampling distributions,

DEFINITION

• *The sampling distribution of F gives all the possible F values along with the $p(F)$ for each value, assuming sampling is random from the population.*

Like the *t* distribution, the *F* distribution varies with *degrees of freedom*. However, the *F* distribution has two values for degrees of freedom, one for the numerator and one for the denominator. As you might guess, we lose 1 degree of freedom for each calculation of variance. Thus,

$$\text{df for the numerator} = \text{df}_1 = n_1 - 1$$

$$\text{df for the denominator} = \text{df}_2 = n_2 - 1$$

FIGURE 15.1

F distribution with 3
degrees of freedom in the
numerator and 16 degrees
of freedom in the
denominator

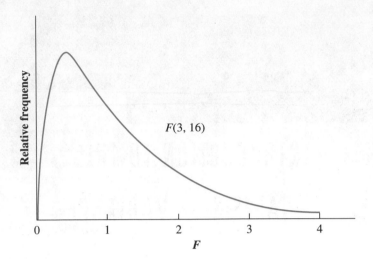

Figure 15.1 shows an F distribution with 3 df in the numerator and 16 df in the denominator. Several features are apparent. First, since F is a ratio of variance estimates, it never has a negative value (s_1^2 and s_2^2 will always be positive). Second, the F distribution is positively skewed. Finally, the median F value is approximately equal to 1.

Like the t test, there is a family of F curves. With the F test, however, there is a different curve for each combination of df_1 and df_2. Table F in Appendix D gives the critical values of F for various combinations of df_1 and df_2. There are two entries for every cell. The light entry gives the critical F value for the 0.05 level. The dark entry gives the critical F value for the 0.01 level. Note that these are one-tailed values for the right-hand tail of the F distribution. To illustrate, Figure 15.2 shows the F distribution for 4 df in the numerator and 20 df in the denominator. From Table F, F_{crit} at the 0.05 level equals 2.87. This means that 5% of the F values are equal to or greater than 2.87. The area containing these values is shown shaded in Figure 15.2.

FIGURE 15.2

Illustration showing
that F_{crit} in Table F is
one-tailed for the
right-hand tail

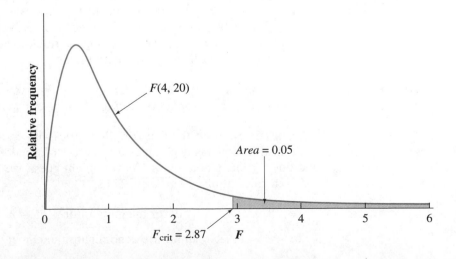

F TEST AND THE ANALYSIS OF VARIANCE (ANOVA)

The *F* test is appropriate in any experiment where the scores can be used to form two independent estimates of the population variance. One quite frequent situation in the behavioral sciences for which the *F* test is appropriate occurs when analyzing the data from experiments that use more than two groups or conditions.

Thus far in the text, we have discussed the most fundamental experiment: the two-group study involving a control group and an experimental group. Although this design is still used frequently, it is more common to encounter experiments that involve three or more groups. A major limitation of the two-group study is that very often two groups are not sufficient to allow a clear interpretation of the findings. For example, in the "thalamus and pain perception" experiment, there were two groups. One received lesions in the thalamus and the other in an area "believed" to be unrelated to pain. The results showed a significantly higher pain threshold for the rats with thalamic lesions. Our conclusion was that lesions of the thalamus increased pain threshold. However, the difference between the two groups could just as well have been due to a lowering of pain threshold as a result of the other lesion rather than a raising of threshold because of the thalamic damage. This ambiguity could have been dispelled if three groups had been run rather than two. The third group would be an unlesioned control group. Comparing the pain threshold of the two lesioned groups with the unlesioned group would help resolve the issue.

Another class of experiments requiring more than two groups involves experiments in which the independent variable is varied as a factor; i.e., a predetermined range of the independent variable is selected, and several values spanning the range are used in the experiment. For example, in the "hormone X and sexual behavior" problem, rather than arbitrarily picking one value of the hormone, the experimenter would probably pick several levels across the range of possible effective values. Each level would be administered to a different group of subjects, randomly sampled from the population. There would be as many groups in the experiment as there are levels of the hormone. This type of experiment has the advantage of allowing the experimenter to determine how the dependent variable changes with several different levels of the independent variable. In this example, the experimenter would find out how mating behavior varies in frequency with different levels of hormone X. Not only does using several levels allow a lawful relationship to emerge if one exists, but in situations where the experimenter is unsure of what single level might be effective, using several levels increases the possibility of a positive result occurring from the experiment.

Granted that it is frequently desirable to do experiments with more than two groups, you may wonder why these experiments aren't analyzed in the usual way. For example, if the experiment used four independent groups, why not simply compare the group means two at a time using the *t* test for independent groups? That is, why not just calculate *t* values comparing group 1 with 2, 3, and 4; 2 with 3 and 4; and 3 with 4?

The answer involves considerations of Type I error. You will recall that, when we set alpha at the 0.05 level, we are in effect saying that we are willing to risk being wrong 5% of the time when we reject H_0. In an experiment with two groups, there would be just one *t* calculation, and we would compare t_{obt} with t_{crit} to see whether t_{obt} fell in the critical region for rejecting H_0. Let's assume alpha = 0.05. The critical value of t at the 0.05 level was originally determined by taking the sampling distribution of t for the appropriate df and locating the t value such that the proportion of the total number of t values that were equal to or more extreme than it equaled 0.05. That is, if we were randomly sampling one t score from the t distribution, the probability it would be $\geq t_{crit}$ is 0.05.

Now what happens when we do an experiment involving many t comparisons, say, 20 of them? We are no longer just sampling one t value from the t distribution but 20. The probability of getting t values equal to or greater than t_{crit} obviously goes up. It is no longer equal to 0.05. The probability of making a Type I error has gone up as a result of doing an experiment with many groups and analyzing the data with more than one comparison.

OVERVIEW OF THE ONE-WAY ANALYSIS OF VARIANCE TECHNIQUE

The analysis of variance is a statistical technique employed to analyze multigroup experiments. Using the F test allows us to make *one* overall comparison that tells whether there is a significant difference between the *means* of the groups. Thus, it avoids the problem of an increased probability of Type I error that occurs when assessing many t values. The analysis of variance, or ANOVA as it is frequently called, is used in both independent groups and repeated measures designs. It is also used when one or more factors (variables) are investigated in the same experiment. In this section, we shall consider the simplest of these designs: the simple randomized-group design. This design is also often referred to as the one-way analysis of variance, independent groups design. A third designation often used is the single-factor experiment, independent groups design.* According to this design, subjects are randomly sampled from the population and then randomly assigned to the conditions, preferably such that there are an equal number of subjects in each condition. There are as many independent groups as there are conditions. If the study is investigating the effect of an independent variable as a factor, then the conditions would be the different levels of the independent variable used. Each group would receive a different level of the independent variable, e.g., a different concentration of hormone X. Thus, in this design, scores from several independent groups are analyzed.

The alternative hypothesis used in the analysis of variance is nondirectional. It states that one or more of the conditions have different effects from at least one of the others on the dependent variable. The null hypothesis states that the different conditions are all equally effective, in which case the scores in each group are random samples from populations having the same mean value. If there are k groups, then the null hypothesis specifies that

$$\mu_1 = \mu_2 = \mu_3 = \cdot\cdot\cdot = \mu_k$$

where μ_1 = Mean of the population from which group 1 is taken
               μ_2 = Mean of the population from which group 2 is taken

---

* The use of ANOVA with repeated measures designs is covered in B. J. Winer, D. R. Brown, and K. M. Michels, *Statistical Principles in Experimental Design,* 3rd ed., McGraw-Hill, New York, 1991.

FIGURE 15.3

Overview of the analysis of variance technique, simple randomized-groups design

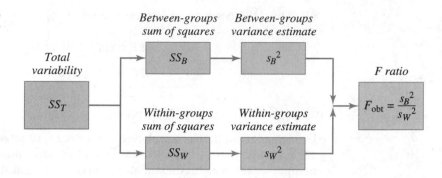

μ_3 = Mean of the population from which group 3 is taken
μ_k = Mean of the population from which group k is taken

Like the t test, the analysis of variance assumes that only the mean of the scores is affected by the independent variable, not the variance. Therefore, the analysis of variance assumes that

$$\sigma_1^2 = \sigma_2^2 = \sigma_3^2 = \cdots = \sigma_k^2$$

where σ_1^2 = Variance of the population from which group 1 is taken
 σ_2^2 = Variance of the population from which group 2 is taken
 σ_3^2 = Variance of the population from which group 3 is taken
 σ_k^2 = Variance of the population from which group k is taken

Essentially, the analysis of variance partitions the total variability of the data (SS_T) into two sources: the variability that exists within each group, called the *within-group sum of squares* (SS_W), and the variability that exists between the groups, called the *between-groups sum of squares* (SS_B) (see Figure 15.3). Each sum of squares is used to form an independent estimate of the H_0 population variance. The estimate based on the within-groups variability is called the *within-groups variance estimate* (s_W^2), and the estimate based on the between-groups variability is called the *between-groups variance estimate* (s_B^2).* Finally, an F ratio is calculated where

$$F_{\text{obt}} = \frac{\text{Between-groups variance estimate } (s_B^2)}{\text{Within-groups variance estimate } (s_W^2)}$$

This process is shown in Figure 15.3. The between-groups variance estimate increases with the magnitude of the independent variable's effect, whereas the within-groups variance estimate is unaffected. Thus, the larger the F ratio, the more unreasonable the null hypothesis becomes. As with the other statistics, we evaluate F_{obt} by comparing it with F_{crit}. If F_{obt} is equal to or exceeds F_{crit}, we reject

---

* s_B^2 is often symbolized MS_B and referred to as "Mean Square Between." Similarly, s_W^2 is often symbolized as MS_W and referred to as "Mean Square Within" or "Mean Square Error."

H_0. Thus, the decision rule states the following:

If $F_{obt} \geq F_{crit}$, reject H_0.

If $F_{obt} < F_{crit}$, retain H_0.

Within-Groups Variance Estimate, s_w^2

One estimate of the H_0 population variance, σ^2, is based on the variability within each group. It is symbolized as s_w^2 and is determined in precisely the same manner as the weighted estimate s_W^2 used in the t test for independent groups. We call it the *within-groups* variance estimate in the analysis of variance to distinguish it from the *between-groups* variance estimate discussed in the next section. You will recall that in the t test for independent groups,

$$s_W^2 = \text{Weighted estimate of } H_0 \text{ population variance, } \sigma^2$$

$$= \text{Weighted average of } s_1^2 \text{ and } s_2^2$$

$$= \frac{SS_1 + SS_2}{(n_1 - 1) + (n_2 - 1)}$$

$$= \frac{SS_1 + SS_2}{N - 2}$$

The analysis of variance utilizes exactly the same estimate except ordinarily we are dealing with three or more groups. Thus, for the analysis of variance,

$$s_W^2 = \text{Within-groups variance estimate}$$

$$= \text{Weighted estimate of } H_0 \text{ population variance, } \sigma^2$$

$$= \text{Weighted average of } s_1^2, s_2^2, s_3^2, \ldots, \text{ and } s_k^2$$

$$= \frac{SS_1 + SS_2 + SS_3 + \cdots + SS_k}{(n_1 - 1) + (n_2 - 1) + (n_3 - 1) + \cdots + (n_k - 1)}$$

where k = Number of groups
     n_k = Number of subjects in group k
    SS_k = Sum of squares of group k

This equation can be simplified to

$$s_W^2 = \frac{SS_1 + SS_2 + SS_3 + \cdots + SS_k}{N - k}$$

conceptual equation for within-groups variance estimate

where $N = n_1 + n_2 + n_3 + \cdots + n_k$

The numerator of this equation is called the *within-groups sum of squares*. It is symbolized by SS_W. The denominator equals the degrees of freedom for the within-groups variance estimate. Since we lose 1 degree of freedom for each sample variance calculated and there are k variances, there are $N - k$ degrees of freedom. Thus,

$$s_W{}^2 = \frac{SS_W}{df_W} \qquad \textit{within-groups variance estimate}$$

where $SS_W = SS_1 + SS_2 + SS_3 + \cdots + SS_k$ *within-groups sum of squares*

$df_W = N - k$ *within-groups degrees of freedom*

This equation for SS_W is fine conceptually, but when actually computing SS_W, it is better to use another equation. This equation is the algebraic equivalent of the conceptual equation, but it is easier to use and leads to fewer rounding errors. The computational equation is given below and will be discussed subsequently when we analyze the data from an experiment.

$$SS_W = \sum^{\substack{\text{all}\\\text{scores}}} X^2 - \left[\frac{(\sum X_1)^2}{n_1} + \frac{(\sum X_2)^2}{n_2} + \frac{(\sum X_3)^2}{n_3} + \cdots + \frac{(\sum X_k)^2}{n_k} \right] \qquad \begin{array}{l}\textit{computational}\\\textit{equation for } SS_W\end{array}$$

Between-Groups Variance Estimate, $s_B{}^2$

The second estimate of the variance of the null-hypothesis populations, σ^2, is based on the variability *between* the groups. It is symbolized by $s_B{}^2$. The null hypothesis states that each group is a random sample from populations where $\mu_1 = \mu_2 = \mu_3 = \cdots = \mu_k$. If the null hypothesis is correct, then we can use the variability between the means of the samples to estimate the variance of these populations, σ^2.

We know from Chapter 12 that, if we take all possible samples of size n from a population and calculate their mean values, the resulting sampling distribution of means has a variance of $\sigma_{\overline{X}}{}^2 = \sigma^2/n$. Solving for σ^2, we arrive at $\sigma^2 = n\sigma_{\overline{X}}{}^2$. If $\sigma_{\overline{X}}{}^2$ can be estimated, we can substitute the estimate in the previous equation to arrive at an independent estimate of σ^2. Since in the actual experiment there are several sample means, we can use the variance of these mean scores to estimate the variance of the full set of sample means, $\sigma_{\overline{X}}{}^2$. Since there are k sample means, we divide by $k - 1$, just as when we have N raw scores we divide by $N - 1$. Thus,

$$s_{\overline{X}}{}^2 = \frac{\sum (\overline{X} - \overline{X}_G)^2}{k - 1} \qquad \textit{estimate of } \sigma_{\overline{X}}{}^2$$

where $\overline{X}_G$ = Grand mean (overall mean of all the scores combined)

k = Number of groups

Using $s_{\overline{X}}{}^2$ for our estimate of $\sigma_{\overline{X}}{}^2$, we arrive at the second independent estimate of σ^2. This estimate is called the *between-groups variance estimate* and is symbolized by $s_B{}^2$. Since $\sigma^2 = n\sigma_{\overline{X}}{}^2$, then

$$s_B{}^2 = \text{Estimate of } \sigma^2 = n s_{\overline{X}}{}^2$$

Substituting for $s_{\overline{X}}^2$,

$$s_B^2 = \frac{n \sum (\overline{X} - \overline{X}_G)^2}{k - 1}$$

Expanding the summation, we arrive at

$$s_B^2 = \frac{n[(\overline{X}_1 - \overline{X}_G)^2 + (\overline{X}_2 - \overline{X}_G)^2 + (\overline{X}_3 - \overline{X}_G)^2 + \cdots + (\overline{X}_k - \overline{X}_G)^2]^*}{k - 1}$$

conceptual equation for the between-groups variance estimate

The numerator of this equation is called the *between-groups sum of squares*. It is symbolized by SS_B. The denominator is the degrees of freedom for the between-groups variance estimate. It is symbolized by df_B. Thus,

$$s_B^2 = \frac{SS_B}{df_B} \quad \textit{between-groups variance estimate}$$

where $\quad SS_B = n[\overline{X}_1 - \overline{X}_G)^2 + (\overline{X}_2 - \overline{X}_G)^2 + (\overline{X}_3 - \overline{X}_G)^2 + \cdots$
$\quad + (\overline{X}_k - \overline{X}_G)^2] \quad \textit{between-groups sum of squares}$

$df_B = k - 1 \quad \textit{between-groups degrees of freedom}$

It should be clear that, as the effect of the independent variable increases, the differences between the sample means increase. This causes $(\overline{X}_1 - \overline{X}_G)^2$, $(\overline{X}_2 - \overline{X}_G)^2, \ldots, (\overline{X}_k - \overline{X}_G)^2$ to increase, which in turn produces an increase in SS_B. Since SS_B is in the numerator, increases in it produce increases in s_B^2. Thus, the between-groups variance estimate (s_B^2) increases with the effect of the independent variable.

This equation for SS_B is fine conceptually, but when actually computing SS_B, it is better to use another equation. As with SS_W, there is a computational equation for SS_B that is the algebraic equivalent of the conceptual equation but is easier to use and leads to fewer rounding errors. The computational equation is given here and will be discussed shortly when we analyze the data from an experiment:

$$SS_B = \left[\frac{(\sum X_1)^2}{n_1} + \frac{(\sum X_2)^2}{n_2} + \frac{(\sum X_3)^2}{n_3} + \cdots + \frac{(\sum X_k)^2}{n_k} \right]$$
$$- \frac{\left(\overset{\text{all}}{\underset{\text{scores}}{\sum}} X \right)^2}{N} \quad \textit{computational equation for } SS_B$$

The F Ratio

We noted above that s_B^2 increases with the effect of the independent variable. However, since an assumption of the analysis of variance is that the independent

---

* This assumes there are n subjects in each group.

variable affects only the mean, not the variance, of each group, the within-groups variance estimate does not change with the effect of the independent variable. Since $F = s_B^2/s_W^2$, F increases with the effect of the independent variable. Thus, the larger the F ratio, the more reasonable it is that the independent variable has had a real effect. Another way of saying this is that s_B^2 is really an estimate of σ^2 plus the effects of the independent variable, whereas s_W^2 is just an estimate of σ^2. Thus,

$$F_{obt} = \frac{s_B^2}{s_W^2} = \frac{\sigma^2 + \text{independent variable effects}}{\sigma^2}$$

The larger F_{obt} becomes, the more reasonable it is that the independent variable has had a real effect. Of course, F_{obt} must be equal to or exceed F_{crit} before H_0 can be rejected. If F_{obt} is less than 1, we don't even need to compare it with F_{crit}. It is obvious the treatment has not had a significant effect, and we can immediately conclude by retaining H_0.

So far, we have been quite theoretical. Now let's do a problem to illustrate the analysis of variance technique.

AN EXPERIMENT

Different Situations and Stress

Suppose you are interested in determining whether certain situations produce differing amounts of stress. You know the amount of the hormone corticosterone circulating in the blood is a good measure of how stressed a person is. You randomly assign 15 students into three groups of 5 each. The students in group 1 have their corticosterone levels measured immediately after returning from vacations (low stress). The students in group 2 have their corticosterone levels measured after they have been in class for a week (moderate stress). The students in group 3 are measured immediately before final exam week (high stress). All measurements are taken at the same time of day. You record the following data. Scores are in milligrams of corticosterone per 100 milliliters of blood.

Group 1, Vacation		Group 2, Class		Group 3, Final Exam	
X_1	X_1^2	X_2	X_2^2	X_3	X_3^2
2	4	10	100	10	100
3	9	8	64	13	169
7	49	7	49	14	196
2	4	5	25	13	169
6	36	10	100	15	225
20	102	40	338	65	859

$$n_1 = 5 \qquad\qquad n_2 = 5 \qquad\qquad n_3 = 5$$

$$\overline{X}_1 = 4.00 \qquad\qquad \overline{X}_2 = 8.00 \qquad\qquad \overline{X}_3 = 13.00$$

$$\sum_{}^{\substack{\text{all}\\\text{scores}}} X = 125 \qquad \sum_{}^{\substack{\text{all}\\\text{scores}}} X^2 = 1299 \qquad \overline{X}_G = \frac{\sum_{}^{\substack{\text{all}\\\text{scores}}} X}{N} = 8.333$$

$$N = 15$$

1. What is the alternative hypothesis?
2. What is the null hypothesis?
3. What is the conclusion? Use $\alpha = 0.05$.

SOLUTION

1. Alternative hypothesis: The alternative hypothesis states that at least one of the situations affects stress differently than at least one of the remaining situations. Therefore, at least one of the means (μ_1, μ_2, or μ_3) differs from at least one of the others.
2. Null hypothesis: The null hypothesis states that the different situations affect stress equally. Therefore, the three sample sets of scores are random samples from populations where $\mu_1 = \mu_2 = \mu_3$.
3. Conclusion, using $\alpha = 0.05$: The conclusion is reached in the same general way as with the other inference tests. First, we calculate the appropriate statistic, in this case F_{obt}, and then we evaluate F_{obt} based on its sampling distribution.

A. Calculating F_{obt}

Step 1: **Calculate the between-groups sum of squares, SS_B.** To calculate SS_B, we shall use the following computational equation:

$$SS_B = \left[\frac{(\sum X_1)^2}{n_1} + \frac{(\sum X_2)^2}{n_2} + \frac{(\sum X_3)^2}{n_3} + \cdots + \frac{(\sum X_k)^2}{n_k} \right]$$

$$- \frac{\left(\overset{\text{all}}{\underset{\text{scores}}{\sum}} X \right)^2}{N} \qquad \textit{computational equation for } SS_B$$

In this problem, since $k = 3$, this equation reduces to

$$SS_B = \left[\frac{(\sum X_1)^2}{n_1} + \frac{(\sum X_2)^2}{n_2} + \frac{(\sum X_3)^2}{n_3} \right] - \frac{\left(\overset{\text{all}}{\underset{\text{scores}}{\sum}} X \right)^2}{N}$$

where $\overset{\text{all}}{\underset{\text{scores}}{\sum}} X^2 = $ Sum of all the scores

Substituting the appropriate values from the data table into this equation, we obtain

$$SS_B = \left[\frac{(20)^2}{5} + \frac{(40)^2}{5} + \frac{(65)^2}{5} \right] - \frac{(125)^2}{15} = 80 + 320 + 845 - 1041.667$$

$$= 203.333$$

Step 2: **Calculate the within-groups sum of squares, SS_W.** The computational equation for SS_W is as follows:

$$SS_W = \overset{\text{all}}{\underset{\text{scores}}{\sum}} X^2 - \left[\frac{(\sum X_1)^2}{n_1} + \frac{(\sum X_2)^2}{n_2} + \frac{(\sum X_3)^2}{n_3} + \cdots \right.$$

$$\left. + \frac{(\sum X_k)^2}{n_k} \right] \qquad \textit{computational equation for } SS_W$$

where $\sum\limits_{}^{\substack{\text{all}\\\text{scores}}} X^2$ = Sum of all the squared scores

Since $k = 3$, for this problem the equation reduces to

$$SS_W = \sum_{}^{\substack{\text{all}\\\text{scores}}} X^2 - \left[\frac{(\sum X_1)^2}{n_1} + \frac{(\sum X_2)^2}{n_2} + \frac{(\sum X_3)^2}{n_3} \right]$$

Substituting the appropriate values into this equation, we obtain

$$SS_W = \sum_{}^{\substack{\text{all}\\\text{scores}}} X^2 - \left[\frac{(\sum X_1)^2}{n_1} + \frac{(\sum X_2)^2}{n_2} + \frac{(\sum X_3)^2}{n_3} \right]$$

$$= 1299 - \left[\frac{(20)^2}{5} + \frac{(40)^2}{5} + \frac{(65)^2}{5} \right]$$

$$= 54$$

Step 3: Calculate the total sum of squares, SS_T. This step is just a check to be sure the calculations in steps 1 and 2 are correct. You will recall that at the beginning of the analysis of variance section, p. 355, we said this technique partitions the total variability into two parts, the within variability and the between variability. The measure of total variability is SS_T, the measure of within variability is SS_W, and the measure of between variability is SS_B. Thus,

$$SS_T = SS_W + SS_B$$

By independently calculating SS_T, we can check to see whether this relationship holds true for the calculations in steps 1 and 2:

$$SS_T = \sum_{}^{\substack{\text{all}\\\text{scores}}} X^2 - \frac{\left(\sum\limits_{}^{\substack{\text{all}\\\text{scores}}} X \right)^2}{N}$$

You will recognize that this equation is quite similar to the sum of squares with each sample, except here we are using the scores of all the samples as a single group. Calculating SS_T, we obtain

$$SS_T = \sum_{}^{\substack{\text{all}\\\text{scores}}} X^2 - \frac{\left(\sum\limits_{}^{\substack{\text{all}\\\text{scores}}} X \right)^2}{N}$$

$$= 1299 - \frac{(125)^2}{15}$$

$$= 257.333$$

Substituting the values of SS_T, SS_W, and SS_B into the equation, we obtain

$$SS_T = SS_W + SS_B$$

$$257.333 = 54 + 203.333$$

$$257.333 = 257.333$$

Note that, if the within sum of squares plus the between sum of squares does not equal the total sum of squares, you've made a calculation error. Go back and check steps 1, 2, and 3 until the equation balances (within rounding error).

Step 4: Calculate the degrees of freedom for each estimate:

$$\mathrm{df}_B = k - 1 = 3 - 1 = 2$$

$$\mathrm{df}_W = N - k = 15 - 3 = 12$$

$$\mathrm{df}_T = N - 1 = 15 - 1 = 14$$

Step 5: Calculate the between-groups variance estimate, $s_B{}^2$. The variance estimates are just the sums of squares divided by their degrees of freedom. Thus,

$$s_B{}^2 = \frac{SS_B}{\mathrm{df}_B} = \frac{203.333}{2} = 101.667$$

Step 6: Calculate the within-groups variance estimate, $s_W{}^2$:

$$s_W{}^2 = \frac{SS_W}{\mathrm{df}_W} = \frac{54}{12} = 4.5$$

Step 7: Calculate F_{obt}. We have calculated two independent estimates of σ^2, the between-variance estimate and the within-variance estimate. The F value is the ratio of $s_B{}^2$ to $s_W{}^2$. Thus,

$$F_{\mathrm{obt}} = \frac{s_B{}^2}{s_W{}^2} = \frac{101.667}{4.5} = 22.59$$

Note that $s_B{}^2$ is always put in the numerator and $s_W{}^2$ in the denominator.

B. Evaluate F_{obt}. Since $s_B{}^2$ is a measure of the effect of the independent variable as well as an estimate of σ^2, it should be larger than $s_W{}^2$, unless chance alone is at work. If $F_{\mathrm{obt}} \leq 1$, it is clear that the independent variable has not had a significant effect and we conclude by retaining H_0 without even bothering to compare F_{obt} with F_{crit}. If $F_{\mathrm{obt}} > 1$, we must compare it with F_{crit}. If $F_{\mathrm{obt}} \geq F_{\mathrm{crit}}$, we reject H_0. From Table F in Appendix D, with $\alpha = .05$, $\mathrm{df}_{\mathrm{numerator}} = 2$, and $\mathrm{df}_{\mathrm{denominator}} = 12$,

$$F_{\mathrm{crit}} = 3.88$$

Note that, in looking up F_{crit} in Table F, it is important to keep the df for the numerator and denominator straight. If by mistake you had entered the table with 2 df for the denominator and 12 df for the numerator, F_{crit} would equal 19.41, which is quite different from 3.88. Since $F_{\mathrm{obt}} > 3.88$, we reject

H_0. The three situations are not all the same in the stress levels they produce. A summary of the solution is shown in Table 15.1.

TABLE 15.1   Summary Table for ANOVA Problem Involving Stress

Source	SS	df	s^2	F_{obt}
Between groups	203.333	2	101.667	22.59*
Within groups	54.000	12	4.500	
Total	257.333	14		

* With $\alpha = 0.05$, $F_{crit} = 3.88$. Therefore, H_0 is rejected.

LOGIC UNDERLYING THE ONE-WAY ANOVA

Now that we have worked through the calculations of an illustrative example, we would like to discuss in more detail the logic underlying the one-way ANOVA. Earlier, we pointed out that the one-way ANOVA partitions the total variability (SS_T) into two parts: the within-groups sum of squares (SS_W) and the between-groups sum of squares (SS_B). We can gain some insight into this partitioning by recognizing that the partitioning is based on the simple idea that the deviation of each score from the grand mean is made up of two parts: the deviation of the score from its own group mean and the deviation of that group mean from the grand mean. Applying this idea to the first score in group 1, we obtain

Deviation of each score from the grand mean		Deviation of the score from its own group mean		Deviation of that group mean from the grand mean
$2 - 8.33$	$=$	$2 - 4.00$	$+$	$4.00 - 8.33$
$X - \overline{X}_G$	$=$	$X - \overline{X}_1$	$+$	$\overline{X}_1 - \overline{X}_G$
$\downarrow$		$\downarrow$		$\downarrow$
SS_T	$=$	SS_W	$+$	SS_B

Note that the term on the left ($X - \overline{X}_G$) when squared and summed over all the scores becomes SS_T. Thus,

$$SS_T = \overset{\overset{\text{all}}{\text{scores}}}{\sum} (X - \overline{X}_G)^2$$

The term in the middle ($X - \overline{X}_1$) when squared and summed for all the scores (of course we must subtract the appropriate group mean from each score) becomes SS_W. Thus,

$$SS_W = SS_1 + SS_2 + SS_3$$
$$= \sum (X - \overline{X}_1)^2 + \sum (X - \overline{X}_2)^2 + \sum (X - \overline{X}_3)^2$$

It is important to note that, since the subjects *within* each group receive the same level of the independent variable, variability among the scores within each group cannot be due to differences in the effect of the independent variable. Thus, the within-group sum of squares (SS_W) is not a measure of the effect of the independent variable. Since $SS_W/df_W = s_W^2$, this means that the within-groups variance estimate (s_W^2) also is not a measure of the real effect of the independent variable. Rather, it provides us with an estimate of the inherent variability of the scores themselves. Thus, s_W^2 is an estimate of σ^2, which is unaffected by treatment differences.

The last term in the equation partitioning the variability of the score 2 from the grand mean is $\overline{X}_1 - \overline{X}_G$. When this term is squared and summed for all the scores, it becomes SS_B. Thus,

$$SS_B = n(\overline{X}_1 - \overline{X}_G)^2 + n(\overline{X}_2 - \overline{X}_G)^2 + n(\overline{X}_3 - \overline{X}_G)^2$$

As discussed previously, SS_B is sensitive to the effect of the independent variable, because the greater the effect of the independent variable, the more the means of each group will differ from each other and, hence, will differ from $\overline{X}_G$. Since $SS_B/df_B = s_B^2$, this means that the between-groups variance estimate (s_B^2) is also sensitive to the real effect of the independent variable. Thus, s_B^2 gives us an estimate of σ^2 plus the effects of the independent variable. Since

$$F_{\text{obt}} = \frac{s_B^2}{s_W^2} = \frac{\sigma^2 + \text{effects of the independent variable}}{\sigma^2}$$

the larger F_{obt} is, the less reasonable is the null-hypothesis explanation. If the independent variable has no effect, then both s_B^2 and s_W^2 are independent estimates of σ^2, and their ratio is distributed as F with df = df_B (numerator) and df_W (denominator). We evaluate the null hypothesis by comparing F_{obt} with F_{crit}. If $F_{\text{obt}} \geq F_{\text{crit}}$, we reject H_0.

Let's try one more problem for practice.

PRACTICE PROBLEM 15.1

A college professor wants to determine the best way to present an important topic to his class. He has the following three choices: (1) he can lecture, (2) he can lecture plus assign supplementary reading, or (3) he can show a film and assign supplementary reading. He decides to do an experiment to evaluate the three options. He solicits 27 volunteers from his class and randomly assigns 9 to each of three conditions. In condition 1, he lectures to the students. In condition 2, he lectures plus assigns supplementary reading. In condition 3, the students see a film on the topic plus receive the same supplementary reading as the students in condition 2. The students are subsequently tested on the material. The following scores (percentage correct) were obtained:

Lecture, Condition 1		Lecture + Reading, Condition 2		Film + Reading, Condition 3	
X_1	X_1^2	X_2	X_2^2	X_3	X_3^2
92	8,464	86	7,396	81	6,561
86	7,396	93	8,649	80	6,400
87	7,569	97	9,409	72	5,184
76	5,776	81	6,561	82	6,724
80	6,400	94	8,836	83	6,889
87	7,569	89	7,921	89	7,921
92	8,464	98	9,604	76	5,776
83	6,889	90	8,100	88	7,744
84	7,056	91	8,281	83	6,889
767	65,583	819	74,757	734	60,088

$$n_1 = 9 \qquad n_2 = 9 \qquad n_3 = 9$$

$$\overline{X}_1 = 85.222 \qquad \overline{X}_2 = 91 \qquad \overline{X}_3 = 81.556$$

$$\overset{\text{all}}{\underset{\text{scores}}{\sum}} X = 2320 \qquad \overset{\text{all}}{\underset{\text{scores}}{\sum}} X^2 = 200,428 \qquad \overline{X}_G = \frac{\overset{\text{all}}{\underset{\text{scores}}{\sum}} X}{N} = 85.926$$

$$N = 27$$

a. What is the overall null hypothesis?
b. What is the conclusion? Use $\alpha = 0.05$.

SOLUTION

a. Null hypothesis: The null hypothesis states that the different methods of presenting the material are equally effective. Therefore, $\mu_1 = \mu_2 = \mu_3$.
b. Conclusion, using $\alpha = 0.05$: To assess H_0, we must calculate F_{obt} and then evaluate it based on its sampling distribution.

A. Calculate F_{obt}.

Step 1: Calculate SS_B:

$$SS_B = \left[\frac{(\sum X_1)^2}{n_1} + \frac{(\sum X_2)^2}{n_2} + \frac{(\sum X_3)^2}{n_3}\right] - \frac{\left(\sum\limits^{\text{all scores}} X\right)^2}{N}$$

$$= \left[\frac{(767)^2}{9} + \frac{(819)^2}{9} + \frac{(734)^2}{9}\right] - \frac{(2320)^2}{27}$$

$$= 408.074$$

Step 2: Calculate SS_W:

$$SS_W = \sum\limits^{\text{all scores}} X^2 - \left[\frac{(\sum X_1)^2}{n_1} + \frac{(\sum X_2)^2}{n_2} + \frac{(\sum X_3)^2}{n_3}\right]$$

$$= 200{,}428 - \left[\frac{(767)^2}{9} + \frac{(819)^2}{9} + \frac{(734)^2}{9}\right]$$

$$= 671.778$$

Step 3: Calculate SS_T:

$$SS_T = \sum\limits^{\text{all scores}} X^2 - \frac{\left(\sum\limits^{\text{all scores}} X\right)^2}{N}$$

$$= 200{,}428 - \frac{(2320)^2}{27}$$

$$= 1079.852$$

This step is a check to see whether SS_B and SS_W were correctly calculated. If so, then $SS_T = SS_B + SS_W$. This check is shown here:

$$SS_T = SS_B + SS_W$$

$$1079.852 = 408.074 + 671.778$$

$$1079.852 = 1079.852$$

Step 4: Calculate df:

$$\text{df}_B = k - 1 = 3 - 1 = 2$$
$$\text{df}_W = N - k = 27 - 3 = 24$$
$$\text{df}_T = N - 1 = 27 - 1 = 26$$

Step 5: Calculate s_B^2.

$$s_B^2 = \frac{SS_B}{\text{df}_B} = \frac{408.074}{2} = 204.037$$

Step 6: Calculate s_W^2:

$$s_W^2 = \frac{SS_W}{\text{df}_W} = \frac{671.778}{24} = 27.991$$

Step 7: Calculate F_{obt}:

$$F_{obt} = \frac{s_B^2}{s_W^2} = \frac{204.037}{27.991} = 7.29$$

B. Evaluate F_{obt}. With $\alpha = 0.05$, $\text{df}_{numerator} = 2$, and $\text{df}_{denominator} = 24$, from Table F,

$$F_{crit} = 3.40$$

Since $F_{obt} > 3.40$, we reject H_0. The methods of presentation are not equally effective. The solution is summarized in Table 15.2.

TABLE 15.2 Summary Table for ANOVA Problem Involving Teaching Methods

Source	SS	df	s^2	F_{obt}
Between groups	408.074	2	204.037	7.29*
Within groups	671.778	24	27.991	
Total	1079.852	26		

* With $\alpha = 0.05$, $F_{crit} = 3.40$. Therefore, H_0 is rejected.

THE RELATIONSHIP BETWEEN THE ANALYSIS OF VARIANCE AND THE t TEST

When a study involves just two independent groups and we are testing the null hypothesis that $\mu_1 = \mu_2$, we can use either the t test for independent groups or the analysis of variance. In such situations, it turns out that $t^2 = F$. To illustrate this point, let's analyze a suitable situation with both the t test and analysis of variance.

EXAMPLE

Dating—Sorority vs Dormitory Women

A sociologist wants to determine whether sorority or dormitory women date more often. He randomly samples 12 women who live in sororities and 12 women who live in dormitories and determines the number of dates they each have during the ensuing month. The following are the results:

Sorority Women		Dormitory Women	
X_1	X_1^2	X_2	X_2^2
8	64	9	81
5	25	7	49
6	36	3	9
4	16	4	16
12	144	4	16
7	49	8	64
9	81	7	49
10	100	5	25
5	25	8	64
3	9	6	36
7	49	3	9
5	25	5	25
81	623	69	443

$$n_1 = 12 \qquad n_2 = 12$$
$$\overline{X}_1 = 6.750 \qquad \overline{X}_2 = 5.750$$

Based on the data in the table, what is his conclusion? Use $\alpha = 0.05_{2\,\text{tail}}$.

SOLUTION

The null hypothesis is that $\mu_1 = \mu_2$. Since this is an independent groups design, we can use either the t test for independent groups or the analysis of variance to evaluate this hypothesis. We shall first analyze the data with the t test, as follows:

$$SS_1 = \sum X_1^2 - \frac{(\sum X_1)^2}{N} \qquad SS_2 = \sum X_2^2 - \frac{(\sum X_2)^2}{N}$$

$$= 623 - \frac{(81)^2}{12} \qquad\qquad = 443 - \frac{(69)^2}{12}$$

$$= 76.250 \qquad\qquad\qquad = 46.250$$

$$t_{obt} = \frac{\bar{X}_1 - \bar{X}_2}{\sqrt{\dfrac{SS_1 + SS_2}{n(n-1)}}} = \frac{6.750 - 5.750}{\sqrt{\dfrac{76.250 + 46.250}{12(11)}}}$$

$$= \frac{1.000}{0.963} = 1.038 = 1.04$$

With $\alpha = 0.05_{2\,tail}$ and df $= N - 2 = 22$, from Table D in Appendix D,

$$t_{crit} = \pm 2.074$$

Since $|t_{obt}| < 2.074$, we retain H_0. We cannot conclude that the sorority or dormitory women date more often.

This problem can also be solved with the analysis of variance. The solution is shown here:

Step 1: **Calculate SS_B:**

$$SS_B = \left[\frac{(\sum X_1)^2}{n_1} + \frac{(\sum X_2)^2}{n_2} \right] - \frac{\left(\overset{\text{all scores}}{\sum X} \right)^2}{N}$$

$$= \left[\frac{(81)^2}{12} + \frac{(69)^2}{12} \right] - \frac{(150)^2}{24}$$

$$= 6.000$$

Step 2: **Calculate SS_W:**

$$SS_W = \overset{\text{all scores}}{\sum} X^2 - \left[\frac{(\sum X_1)^2}{n_1} + \frac{(\sum X_2)^2}{n_2} \right]$$

$$= 1066 - \left[\frac{(81)^2}{12} + \frac{(69)^2}{12} \right]$$

$$= 122.500$$

Step 3: **Calculate SS_T:**

$$SS_T = \overset{\text{all scores}}{\sum} X^2 - \frac{\left(\overset{\text{all scores}}{\sum X} \right)^2}{N}$$

$$= 1066 - \frac{(150)^2}{24}$$

$$= 128.500$$

As a check,

$$SS_T = SS_B + SS_W$$
$$128.500 = 6.000 + 122.500$$
$$128.500 = 128.500$$

Step 4: Calculate df:

$$df_B = k - 1 = 2 - 1 = 1$$
$$df_W = N - k = 24 - 2 = 22$$
$$df_T = N - 1 = 24 - 1 = 23$$

Step 5: Calculate $s_B{}^2$:

$$s_B{}^2 = \frac{SS_B}{df_B} = \frac{6.000}{1} = 6.000$$

Step 6: Calculate $s_W{}^2$:

$$s_W{}^2 = \frac{SS_W}{df_W} = \frac{122.500}{22} = 5.568$$

Step 7: Calculate F_{obt}:

$$F_{obt} = \frac{s_B{}^2}{s_W{}^2} = \frac{6.000}{5.568} = 1.078 = 1.08$$

With $\alpha = 0.05$, $df_{numerator} = 1$, and $df_{denominator} = 22$, from Table F,

$$F_{crit} = 4.30$$

Since $F_{obt} < 4.30$, we retain H_0. We cannot conclude that sorority or dormitory women date more often.

Note that we have reached the same conclusion with both the t test and the F test. A comparison of the two obtained statistics and the associated critical values are shown here:

$$t_{obt}{}^2 = F_{obt} \qquad t_{crit}{}^2 = F_{crit}$$
$$(1.04)^2 = 1.08 \qquad (2.074)^2 = 4.30$$

Thus, in the independent groups design, when there are two groups and we are testing the null hypothesis that $\mu_1 = \mu_2$, $t_{obt}{}^2 = F$ and $t_{crit}{}^2 = F_{crit}$. Therefore, both tests give the same results.

ASSUMPTIONS UNDERLYING THE ANALYSIS OF VARIANCE

The assumptions underlying the analysis of variance are similar to those of the t test for independent groups:

1. The populations from which the samples were taken are normally distributed.

2. The samples are drawn from populations of equal variances. As pointed out in Chapter 14 in connection with the *t* test for independent groups, this assumption is called the homogeneity of variance assumption. The analysis of variance also assumes homogeneity of variance.*

Like the *t* test, the analysis of variance is a robust test. It is minimally affected by violations of population normality. It is also relatively insensitive to violations of homogeneity of variance provided the samples are of equal size.†

TWO-WAY ANALYSIS OF VARIANCE

Thus far, we have discussed the most elementary analysis of variance design. We have called it the simple randomized-groups design, the one-way analysis of variance, independent groups design, or the single-factor experiment, independent groups design. The characteristics of this design are that there is only one independent variable (one factor) that is being investigated, there are several levels of the independent variable (several conditions) represented, and subjects are randomly assigned to each condition.

Actually, the analysis of variance design is not limited to single-factor experiments. In fact, the effect of many different factors may be investigated at the same time in one experiment. Such experiments are called factorial experiments.

DEFINITION

- *A **factorial experiment** is one in which the effects of two or more factors are assessed in one experiment. In a factorial experiment, the treatments used are combinations of the levels of the factors.*

In this section, we shall acquaint you in a *qualitative* way with the main features of the most elementary factorial design: the two-way analysis of variance, independent groups design, fixed effects model. For anyone who wishes to explore this design in depth, we have included a more detailed quantitative presentation in Chapter 17.

The two-way analysis of variance is a bit more complicated than the one-way design. However, we get a lot more information from the two-way design.

*Basically, the **two-way analysis of variance** allows us in one experiment to evaluate the effect of two independent variables and the interaction between them.*

To illustrate this design, suppose a professor in physical education conducts an experiment to compare the effects on nighttime sleep of different intensities of exercise and the time of day when the exercise is done. For this example, let's assume that there are two levels of exercise (light and heavy) and two times of day (morning and evening). The experiment is shown diagrammatically in Figure 15.4. From this figure, we can see that there are two factors (or independent

* See Chapter 14 footnote * on p. 339. Some statisticians would also limit the use of ANOVA to data that are interval or ratio in scaling. For a discussion of this point, see the references in the Chapter 2 footnote on p. 27.
† For an extended discussion of these points, see G. V. Glass, P. D. Peckham, and J. R. Sanders, "Consequences of Failure to Meet the Assumptions Underlying the Use of Analysis of Variance and Covariance," *Review of Educational Research, 42* (1972), 237–288.

FIGURE 15.4

Schematic diagram of two-way analysis of variance example involving exercise intensity and time of day

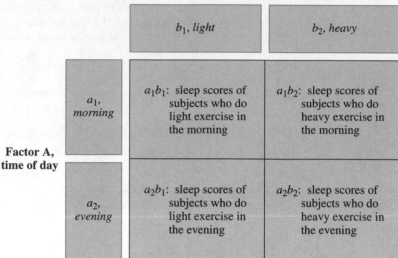

Factor B, exercise intensity

	b_1, *light*	b_2, *heavy*
a_1, *morning*	a_1b_1: sleep scores of subjects who do light exercise in the morning	a_1b_2: sleep scores of subjects who do heavy exercise in the morning
a_2, *evening*	a_2b_1: sleep scores of subjects who do light exercise in the evening	a_2b_2: sleep scores of subjects who do heavy exercise in the evening

Factor A, time of day

variables): factor A, which is *time of day,* and factor B, which is *exercise intensity.* Each factor has two levels. Thus, this design is referred to as a 2 × 2 (read "two by two") design where each number stands for a factor and the magnitude of the number designates the number of levels within the factor. For example, if factor A had three levels, then the experiment would be called a 3 × 2 design. In a 2 × 4 × 3 design, there would be three factors having, respectively, two, four, and three levels. In the present example, there are two factors each having two levels. This results in four cells or conditions: a_1b_1 (morning—light exercise), a_1b_2 (morning—heavy exercise), a_2b_1 (evening—light exercise), and a_2b_2 (evening—heavy exercise). Since this is an independent groups design, subjects would be randomly assigned to each of the cells so that a different group of subjects occupies each cell. Since the levels of each factor were systematically chosen by the experimenter rather than being randomly chosen, this is called a *fixed effects* design.

There are three analyses done in this design. First, we want to determine whether factor A has a significant effect, disregarding the effect of factor B. In this illustration, we are interested in determining whether "time of day" makes a difference in the effect of exercise on sleep, disregarding the effect of "exercise intensity." Second, we want to determine whether factor B has a significant effect, without considering the effect of factor A. For this experiment, we are interested in determining whether the intensity of exercise makes a difference in sleep activity, disregarding the effect of time of day. Finally, we want to determine whether there is an interaction between factors A and B. In the present experiment, we want to determine whether there is an interaction between time of day and intensity of exercise in their effect on sleep.

DEFINITION

• *The effect of factor A (averaged over the levels of factor B) and the effect of factor B (averaged over the levels of factor A) are called* **main effects.** *An* **interaction effect** *occurs when the effect of one factor is not the same at all levels of the other factor.*

Figure 15.5 shows some possible outcomes of this experiment. In part (a), there are no significant effects. In part (b), there is a significant main effect for time of day but no effect for intensity of exercise and no interaction. Thus, the subjects get significantly more sleep if the exercise is done in the morning rather than in the evening. However, it doesn't seem to matter if the exercise is light or heavy. In part (c), there is a significant main effect for intensity of exercise but no effect for time of day and no interaction. In this example, heavy exercise results in significantly more sleep than light exercise, and it doesn't matter whether the exercise is done in the morning or evening—the effect appears to be the same. Part (d) shows a significant main effect for intensity of exercise and time of day, with no interaction effect.

Both parts (e) and (f) show significant interaction effects. As stated previously, the essence of an interaction is that the effect of one factor is not the same at

FIGURE 15.5

Some possible outcomes of the experiment investigating the effects of intensity of exercise and time of day

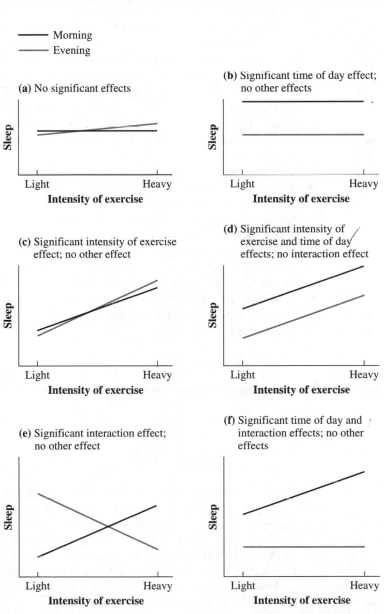

all levels of the other factor. This means that, when an interaction occurs between factors A and B, the differences in the dependent variable due to changes in one factor are not the same for each level of the other factor. In part (e), there is a significant interaction effect between intensity of exercise and time of day. The effect of different intensities of exercise is not the same for all levels of time of day. Thus, if the exercise is done in the evening, light exercise results in significantly more sleep than heavy exercise. On the other hand, if the exercise is done in the morning, light exercise results in significantly less sleep than heavy exercise. In part (f), there is a significant main effect for time of day and a significant interaction effect. Thus, when the exercise is done in the morning, it results in significantly more sleep than when done in the evening, regardless of whether it is light or heavy exercise. In addition to this main effect, there is an interaction between the intensity of exercise and the time of day. Thus, there is no difference in the effect of the two intensities when the exercise is done in the evening, but when done in the morning, heavy exercise results in more sleep than light exercise.

In analyzing the data from a two-way analysis of variance design, we determine four variance estimates: s_W^2, s_R^2, s_C^2, and s_{RC}^2. The estimate s_W^2 is the *within-cells variance estimate* and corresponds to the within-groups variance estimate used in the one-way ANOVA. It becomes the standard against which each of the other estimates is compared. The other estimates are sensitive to the effects of the independent variables. The estimate s_R^2 is called the *row variance estimate*. It is based on the variability of the row means (see Figure 15.4) and, hence, is sensitive to the effects of variable A. The estimate s_C^2 is called the *column variance estimate*. It is based on the variability of the column means and, hence, is sensitive to the effects of variable B. The estimate s_{RC}^2 is the *row × column* (read "row by column") *variance estimate*. It is based on the variability of the cell means and, hence, is sensitive to the interaction effects of variables A and B. If variable A has no effect, s_R^2 is an independent estimate of σ^2. If variable B has no effect, then s_C^2 is an independent estimate of σ^2. Finally, if there is no interaction between variables A and B, s_{RC}^2 is also an independent estimate of σ^2. Thus, the estimates s_R^2, s_C^2, and s_{RC}^2 are analogous to the between-groups variance estimate of the one-way design. To test for significance, three F ratios are formed:

For variable A,
$$F_{\text{obt}} = \frac{s_R^2}{s_W^2}$$

For variable B,
$$F_{\text{obt}} = \frac{s_C^2}{s_W^2}$$

For the interaction between A and B,
$$F_{\text{obt}} = \frac{s_{RC}^2}{s_W^2}$$

Each F_{obt} value is evaluated against F_{crit} as in the one-way analysis. For the rows comparison, if $F_{\text{obt}} \geq F_{\text{crit}}$, there is a significant main effect for factor A. If $F_{\text{obt}} \geq F_{\text{crit}}$ for the columns comparison, there is a significant main effect for factor B. Finally, if $F_{\text{obt}} \geq F_{\text{crit}}$ for the row × column comparison, there is a significant interaction effect. Thus, there are many similarities between the one-way and two-way designs. The biggest difference is that, with a two-way design, we can do essentially two one-way experiments plus we are able to evaluate the interaction between the two independent variables.

MUCH ADO ABOUT ALMOST NOTHING

In a recent magazine advertisement placed by Rawlings Golf Company, Rawlings claims to have developed a new golf ball that travels a greater distance. The ball is called Tony Penna DB (DB stands for distance ball). To its credit, Rawlings not only offered terms like *high rebound core, Surlyn cover, centrifugal action,* etc., to explain why it is reasonable to believe that its ball would travel farther, but hired a consumer testing institute to conduct an experiment to determine whether, in fact, the Tony Penna DB ball does travel farther. In this experiment, six different brands of balls were evaluated. Fifty-one golfers each hit 18 new balls (3 of each brand) off a driving tee, with a driver. The mean distance traveled for each ball was reported as follows:

1. Tony Penna DB   254.57 yd
2. Titleist Pro Trajectory 252.50 yd
3. Wilson Pro Staff   249.24 yd
4. Titleist DT   249.16 yd
5. Spalding Top-Flite   247.12 yd
6. Dunlop Blue Max   244.22 yd

Although no inference testing was reported, the ad concludes, "as you can see, while we can't promise *you* 250 yards off the tee, we can offer you a competitive edge, if only a yard or two. But an edge is an edge." Since you are by now thoroughly grounded in inferential statistics, how do you respond to this ad?

ANSWER First, I think you should commend the company on conducting evaluative research that compares its product with competitors' on a very important dependent variable. It is to be further commended in engaging an impartial organization to conduct the research. Finally, it is to be commended for reporting the results and calling readers' attention to the fact that the differences between balls are quite small (although, admittedly, the wording of the ad tries to achieve a somewhat different result).

A major criticism of this ad (and an old friend by now) is that we have not been told whether these results are statistically significant. Without establishing this point, the most reasonable explanation of the differences may be "chance." Of course, if chance is the correct explanation, then using the Tony Penna DB ball won't even give you a yard or two advantage! Before we can take the superiority claim seriously, the manufacturer must report that the differences were statistically significant. Without this statement, as a general rule, I believe we should assume the differences were tested and were not significant, in which case chance alone remains a reasonable explanation of the data. (By the way, what inference test would you have used? Did you answer ANOVA? Nice going!)

For the sake of my second point, let's say the appropriate inference testing has been done, and the data are statistically significant. We still need to ask: *"So what? Even if the results are statistically significant, is the magnitude of the effect worth bothering about?"* Regarding the difference in yardage between the first two brands, I think the answer is "no." Even if I were an avid golfer, I fail to see how a

(continues)

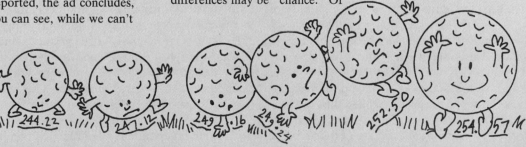

MUCH ADO ABOUT ALMOST NOTHING (*continued*)

yard or two would make any practical difference in my golf game. In all likelihood, my 18-hole score would not change by even one stroke, regardless of which ball I used. On the other hand, if I had been using a Dunlop Blue Max ball, these results would cause me to try one of the top 2 brands. Regarding the

3rd-, 4th-, and 5th-place brands, a reasonable person could go either way. If there were no difference in cost, I think I would switch to one of the first two brands, on a trial basis.

In summary, there are two points I have tried to make. The first is that *product claims of superiority based on sample data*

should report whether the results are statistically significant. The second is that *"statistical significance" and "importance" are different issues.* Once statistical significance has been established, we must look at the magnitude of the effect to see if it is large enough to warrant changing our behavior.

SUMMARY

In this chapter, we have discussed the F test and the analysis of variance. The F test is fundamentally the ratio of two independent variance estimates of the same population variance, σ^2. The F distribution is a family of curves that varies with degrees of freedom. Since F_{obt} is a ratio, there are two values for degrees of freedom, one for the numerator and one for the denominator. The F distribution (1) is positively skewed, (2) has no negative values, and (3) has a median approximately equal to 1, depending on the ns of the estimates.

The analysis of variance technique is used in conjunction with experiments involving more than two independent groups. Basically, it allows the means of the various groups to be compared in one overall evaluation, thus avoiding the inflated probability of making a Type I error when doing many t tests. In the one-way analysis of variance, the total variability of the data (SS_T) is partitioned into two parts: the variability that exists within each group, called the within-groups sum of squares (SS_W), and the variability that exists between the groups, called the between-groups sum of squares (SS_B). Each sum of squares is used to form an independent estimate of the variance of the null hypothesis populations. Finally, an F ratio is calculated where the between-groups variance esti-

mate ($s_B{}^2$) is in the numerator and the within-groups variance estimate ($s_W{}^2$) is in the denominator. Since the between-groups variance estimate increases with the effect of the independent variable and the within-groups variance estimate remains constant, the larger the F ratio, the more unreasonable the null hypothesis becomes. We evaluate F_{obt} by comparing it with F_{crit}. If $F_{obt} \geq F_{crit}$, we reject the null hypothesis and conclude that at least one of the conditions differs significantly from at least one of the other conditions.

Next, we discussed the assumptions underlying the analysis of variance. There are two assumptions: (1) the populations from which the samples are drawn should be normal; and (2) there should be homogeneity of variance. The F test is robust with regard to violations of normality and homogeneity of variance.

Finally, we presented a qualitative discussion of the two-way analysis of variance, independent groups design. Like the one-way design, in the two-way design, subjects are randomly assigned to the conditions. However, the two-way design allows us to investigate two independent variables and the interaction between them in one experiment. The effect of either independent variable (averaged over the levels of the other variable) is called a main effect. An interaction occurs when the effect of one of the variables is not the same at each level of the other variable. In this design, four variance estimates are determined: s_W^2, s_R^2, s_C^2, and s_{RC}^2. The first estimate, s_W^2, is the within-cells variance estimate of σ^2 and is not affected by the treatments. It is the standard against which the other variance estimates are compared. The other three variance estimates are sensitive to the effects of the two independent variables and their interaction. If there are no main effects and no interaction, then each variance is an inde-

pendent estimate of σ^2. To test whether there are any main effects, $F_{obt} = s_R^2/s_W^2$ and $F_{obt} = s_C^2/s_W^2$ are calculated. To test whether there is an interaction, $F_{obt} = s_{RC}^2/s_W^2$ is determined. These F ratios are then compared with F_{crit}. If $s_R^2/s_W^2 \geq F_{crit}$, H_0 is rejected and we concluded that factor A has a significant effect. If $s_C^2/s_W^2 \geq F_{crit}$, we reject H_0 and conclude that factor B has a significant effect. If $s_{RC}^2/s_W^2 \geq F_{crit}$, we conclude there is a significant interaction effect.

IMPORTANT TERMS

Analysis of variance (p. 353)
Between-groups sum of squares (SS_B) (p. 358)
Between-groups variance estimate (s_B^2) (p. 357)
Column variance estimate (s_C^2) (p. 374)
F test (p. 351)
F_{crit} (p. 352)
Grand mean ($\overline{X}_G$) (p. 357)
Interaction (p. 371)
Main effect (p. 372)

One-way analysis of variance, independent groups design (p. 354)
Row × column variance estimate (s_{RC}^2) (p. 374)
Row variance estimate (s_R^2) (p. 374)
Simple randomized-groups design (p. 354)
Single-factor experiment, independent groups design (p. 354)

Total variability (SS_T) (p. 355)
Two-way analysis of variance (p. 371)
Within-cells variance estimate (s_W^2) (p. 374)
Within-groups sum of squares (SS_W) (p. 356)
Within-groups variance estimate (s_W^2) (p. 356)

QUESTIONS AND PROBLEMS

1. Identify or define the terms in the "Important Terms" section.
2. What are the characteristics of the F distribution?
3. What advantages are there in doing experiments with more than two groups or conditions?
4. When doing an experiment with many groups, what is the problem with doing t tests between all possible groups without any correction? Why does use of the analysis of variance avoid that problem?
5. The analysis of variance technique analyzes the variability of the data. Yet a significant F value indicates that there is at least one significant mean difference between the conditions. How does analyzing the variability of the data allow conclusions about the means of the conditions?
6. What are the steps in forming an F ratio in using the one-way analysis of variance technique?
7. In the analysis of variance, if F_{obt} is less than 1, we don't even need to compare it with F_{crit}. It is obvious that the independent variable has not had a significant effect. Why is this so?
8. What are the assumptions underlying the analysis of variance?

9. The analysis of variance is a nondirectional technique, and yet it uses a one-tailed evaluation. Is this statement correct? Explain.
10. What are the advantages of the two-way ANOVA compared to the one-way ANOVA?
11. What is a factorial experiment?
12. In the two-way ANOVA, what is a *main* effect? What is an *interaction*? Is it possible to have a main effect without an interaction? An interaction without a main effect? Explain.
13. Find F_{crit} for the following situations:
    a. df(numerator) = 2, df(denominator) = 16, $\alpha = 0.05$
    b. df(numerator) = 3, df(denominator) = 36, $\alpha = 0.05$
    c. df(numerator) = 3, df(denominator) = 36, $\alpha = 0.01$
    What happens to F_{crit} as the degrees of freedom increase and alpha is held constant? What happens to F_{crit} when the degrees of freedom are held constant and alpha is made more stringent?
14. Verify that $F = t^2$ when there are just two independent groups by calculating F_{obt} for the data of Practice Problem 14.4.

15. The accompanying table is a one-way, independent-groups ANOVA summary table with part of the material missing.

Source	SS	df	s^2	F_{obt}
Between groups	1253.68	3		
Within groups				
Total	5016.40	39		

a. Filling in the missing values.
b. How many groups are there in the experiment?
c. Assuming an equal number of subjects in each group, how many subjects are there in each group?
d. What is the value of F_{crit}, using $\alpha = 0.05$?
e. Is there a significant effect?

16. Assume you are a nutritionist who has been asked to determine whether there is a difference in sugar content among the three leading brands of breakfast cereal (brands A, B, and C). To assess the amount of sugar in the cereals, you randomly sample six packages of each brand and chemically determine their sugar content. The following percentages of sugar were found:

Breakfast Cereal

A	B	C
1	7	5
4	5	4
3	3	4
3	6	5
2	4	7
5	7	8

a. Using the conceptual equations of the one-way ANOVA, determine whether any of the brands differ in sugar content. Use $\alpha = 0.05$.
b. Same as part **a**, except use the computational equations. Which do you prefer? Why?

17. A sleep researcher conducts an experiment to determine whether sleep loss affects the ability to maintain sustained attention. Fifteen individuals are randomly divided into the following three groups of 5 subjects each: group 1, which gets the normal amount of sleep (7–8 hours); group 2, which is sleep-deprived for 24 hours; and group 3, which is sleep-deprived for 48 hours. All three groups are tested on the same auditory vigilance task. Subjects are presented with half-second tones spaced at irregular intervals over a 1-hour duration. Occasionally, one of the tones is slightly shorter than the rest. The subject's task is to detect the shorter tones. The following percentages of correct detections were observed:

Normal Sleep	Sleep-Deprived for 24 Hours	Sleep-Deprived for 48 Hours
85	60	60
83	58	48
76	76	38
64	52	47
75	63	50

a. Determine whether there is an overall effect for sleep deprivation, using the conceptual equations of the one-way ANOVA. Use $\alpha = 0.05$.
b. Same as part **a**, except use the computational equations.
c. Which do you prefer? Why?

18. To test whether memory changes with age, a researcher conducts an experiment in which there are four groups of six subjects each. The groups differ according to the age of the subjects. In group 1, the subjects are each 30 years old; group 2, 40 years old; group 3, 50 years old; and group 4, 60 years old. Assume that the subjects are all in good health and that the groups are matched on other important variables such as years of education, IQ, gender, motivation, etc. Each subject is shown a series of nonsense syllables (a meaningless combination of three letters such as DAF or FUM) at a rate of one syllable every 4 seconds. The series is shown twice after which the subjects are asked to write down as many of the syllables as they can remember. The number of syllables remembered by each subject is shown here:

30 Years Old	40 Years Old	50 Years Old	60 Years Old
14	12	17	13
13	15	14	10
15	16	14	7
17	11	9	8
12	12	13	6
10	18	15	9

Use the analysis of variance with $\alpha = 0.05$ to determine whether age has an effect on memory.

19. Assume you are employed by a consumer-products rating service and your assignment is to assess car batteries. For this part of your investigation, you want to determine whether there is a difference in useful life among the top-of-the-line car batteries produced by three manufacturers (A, B, and C). To provide the database for your assessment, you randomly sample four batteries from each manufacturer and run them through laboratory tests that allow you to determine the useful life of each battery. The following are the results given in months of useful battery life:

Battery Manufacturer

A	B	C
52	46	44
57	52	53
55	51	50
59	50	51

Use the analysis of variance with $\alpha = 0.05$ to determine whether there is a difference among these three top-of-the-line brands of batteries.

20. In Chapter 14, an illustrative experiment involved investigating the effect of hormone X on sexual behavior. Although we presented only two concentrations in that problem, let's assume the experiment actually involved four different concentrations of the hormone. The full data are shown here, where the concentrations are arranged in ascending order; i.e., 0 concentration is where there is zero amount of hormone X (this is the placebo group), and concentration 3 represents the highest amount of the hormone:

Concentration of Hormone X

0	1	2	3
5	4	8	13
6	5	10	10
3	6	12	9
4	4	6	12
7	5	6	12
8	7	7	14
6	7	9	9
5	8	8	13
4	4	7	10
8	8	11	12

Using the analysis of variance with $\alpha = 0.05$, determine whether hormone X affects sexual behavior.

21. A clinical psychologist is interested in evaluating the effectiveness of the following three techniques for treating mild depression: cognitive restructuring, assertiveness training, and an exercise/nutrition program. Forty undergraduate students suffering from mild depression are randomly sampled from the university counseling center's waiting list and randomly assigned 10 each to the three techniques previously mentioned, and the remaining 10 to a placebo control group. Treatment is conducted for 10 weeks, after which depression is measured using the Beck Depression Scale. The post-treatment depression scores are given here. Higher scores indicate greater depression.

Treatment

Placebo	Cognitive Restructuring	Assertiveness Training	Exercise/ Nutrition
27	10	16	26
16	8	18	24
18	14	12	17
26	16	15	23
18	18	9	25
28	8	13	22
25	12	17	16
20	14	20	15
24	9	21	18
26	7	19	23

a. What is the overall null hypothesis?
b. Using $\alpha = 0.05$, what do you conclude?

16 | Multiple Comparisons

Introduction

In Chapter 15, we discussed the one-way and two-way analyses of variance when used with an independent groups design. We pointed out that the analysis of variance is a statistical technique employed in multigroup experiments that allows us to determine with one overall test (an F test) whether the independent variable has had a significant effect. In the one-way ANOVA, there is only one independent variable to evaluate; in the two-way ANOVA, there are two independent variables and their interaction to evaluate. For the sake of clarity, we shall confine our discussion of multiple comparisons to the one-way ANOVA. However, similar considerations also apply to the two-way situation.

In the one-way ANOVA, a significant F value indicates that all the conditions do not have the same effect on the dependent variable. For example, in the illustrative experiment presented in Chapter 15 (which investigated the amount of stress produced by three situations), a significant F value was obtained, and we concluded that the three situations were not the same in the stress levels they produced. For pedagogical reasons, we ended the analysis at this conclusion. However, in actual practice, the analysis does not ordinarily stop at this point. Usually, we are also interested in determining which of the conditions differ from each other. A significant F value tells us that at least one condition differs from at least one of the others. It is also possible that they are all different or any combination in between may be true. To determine which conditions differ, multiple comparisons between pairs of group means are usually made. In this chapter, we shall discuss two types of comparisons that may be made: *a priori* comparisons and *a posteriori* comparisons.

A Priori, or Planned, Comparisons

A priori, or planned, comparisons are planned in advance of the experiment and often arise from predictions based on theory and prior research. With *a priori*, or planned, comparisons, we do not correct for the higher probability of a Type I error that arises due to multiple comparisons, as is done with the *a posteriori* methods. This correction, which we shall cover in the next section, in effect makes it harder for the null hypothesis to be rejected. When doing *a priori,* or planned, comparisons, statisticians do not agree on whether the comparisons must be orthogonal, i.e., independent.* We have followed the position taken by Keppel and Winer that planned comparisons need not be orthogonal so long as they flow meaningfully and logically from the experimental design and are few in number.†

In doing planned comparisons, the *t* test for independent groups is used. We could calculate t_{obt} in the usual way. For example, in comparing conditions 1 and 2, we could use the equation

$$t_{obt} = \frac{\overline{X}_1 - \overline{X}_2}{\sqrt{\left(\dfrac{SS_1 + SS_2}{n_1 + n_2 - 2}\right)\left(\dfrac{1}{n_1} + \dfrac{1}{n_2}\right)}}$$

However, remembering that $(SS_1 + SS_2)/(n_1 + n_2 - 2)$ is an estimate of σ^2 based on the within variance of the two groups, we can use a better estimate since we have three or more groups in the experiment. Instead of $(SS_1 + SS_2)/(n_1 + n_2 - 2)$, we can use the within variance estimate s_W^2, which is based on all of the groups. Thus,

$$t_{obt} = \frac{\overline{X}_1 - \overline{X}_2}{\sqrt{s_W^2\left(\dfrac{1}{n_1} + \dfrac{1}{n_2}\right)}}$$ *t equation for a priori, or planned, comparisons, general equation*

With $n_1 = n_2 = n$,

$$t_{obt} = \frac{\overline{X}_1 - \overline{X}_2}{\sqrt{2s_W^2/n}}$$ *t equation for a priori, or planned, comparisons with equal n in the two groups*

Let's apply this to the stress example presented in Chapter 15. For convenience, we have shown the data and ANOVA solution in the table:

Group 1 Vacation		Group 2 Class		Group 3 Final Exam	
X_1	X_1^2	X_2	X_2^2	X_3	X_3^2
2	4	10	100	10	100
3	9	8	64	13	169
7	49	7	49	14	196
2	4	5	25	13	169
6	36	10	100	15	225
20	102	40	338	65	859

---

* For a discussion of this point, see G. Keppel, *Design and Analysis,* Prentice-Hall, Englewood Cliffs, N.J., 1973, pp. 92–93.

† See Note 16.1 for a discussion of orthogonal comparisons.

$$n_1 = 5 \qquad\qquad n_2 = 5 \qquad\qquad n_3 = 5$$

$$\overline{X}_1 = 4.00 \qquad\qquad \overline{X}_2 = 8.00 \qquad\qquad \overline{X}_3 = 13.00$$

$$\overset{\text{all scores}}{\sum} X = 125 \qquad \overset{\text{all scores}}{\sum} X^2 = 1299 \qquad \overline{X}_G = \frac{\overset{\text{all scores}}{\sum} X}{N} = 8.333$$

$$N = 15$$

Source	SS	df	s^2	F_{obt}
Between groups	203.333	2	101.667	22.59*
Within groups	54	12	4.5	
Total	257.333	14		

* With $\alpha = 0.05$, $F_{\text{crit}} = 3.88$. Therefore, H_0 is rejected.

Suppose we have the *a priori* hypothesis based on theoretical grounds that the effect of condition 3 will be different from the effect of both conditions 1 and 2. Therefore, prior to collecting any data, we have planned to compare the scores of group 3 with those of group 1 and group 2. To perform the planned comparisons, we first calculate the appropriate t_{obt} values and then compare them with t_{crit}. The calculations are as follows:

Group 1 and Group 3:

$$t_{\text{obt}} = \frac{\overline{X}_1 - \overline{X}_3}{\sqrt{2s_W^2/n}} = \frac{4.00 - 13.00}{\sqrt{2(4.50)/5}} = -6.71$$

Group 2 and Group 3:

$$t_{\text{obt}} = \frac{\overline{X}_2 - \overline{X}_3}{\sqrt{2s_W^2/n}} = \frac{8.00 - 13.00}{\sqrt{2(4.50)/5}} = -3.73$$

Are any of these t values significant? The value of t_{crit} is found from Table D in Appendix D, using the degrees of freedom for s_W^2. Thus, with df $= N - k = 12$ and $\alpha = 0.05_{2\,\text{tail}}$,

$$t_{\text{crit}} = \pm 2.18$$

Since both of the obtained t scores have absolute values greater than 2.18, we conclude that condition 3 differs significantly from conditions 1 and 2.

A Posteriori, or *Post Hoc*, Comparisons

When the comparisons are not planned in advance, we must use an *a posteriori* test. These comparisons usually arise after the experimenter sees the data and picks groups with mean scores that are far apart, or else they arise from doing all the comparisons possible with no theoretical *a priori* basis. Since these comparisons were not planned before the experiment, we must correct for the inflated probability values that occur when doing multiple comparisons, as discussed in the introduction.

There are many methods available for achieving this correction.* The topic is fairly complex, and it is beyond the scope of this text to present all of the methods. However, we shall present two of the most commonly accepted methods: a method devised by Tukey called the HSD (honestly significant difference) test and the Newman–Keuls test. Both of these tests are *post hoc* multiple comparison tests. They maintain the Type I error rate at α while making all possible comparisons between pairs of sample means.

You will recall that the problem with doing multiple *t* test comparisons is that the critical values of *t* were derived under the assumption that there are only two samples whose means are to be compared. This would be accomplished by performing one *t* test. When there are many samples and hence more than one comparison, the sampling distribution of *t* is no longer appropriate. In fact, if it were to be used, the actual probability of making a Type I error would greatly exceed alpha, particularly if many comparisons were made. Both the Tukey and Newman–Keuls method avoid this difficulty by using sampling distributions based on comparing the means of many samples rather than just two. These distributions, called the *Q* or Studentized range distributions, were developed by randomly taking *k* samples of equal *n* from the same population (rather than just 2, as with the *t* test) and determining the difference between the highest and lowest sample means. The differences were then divided by $\sqrt{s_W^2/n}$, producing distributions that were like the *t* distributions except that these provide the basis for making multiple comparisons, not just a single comparison as in the *t* test. The 95th and 99th percentile points for the *Q* distribution are given in Table G in Appendix D. These values are the critical values of *Q* for the 0.05 and 0.01 alpha levels. As you might guess, the critical values depend on the number of sample means and the degrees of freedom associated with s_W^2.

In discussing the HSD and Newman–Keuls tests, it is useful to distinguish between two aspects of Type I errors: the *experiment-wise* error rate and the *comparison-wise* error rate.

DEFINITION

• *The experiment-wise error rate is defined as the probability of making one or more Type I errors for the full set of possible comparisons in an experiment. The comparison-wise error rate is the probability of making a Type I error for any of the possible comparisons.*

As we shall see in the following sections, the HSD test and the Newman–Keuls test differ in which of these rates they maintain equal to alpha.

THE TUKEY HONESTLY SIGNIFICANT DIFFERENCE (HSD) TEST

The Tukey Honestly Significant Difference test is designed to compare all possible pairs of means while maintaining the Type I error for making the complete

---

* For a detailed discussion of these methods, see R. E. Kirk, *Experimental Design,* 3rd ed., Brooks/Cole, 1995, pp. 144–159.

set of comparisons at α. Thus, the HSD test maintains the experiment-wise Type I error rate at α. The statistic calculated for this test is Q. It is defined by the following equation:

$$Q_{obt} = \frac{\overline{X}_i - \overline{X}_j}{\sqrt{s_w^2/n}}$$

where $\overline{X}_i$ = Larger of the two means being compared
$\overline{X}_j$ = Smaller of the two means being compared
s_w^2 = Within-groups variance estimate
n = Number of subjects in each group

Note that in calculating Q_{obt} the smaller mean is always subtracted from the larger mean. This always makes Q_{obt} positive. Otherwise, the Q statistic is very much like the t statistic, except it uses the Q distributions rather than the t distributions. To use the statistic, we calculate Q_{obt} for the desired comparisons and compare Q_{obt} with Q_{crit}, determined from Table G. The decision rule states that if $Q_{obt} \geq Q_{crit}$, reject H_0. If not, then retain H_0.

To illustrate the use of the HSD test, we shall apply it to the data of the stress experiment. For the sake of illustration, we shall assume that all three comparisons are desired. There are two steps in using the HSD test. First, we must calculate the Q_{obt} value for each comparison and then compare each value with Q_{crit}. The calculations for Q_{obt} are as follows:

Group 2 and Group 1:

$$Q_{obt} = \frac{\overline{X}_2 - \overline{X}_1}{\sqrt{s_w^2/n}} = \frac{8.00 - 4.00}{\sqrt{4.50/5}} = \frac{4.00}{0.949} = 4.21$$

Group 3 and Group 1:

$$Q_{obt} = \frac{\overline{X}_3 - \overline{X}_1}{\sqrt{s_w^2/n}} = \frac{13.00 - 4.00}{\sqrt{4.50/5}} = \frac{9.00}{0.949} = 9.48$$

Group 3 and Group 2:

$$Q_{obt} = \frac{\overline{X}_3 - \overline{X}_2}{\sqrt{s_w^2/n}} = \frac{13.00 - 8.00}{\sqrt{4.50/5}} = \frac{5.00}{0.949} = 5.27$$

The next step is to compare the Q_{obt} values with Q_{crit}. The value of Q_{crit} is determined from Table G. To locate the appropriate value, we must know the df, the alpha level, and k. The df are the degrees of freedom associated with s_w^2. In this experiment, df = 12. As mentioned earlier, k stands for the number of groups in the experiment. In the present experiment $k = 3$. For this experiment, alpha was set at 0.05. From Table G, with df = 12, $k = 3$, and $\alpha = 0.05$, we obtain

$$Q_{crit} = 3.77$$

Since $Q_{obt} > 3.77$ for each comparison, we reject H_0 in each case and conclude that $\mu_1 \neq \mu_2 \neq \mu_3$. All three conditions differ in stress-inducing value. The solution is summarized in Table 16.1.

TABLE 16.1 *Post Hoc* Individual Comparisons Analysis of the Stress Experiment Using Tukey's HSD Test

	Group			Calculation
	1	**2**	**3**	
$\overline{X}$	4.00	8.00	13.00	Groups 2 and 1:
$\overline{X}_i - \overline{X}_j$		4.00	9.00	
			5.00	$Q_{obt} = \dfrac{\overline{X}_2 - \overline{X}_1}{\sqrt{s_w^2/n}} = \dfrac{8.00 - 4.00}{\sqrt{4.50/5}} = 4.21$
Q_{obt}		4.21*	9.48*	Groups 3 and 1:
			5.27*	
Q_{crit}		3.77	3.77	$Q_{obt} = \dfrac{\overline{X}_3 - \overline{X}_1}{\sqrt{s_w^2/n}} = \dfrac{13.00 - 4.00}{\sqrt{4.50/5}} = 9.48$
df = 12			3.77	
k = 3				Groups 3 and 2:
α = 0.05				
				$Q_{obt} = \dfrac{\overline{X}_3 - \overline{X}_2}{\sqrt{s_w^2/n}} = \dfrac{13.00 - 8.00}{\sqrt{4.50/5}} = 5.27$

* Reject H_0.

The Newman–Keuls Test

The Newman–Keuls test is also a *post hoc* test that allows us to make all possible pairwise comparisons among the sample means. It is like the HSD test in that Q_{obt} is calculated for each comparison and compared with Q_{crit} to evaluate the null hypothesis. It differs from the HSD test in that it maintains the Type I error rate at α *for each comparison,* rather than for the entire set of comparisons. Thus:

The Newman–Keuls test maintains the comparison-wise error rate at α, whereas the HSD test maintains the experiment-wise error rate at α.

To keep the comparison-wise error rate at α, the Newman–Keuls method varies the value of Q_{crit} for each comparison. The value of Q_{crit} used for any comparison is given by the sampling distribution of Q for the number of groups having means encompassed by $\overline{X}_i$ and $\overline{X}_j$ after all the means have been rank-ordered. This number is symbolized by r to distinguish it from k, which symbolizes the total number of groups in the experiment.

To illustrate the Newman–Keuls method, we shall use it to analyze the data from the stress experiment. The first step is to rank-order the means. This has been done in Table 16.2. Next, Q_{obt} is calculated for each comparison. The calculations and Q_{obt} values are also shown in Table 16.2. The next step is to determine the values of Q_{crit}. Note that to keep the Type I error rate for each comparison at α, there is a different critical value for each comparison, depending on r. Recall that r for any comparison equals the number of groups having means that are encompassed by $\overline{X}_i$ and $\overline{X}_j$, after the means have been rank-ordered. Thus, for the comparison between groups 3 and 1, $\overline{X}_i = \overline{X}_3$ and $\overline{X}_j = \overline{X}_1$. When the means are rank-ordered, there are three groups (groups 1, 2, and 3) whose means are encompassed by $\overline{X}_3$ and $\overline{X}_1$. Thus, $r = 3$ for this comparison. For the comparison between groups 3 and 2, $\overline{X}_i = \overline{X}_3$ and $\overline{X}_j = \overline{X}_2$. After rank-ordering

TABLE 16.2   *Post Hoc* Individual Comparisons Analysis of the Stress Experiment Using the Newman–Keuls Test

	Group 1	2	3	Calculation
$\overline{X}$	4.00	8.00	13.00	Groups 3 and 1:
$\overline{X}_i - \overline{X}_j$		4.00	9.00	$Q_{obt} = \dfrac{\overline{X}_3 - \overline{X}_1}{\sqrt{s_W^2/n}} = \dfrac{13.00 - 4.00}{\sqrt{4.50/5}} = 9.48$
			5.00	
Q_{obt}		4.21*	9.48*	Groups 3 and 2:
			5.27*	
Q_{crit}		3.08	3.77	$Q_{obt} = \dfrac{\overline{X}_3 - \overline{X}_2}{\sqrt{s_W^2/n}} = \dfrac{13.00 - 8.00}{\sqrt{4.50/5}} = 5.27$
df = 12			3.08	
$r = 2, 3$				Groups 2 and 1:
$\alpha = 0.05$				$Q_{obt} = \dfrac{\overline{X}_2 - \overline{X}_1}{\sqrt{s_W^2/n}} = \dfrac{8.00 - 4.00}{\sqrt{4.50/5}} = 4.21$

* Reject H_0.

all the means, $\overline{X}_3$ and $\overline{X}_2$ are directly adjacent so that there are only two means encompassed by $\overline{X}_3$ and $\overline{X}_2$, namely, $\overline{X}_2$ and $\overline{X}_3$. Thus, $r = 2$ for this comparison. The same holds true for the comparison between groups 2 and 1. For this comparison, $\overline{X}_i = \overline{X}_2$, $\overline{X}_j = \overline{X}_1$, and $r = 2$ (the two encompassed means are $\overline{X}_2$ and $\overline{X}_1$).

We are now ready to determine Q_{crit} for each comparison. The value of Q_{crit} is found in Table G, using the appropriate values for df, r, and α. The degrees of freedom are the df for s_W^2, which equal 12, and we shall use the same α level. Thus, $\alpha = 0.05$. For the comparison between groups 1 and 3, with df = 12, $r = 3$, and $\alpha = 0.05$, $Q_{crit} = 3.77$. For the comparisons between groups 1 and 2, and groups 2 and 3, with df = 12, $r = 2$, and $\alpha = 0.05$, $Q_{crit} = 3.08$. These values are also shown in Table 16.2.

The final step is to compare Q_{obt} with Q_{crit}. In making these comparisons, we follow the rule that we begin with the largest Q_{obt} value in the table (the upper right-hand corner of the Q_{obt} values) and compare it with the appropriate Q_{crit}. If it is significant, we proceed to the left one step and compare the next Q_{obt} with the corresponding Q_{crit}. We continue in this manner until we finish the row or until we reach a nonsignificant Q_{obt}. In the latter case, all the remaining Q_{obt} values left of this first nonsignificant Q_{obt} are considered nonsignificant. When the row is finished or when a nonsignificant Q_{obt} is reached, we drop to the next row and begin again at the rightmost Q_{obt} value. We continue in this manner until all the Q_{obt} values have been evaluated.

In the present example, we begin with 9.48 and compare it with 3.77. It is significant. Next, 4.21 is compared with 3.08. It is also significant. Since that comparison ends the row, we drop a row and compare 5.27 and 3.08. Again the Q_{obt} value is significant. Thus, we are able to reject H_0 in each comparison. Therefore, we end the analysis by concluding that $\mu_1 \neq \mu_2 \neq \mu_3$. The solution is summarized in Table 16.2.

Now let's do a practice problem.

PRACTICE PROBLEM 16.1

Using the data of Practice Problem 15.1 (page 364),

a. Test the planned comparisons that (1) lecture + reading and lecture have different effects and (2) that lecture + reading and film + reading have different effects. Use $\alpha = 0.05_{2\,tail}$.
b. Make all possible *post hoc* comparisons using the HSD test. Use $\alpha = 0.05$.
c. Make all possible *post hoc* comparisons using the Newman–Keuls test. Use $\alpha = 0.05$.

SOLUTION

For convenience, the data and the ANOVA solution are shown again.

Condition 1 Lecture		Condition 2 Lecture + Reading		Condition 3 Film + Reading	
X_1	X_1^2	X_2	X_2^2	X_3	X_3^2
92	8,464	86	7,396	81	6,561
86	7,396	93	8,649	80	6,400
87	7,569	97	9,409	72	5,184
76	5,776	81	6,561	82	6,724
80	6,400	94	8,836	83	6,889
87	7,569	89	7,921	89	7,921
92	8,464	98	9,604	76	5,776
83	6,889	90	8,100	88	7,744
84	7,056	91	8,281	83	6,889
767	65,583	819	74,757	734	60,088

$n_1 = 9$ $n_2 = 9$ $n_3 = 9$

$\overline{X}_1 = 85.222$ $\overline{X}_2 = 91$ $\overline{X}_3 = 81.556$

$$\overset{\text{all scores}}{\sum} X = 2320 \qquad \overset{\text{all scores}}{\sum} X^2 = 200{,}428 \qquad \overline{X}_G = \frac{\overset{\text{all scores}}{\sum} X}{N}$$

$$N = 27$$

$$= 85.926$$

Source	SS	df	s^2	F_{obt}
Between groups	408.074	2	204.037	7.29*
Within groups	671.778	24	27.991	
Total	1079.852	26		

* With $\alpha = 0.05$, $F_{crit} = 3.40$. Therefore, H_0 is rejected.

a. Planned comparisons: The comparisons are as follows:

Lecture + reading and lecture (condition 2 and condition 1):

$$t_{obt} = \frac{\overline{X}_2 - \overline{X}_1}{\sqrt{2s_W^2/n}} = \frac{91 - 85.222}{\sqrt{2(27.991)/9}} = 2.32$$

Lecture + reading and film + reading (condition 2 and condition 3):

$$t_{obt} = \frac{\overline{X}_2 - \overline{X}_3}{\sqrt{2s_W^2/n}} = \frac{91 - 81.556}{\sqrt{2(27.991)/9}} = 3.79$$

To evaluate these values of t_{obt}, we must determine t_{crit}. From Table D, with $\alpha = 0.05_{2\,tail}$ and df = 24,

$$t_{crit} = \pm 2.064$$

Since $|t_{obt}| > 2.064$ in both comparisons, we reject H_0 in each case and conclude that $\mu_1 \neq \mu_2$ and $\mu_2 \neq \mu_3$. By using *a priori* tests, lecture + reading appears to be the most effective method.

b. *Post hoc* comparisons using the HSD test: With the HSD test, Q_{obt} is determined for each comparison and then evaluated against Q_{crit}. The value of Q_{crit} is the same for each comparison and is such that the experiment-wise error rate is maintained at α. Although it is not necessary for this test, we have first rank-ordered the means for comparison purposes with the Newman–Keuls test. They are shown in Table 16.3. The calculations for Q_{obt} are as follows:

Lecture (1) *and Film + reading* (3):

$$Q_{obt} = \frac{\overline{X}_1 - \overline{X}_3}{\sqrt{s_W^2/n}} = \frac{85.222 - 81.556}{\sqrt{27.991/9}} = \frac{3.666}{1.764} = 2.08$$

Lecture + reading (2) *and Film + reading* (3):

$$Q_{obt} = \frac{\overline{X}_2 - \overline{X}_3}{\sqrt{s_W^2/n}} = \frac{91 - 81.556}{\sqrt{27.991/9}} = \frac{9.444}{1.764} = 5.35$$

Lecture + reading (2) *and Lecture* (1):

$$Q_{obt} = \frac{\bar{X}_2 - \bar{X}_1}{\sqrt{s_w^2/n}} = \frac{91 - 85.222}{\sqrt{27.991/9}} = \frac{5.778}{1.764} = 3.28$$

Next we must determine Q_{crit}. From Table G, with df = 24, k = 3, and α = 0.05, we obtain

$$Q_{crit} = 3.53$$

Comparing the three values of Q_{obt} with Q_{crit}, we find that only the comparison between film + reading and lecture + reading is significant. (For this comparison $Q_{obt} > 3.53$, whereas for the others $Q_{obt} < 3.53$.) Thus, on the basis of the HSD test, we may reject H_0 for the lecture + readiing and film + reading comparison (conditions 2 and 3). Lecture + reading appears to be more effective than film + reading. However, we cannot reject H_0 with regard to the other comparisons. The results are summarized in Table 16.3.

TABLE 16.3   *Post Hoc* Analysis of the Teaching Methods Experiment Using Tukey's HSD Test

	Condition			
	3	**1**	**2**	**Calculation**
$\bar{X}$	81.556	85.222	91	Conditions 1 and 3:
$\bar{X}_i - \bar{X}_j$		3.666	9.444	$Q_{obt} = \dfrac{\bar{X}_1 - \bar{X}_3}{\sqrt{s_w^2/n}} = \dfrac{85.222 - 81.556}{\sqrt{27.991/9}} = 2.08$
			5.778	
Q_{obt}		2.08	5.35*	Conditions 2 and 3:
			3.28	
Q_{crit} = 3.53		3.53	3.53	$Q_{obt} = \dfrac{\bar{X}_2 - \bar{X}_3}{\sqrt{s_w^2/n}} = \dfrac{91 - 81.556}{\sqrt{27.991/9}} = 5.35$
df = 24			3.53	
k = 3				Conditions 2 and 1:
α = 0.05				$Q_{obt} = \dfrac{\bar{X}_2 - \bar{X}_1}{\sqrt{s_w^2/n}} = \dfrac{91 - 85.222}{\sqrt{27.991/9}} = 3.28$

* Reject H_0.

c. *Post hoc* comparisons using the Newman–Keuls test: As with the HSD test, Q_{obt} is calculated for each comparison and then evaluated against Q_{crit}. However, with the Newman–Keuls test, the value of Q_{crit} changes with each comparison so as to keep the comparison-wise error rate at α. First, the means are rank-ordered from lowest to highest. This is shown in Table 16.4. Then, Q_{obt} is calculated for each comparison. These calculations, as well as the Q_{obt} values, have been entered in Table 16.4.

Next, Q_{crit} for each comparison is determined from Table G. The values of Q_{crit} depend on α, df for s_W^2, and r, where r equals the number of groups having means that are encompassed by $\overline{X}_i$ and $\overline{X}_j$ after the means have been rank-ordered. Thus, for the comparison between $\overline{X}_2$ and $\overline{X}_1$, $r = 2$; between $\overline{X}_2$ and $\overline{X}_3$, $r = 3$; and between $\overline{X}_1$ and $\overline{X}_3$, $r = 2$. For this experiment, df = 24 and $\alpha = 0.05$. The values of Q_{crit} for each comparison have been entered in Table 16.4. Comparing Q_{obt} with Q_{crit}, starting with the highest Q_{obt} value in the first row and proceeding to the left, we find that we can reject H_0 for the comparison between conditions 2 and 3, but not for the comparison between conditions 1 and 3. Dropping down to the next row, since $3.28 > 2.92$, we can also reject H_0 for the comparison between conditions 2 and 1. Thus, based on the Newman–Keuls test, it appears that lecture + reading is superior to both lecture alone and film + reading.

TABLE 16.4   *Post Hoc* Analysis of the Teaching Methods Experiment Using the Newman–Keuls Test

| | Condition | | | |
	3	1	2	Calculation
$\overline{X}$	81.556	85.222	91	Conditions 1 and 3:
$\overline{X}_i - \overline{X}_j$		3.666	9.444	
			5.778	$Q_{obt} = \dfrac{\overline{X}_1 - \overline{X}_3}{\sqrt{s_W^2/n}} = \dfrac{85.222 - 81.556}{\sqrt{27.991/9}} = 2.08$
Q_{obt}		2.08	5.35*	Conditions 2 and 3:
			3.28*	
Q_{crit}		2.92	3.53	$Q_{obt} = \dfrac{\overline{X}_2 - \overline{X}_3}{\sqrt{s_W^2/n}} = \dfrac{91 - 81.556}{\sqrt{27.991/9}} = 5.35$
df = 24			2.92	
$r = 2, 3$				Conditions 2 and 1:
$\alpha = 0.05$				$Q_{obt} = \dfrac{\overline{X}_2 - \overline{X}_1}{\sqrt{s_W^2/n}} = \dfrac{91 - 85.222}{\sqrt{27.991/9}} = 3.28$

* Reject H_0.

HSD and Newman–Keuls Tests with Unequal n

As pointed out previously, both the HSD and the Newman–Keuls tests are appropriate when there are an equal number of subjects in each group. If the ns are unequal, these tests still can be used, provided the ns do not differ greatly. To use the HSD or the Newman–Keuls tests with unequal n, we calculate the harmonic mean ($\tilde{n}$) of the various ns and use it in the denominator of the Q equation. The equation for $\tilde{n}$ is

$$\tilde{n} = \frac{k}{(1/n_1) + (1/n_2) + (1/n_3) + \cdots + (1/n_k)} \qquad \textit{harmonic mean}$$

where k = Number of groups
n_k = Number of subjects in the kth group

Suppose in the stress experiment that $n_1 = 5$, $n_2 = 7$, and $n_3 = 8$. Then

$$\tilde{n} = \frac{k}{(1/n_1) + (1/n_2) + (1/n_3)} = \frac{3}{\frac{1}{5} + \frac{1}{7} + \frac{1}{8}} = 6.41$$

COMPARISON BETWEEN PLANNED COMPARISONS, TUKEY'S HSD, AND THE NEWMAN–KEULS TESTS

Since planned comparisons do not correct for an increased probability of making a Type I error, they are more powerful than either of the *post hoc* tests we have discussed. This is the method of choice when applicable. It is important to note, however, that planned comparisons should be relatively few in number and should flow meaningfully and logically from the experimental design.

Deciding between Tukey's HSD and the Newman–Keuls tests really depends on one's philosophy. Since the HSD test keeps the Type I error rate at α for the entire set of comparisons, whereas the Newman–Keuls test maintains the Type I error rate at α for each comparison, the Newman–Keuls test has a somewhat higher experiment-wise Type I error rate than the HSD test (although still considerably lower than when making no adjustment at all). Because it uses a less stringent experiment-wise α level than the HSD test, the Newman–Keuls test is more powerful. An example of this occurred in the last experiment we analyzed. With Newman–Keuls, we were able to reject H_0 for the comparisons between conditions 2 and 3 and between conditions 2 and 1. With the HSD test we rejected H_0 only for the comparison between conditions 2 and 3. Thus, there is a trade-off. Newman–Keuls has a higher experiment-wise Type I error rate but a lower Type II error rate. As between the two tests, a conservative experimenter would probably choose Tukey's HSD test, whereas a more liberal researcher would probably prefer the Newman–Keuls test. Of course, if the consequences of making a Type I error were much greater than making a Type II error or vice versa, the researcher would choose the test that minimized the appropriate error rate.

SUMMARY

In experiments using the ANOVA technique, a significant F value indicates that the conditions are not all equal in their effects. To determine which conditions differ from each other, multiple comparisons between pairs of group means are usually performed. There are two approaches to doing multiple comparisons: *a priori,* or planned, comparisons and *a posteriori,* or *post hoc,* comparisons.

In the *a priori* approach, there are between-group comparisons that have been planned in advance of collecting the data. These may be done in the usual way, regardless of whether the obtained F value is significant, by calculating t_{obt} for the two groups and evaluating t_{obt} by comparing it with t_{crit}. In conducting the analysis, we use the within-groups variance estimate calculated in doing the analysis of variance.

Since this estimate is based on more groups than the two-group estimate employed in the ordinary t test, it is more accurate. There is no correction necessary for multiple comparisons. However, statisticians do not agree on whether the comparisons should be orthogonal. We have followed the view that *a priori* comparisons need not be orthogonal so long as they flow meaningfully and logically from the experimental design and are few in number.

A posteriori, or *post hoc,* comparisons were not planned prior to conducting the experiment and arise after looking at the data. As a result, we must be very careful about Type I error considerations. *Post hoc* comparisons must be made with a method that corrects for the inflated Type I error probability. There are many methods that do this.

In this chapter, we have described Tukey's HSD test and the Newman–Keuls test. Both of these tests maintain the Type I error rate at α while making all possible comparisons between pairs of sample means. The HSD test keeps the experiment-wise error rate at α, whereas the Newman–Keuls test keeps the comparison-wise error rate at α. Both tests use the Q or studentized range statistic. As with the t test, Q_{obt} is calculated for each comparison and evaluated against Q_{crit} determined from the sampling distribution of Q. If $Q_{obt} \geq Q_{crit}$, the null hypothesis is rejected.

IMPORTANT TERMS

A posteriori comparisons (p. 383)
A priori comparisons (p. 381)
Comparison-wise error rate (p. 384)
Experiment-wise error rate (p. 384)
Newman–Keuls test (p. 386)
Planned comparisons (p. 381)
Post hoc comparisons (p. 383)
Q_{crit} (p. 385)
Q_{obt} (p. 385)
Tukey's HSD test (p. 384)

QUESTIONS AND PROBLEMS

1. Identify or define the terms in the "Important Terms" section.
2. Explain why we must correct for doing multiple comparisons when doing *post hoc* comparisons.
3. How do planned comparisons, *post hoc* comparisons using the HSD test, and *post hoc* comparisons using the Newman–Keuls test differ with regard to
   a. Power?
   b. The probability of making a Type I error? Explain.
4. What are the Q or Studentized range distributions? How do they avoid the problem of inflated Type I errors that result from doing multiple comparisons with the t distribution?
5. In doing planned comparisons, it is better to use s_W^2 from the ANOVA rather than the weighted variance estimate from the two groups being compared. Is this statement correct? Why?
6. Refer to Chapter 15, Problem 16. In that problem, the ANOVA yielded a significant F, indicating that at least one of the cereals differed in sugar content.

   a. Do a *post hoc* analysis on each pair of means using the Tukey HSD test with $\alpha = 0.05$ to determine which cereals are different in sugar content.
   b. Same as part **a**, but use the Newman–Keuls test.
7. Refer to Chapter 15, Problem 17. In this problem, F_{obt} was significant, indicating that at least one of the conditions differed in its effect on the ability to maintain sustained attention.
   a. Do a planned comparison between the means of the 48-hour sleep-deprived group and the normal sleep group to see whether these conditions differ in their effect on the ability to maintain sustained attention. Use $\alpha = 0.05_{2\,tail}$. What do you conclude?
   b. Do *post hoc* comparisons, comparing each pair of means using the Newman–Keuls test and $\alpha = 0.05_{2\,tail}$. What do you conclude?
   c. Same as part **b**, but use the HSD test. Compare your answers to parts **b** and **c**. Explain any difference.
8. Refer to Chapter 15, Problem 18. The ANOVA

for this problem yielded a significant F value, indicating that at least one of the age conditions differed in its effect on memory.
  a. Using planned comparisons with $\alpha = 0.05_{2\,tail}$, compare the means of the 60-year-old and the 30-year-old groups. What do you conclude?
  b. Use the Newman–Keuls test with $\alpha = 0.05_{2\,tail}$ to compare all possible pairs of means. What do you conclude?
9. Refer to Chapter 15, Problem 19. Suppose you are asked to make a recommendation regarding the batteries based on useful life. Use the HSD test with $\alpha = 0.05_{2\,tail}$ to help you with your decision.

10. Refer to Chapter 15, Problem 20.
  a. Using planned comparisons with $\alpha = 0.05_{2\,tail}$, compare the mean of concentration 3 with that of concentration 0. What do you conclude?
  b. Using the Newman–Keuls test with $\alpha = 0.05_{2\,tail}$, compare all possible pairs of means. What do you conclude?
  c. Same as part **b**, except use the HSD test.
11. Refer to Chapter 15, Problem 21. Do *post hoc* comparisons, using the Tukey HSD test, with $\alpha = 0.05_{2\,tail}$. What do you conclude?

NOTES

16.1 *Orthogonal comparisons.* When making comparisons between the means of the groups, we can represent any comparison as a weighted sum of the means. For example, if there are four means, we can represent the comparison between any of the groups as the weighted sum of $\bar{X}_1, \bar{X}_2, \bar{X}_3$, and $\bar{X}_4$. Thus, if we are evaluating $\bar{X}_1 - \bar{X}_2$, the weighted sum would be $(1)\bar{X}_2 + (-1)\bar{X}_2 + (0)\bar{X}_3 + (0)\bar{X}_4$, where 1, −1, 0, and 0 are the weights. If we are evaluating $\bar{X}_3 - \bar{X}_4$, the weighted sum would be $(0)\bar{X}_1 + (0)\bar{X}_2 + (1)\bar{X}_3 + (-1)\bar{X}_4$, where 0, 0, 1, and −1 are the weights. *In general, two comparisons are said to be orthogonal if the sum of the products of the two weights for each mean equals zero.* Using this information, let's now determine whether the above two comparisons are orthogonal. The appropriate paired weights have been multiplied and summed, as follows:

Comparison	Weighted Sum	Weights
$\bar{X}_1 - \bar{X}_2$	$(1)\bar{X}_1 + (-1)\bar{X}_2 + (0)\bar{X}_3 + (0)\bar{X}_4$	1, −1, 0, 0
$\bar{X}_3 - \bar{X}_4$	$(0)\bar{X}_1 + (0)\bar{X}_2 + (1)\bar{X}_3 + (-1)\bar{X}_4$	0, 0, 1, −1
	$(1)(0) + (-1)(0) + (0)(1) + (0)(-1) = 0$	

Since the sum of the products of the two weights for each mean equals zero, these two comparisons are orthogonal, i.e., independent. In general, if there are k groups in an experiment, there are $k - 1$ independent comparisons possible.*

---

* For a more detailed discussion of orthogonal comparisons, see R. E. Kirk, *Experimental Design,* 3rd. ed., Brooks/Cole, 1995, pp. 115–118.

17

INTRODUCTION TO TWO-WAY ANALYSIS OF VARIANCE

INTRODUCTION

The two-way analysis of variance, independent groups, fixed-effects design was discussed in a qualitative way in Chapter 15. In this chapter, we shall present a more detailed quantitative discussion of the data analysis for this design. This discussion will assume that the reader is already familiar with the appropriate material in Chapter 15.

OVERVIEW OF THE TWO-WAY ANALYSIS OF VARIANCE

In the one-way analysis of variance, the total sum of squares is partitioned into two components: the within-groups sum of squares and the between-groups sum of squares. These two components are divided by the appropriate degrees of freedom to form two variance estimates: the within-groups variance estimate (s_W^2) and the between-groups variance estimate (s_B^2). If the null hypothesis is correct, then both estimates are estimates of the null-hypothesis population variance (σ^2), and the ratio s_B^2/s_W^2 will be distributed as F. If the independent variable has a real effect, then s_B^2 will tend to be larger than otherwise, and so will the F ratio. Thus, the larger the F ratio, the more unreasonable the null hypothesis becomes. When $F_{\text{obt}} \geq F_{\text{crit}}$, we reject H_0 as being too unreasonable to entertain as an explanation of the data.

The situation is quite similar in the two-way analysis of variance. However, in the two-way analysis of variance, we partition the total sum of squares (SS_T) into four components: the within-cells sum of squares (SS_W), the row sum of squares (SS_R), the column sum of squares (SS_C), and the row × column sum of squares (SS_{RC}). This partitioning is shown in Figure 17.1. When these sums of squares are divided by the appropriate degrees of freedom, they form four

FIGURE 17.1   Overview of two-way analysis of variance technique, independent-groups design

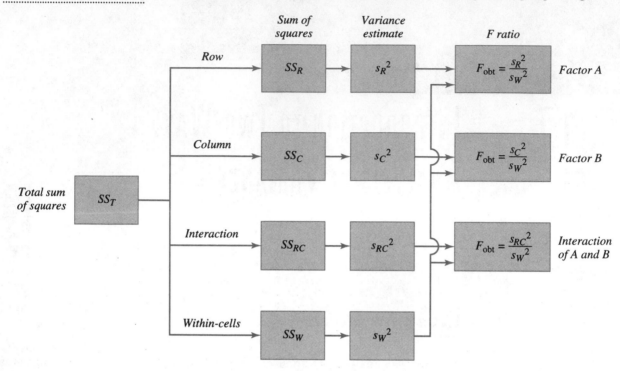

variance estimates. These estimates are the within-cells variance estimate (s_W^2), the row variance estimate (s_R^2), the column variance estimate (s_C^2), and the row × column variance estimate (s_{RC}^2). In discussing each of these variance estimates, it will be useful to refer to Figure 17.2, which shows the notation and general layout of data for a two-way analysis of variance, independent groups design. We have assumed in the following discussion that the number of subjects in each cell is the same.

Within-Cells Variance Estimate (s_W^2)

This estimate is derived from the variability of the scores within each cell. Since all the subjects within each cell receive the same level of variables A and B, the variability of their scores cannot be due to treatment differences. The within-cells variance estimate is analogous to the within-groups variance estimate used in the one-way analysis of variance. It is a measure of the inherent variability of the scores and, hence, gives us an estimate of the null-hypothesis population variance (σ^2). It is the yardstick against which we compare each of the other variance estimates. In equation form,

$$s_W^2 = \frac{SS_W}{df_W} \qquad \textit{equation for within-cells variance estimate}$$

where SS_W = Within-cells sum of squares
df_W = Within-cells degrees of freedom

FIGURE 17.2

FIGURE 17.2

Notation and general layout of data for a two-way analysis of variance design

	Levels of factor B				
	b_1	b_2		b_c	
a_1	Cell 11 — — — —	Cell 12 — — — —	— — — — —	Cell 1c — — — —	$\displaystyle\sum_{1}^{\text{row}} X$
a_2	Cell 21 — — — —	— — — — —	— — — — —	— — — — —	$\displaystyle\sum_{2}^{\text{row}} X$
	— — — — —	— — — — —	— — — — —	— — — — —	
a_r	Cell r1 — — — —	— — — — —	— — — — —	Cell rc — — — —	$\displaystyle\sum_{r}^{\text{row}} X$
	$\displaystyle\sum_{1}^{\text{col.}} X$	$\displaystyle\sum_{2}^{\text{col.}} X$		$\displaystyle\sum_{c}^{\text{col.}} X$	

Levels of factor A

where a_1 = first level of factor A

a_r = last level of factor A

b_1 = first level of factor B

b_c = last level of factor B

$\displaystyle\sum_{1}^{\text{row}} X$ = sum of scores for row 1

$\displaystyle\sum_{r}^{\text{row}} X$ = sum of scores for row r

$\displaystyle\sum_{1}^{\text{col.}} X$ = sum of scores for column 1

$\displaystyle\sum_{c}^{\text{col.}} X$ = sum of scores for column c

r = number of rows

c = number of columns

N = total number of scores

n_{cell} = number of scores in each cell

n_{row} = number of scores in each row

$n_{col.}$ = number of scores in each column

The within-cells sum of squares (SS_W) is just the sum of squares within each cell added together. Conceptually,

$$SS_W = SS_{11} + SS_{12} + \cdots + SS_{rc} \qquad \text{within-cells sum of squares, conceptual equation}$$

where SS_{11} = Sum of squares for the scores in the cell defined by the intersection of row 1 and column 1

SS_{12} = Sum of squares for the scores in the cell defined by the intersection of row 1 and column 2

SS_{rc} = Sum of squares for the scores in the cell defined by the intersection of row r and column c; this is the last cell in the matrix

As has been the case so often previously, the conceptual equation is not the best equation to use for computational purposes. The computational equation is given here:

$$SS_W = \sum^{\substack{\text{all} \\ \text{scores}}} X^2 - \left[\frac{\left(\sum^{\substack{\text{cell} \\ 11}} X \right)^2 + \left(\sum^{\substack{\text{cell} \\ 12}} X \right)^2 + \cdots + \left(\sum^{\substack{\text{cell} \\ rc}} X \right)^2}{n_{\text{cell}}} \right]$$

within-cells sum of squares, computational equation

Note the similarity of these equations to the comparable equations for the within-group variance estimate used in the one-way ANOVA. The only difference is that in the two-way ANOVA, summation is with regard to the cells, whereas in the one-way ANOVA, summation is with regard to the groups.

In computing SS_W, there are n deviation scores for each cell. Therefore, there are $n - 1$ degrees of freedom contributed by each cell. Since we sum over all cells in calculating SS_W, the within-cells degrees of freedom equal $n - 1$ times the number of cells. If we let r equal the number of rows and c equal the number of columns, then rc equals the number of cells. Therefore, the within-cells degrees of freedom equal $rc(n - 1)$. Thus,

$$\text{df}_W = rc(n - 1) \qquad \text{within-cells degrees of freedom}$$

where r = Number of rows
c = Number of columns

Row Variance Estimate ($s_R{}^2$)

This estimate is based on the differences among the row means. It is analogous to the between-groups variance estimate ($s_B{}^2$) in the one-way ANOVA. You will recall that $s_B{}^2$ is an estimate of σ^2 plus the effect of the independent variable. Similarly, the row variance estimate ($s_R{}^2$) in the two-way ANOVA is an estimate of σ^2 plus the effect of factor A. If factor A has no effect, then the population row means are equal ($\mu_{a_1} = \mu_{a_2} = \cdots = \mu_{a_r}$), and the differences among sample row means will just be due to random sampling from identical populations. In this case, $s_R{}^2$ becomes an estimate of just σ^2 alone. If factor A has an effect, then the differences among the row means, and hence $s_R{}^2$, will tend to be larger than otherwise. In equation form,

$$s_R^2 = \frac{SS_R}{df_R} \qquad \textit{equation for the row variance estimate}$$

where SS_R = Row sum of squares
df_R = Row degrees of freedom

The row sum of squares is very similar to the between-groups sum of squares in the one-way ANOVA. The only difference is that with the row sum of squares we use the row means, whereas the between-groups sum of squares used the group means. The conceptual equation for SS_R follows. Note that in computing row means, all the scores in a given row are combined and averaged. This is referred to as computing the row means "averaged over the columns" (see Figure 17.2). Thus, the row means are arrived at by averaging over the columns:

$$SS_R = n_{\text{row}}[(\overline{X}_{\text{row}\,1} - \overline{X}_G)^2 + (\overline{X}_{\text{row}\,2} - \overline{X}_G)^2 + \cdots + (\overline{X}_{\text{row}\,r} - \overline{X}_G)^2]$$

conceptual equation for the row sum of squares

where $$\overline{X}_{\text{row}\,1} = \frac{\sum\limits^{\text{row}\,1} X}{n_{\text{row}\,1}}$$

$$\overline{X}_{\text{row}\,r} = \frac{\sum\limits^{\text{row}\,r} X}{n_{\text{row}\,r}}$$

$\overline{X}_G$ = Grand mean

From the conceptual equation, it is easy to see that SS_R increases with the effect of variable A. As the effect of variable A increases, the row means become more widely separated, which in turn causes $(\overline{X}_{\text{row}\,1} - \overline{X}_G)^2$, $(\overline{X}_{\text{row}\,2} - \overline{X}_G)^2$, ..., $(\overline{X}_{\text{row}\,r} - \overline{X}_G)^2$ to increase. Since these terms are in the numerator, SS_R increases. Of course, if SS_R increases, so does s_R^2.

In calculating SS_R, there are r deviation scores. Thus, the row degrees of freedom equal $r - 1$. In equation form,

$$df_R = r - 1 \qquad \textit{row degrees of freedom}$$

Recall that the between-groups degrees of freedom $(df_B) = k - 1$ for the one-way ANOVA. The row degrees of freedom are quite similar except we are using rows rather than groups.

Again the conceptual equation turns out not to be the best equation to use for computing SS_R. The computational equation is given here:

$$SS_R = \left[\frac{\left(\sum\limits^{\text{row}\,1} X\right)^2 + \left(\sum\limits^{\text{row}\,2} X\right)^2 + \cdots + \left(\sum\limits^{\text{row}\,r} X\right)^2}{n_{\text{row}}} \right] - \frac{\left(\sum\limits^{\text{all scores}} X\right)^2}{N}$$

computational equation for the row sum of squares

Column Variance Estimate (s_C^2)

This estimate is based on the differences among the column means. It is exactly the same as s_R^2, except that it uses the column means rather than the row means. Since factor B affects the column means, the column variance estimate (s_C^2) is an estimate of σ^2 plus the effects of factor B. If the levels of factor B have no differential effect, then the population column means are equal ($\mu_{b_1} = \mu_{b_2} = \mu_{b_3} = \cdots = \mu_{b_c}$), and the differences among the sample column means are due to random sampling from identical populations. In this case, s_C^2 will be an estimate of σ^2 alone. If factor B has an effect, then the differences among the column means, and hence s_C^2, will tend to be larger than otherwise.

The equation for s_C^2 is

$$s_C^2 = \frac{SS_C}{df_C} \qquad column\ variance\ estimate$$

where SS_C = Column sum of squares
df_C = Column degrees of freedom

The column sum of squares is also very similar to the row sum of squares. The only difference is that we use the column means in calculating the column sum of squares rather than the row means. The conceptual equation for SS_C is shown here. Note that, in computing the column means, all the scores in a given column are combined and averaged. Thus, the column means are arrived at by averaging over the rows.

$$SS_C = n_{col.}[(\overline{X}_{col.\ 1} - \overline{X}_G)^2 + (\overline{X}_{col.\ 2} - \overline{X}_G)^2 + \cdots + (\overline{X}_{col.\ c} - \overline{X}_G)^2]$$

$$conceptual\ equation\ for\ the$$
$$column\ sum\ of\ squares$$

where $$\overline{X}_{col.\ 1} = \frac{\overset{col.}{\overset{1}{\sum}} X}{n_{col.\ 1}}$$

$$\overline{X}_{col.\ c} = \frac{\overset{col.}{\overset{c}{\sum}} X}{n_{col.\ c}}$$

Again, we can see from the conceptual equation that SS_C increases with the effect of variable B. As the effect of variable B increases, the column means become more widely spaced, which in turn causes $(\overline{X}_{col.\ 1} - \overline{X}_G)^2$, $(\overline{X}_{col.\ 2} - \overline{X}_G)^2$, $\ldots$, $(\overline{X}_{col.\ c} - \overline{X}_G)^2$ to increase. Since these terms are in the numerator of the equation for SS_C, the result is an increase in SS_C. Of course, an increase in SS_C results in an increase in s_C^2.

Since there are c deviation scores used in calculating SS_C, the column degrees of freedom equal $c - 1$. Thus,

$$df_C = c - 1 \qquad column\ degrees\ of\ freedom$$

The computational equation for SS_C is

$$SS_C = \left[\frac{\left(\overset{\text{col.}\ 1}{\sum X}\right)^2 + \left(\overset{\text{col.}\ 2}{\sum X}\right)^2 + \cdots + \left(\overset{\text{col.}\ c}{\sum X}\right)^2}{n_{\text{col.}}}\right] - \frac{\left(\overset{\text{all}\ \text{scores}}{\sum X}\right)^2}{N}$$

*computational equation for
the column sum of squares*

Row × Column Variance Estimate (s_{RC}^2)

In Chapter 15, we pointed out that an interaction exists when the effect of one of the variables is not the same at all levels of the other variable. Another way of saying this is that an interaction exists when the effect of the combined action of the variables is different from that which would be predicted by the individual effects of the variables. To illustrate this point, consider Figure 15.5(f), p. 373, where there is an interaction between the time of day and the intensity of exercise. An interaction exists because the sleep score for heavy exercise done in the morning is higher than would be predicted based on the individual effects of the time of day and intensity of exercise variables. If there were no interaction, then we would expect the lines to be parallel. The intensity of exercise variable would have the same effect when done in the evening as when done in the morning.

The row × column variance estimate (s_{RC}^2) is used to evaluate the interaction of variables A and B. As such, it is based on the differences among the cell means beyond that which is predicted by the individual effects of the two variables. The row × column variance estimate is an estimate of σ^2 plus the interaction of A and B. If there is no interaction and any main effects are removed, then the population cell means are equal ($\mu_{a_1 b_1} = \mu_{a_1 b_2} = \cdots = \mu_{a_r b_c}$), and differences among cell means must be due to random sampling from identical populations. In this case, s_{RC}^2 will be an estimate of σ^2 alone. If there is an interaction between factors A and B, then the differences among the cell means and, hence s_{RC}^2, will tend to be higher than otherwise.

The equation for s_{RC}^2 is

$$s_{RC}^2 = \frac{SS_{RC}}{\text{df}_{RC}} \qquad row \times column\ variance\ estimate$$

where SS_{RC} = Row × column sum of squares
df_{RC} = Row × column degrees of freedom

The row × column sum of squares is equal to the variability of the cell means when the variability due to the individual effects of factors A and B has been removed. Both the conceptual and computational equations are given here:

$$SS_{RC} = n_{\text{cell}}[(\overline{X}_{\text{cell}\ 11} - \overline{X}_G)^2 + (\overline{X}_{\text{cell}\ 12} - \overline{X}_G)^2 + \cdots + (\overline{X}_{\text{cell}\ rc} - \overline{X}_G)^2]$$

$$- SS_R - SS_C \qquad \begin{array}{l} \textit{conceptual equation for the} \\ \textit{row} \times \textit{column sum of squares} \end{array}$$

$$SS_{RC} = \left[\frac{\left(\overset{\text{cell}}{\underset{11}{\sum}} X \right)^2 + \left(\overset{\text{cell}}{\underset{12}{\sum}} X \right)^2 + \cdots + \left(\overset{\text{cell}}{\underset{rc}{\sum}} X \right)^2}{n_{\text{cell}}} \right] - \frac{\left(\overset{\text{all}}{\underset{\text{scores}}{\sum}} X \right)^2}{N}$$

$$- SS_R - SS_C \qquad \begin{array}{l} \textit{computational equation for the} \\ \textit{row} \times \textit{column sum of squares} \end{array}$$

The degrees of freedom for the row × column variance estimate equal $(r - 1)(c - 1)$. Thus,

$$\text{df}_{RC} = (r - 1)(c - 1) \qquad \textit{row} \times \textit{column degrees of freedom}$$

Computing F Ratios

Once the variance estimates have been determined, they are used in conjunction with s_W^2 to form F ratios (see Figure 17.1) to test the main effects of the variables and their interaction. The following three F ratios are computed:

To test the main effect of variable A (row effect):

$$F_{\text{obt}} = \frac{s_R^2}{s_W^2} = \frac{\sigma^2 + \text{effects of variable A}}{\sigma^2}$$

To test the main effect of variable B (column effect):

$$F_{\text{obt}} = \frac{s_C^2}{s_W^2} = \frac{\sigma^2 + \text{effects of variable B}}{\sigma^2}$$

To test the interaction of variables A and B (row × column effect):

$$F_{\text{obt}} = \frac{s_{RC}^2}{s_W^2} = \frac{\sigma^2 + \text{interaction effects of A and B}}{\sigma^2}$$

The ratio s_R^2/s_W^2 is used to test the main effect of variable A. If variable A has no main effect, s_R^2 is an independent estimate of σ^2, and s_R^2/s_W^2 is distributed as F with degrees of freedom equal to df_R and df_W. If variable A has a main effect, s_R^2 will be larger than otherwise, and the F_{obt} value for rows will increase.

The ratio s_C^2/s_W^2 is used to test the main effect of variable B. If this variable has no main effect, then s_C^2 is an independent estimate of σ^2, and s_C^2/s_W^2 is distributed as F with degrees of freedom equal to df_C and df_W. If variable B has a main effect, s_C^2 will be larger than otherwise, causing an increase in the F_{obt} value for columns.

The interaction between A and B is tested using the ratio s_{RC}^2/s_W^2. If there is no interaction, s_R^2 is an independent estimate of σ^2, and s_{RC}^2/s_W^2 is distributed as F with degrees of freedom equal to df_{RC} and df_W. If there is an interaction, s_{RC}^2 will be larger than otherwise, causing an increase in the F_{obt} value for interaction.

The main effect of each variable and their interaction are tested by comparing the appropriate F_{obt} value with F_{crit}. F_{crit} is found in Table F in Appendix D, using α and the degrees of freedom of the F value being evaluated. The decision rule is the same as with the one-way ANOVA, namely,

If $F_{obt} \geq F_{crit}$, reject H_0.   *decision rule for evaluating H_0 in two-way ANOVA*

We are now ready to analyze the data from an illustrative example.

AN EXPERIMENT

Effect of Exercise on Sleep

Let's assume a professor in physical education conducts an experiment to compare the effects on nighttime sleep of different amounts of exercise and the time of day when the exercise is done. The experiment uses a fixed effects, 3×2 factorial design with independent groups. There are three levels of exercise (light, moderate, and heavy) and two times of day (morning and evening). Thirty-six college students who are in good physical condition are randomly assigned to the six cells such that there are 6 subjects per cell. The subjects who do heavy exercise jog for 3 miles, the subjects who do moderate exercise jog for 1 mile, and the subjects in the light exercise condition jog for only $\frac{1}{4}$ mile. Morning exercise is done at 7:30 A.M., whereas evening exercise is done at 7:00 P.M. Each subject exercises once, and the number of hours slept that night is recorded. The following data are obtained:

Time of Day	Exercise			
	Light	*Moderate*	*Heavy*	
Morning	6.5 7.4	7.4 7.3	8.0 7.6	$\Sigma X = 129.20$
	7.3 7.2	6.8 7.6	7.7 6.6	$\Sigma X^2 = 930.50$
	6.6 6.8	6.7 7.4	7.1 7.2	$n = 18$
	$\bar{X} = 6.97$	$\bar{X} = 7.20$	$\bar{X} = 7.37$	$\bar{X} = 7.18$
Evening	7.1 7.7	7.4 8.0	8.2 8.7	$\Sigma X = 147.20$
	7.9 7.5	8.1 7.6	8.5 9.6	$\Sigma X^2 = 1212.68$
	8.2 7.6	8.2 8.0	9.5 9.4	$n = 18$
	$\bar{X} = 7.67$	$\bar{X} = 7.88$	$\bar{X} = 8.98$	$\bar{X} = 8.18$
	$\Sigma X = 87.80$	$\Sigma X = 90.50$	$\Sigma X = 98.10$	$\Sigma X = 276.40$
	$\Sigma X^2 = 645.30$	$\Sigma X^2 = 685.07$	$\Sigma X^2 = 812.81$	$\Sigma X^2 = 2143.18$
	$n = 12$	$n = 12$	$n = 12$	$N = 36$
	$\bar{X} = 7.32$	$\bar{X} = 7.54$	$\bar{X} = 8.18$	

1. What are the null hypotheses for this experiment?
2. Using $\alpha = 0.05$, what do you conclude?

SOLUTION

1. Null hypothesis:
   a. *For the A variable (main effect).* The time of day when exercise is done does not affect nighttime sleep. The population row means for morning and evening exercise averaged over the different levels of exercise are equal ($\mu_{a_1} = \mu_{a_2}$).
   b. *For the B variable (main effect).* The different levels of exercise have the same effect on nighttime sleep. The population column means for light, medium, and heavy exercise averaged over time of day conditions are equal ($\mu_{b_1} = \mu_{b_2} = \mu_{b_3}$).

c. *For the interaction between A and B.* There is no interaction between time of day and level of exercise. With any main effects removed, the population cell means are equal ($\mu_{a_1b_1} = \mu_{a_1b_2} = \mu_{a_1b_3} = \mu_{a_2b_1} = \mu_{a_2b_2} = \mu_{a_2b_3}$).

2. Conclusion, using $\alpha = 0.05$:
   a. Calculate F_{obt} for each hypothesis:

Step 1: Calculate the row sum of squares, SS_R:

$$SS_R = \left[\frac{\left(\overset{row\,1}{\sum} X\right)^2 + \left(\overset{row\,2}{\sum} X\right)^2}{n_{row}}\right] - \frac{\left(\overset{all\,scores}{\sum} X\right)^2}{N}$$

$$= \left[\frac{(129.2)^2 + (147.2)^2}{18}\right] - \frac{(276.4)^2}{36}$$

$$= 9.000$$

Step 2: Calculate the column sum of squares, SS_C:

$$SS_C = \left[\frac{\left(\overset{col.\,1}{\sum} X\right)^2 + \left(\overset{col.\,2}{\sum} X\right)^2 + \left(\overset{col.\,3}{\sum} X\right)^2}{n_{col.}}\right] - \frac{\left(\overset{all\,scores}{\sum} X\right)^2}{N}$$

$$= \left[\frac{(87.8)^2 + (90.5)^2 + (98.1)^2}{12}\right] - \frac{(276.4)^2}{36}$$

$$= 4.754$$

Step 3: Calculate the row × column sum of squares, SS_{RC}:

$$SS_{RC} = \left[\frac{\left(\overset{cell\,11}{\sum} X\right)^2 + \left(\overset{cell\,12}{\sum} X\right)^2 + \left(\overset{cell\,13}{\sum} X\right)^2 + \left(\overset{cell\,21}{\sum} X\right)^2 + \left(\overset{cell\,22}{\sum} X\right)^2 + \left(\overset{cell\,23}{\sum} X\right)^2}{n_{cell}}\right]$$

$$- \frac{\left(\overset{all\,scores}{\sum} X\right)^2}{N} - SS_R - SS_C$$

$$= \left[\frac{(41.8)^2 + (43.2)^2 + (44.2)^2 + (46.0)^2 + (47.3)^2 + (53.9)^2}{6}\right] - \frac{(276.4)^2}{36}$$

$$- 9.000 - 4.754$$

$$= 1.712$$

Step 4: Calculate the within-cells sum of squares, SS_W:

$$SS_W = \overset{all\,scores}{\sum} X^2$$

$$- \left[\frac{\left(\overset{cell\,11}{\sum} X\right)^2 + \left(\overset{cell\,12}{\sum} X\right)^2 + \left(\overset{cell\,13}{\sum} X\right)^2 + \left(\overset{cell\,21}{\sum} X\right)^2 + \left(\overset{cell\,22}{\sum} X\right)^2 + \left(\overset{cell\,23}{\sum} X\right)^2}{n_{cell}}\right]$$

$$= 2143.18 - \left[\frac{(41.8)^2 + (43.2)^2 + (44.2)^2 + (46.0)^2 + (47.3)^2 + (53.9)^2}{6}\right]$$

$$= 5.577$$

Step 5. **Calculate the total sum of squares, SS_T.** This step is a check to be sure the previous calculations are correct. Once we calculate SS_T, we can use the following equation to check the other calculations:

$$SS_T = SS_R + SS_C + SS_{RC} + SS_W$$

First, we must calculate SS_T:

$$SS_T = \sum_{}^{\substack{all \\ scores}} X^2 - \frac{\left(\sum_{}^{\substack{all \\ scores}} X \right)^2}{N}$$

$$= 2143.18 - \frac{(276.4)^2}{36}$$

$$= 21.042$$

Substituting the obtained values of SS_T, SS_R, SS_{RC}, and SS_W into the partitioning equation for SS_T, we obtain

$$SS_T = SS_R + SS_C + SS_{RC} + SS_W$$

$$21.042 = 9.000 + 4.754 + 1.712 + 5.577$$

$$21.042 = 21.043$$

The equation checks within rounding accuracy. Therefore, we can assume our calculations up to this point are correct.

Step 6: **Calculate the degrees of freedom for each variance estimate:**

$$df_R = r - 1 = 2 - 1 = 1$$

$$df_C = c - 1 = 3 - 1 = 2$$

$$df_{RC} = (r - 1)(c - 1) = 2(1) = 2$$

$$df_W = rc(n_{cell} - 1) = 2(3)(5) = 30$$

$$df_T = N - 1 = 35$$

Note that

$$df_T = df_R + df_C + df_{RC} + df_W$$

$$35 = 1 + 2 + 2 + 30$$

$$35 = 35$$

Step 7: **Calculate the variance estimates s_R^2, s_C^2, s_{RC}^2, and s_W^2.** Each variance estimate is equal to the sum of squares divided by the appropriate degrees of freedom. Thus,

$$\text{Row variance estimate} = s_R^2 = \frac{SS_R}{df_R} = \frac{9.000}{1} = 9.000$$

$$\text{Column variance estimate} = s_C^2 = \frac{SS_C}{df_C} = \frac{4.754}{2} = 2.377$$

$$\text{Row} \times \text{column variance estimate} = s_{RC}^2 = \frac{SS_{RC}}{df_{RC}} = \frac{1.712}{2} = 0.856$$

$$\text{Within-cells variance estimate} = s_W^2 = \frac{SS_W}{df_W} = \frac{5.577}{30} = 0.186$$

Step 8: Calculate the F ratios. For the row effect,

$$F_{\text{obt}} = \frac{s_R^2}{s_W^2} = \frac{9.000}{0.186} = 48.42$$

For the column effect,

$$F_{\text{obt}} = \frac{s_C^2}{s_W^2} = \frac{2.377}{0.186} = 12.78$$

For the row $\times$ column interaction effect,

$$F_{\text{obt}} = \frac{s_{RC}^2}{s_W^2} = \frac{0.856}{0.186} = 4.60$$

b. Evaluate the F_{obt} values:

For the row effect. From Table F, with $\alpha = 0.05$, $df_{\text{numerator}} = df_R = 1$, and $df_{\text{denominator}} = df_W = 30$, $F_{\text{crit}} = 4.17$. Since F_{obt} (48.39) > 4.17, we reject H_0 with respect to the A variable, which in this experiment is time of day. There is a significant effect for time of day.

For the column effect. From Table F, with $\alpha = 0.05$, $df_{\text{numerator}} = df_C = 2$, and $df_{\text{denominator}} = df_W = 30$, $F_{\text{crit}} = 3.32$. Since F_{obt} (12.78) > 3.32, we reject H_0 with respect to the B variable, which in this experiment is amount of exericse. There is a significant main effect for amount of exercise.

For the row $\times$ column interaction effect. From Table F, with $\alpha = 0.05$, $df_{\text{numerator}} = df_{RC} = 2$, and $df_{\text{denominator}} = df_W = 30$, $F_{\text{crit}} = 3.32$. Since F_{obt} (4.60) > 3.32, we reject H_0 regarding the interaction of variables A and B. There is a significant interaction between the amount of exercise and the time of day when the exercise is done.

The analysis is summarized in Table 17.1.

TABLE 17.1   ANOVA Table for Exercise and Time of Day Experiment

Source	SS	df	s^2	F_{obt}	F_{crit}
Rows (time of day)	9.000	1	9.000	48.42*	4.17
Columns (exercise)	4.754	2	2.377	12.79*	3.32
Rows × columns	1.712	2	0.856	4.60*	3.32
Within cells	5.577	30	0.186		
Total	21.042	35			

* Since $F_{\text{obt}} > F_{\text{crit}}$, H_0 is rejected.

Interpreting the Results

In the preceding analysis, we have rejected the null hypothesis for both the row and column effects. A significant effect for rows indicates that variable A has had a significant main effect. The differences between the row means, averaged

FIGURE 17.3

Cell means from the
exercise and sleep
experiment

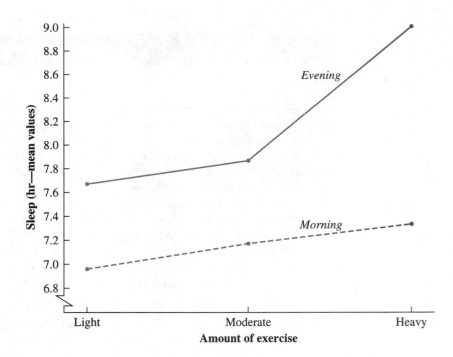

over the columns, were too great to attribute to random sampling from populations where $\mu_{a_1} = \mu_{a_2}$. In the present experiment, the significant row effect indicates that there was a significant main effect for the time of day factor. The
differences between the means for the time of day conditions averaged over the
amount of exercise conditions were too great to attribute to chance. We have
plotted the mean of each cell in Figure 17.3. From this figure, it can be seen that
evening exercise resulted in greater sleep than morning exercise.

A significant effect for columns indicates that variable B has had a significant
main effect—that the differences among the column means, computed by averaging over the rows, was too great to attribute to random sampling from the
null-hypothesis population. In the present experiment, the significant effect for
columns tells us that the differences between the means of the three exercise
conditions computed by averaging over the time of day conditions were too great
to attribute to random sampling fluctuations. From Figure 17.3, it can be seen
that the effect of increasing the amount of exercise averaged over the time of
day conditions was to increase the amount of sleep.

The results of this experiment also showed a significant row × column interaction effect. As discussed previously, a significant interaction effect indicates that
the effects of one of the factors on the dependent variable are not the same at
all the levels of the other factor. Plotting the mean for each cell is particularly
helpful for interpreting an interaction effect. From Figure 17.3, we can see that
the increase in the amount of sleep is about the same in going from light to
moderate exercise whether the exercise is done in the morning or evening.
However, the difference in the amount of sleep in going from moderate to heavy
exercise varies depending on whether the exercise is done in the morning or
evening. Heavy exercise results in a much greater increase in the amount of
sleep when the exercise is done in the evening than when done in the morning.

Let's do another problem for practice.

PRACTICE PROBLEM 17.1

A statistics professor conducts an experiment to compare the effectiveness of two methods of teaching his course. Method I is the usual way he teaches the course: lectures, homework assignments, and a final exam. Method II is the same as method I, except that students receiving method II get 1 additional hour per week in which they solve illustrative problems under the guidance of the professor. Since the professor is also interested in how the methods affect students of differing mathematical abilities, volunteers for the experiment are subdivided according to mathematical ability into superior, average, and poor groups. Five students from each group are randomly assigned to method I and 5 students from each group to method II. At the end of the course, all 30 students take the same final exam. The following final exam scores resulted:

Mathematical Ability	Method I (1)		Method II (2)	
Superior (1)	39*	41	49	47
	48	42	47	48
	44		43	
Average (2)	43	36	38	46
	40	35	45	44
	42		42	
Poor (3)	30	33	37	41
	29	36	34	33
	37		40	

* Scores are the number of points received out of a total of 50 possible points.

a. What are the null hypotheses for this experiment?
b. Using $\alpha = .05$, what do you conclude?

SOLUTION

a. Null hypotheses:
   1. *For the A variable (main effect).* The three levels of mathematical ability do not differentially affect final exam scores in this course. The population row means for the three levels of mathematical ability averaged over teaching methods are equal ($\mu_{a_1} = \mu_{a_2} = \mu_{a_3}$).

2. *For the B variable (main effect)*. Teaching methods I and II are equal in their effects on final exam scores in this course. The population column means for teaching methods I and II averaged over the three levels of mathematical ability are equal ($\mu_{b_1} = \mu_{b_2}$).

3. *For the interaction between variables A and B*. There is no interaction effect between variables A and B. With any main effects removed, the population cell means are equal ($\mu_{a_1b_1} = \mu_{a_1b_2} = \mu_{a_2b_1} = \mu_{a_2b_2} = \mu_{a_3b_1} = \mu_{a_3b_2}$).

b. Conclusion, using $\alpha = 0.05$:

1. Calculating F_{obt}:

Step 1: Calculate SS_R:

$$SS_R = \left[\frac{\left(\overset{\text{row}_1}{\sum} X \right)^2 + \left(\overset{\text{row}_2}{\sum} X \right)^2 + \left(\overset{\text{row}_3}{\sum} X \right)^2}{n_{\text{row}}} \right] - \frac{\left(\overset{\text{all scores}}{\sum} X \right)^2}{N}$$

$$= \frac{(448)^2 + (411)^2 + (350)^2}{10} - \frac{(1209)^2}{30} = 489.800$$

Step 2: Calculate SS_C:

$$SS_C = \left[\frac{\left(\overset{\text{col.}_1}{\sum} X \right)^2 + \left(\overset{\text{col.}_2}{\sum} X \right)^2}{n_{\text{col.}}} \right] - \frac{\left(\overset{\text{all scores}}{\sum} X \right)^2}{N}$$

$$= \frac{(575)^2 + (634)^2}{15} - \frac{(1209)^2}{30} = 116.033$$

Step 3: Calculate SS_{RC}:

$$SS_{RC} = \left[\frac{\left(\overset{\text{cell}_{11}}{\sum} X \right)^2 + \left(\overset{\text{cell}_{12}}{\sum} X \right)^2 + \left(\overset{\text{cell}_{21}}{\sum} X \right)^2 + \left(\overset{\text{cell}_{22}}{\sum} X \right)^2 + \left(\overset{\text{cell}_{31}}{\sum} X \right)^2 + \left(\overset{\text{cell}_{32}}{\sum} X \right)^2}{n_{\text{cell}}} \right]$$

$$- \frac{\left(\overset{\text{all scores}}{\sum} X \right)^2}{N} - SS_R - SS_C$$

$$= \left[\frac{(214)^2 + (234)^2 + (196)^2 + (215)^2 + (165)^2 + (185)^2}{5} \right] - \frac{(1209)^2}{30}$$

$$- 489.8 - 116.083 = 49{,}328.6 - 48{,}722.7 - 489.8 - 116.033 = 0.067$$

Step 4: Calculate SS_W:

$$SS_W = \sum^{\substack{\text{all}\\\text{scores}}} X^2$$

$$-\left[\frac{\left(\sum^{\substack{\text{cell}\\11}} X\right)^2 + \left(\sum^{\substack{\text{cell}\\12}} X\right)^2 + \left(\sum^{\substack{\text{cell}\\21}} X\right)^2 + \left(\sum^{\substack{\text{cell}\\22}} X\right)^2 + \left(\sum^{\substack{\text{cell}\\31}} X\right)^2 + \left(\sum^{\substack{\text{cell}\\32}} X\right)^2}{n_{\text{cell}}}\right]$$

$$= 49{,}587 - \left[\frac{(214)^2 + (234)^2 + (196)^2 + (215)^2 + (165)^2 + (185)^2}{5}\right]$$

$$= 49{,}587 - 49{,}328.6 = 258.4$$

Step 5: Calculate SS_T. This step is to check the previous calculations:

$$SS_T = \sum^{\substack{\text{all}\\\text{scores}}} X^2 - \frac{\left(\sum^{\substack{\text{all}\\\text{scores}}} X\right)^2}{N}$$

$$= 49{,}587 - \frac{(1209)^2}{30}$$

$$= 864.3$$

$$SS_T = SS_R + SS_C + SS_{RC} + SS_W$$

$$864.3 = 489.8 + 116.033 + 0.067 + 258.4$$

$$864.3 = 864.3$$

Since the partitioning equation checks, we can assume our calculations thus far are correct.

Step 6: Calculate df:

$$df_R = r - 1 = 3 - 1 = 2$$

$$df_C = c - 1 = 2 - 1 = 1$$

$$df_{RC} = (r - 1)(c - 1) = (3 - 1)(2 - 1) = 2$$

$$df_W = rc(n_{\text{cell}} - 1) = 6(4) = 24$$

$$df_T = N - 1 = 29$$

Step 7: Calculate s_R^2, s_C^2, s_{RC}^2, and s_W^2:

$$s_R^2 = \frac{SS_R}{df_R} = \frac{489.8}{2} = 244.9$$

$$s_C^2 = \frac{SS_C}{df_C} = \frac{116.033}{1} = 116.033$$

$$s_{RC}^2 = \frac{SS_{RC}}{df_{RC}} = \frac{0.067}{2} = 0.034$$

$$s_W^2 = \frac{SS_W}{df_W} = \frac{258.4}{24} = 10.767$$

Step 8: Calculate F_{obt}. For the row effect,

$$F_{obt} = \frac{s_R^2}{s_W^2} = \frac{244.9}{10.767} = 22.75$$

For the column effect,

$$F_{obt} = \frac{s_C^2}{s_W^2} = \frac{116.033}{10.767} = 10.78$$

For the row $\times$ column effect,

$$F_{obt} = \frac{s_{RC}^2}{s_W^2} = \frac{0.034}{10.767} = 0.00$$

2. Evaluate the F_{obt} values:

For the row effect. From Table F, with $\alpha = 0.05$, $df_{numerator} = df_R = 2$, and $df_{denominator} = df_W = 24$, $F_{crit} = 3.40$. Since F_{obt} (22.75) > 3.40, we reject H_0 with respect to the A variable. There is a significant effect for mathematical ability.

For the column effect. From Table F, with $\alpha = 0.05$, $df_{numerator} = df_C = 1$, and $df_{denominator} = df_W = 24$, $F_{crit} = 4.26$. Since F_{obt} (10.78) > 4.26, we reject H_0 with respect to the B variable. There is a significant main effect for teaching methods.

For the row $\times$ column interaction effect. Since F_{obt} (0.00) < 1, we retain H_0 and conclude that the data do not support the hypothesis that there is an interaction between mathematical ability and teaching methods.

The solution to this problem is summarized in Table 17.2.

TABLE 17.2   ANOVA Table for Mathematical Ability and Teaching Method Experiment

Source	SS	df	s^2	F_{obt}	F_{crit}
Rows (mathematical ability)	489.800	2	244.900	22.75*	3.40
Columns (teaching method)	116.033	1	116.033	10.78*	4.26
Rows × columns	0.067	2	0.034	0.00	3.40
Within cells	258.400	24	10.767		
Total	864.300	29			

* Since $F_{obt} > F_{crit}$, H_0 is rejected.

Interpreting the Results of Practice Problem 17.1   In the preceding analysis, we rejected the null hypothesis for both the row and column effects. Rejecting H_0 for rows means that there was a significant main effect for variable A, mathematical ability. The differences among the means for the different levels of mathematical ability averaged over teaching methods were too great to attribute to chance. The mean of each cell has been plotted in Figure 17.4. From this figure, it can be seen that increasing the level of mathematical ability results in increased final exam scores.

Rejecting H_0 for columns indicates that there was a significant main effect for the B variable, teaching methods. The difference between the means for teaching method I and teaching method II averaged over mathematical ability was too great to attribute to random sampling fluctuations. From Figure 17.4, we can see that method II was superior to method I.

In this experiment, there was no significant interaction effect. This means that, within the limits of sensitivity of this experiment, the effect of each variable

FIGURE 17.4

Cell means from the teaching method and mathematical ability experiment

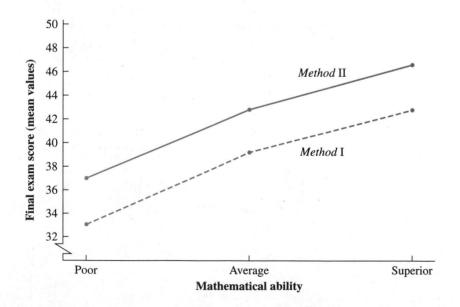

was the same over all levels of the other variable. This can be most clearly seen by viewing Figure 17.4 with regard to variable A. The lack of a significant interaction effect indicates that the effect of different levels of mathematical ability on final exam scores was the same for teaching methods I and II. This results in parallel lines when the means of the cells are plotted (see Figure 17.4). *In fact, it is a general rule that, when the lines are parallel in a graph of the individual cell means, you can be sure there is no interaction effect.* For there to be an interaction effect, the lines must deviate significantly from being parallel. In this regard, it will be useful to review Figure 15.5 to see whether you can determine which graphs show interaction effects.*

PRACTICE PROBLEM 17.2

A clinical psychologist is interested in the effect that anxiety has on the ability of individuals to learn new material. She is also interested in whether the effect of anxiety depends on the difficulty of the new material. An experiment is conducted in which there are three levels of anxiety (high, medium, and low) and three levels of difficulty (high, medium, and low) for the material which is to be learned. Out of a pool of volunteers, 15 low-anxious, 15 medium-anxious, and 15 high-anxious subjects are selected and randomly assigned 5 each to the three difficulty levels. Each subject is given $\frac{1}{2}$ hour to learn the new material, after which the subjects are tested to determine the amount learned.

The following data are collected:

Difficulty of Material	Anxiety					
	Low		Medium		High	
Low	18*	17	18	18	18	17
	20	16	19	15	16	18
	17		17		19	
Medium	18	14	18	17	14	15
	17	16	18	15	17	12
	14		14		16	
High	11	6	15	12	9	8
	10	10	13	11	7	8
	8		12		5	

* Each score is the total points obtained out of a possible 20 points.

* Of course you can't tell whether the interaction is significant without doing a statistical analysis.

a. What are the null hypotheses?
b. Using $\alpha = 0.05$, what do you conclude?

SOLUTION

a. Null hypothesis:
  1. *For variable A (main effect).* The null hypothesis states that the difficulty of the material has no effect on the amount learned. The population row means for low, medium, and high difficulty levels averaged over anxiety levels are equal ($\mu_{a_1} = \mu_{a_2} = \mu_{a_3}$).
  2. *For variable B (main effect).* The null hypothesis states that anxiety level has no effect on the amount learned. The population column means for low, medium, and high anxiety levels averaged over difficulty levels are equal ($\mu_{b_1} = \mu_{b_2} = \mu_{b_3}$).
  3. *For the interaction between variables A and B.* The null hypothesis states that there is no interaction between difficulty of material and anxiety. With any main effects removed, the population cell means are equal ($\mu_{a_1b_1} = \mu_{a_1b_2} = \cdots = \mu_{a_3b_3}$).
b. Conclusion, using $\alpha = 0.05$:
  1. Calculate F_{obt}:

Step 1: Calculate SS_R:

$$SS_R = \left[\frac{\left(\overset{\text{row}\,1}{\sum X}\right)^2 + \left(\overset{\text{row}\,2}{\sum X}\right)^2 + \left(\overset{\text{row}\,3}{\sum X}\right)^2}{n_{\text{row}}}\right] - \frac{\left(\overset{\text{all scores}}{\sum X}\right)^2}{N}$$

$$= \frac{(263)^2 + (235)^2 + (145)^2}{15} - \frac{(643)^2}{45} = 506.844$$

Step 2: Calculate SS_C:

$$SS_C = \left[\frac{\left(\overset{\text{col.}\,1}{\sum X}\right)^2 + \left(\overset{\text{col.}\,2}{\sum X}\right)^2 + \left(\overset{\text{col.}\,3}{\sum X}\right)^2}{n_{\text{col.}}}\right] - \frac{\left(\overset{\text{all scores}}{\sum X}\right)^2}{N}$$

$$= \frac{(212)^2 + (232)^2 + (199)^2}{15} - \frac{(643)^2}{45}$$

$$= 36.844$$

Step 3: Calculate SS_{RC}:

$$SS_{RC} = \left[\frac{\left(\overset{\text{cell}}{\underset{11}{\sum}} X\right)^2 + \left(\overset{\text{cell}}{\underset{12}{\sum}} X\right)^2 + \cdots + \left(\overset{\text{cell}}{\underset{33}{\sum}} X\right)^2}{n_{\text{cell}}} \right] - \frac{\left(\overset{\text{all}}{\overset{\text{scores}}{\sum}} X\right)^2}{N}$$

$$- SS_R - SS_C$$

$$= \left[\frac{(88)^2 + (87)^2 + (88)^2 + (79)^2 + (82)^2 + (74)^2 + (45)^2 + (63)^2 + (37)^2}{5} \right]$$

$$- \frac{(643)^2}{45} - 506.844 - 36.844$$

$$= 40.756$$

Step 4: Calculate SS_W:

$$SS_W = \overset{\text{all}}{\overset{\text{scores}}{\sum}} X^2 - \left[\frac{\left(\overset{\text{cell}}{\underset{11}{\sum}} X\right)^2 + \left(\overset{\text{cell}}{\underset{12}{\sum}} X\right)^2 + \cdots + \left(\overset{\text{cell}}{\underset{33}{\sum}} X\right)^2}{n_{\text{cell}}} \right]$$

$$= 9871$$

$$- \left[\frac{(88)^2 + (87)^2 + (88)^2 + (79)^2 + (82)^2 + (74)^2 + (45)^2 + (63)^2 + (37)^2}{5} \right]$$

$$= 98.800$$

Step 5: Calculate SS_T. This step is a check on the previous calculations:

$$SS_T = \overset{\text{all}}{\overset{\text{scores}}{\sum}} X^2 - \frac{\left(\overset{\text{all}}{\overset{\text{scores}}{\sum}} X\right)^2}{N} = 9871 - \frac{(643)^2}{45}$$

$$= 683.244$$

$$SS_T = SS_R + SS_C + SS_{RC} + SS_W$$

$$683.224 = 506.844 + 36.844 + 40.756 + 98.800$$

$$683.224 = 683.224$$

Since the partitioning equation checks, we can assume our calculations thus far are correct.

Step 6: Calculate df:

$$df_R = r - 1 = 3 - 1 = 2$$

$$df_C = c - 1 = 3 - 1 = 2$$

$$df_{RC} = (r - 1)(c - 1) = 2(2) = 4$$

$$df_W = rc(n_{cell} - 1) = 3(3)(5 - 1) = 36$$

$$df_T = N - 1 = 45 - 1 = 44$$

Step 7: Calculate s_R^2, s_C^2, s_{RC}^2, and s_W^2.

$$s_R^2 = \frac{SS_R}{df_R} = \frac{506.844}{2} = 253.422$$

$$s_C^2 = \frac{SS_C}{df_C} = \frac{36.844}{2} = 18.442$$

$$s_{RC}^2 = \frac{SS_{RC}}{df_{RC}} = \frac{40.756}{4} = 10.189$$

$$s_W^2 = \frac{SS_W}{df_W} = \frac{98.800}{36} = 2.744$$

Step 8: Calculate F_{obt}. For the row effect,

$$F_{obt} = \frac{s_R^2}{s_W^2} = \frac{253.422}{2.744} = 92.35$$

For the column effect,

$$F_{obt} = \frac{s_C^2}{s_W^2} = \frac{18.442}{2.744} = 6.72$$

For the row × column interaction effect,

$$F_{opt} = \frac{s_{RC}^2}{s_W^2} = \frac{10.189}{2.744} = 3.71$$

2. Evaluate F_{obt}:

For the row effect. With $\alpha = 0.05$, $df_{numerator} = df_R = 2$, and $df_{denominator} = df_W = 36$, from Table F, $F_{crit} = 3.26$. Since F_{obt} (92.35) > 3.26, we reject H_0 for the A variable. There is a significant main effect for difficulty of material.

For the column effect. With $\alpha = 0.05$, $df_{numerator} = df_C = 2$, and $df_{denominator} = df_W = 36$, from Table F, $F_{crit} = 3.26$. Since F_{obt} (6.72) > 3.26, H_0 is rejected for the B variable. There is a significant main effect for anxiety level.

For the row × column effect. With $\alpha = 0.05$, $df_{numerator} = df_{RC} = 4$, and $df_{denominator} = df_W = 36$, from Table F, $F_{crit} = 2.63$. Since F_{obt} (3.71) > 2.63, H_0 is rejected. There is a significant interaction between difficulty of material and anxiety level.

The solution is summarized in Table 17.3.

TABLE 17.3   ANOVA Table for Difficulty of Material and Anxiety Level Experiment

Source	SS	df	s^2	F_{obt}	F_{crit}
Rows (difficulty of material)	506.844	2	253.422	92.35*	3.26
Columns (anxiety level)	36.844	2	18.442	6.72*	3.26
Rows × columns	40.756	4	10.189	3.71*	2.63
Within cells	98.800	36	2.744		
Total	683.244	44			

* Since $F_{obt} > F_{crit}$, H_0 is rejected.

Interpreting the Results of Practice Problem 17.2   In the preceding analysis, there was a significant main effect for both difficulty of material and anxiety level. The significant main effect for difficulty of material indicates that the differences among the means for the three difficulty levels averaged over anxiety levels were too great to attribute to chance. The cell means have been plotted in Figure 17.5. From this figure, it can be seen that increasing the difficulty of the material results in lower mean values when the scores are averaged over anxiety levels.

FIGURE 17.5

Cell means from the difficulty of material and anxiety level experiment

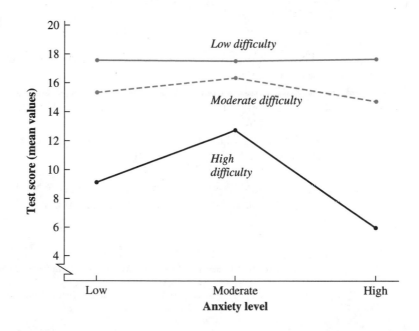

The significant main effect for anxiety level is more difficult to interpret. Of course, at the operational level, this main effect tells us that the differences among the means for the three levels of anxiety when averaged over difficulty levels were too great to attribute to chance. However, beyond this, the interpretation is not clear because of the interaction between the two variables. From Figure 17.5, we can see that the effect of different anxiety levels depends on the difficulty of the material. At the low level of difficulty, differences in anxiety level seem to have no effect on the test scores. However, for the other two difficulty levels, differences in anxiety levels do affect performance. The interaction is a complicated one such that both low and high levels of anxiety seem to interfere with performance when compared to moderate anxiety. This is an example of the inverted U-shaped curve that occurs frequently in psychology when relating performance and arousal levels.

MULTIPLE COMPARISONS

In the three examples we have just analyzed, we have ended the analyses by evaluating the F_{obt} values. In actual practice, the analysis is usually carried further, by doing multiple comparisons on the appropriate pairs of means. For example, in Practice Problem 17.2, there was a significant F_{obt} value for difficulty of material. The next step ordinarily is to determine which difficulty levels are significantly different from each other. Conceptually, this topic is very similar to that which we presented in Chapter 16 when discussing multiple comparisons in conjunction with the one-way ANOVA. One main difference is that in the two-way ANOVA we are often evaluating pairs of *row* means or *column* means rather than pairs of group means. Further exposition of this topic is beyond the scope of this textbook.*

ASSUMPTIONS UNDERLYING THE TWO-WAY ANOVA

The assumptions underlying the two-way ANOVA are the same as for the one-way ANOVA:

1. The population from which the samples were taken are normally distributed.
2. The population variances for each of the cells are equal. This is the *homogeneity of variance* assumption.

As with the one-way ANOVA, the two-way ANOVA is robust with regard to violations of these assumptions, provided the samples are of equal size.†

---

* The interested reader should consult Winer et al., *Statistical Principles in Experimental Design,* 3rd ed., McGraw-Hill, New York, 1991.
† Some statisticians also require the data to be interval or ratio in scaling. For a discussion of this point, see the footnoted references in Chapter 2, p. 27.

SUMMARY

The two-way analysis of variance is very similar to the one-way ANOVA. However, in the two-way ANOVA, the total sum of squares (SS_T) is partitioned into four components: the within-cells sum of squares (SS_W), the row sum of squares (SS_R), the column sum of squares (SS_C), and the row × column sum of squares (SS_{RC}). When these sums of squares are divided by the appropriate degrees of freedom, they form four variance estimates: the within-cells variance estimate (s_W^2), the row variance estimate (s_R^2), the column variance estimate (s_C^2), and the row × column variance estimate (s_{RC}^2).

The within-cells variance estimate (s_W^2) is the yardstick against which the other variance estimates are compared. Since all the subjects within each cell receive the same level of variables A and B, the within-cells variability cannot be due to treatment differences. Rather, it is a measure of the inherent variability of the scores and, hence, gives us an estimate of the null-hypothesis population variance (σ^2). The row variance estimate (s_R^2) is based on the differences among the row means. It is an estimate of σ^2 + the effect of factor A. It is used to evaluate the main effect of variable A. The column variance estimate (s_C^2) is based on the differences among the column means. It is an estimate of σ^2 + the effect of factor B and is used to evaluate the main effect of variable B. The row × column variance estimate (s_{RC}^2) is based on the differences among the cell means beyond that which is predicted by the individual effects of the two variables. It is an estimate of σ^2 + the interaction of A and B. As such, it is used to evaluate the interaction of variables A and B.

In addition to presenting the conceptual basis for the two-way ANOVA, equations for computing each of the four variance estimates were developed, and several illustrative examples were given for practice in using the two-way ANOVA technique. It was further pointed out that multiple comparison techniques similar to those used with the one-way ANOVA are used with the two-way ANOVA. Finally, the assumptions underlying the two-way ANOVA were presented.

IMPORTANT TERMS

Column degrees of freedom (df_C) (p. 400)
Column sum of squares (SS_C) (p. 400)
Column variance estimate (s_C^2) (p. 400)
Row degrees of freedom (df_R) (p. 399)
Row sum of squares (SS_R) (p. 399)
Row × column degrees of freedom (df_{RC}) (p. 401)
Row × column sum of squares (SS_{RC}) (p. 401)
Row × column variance estimate (s_{RC}^2) (p. 401)
Row variance estimate (s_R^2) (p. 398)
Within-cells degrees of freedom (df_W) (p. 396)
Within-cells sum of squares (SS_W) (p. 396)
Within-cells variance estimate (s_W^2) (p. 396)

QUESTIONS AND PROBLEMS

1. Define or identify each of the terms in the "Important Terms" section.
2. In the two-way ANOVA, the total sum of squares is partitioned into four components. What are the four components?
3. Why is the within-cells variance estimate used as the yardstick against which the other variance estimates are compared?
4. The four variance estimates (s_R^2, s_C^2, s_{RC}^2, and s_W^2) are also referred to as *mean squares*. Can you explain why?
5. If the A variable's effect increased, what would you expect would happen to the differences among the row means? What would happen to s_R^2? Explain. Assuming there is no interaction, what would happen to the differences among the column means?
6. If the B variable's effect increased, what would happen to the differences among the column means? What would happen to s_C^2? Explain. Assuming there is no interaction, what would happen to the differences among the row means?

7. What are the assumptions underlying the two-way ANOVA, independent groups design?

8. It is theorized that repetition aids recall and that the learning of material can interfere with the recall of previously learned material. A professor interested in human learning and memory conducts a 2 × 3 factorial experiment to investigate the effects of these two variables on recall. The material to be recalled consists of a list of 16 nonsense syllable pairs. The pairs are presented, one at a time, for 4 seconds, cycling through the entire list, before the first pair is shown again. There are three levels of repetition: level 1, in which each pair is shown 4 times; level 2, in which each pair is shown 8 times; and level 3, in which each pair is shown 12 times. After being presented the list the requisite number of times and prior to testing for recall, each subject is required to learn some intervening material. The intervening material is of two types: type 1, which consists of number pairs, and type 2, which consists of nonsense syllable pairs. After the intervening material has been presented, the subjects are tested for recall of the original list of 16 nonsense syllable pairs. Thirty-six college freshmen serve as subjects. They are randomly assigned so that there are 6 per cell. The following scores are recorded; each is the number of syllable pairs from the original list correctly recalled.

	Number of Repetitions					
Intervening Material	4 Times		8 Times		12 Times	
Number pairs	10	11	16	12	16	14
	12	15	11	15	16	13
	14	10	13	14	15	16
Nonsense syllable pairs	8	7	11	13	14	12
	4	5	9	10	16	15
	5	6	8	9	12	13

a. What are the null hypotheses for this experiment?

b. Using $\alpha = 0.05$, what do you conclude? Plot a graph of the cell means to help you interpret the results.

9. Assume you have just accepted a position as chief scientist for a leading agricultural company. Your first assignment is to make a recommendation concerning the best type of grass to grow in the Pacific Northwest and the best fertilizer for it. To provide the database for your recommendation, having just graduated *summa cum laude* in statistics, you decide to conduct an experiment involving a factorial independent groups design. Since there are three types of grass and two fertilizers under active consideration, the experiment you conduct is 2 × 3 factorial, where the A variable is the type of fertilizer and the B variable is the type of grass. In your field station, you duplicate the soil and the climate of the Pacific Northwest. Next you divide the soil into 30 equal areas and randomly set aside 5 for each combination of treatments. Next, you fertilize the areas with the appropriate fertilizer and plant in each area the appropriate grass seed. Thereafter, all areas are treated alike. When the grass has grown sufficiently, you determine the number of grass blades per square inch in each area. Your recommendation is based on this dependent variable. The "denser" the grass is, the better. The following scores are obtained:

	Number of Grass Blades per Square Inch					
Fertilizer	Red Fescue		Kentucky Blue		Green Velvet	
Type 1	14	15	15	17	20	19
	16	17	12	18	15	22
	10		11		25	
Type 2	11	7	10	6	15	11
	11	8	8	13	18	10
	14		12		19	

a. What are the null hypotheses for this experiment?

b. Using $\alpha = 0.05$, what are your conclusions? Draw a graph of the cell means to aid in interpreting the results.

10. A sleep researcher conducts an experiment to determine whether a hypnotic drug called Drowson, which is advertised as a remedy for insomnia,

actually does promote sleep. In addition, the researcher is interested in whether a tolerance to the drug develops with chronic use. The design of the experiment is a 2×2 factorial independent groups design. One of the variables is the concentration of Drowson. There are two levels: (1) zero concentration (placebo) and (2) the manufacturer's minimum recommended dosage. The other variable concerns the previous use of Drowson. Again there are two levels: (1) subjects with no previous use and (2) chronic users. Sixteen individuals with sleep-onset insomnia (difficulty in falling asleep) who have had no previous use of Drowson are randomly assigned to the two concentration conditions such that there are 8 subjects in each condition. Sixteen chronic users of Drowson are also assigned randomly to the two conditions, 8 subjects per condition. All subjects take their prescribed "medication" for three consecutive nights, and the time to fall asleep is recorded. The scores shown in the following table are the mean times in minutes to fall asleep for each subject, averaged over the 3 days:

Concentration of Drowson

Previous Use	Placebo		Minimum Recommended Dosage	
No previous use	45	53	30	47
	48	58	33	35
	62	55	40	31
	70	64	50	39
Chronic users	47	68	52	46
	52	64	60	49
	55	58	58	50
	62	59	68	55

a. What are the null hypotheses for this experiment?
b. Using $\alpha = 0.05$, what do you conclude? Plot a graph of the cell means to help you interpret the results.

18

CHI-SQUARE AND OTHER NONPARAMETRIC TESTS

INTRODUCTION: DISTINCTION BETWEEN PARAMETRIC AND NONPARAMETRIC TESTS

Statistical inference tests are often classified as to whether they are parametric or nonparametric. You will recall from our discussion in Chapter 1 that a parameter is a characteristic of a population. A parametric inference test is one that depends considerably on population characteristics, or parameters, for its use. The z test, t test, and F test are examples of parametric tests. The z test, for instance, requires that we specify the mean and standard deviation of the null-hypothesis population, as well as requiring that the population scores must be normally distributed for small Ns. The t test for single samples has the same requirements, except that we don't specify σ. The t tests for two samples or conditions (correlated t or independent t) both require that the population scores be normally distributed when the samples are small. The independent t test further requires that the population variances be equal. The analysis of variance has requirements quite similar to the independent t test.

Although all inference tests depend on population characteristics to some extent, the requirements of nonparametric tests are minimal. The sign test and the Mann–Whitney U test are examples of nonparametric inference tests. For both of these tests, the shape of the population is unimportant. For example, it is not necessary that the samples be random samples from normally distributed populations as with the t and F tests. For the Mann–Whitney U test, for example, all that is necessary is that the sample scores be random samples from populations having the same distributions. For this reason, nonparametric inference tests are sometimes referred to as *distribution-free tests.*

Since nonparametric inference tests have fewer requirements or assumptions about population characteristics, the question arises as to why not use them all the time and forget about parametric tests. The answer is twofold. First, many of the parametric inference tests are robust with regard to violations of underlying assumptions. You will recall that a test is robust if violations in the assumptions do not greatly disturb the sampling distribution of its statistic. Thus, the t test is robust regarding the violation of normality in the population. Even though, theoretically, normality in the population is required with small samples, it turns out empirically that unless the departures from normality are substantial, the sampling distribution of t remains essentially the same. Thus, the t test can be used with data even though the data violate the assumptions of normality.

The mean reasons for preferring parametric to nonparametric tests are that, in general, they are more powerful and more versatile than nonparametric tests. We saw an example of the higher power of parametric tests when we compared the t test with the sign test for correlated groups. The factorial design discussed in Chapters 15 and 17 provides a good example of the versatility of parametric statistics. With this design, we can test two, three, four, or more variables and their interactions. No comparable statistical technique exists with nonparametric statistics.

As a general rule, investigators will use parametric tests whenever possible. However, when there is an extreme violation of an assumption of the parametric test or if the investigator believes the scaling of the data makes the parametric test inappropriate, a nonparametric inference test will be employed. We have already presented two nonparametric tests, the sign test and the Mann–Whitney U test. In the remaining sections of this chapter, we shall present three more nonparametric tests: chi-square, the Wilcoxon matched-pairs signed ranks test, and the Kruskal–Wallis test.*

CHI-SQUARE (χ^2)

Single-Variable Experiments

Thus far, we have presented inference tests used primarily in conjunction with ordinal, interval, or ratio data. But what about nominal data? Experiments involving nominal data occur fairly often, particularly in social psychology. You will recall that with this type of data, observations are grouped into several discrete, mutually exclusive categories, and one counts the frequency of occurrence in each category. The inference test most often used with nominal data is a nonparametric test called *chi-square* (χ^2). As has been our procedure throughout the text, we shall begin our discussion of chi-square with an experiment.

AN EXPERIMENT

Preference for Different Brands of Light Beer

Suppose you are interested in determining whether there is a difference among beer drinkers living in the Puget Sound area in their preference for different brands of light beer. You decide to conduct an experiment in which you randomly sample 150 beer

---

* Although we cover several nonparametric tests, there are many more. The interested reader should consult S. Siegel and N. Castellan, Jr., *Nonparametric Statistics for the Behavioral Sciences,* McGraw-Hill, New York, 1988, or W. Daniel, *Applied Nonparametric Statistics,* 2nd ed., PWS-Kent, Boston, 1990.

drinkers and let them taste the three leading brands. Assume all the precautions of good experimental design are followed, like not disclosing the names of the brands to the subjects and so forth. The resulting data are presented in the table:

Brand A	Brand B	Brand C	Total
1 45	2 40	3 65	150

SOLUTION

The entries in each cell are the number or frequency of subjects appropriate to that cell. Thus, 45 subjects preferred brand A (cell 1); 40, brand B (cell 2); and 65, brand C (cell 3). Can we conclude from these data that there is a difference in preference in the population? The null hypothesis for this experiment states that there is no difference in preference among the brands in the population. More specifically, in the population, the proportion of individuals favoring brand A is equal to the proportion favoring brand B, which is equal to the proportion favoring brand C. Referring to the table, it is clear that in the sample the number of individuals preferring each brand is different. However, it doesn't necessarily follow that there is a difference in the population. Isn't it possible that these scores could be due to random sampling from a population of beer drinkers in which the proportion of individuals favoring each brand is equal? Of course, the answer is "yes." Chi-square allows us to evaluate this possibility.

Computation of χ^2_{obt}

To calculate χ^2_{obt}, we must first determine the frequency we would expect to get in each cell if sampling is random from the null-hypothesis population. These frequencies are called *expected frequencies* and are symbolized by f_e. The frequencies actually obtained in the experiment are called *observed frequencies* and are symbolized by f_o. Thus,

f_e = Expected frequency under the assumption sampling is random from the null-hypothesis population

f_o = Observed frequency in the sample

It should be clear that the closer the observed frequency of each cell is to the expected frequency for that cell, the more reasonable is H_0. On the other hand, the greater the difference between f_o and f_e, the more reasonable H_1 becomes.

After determining f_e for each cell, we obtain the difference between f_o and f_e, square the difference, and divide by f_e. In symbolic form, $(f_o - f_e)^2/f_e$ is computed for each cell. Finally, we sum the resultant values from each of the cells. In equation form,

$$\chi^2_{\text{obt}} = \sum \frac{(f_o - f_e)^2}{f_e} \qquad \textit{equation for calculating } \chi^2$$

where f_o = Observed frequency in the cell

f_e = Expected frequency in the cell, and Σ is over all the cells

From this equation, you can see that χ^2 is basically a measure of how different the observed frequencies are from the expected frequencies.

To calculate the value of χ^2_{obt} for the present experiment, we must determine f_e for each cell. The values of f_o are given in the table. If the null hypothesis is true, then the proportion of beer drinkers in the population that prefers brand A is equal to the proportion that prefers brand B, which in turn is equal to the proportion that prefers brand C. This means that one-third of the population must prefer brand A; one third, brand B; and one-third, brand C. Therefore, if the null hypothesis is true, we would expect one-third of the individuals in the population and, hence, in the sample to prefer brand A, one-third to prefer brand B, and one-third to prefer brand C. Since there are 150 subjects in the sample, f_e for each cell = $\frac{1}{3}$(150) = 50. We have redrawn the data table below and entered the f_e values in parentheses:

Brand A	Brand B	Brand C	Total
45	40	65	150
(50)	(50)	(50)	(150)

Now that we have determined the value of f_e for each cell, we can calculate χ^2_{obt}. All we need do is sum the values of $(f_o - f_e)^2/f_e$ for each cell. Thus,

$$\chi^2_{obt} = \sum \frac{(f_o - f_e)^2}{f_e}$$

$$= \frac{(45 - 50)^2}{50} + \frac{(40 - 50)^2}{50} + \frac{(65 - 50)^2}{50}$$

$$= 0.50 + 2.00 + 4.50 = 7.00$$

Evaluation of χ^2_{obt}

The theoretical sampling distribution of χ^2 is shown in Figure 18.1. The χ^2 distribution consists of a family of curves that, like the t distribution, vary with degrees of freedom. For the lower degrees of freedom, the curves are positively skewed. The degrees of freedom are determined by the number of f_o scores that are free to vary. In the present experiment, two of the f_o scores are free to vary. Once two of the f_o scores are known, the third f_o score is fixed, since the sum of the three f_o scores must equal N. Therefore, df = 2. In general, with experiments involving just one variable, there are $k - 1$ degrees of freedom, where k equals the number of groups or categories. When we take up the use of χ^2 with contingency tables, there will be another equation for determining degrees of freedom. We shall discuss it when the topic arises.

Table H in Appendix D gives the critical values of χ^2 for different alpha levels. Since χ^2 is basically a measure of the overall discrepancy between f_o and f_e,

FIGURE 18.1

Distribution of χ^2 for
various degrees of
freedom

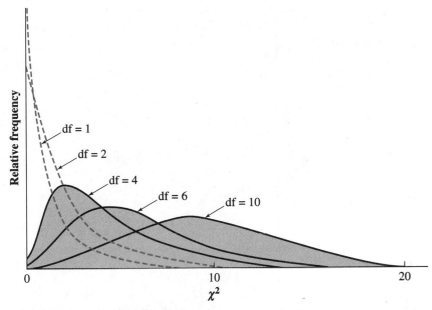

From *Design and Analysis of Experiments in Psychology and Education* by E. F. Lindquist.
Copyright © 1953 Houghton Mifflin Company. Reproduced by permission.

it follows that the larger the discrepancy between the observed and expected
frequencies, the larger will be the value of χ^2_{obt}. Therefore, the larger the value
of χ^2_{obt}, the more unreasonable is the null hypothesis. As with the t and F tests,
if χ^2_{obt} falls within the critical region for rejection, then we reject the null hypothesis. The decision rule states the following:

$$\text{If } \chi^2_{obt} \geq \chi^2_{crit}, \text{reject } H_0.$$

It should be noted that in calculating χ^2_{obt} it doesn't matter whether f_o is greater
or less than f_e. The difference is *squared,* divided by f_e, and added to the other
cells to obtain χ^2_{obt}. Since the direction of the difference is immaterial, the χ^2 test
is a nondirectional test.* Further, since each difference adds to the value of
χ^2, the critical region for rejection always lies under the right-hand tail of the
χ^2 distribution.

In the present experiment, we determined that $\chi^2_{obt} = 7.00$. To evaluate it, we
must determine χ^2_{crit} to see if χ^2_{obt} falls into the critical region for rejection of H_0.
From Table H, with df = 2 and $\alpha = 0.05$,

$$\chi^2_{crit} = 5.991$$

---

* Please see Note 18.1 for further discussion of this point.

FIGURE 18.2

Evaluation of χ^2_{obt} for the light beer drinking problem, df = 2 and $\alpha = 0.05$

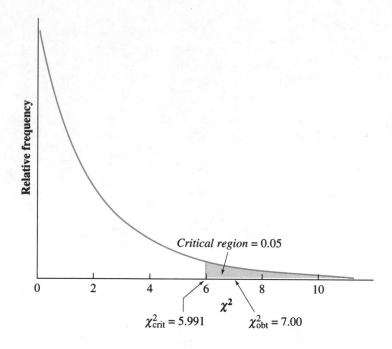

Figure 18.2 shows the χ^2 distribution with df = 2 and the critical region for $\alpha = 0.05$. Since $\chi^2_{obt} > 5.991$, it falls within the critical region, and we reject H_0. There is a difference in the population regarding preference for the three brands of light beer tested. It appears as though brand C is the favored brand.

Let's try one more problem for practice.

PRACTICE PROBLEM 18.1

A political scientist believes that, in recent years, the ethnic composition of the city in which he lives has changed. The most current figures (collected a few years ago) show that the inhabitants were 53% Norwegian, 32% Swedish, 8% Irish, 5% German, and 2% Italian. (Note that nationalities with percentages under 2% have not been included.) To test his belief, a random sample of 750 inhabitants is taken, and the results are shown in the following table:

Norwegian	Swedish	Irish	German	Italian	Total
1 399	2 193	3 63	4 82	5 13	750

a. What is the null hypothesis?
b. What do you conclude? Use $\alpha = 0.05$.

SOLUTION

The solution is shown in Table 18.1.

TABLE 18.1
....

a. Null hypothesis: The ethnic composition of the city has not changed. Therefore, the sample of 750 individuals is a random sample from a population in which 53% are Norwegian, 32% Swedish, 8% Irish, 5% German, and 2% Italian.
b. Conclusion, using $\alpha = 0.05$:

Step 1: **Calculate the appropriate statistic.** The appropriate statistic is χ^2_{obt}. The calculations are shown here:

Cell No.	f_o	f_e	$\dfrac{(f_o - f_e)^2}{f_e}$
1	399	$0.53(750) = 397.5$	$\dfrac{(399 - 397.5)^2}{397.5} = 0.006$
2	193	$0.32(750) = 240.0$	$\dfrac{(193 - 240)^2}{240} = 9.204$
3	63	$0.08(750) = 60.0$	$\dfrac{(63 - 60)^2}{60} = 0.150$
4	82	$0.05(750) = 37.5$	$\dfrac{(82 - 37.5)^2}{37.5} = 52.807$
5	13	$0.02(750) = 15.0$	$\dfrac{(13 - 15)^2}{15} = 0.267$

$$\chi^2_{obt} = \sum \frac{(f_o - f_e)^2}{f_e} = 62.434$$
$$= 62.43$$

Step 2: **Evaluate the statistic.** Degrees of freedom = $5 - 1 = 4$. With df = 4 and $\alpha = 0.05$, from Table H,

$$\chi^2_{crit} = 9.488$$

Since $\chi^2_{obt} > 9.488$, we reject H_0. The ethnic composition of the city appears to have changed. There has been an increase in the proportion of Germans and a decrease in the Swedish.

TEST OF INDEPENDENCE BETWEEN TWO VARIABLES

One of the main uses of χ^2 is in determining whether two categorical variables are independent or are related. To illustrate, let's consider the following example.

AN EXPERIMENT

Political Affiliation and Attitude

Suppose a bill that proposes to lower the legal age for drinking to 18 is pending before the state legislature. A political scientist, living in the state, is interested in determining whether there is a relationship between political affiliation and attitude toward the bill. A random sample of 200 registered Republicans and 200 registered Democrats is sent letters, explaining the scientist's interest and asking the recipients whether they are in favor of the bill, are undecided, or are against the bill. Strict confidentiality is assured. A self-addressed envelope is included to facilitate responding. Answers are received from all 400 Republicans and Democrats. The results are shown in the 2 × 3 table:

	Attitude			Row Marginal
	For	*Undecided*	*Against*	
Republican	68	22	110	200
Democrat	92	18	90	200
Column Marginal	160	40	200	400

The entries in each cell are the frequency of subjects appropriate to the cell. For example, with the Republicans, 68 are for the bill, 22 are undecided, and 110 are against. With the Democrats, 92 are for the bill, 18 are undecided, and 90 are against. This type of table is called a *contingency table*.

DEFINITION

- *A **contingency table** is a two-way table showing the contingency between two variables where the variables have been classified into mutually exclusive categories and the cell entries are frequencies.*

Note that in constructing a contingency table, it is essential that the categories be mutually exclusive. Thus, if an entry is appropriate for one of the cells, the categories must be such that it cannot appropriately be entered in any other cell.

This contingency table contains the data bearing on the contingency between political affiliation and attitude toward the bill. The null hypothesis states that there is no contingency between the variables in the population. For this example, H_0 states that, in the population, attitude toward the bill and political affiliation are **independent.** If this is true, then both the Republicans and Democrats in the population should have the same proportion of individuals "for," "undecided," and "against" the bill. It is clear that in the contingency table the frequencies in these three columns are different for Republicans than for Democrats. The null hypothesis states that these frequencies are due to random sampling from a population in which the **proportion** of Republicans is equal to the **proportion** of the Democrats in each of the categories. The alternative hypothesis is that Republicans and Democrats do differ in their attitudes toward the bill. If so, then in the population, the proportions would be different.

Computation of χ^2_{obt}

To test the null hypothesis, we must calculate χ^2_{obt} and compare it with χ^2_{crit}. With experiments involving two variables, the most difficult part of the process is in determining f_e for each cell. As discussed, the null hypothesis states that, in the population, the proportion of Republicans for each category is the same as the proportion of Democrats. If we knew these proportions, we could just multiply them by the number of Republicans or Democrats in the sample to find f_e for each cell. For example, suppose that, if H_0 is true, the proportion of Republicans in the population against the bill equals 0.50. To find f_e for that cell, all we would have to do is multiply 0.50 by the number of Republicans in the sample. Thus, for the "Republican-against" cell, f_e would equal 0.50(200) = 100.

Since we do not know the population proportions, we estimate them from the sample. In the present experiment, 160 Republicans and Democrats out of 400 were for the bill, 40 out of 400 were undecided, and 200 out of 400 were against the bill. Since the null hypothesis assumes independence between political party and attitude, we can use these sample proportions as our estimates of the null-hypothesis population proportions. Then, we can use these estimates to calculate the expected frequencies. Our estimates for the null-hypothesis population proportions are as follows:

$$\begin{array}{l} \text{Estimated } H_0 \text{ population} \\ \text{proportion for the bill} \end{array} = \frac{\text{Number of subjects for the bill}}{\text{Total number of subjects}} = \frac{160}{400}$$

$$\begin{array}{l} \text{Estimated } H_0 \text{ population} \\ \text{proportion undecided} \end{array} = \frac{\text{Number of subjects undecided}}{\text{Total number of subjects}} = \frac{40}{400}$$

$$\begin{array}{l} \text{Estimated } H_0 \text{ population} \\ \text{proportion against} \\ \text{the bill} \end{array} = \frac{\text{Number of subjects against}}{\text{Total number of subjects}} = \frac{200}{400}$$

Using these estimates to calculate the expected frequencies, we obtain the following values for f_e:

For the Republican-for cell (cell 1 in the table on p. 432):

$$f_e = \left(\begin{array}{c} \text{Estimated } H_0 \text{ population} \\ \text{proportion for the bill} \end{array} \right) \left(\begin{array}{c} \text{Total number of} \\ \text{Republicans} \end{array} \right) = \frac{160}{400}(200) = 80$$

For the Republican-undecided cell (cell 2):

$$f_e = \left(\begin{array}{c} \text{Estimated } H_0 \text{ population} \\ \text{proportion undecided} \end{array} \right) \left(\begin{array}{c} \text{Total number of} \\ \text{Republicans} \end{array} \right) = \frac{40}{400}(200) = 20$$

For the Republican-against cell (cell 3):

$$f_e = \left(\begin{array}{c}\text{Estimated } H_0 \text{ population} \\ \text{proportion against} \\ \text{the bill}\end{array}\right)\left(\begin{array}{c}\text{Total number of} \\ \text{Republicans}\end{array}\right) = \frac{200}{400}(200) = 100$$

For the Democrat-for cell (cell 4):

$$f_e = \left(\begin{array}{c}\text{Estimated } H_0 \text{ population} \\ \text{proportion for the bill}\end{array}\right)\left(\begin{array}{c}\text{Total number of} \\ \text{Democrats}\end{array}\right) = \frac{160}{400}(200) = 80$$

For the Democrat-undecided cell (cell 5):

$$f_e = \left(\begin{array}{c}\text{Estimated } H_0 \text{ population} \\ \text{proportion undecided}\end{array}\right)\left(\begin{array}{c}\text{Total number of} \\ \text{Democrats}\end{array}\right) = \frac{40}{400}(200) = 20$$

For the Democrat-against cell (cell 6):

$$f_e = \left(\begin{array}{c}\text{Estimated } H_0 \text{ population} \\ \text{proportion against} \\ \text{the bill}\end{array}\right)\left(\begin{array}{c}\text{Total number of} \\ \text{Democrats}\end{array}\right) = \frac{200}{400}(200) = 100$$

For convenience, the 2×3 contingency table has been redrawn here, and the f_e values entered within parentheses in the appropriate cells:

	Attitude			Row
	For	*Undecided*	*Against*	Marginal
Republican	1 68 (80)	2 22 (20)	3 110 (100)	200
Democrat	4 92 (80)	5 18 (20)	6 90 (100)	200
Column Marginal	160	40	200	400

The same values for f_e can also be found directly by multiplying the "marginals" for that cell and dividing by N. The marginals are the row and column totals lying outside the table. For example, the marginals for cell 1 are 160 (column total) and 200 (row total). Let's use this method to find f_e for each cell. Multiplying the marginals and dividing by N, we obtain

$$f_e \text{ (cell 1)} = \frac{160(200)}{400} = 80$$

$$f_e \text{ (cell 2)} = \frac{40(200)}{400} = 20$$

$$f_e \text{ (cell 3)} = \frac{200(200)}{400} = 100$$

$$f_e \text{ (cell 4)} = \frac{160(200)}{400} = 80$$

$$f_e \text{ (cell 5)} = \frac{40(200)}{400} = 20$$

$$f_e \text{ (cell 6)} = \frac{200(200)}{400} = 100$$

These values are, of course, the same ones we arrived at previously. Although using the marginals doesn't give much insight into why f_e should be that value, from a practical standpoint it is the best way to calculate f_e for the various cells. You should note that a good check to make sure your calculations of f_e are correct is to see whether the row and column totals of f_e equal the row and column marginals.

Once f_e for each cell has been determined, the next step is to calculate χ^2_{obt}. As before, this is done by summing $(f_o - f_e)^2/f_e$ for each cell. Thus, for the present experiment,

$$\chi^2_{obt} = \sum \frac{(f_o - f_e)^2}{f_e}$$

$$= \frac{(68 - 80)^2}{80} + \frac{(22 - 20)^2}{20} + \frac{(110 - 100)^2}{100} + \frac{(92 - 80)^2}{80}$$

$$+ \frac{(18 - 20)^2}{20} + \frac{(90 - 100)^2}{100}$$

$$= 1.80 + 0.20 + 1.00 + 1.80 + 0.20 + 1.00 = 6.00$$

Evaluation of χ^2_{obt}

To evaluate χ^2_{obt}, we must compare it with χ^2_{crit} for the appropriate df. As discussed previously, the degrees of freedom are equal to the number of f_o scores that are free to vary while keeping the totals constant. In the two-variable experiment, we must keep both the column and row marginals at the same values. Thus, the degrees of freedom for experiments involving a contingency between two variables are equal to the number of f_o scores that are free to vary while at the same time keeping the column and row marginals the same. In the case of a 2 × 3 table, there are only 2 degrees of freedom. Only two f_o scores are free to vary, and all the remaining f_o and f_e scores are fixed.

To illustrate, consider the 2 × 3 table shown here:

1	2	3	
68	22		200
4 5		6	200
160	40	200	400

If we fill in any two f_o scores, all the remaining f_o scores are fully determined, provided the marginals are kept at the same values. For example, in the table, we have filled in the f_o scores for cells 1 and 2. Note that all the other scores are fixed in value once two f_o scores are given; e.g., the f_o score for cell 3 must be 110 [200 − (68 + 22)].

There is also an equation to calculate the df for contingency tables. It states

$$df = (r − 1)(c − 1)$$

where r = Number of rows
c = Number of columns

Applying the equation to the present experiment, we obtain

$$df = (r − 1)(c − 1) = (2 − 1)(3 − 1) = 2$$

The χ^2 test is not limited to 2 × 3 tables. It can be used with contingency tables containing any number of rows and columns. This equation is perfectly general and applies to all contingency tables. Thus, if we did an experiment involving two variables and had four rows and six columns in the table, df = $(r − 1)$ $(c − 1) = (4 − 1)(6 − 1) = 15$.

Returning to the evaluation of the present experiment, let's assume $\alpha = 0.05$. With df = 2 and $\alpha = 0.05$, from Table H,

$$\chi^2_{crit} = 5.991$$

Since $\chi^2_{obt} > 5.991$, we reject H_0. Political affiliation is related to attitude toward the bill. The Democrats appear to be more favorably disposed toward the bill than the Republicans. The complete solution is shown in Table 18.2.

TABLE 18.2   Solution to Political Affiliation and Attitude Problem

a. Null hypothesis: Political affiliation and attitude toward the bill are independent. The frequency obtained in each cell is due to random sampling from a population where the proportions of Republicans and Democrats that are for, undecided about, and against the bill are equal.

b. Conclusion, using $\alpha = 0.05$:

Step 1: **Calculate the appropriate statistic.** The appropriate statistic is χ^2_{obt}. The data are shown on p. 432. The calculations are shown in the table at the top of the facing page.

Step 2: **Evaluate the statistic.** Degrees of freedom = $(r − 1)(c − 1) = (2 − 1)(3 − 1) = 2$. With df = 2 and $\alpha = 0.05$, from Table H,

$$\chi^2_{crit} = 5.991$$

Since $\chi^2_{obt} > 5.991$, we reject H_0. Political affiliation and attitude toward the bill are related. Democrats appear to favor the bill more than Republicans.

Cell No.	f_o	f_e	$\dfrac{(f_o - f_e)^2}{f_e}$
1	68	$\dfrac{160(200)}{400} = 80$	$\dfrac{(68 - 80)^2}{80} = 1.80$
2	22	$\dfrac{40(200)}{400} = 20$	$\dfrac{(22 - 20)^2}{20} = 0.20$
3	110	$\dfrac{200(200)}{400} = 100$	$\dfrac{(110 - 100)^2}{100} = 1.00$
4	92	$\dfrac{160(200)}{400} = 80$	$\dfrac{(92 - 80)^2}{80} = 1.80$
5	18	$\dfrac{40(200)}{400} = 20$	$\dfrac{(18 - 20)^2}{20} = 0.20$
6	90	$\dfrac{200(200)}{400} = 100$	$\dfrac{(90 - 100)^2}{100} = 1.00$

$$\chi^2_{obt} = \sum \frac{(f_o - f_e)^2}{f_e} = 6.00$$

Let's try a problem for practice.

PRACTICE PROBLEM 18.2

A university is considering implementing one of the following three grading systems: (1) All grades are pass–fail, (2) all grades are on the 4.0 system, and (3) 90% of the grades are on the 4.0 system and 10% are pass–fail. A survey is taken to determine whether there is a relationship between undergraduate major and grading system preference. A random sample of 200 students with engineering majors, 200 students with arts and science majors, and 100 students with fine arts majors is selected. Each student is asked which of the three grading systems he or she prefers. The results are shown below in the 3 × 3 contingency table:

	Grading System			Row Marginal
	Pass–Fail	4.0 and Pass–Fail	4.0	
Fine arts	1 26	2 55	3 19	100
Arts & sciences	4 24	5 118	6 58	200
Engineering	7 20	8 112	9 68	200
Column Marginal	70	285	145	500

a. What is the null hypothesis?

b. What do you conclude? Use $\alpha = 0.05$.

SOLUTION

The solution is shown in Table 18.3.

TABLE 18.3

a. Null hypothesis: Undergraduate major and grading system preference are independent. The frequency obtained in each cell is due to random sampling from a population where the proportions of fine arts, arts and sciences, and engineering majors who prefer each grading system are the same.

b. Conclusion, using $\alpha = 0.05$:

Step 1: **Calculate the appropriate statistic.** The data are shown in the table. The appropriate statistic is χ^2_{obt}. Before calculating χ^2_{obt}, we must first calculate f_e for each cell. The values of f_e were found using the marginals.

Cell No.	f_o	f_e	$\dfrac{(f_o - f_e)^2}{f_e}$
1	26	$\dfrac{70(100)}{500} = 14$	$\dfrac{(26-14)^2}{14} = 10.286$
2	55	$\dfrac{285(100)}{500} = 57$	$\dfrac{(55-57)^2}{57} = 0.070$
3	19	$\dfrac{145(100)}{500} = 29$	$\dfrac{(19-29)^2}{29} = 3.448$
4	24	$\dfrac{70(200)}{500} = 28$	$\dfrac{(24-28)^2}{28} = 0.571$
5	118	$\dfrac{285(200)}{500} = 114$	$\dfrac{(118-114)^2}{114} = 0.140$
6	58	$\dfrac{145(200)}{500} = 58$	$\dfrac{(58-58)^2}{58} = 0.000$
7	20	$\dfrac{70(200)}{500} = 28$	$\dfrac{(20-28)^2}{28} = 2.286$
8	112	$\dfrac{285(200)}{500} = 114$	$\dfrac{(112-114)^2}{114} = 0.035$
9	68	$\dfrac{145(200)}{500} = 58$	$\dfrac{(68-58)^2}{58} = \underline{1.724}$

$$\chi^2_{obt} = \sum \frac{(f_o - f_e)^2}{f_e} = 18.560$$

$$= 18.56$$

Step 2: **Evaluate the statistic.** Degrees of freedom $= (r-1)(c-1) = (3-1)(3-1) = 4$. With df $= 4$ and $\alpha = 0.05$, from Table H,

$$\chi^2_{crit} = 9.488$$

Since $\chi^2_{obt} > 9.488$, we reject H_0. Undergraduate major and grading system preference are related.

In trying to determine what the differences in preference were between the groups (since the number of subjects differ considerably for the fine arts majors), it is necessary to convert the frequency entries into proportions. These proportions are shown in the following table:

	Pass–Fail	4.0 and Pass–Fail	4.0
Fine arts	0.26	0.55	0.19
Arts & sciences	0.12	0.59	0.29
Engineering	0.10	0.56	0.34

From this table, it appears that the differences between groups are in their preferences for the all pass–fail or all 4.0 grading systems. The fine arts students show a higher proportion favoring the pass–fail system rather than the all 4.0 system, whereas the arts and sciences and engineering students show the reverse pattern. All groups show about the same proportions favoring the system advocating a combination of 4.0 and pass–fail grades.

Let's try one more problem for practice.

PRACTICE PROBLEM 18.3

A social psychologist is interested in determining whether there is a relationship between the education level of parents and the number of children they have. Accordingly, a survey is taken, and the following results are obtained:

No. of Children

	Two or Less	More than Two	Row Marginal
College education	[1] 53	[2] 22	75
High school education only	[3] 37	[4] 38	75
Column Marginal	90	60	150

a. What is the null hypothesis?
b. What is the conclusion? Use $\alpha = 0.05$.

SOLUTION

The solution is shown in Table 18.4.

TABLE 18.4

a. Null hypothesis: The educational level of parents and the number of children they have are independent. The frequency obtained in each cell is due to random sampling from a population where the proportions of college-educated and only high-school-educated parents that have (1) two or less and (2) more than two children are equal.

b. Conclusion, using $\alpha = 0.05$:

Step 1: Calculate the appropriate statistic. The data are shown in the table. The appropriate statistic is χ^2_{obt}. The calculations follow.

Cell No.	f_o	f_e	$\dfrac{(f_o - f_e)^2}{f_e}$
1	53	$\dfrac{90(75)}{150} = 45$	$\dfrac{(53-45)^2}{45} = 1.422$
2	22	$\dfrac{60(75)}{150} = 30$	$\dfrac{(22-30)^2}{30} = 2.133$
3	37	$\dfrac{90(75)}{150} = 45$	$\dfrac{(37-45)^2}{45} = 1.422$
4	38	$\dfrac{60(75)}{150} = 30$	$\dfrac{(38-30)^2}{30} = 2.133$

$$\chi^2_{\text{obt}} = \sum \frac{(f_o - f_e)^2}{f_e} = 7.11$$

> **Step 2:** **Evaluate the statistic.** Degrees of freedom = $(r - 1)(c - 1) =$ $(2 - 1)(2 - 1) = 1$. With df = 1 and $\alpha = 0.05$, from Table H,
>
> $$\chi^2_{\text{crit}} = 3.841$$
>
> Since $\chi^2_{\text{obt}} > 3.841$, we reject H_0. The educational level of parents and the number of children they have are related.

Assumptions Underlying χ^2

A basic assumption in using χ^2 is that there is independence between each observation recorded in the contingency table. This means that each subject can only have one entry in the table. It is not permissible to take several measurements on the same subject and enter them as separate frequencies in the same or different cells. This error would produce a larger N than there are independent observations.

A second assumption is that the sample size must be large enough so that the *expected* frequency in each cell is at least 5 for tables where r or c is greater than 2. If the table is a 1×2 or 2×2 table, then each expected frequency should be at least 10. If the sample size is small enough to result in expected frequencies that violate these requirements, then the actual sampling distribution of χ^2 deviates considerably from the theoretical one, and the probability values given in Table H do not apply. If the experiment involves a 2×2 contingency table and the data violate this assumption, Fisher's exact probability test should be used.*

Although χ^2 is used frequently when the data are only of nominal scaling, it is not limited to nominal data. Chi-square can be used with ordinal, interval, and ratio data. However, regardless of the actual scaling, the data must be reduced to mutually exclusive categories and appropriate frequencies before χ^2 can be employed.

THE WILCOXON MATCHED-PAIRS SIGNED RANKS TEST

The Wilcoxon signed ranks test is used in conjunction with the correlated groups design with data that are at least ordinal in scaling. It is a relatively powerful test sometimes used in place of the t test for correlated groups when there is an extreme violation of the normality assumption or when the data are not of appropriate scaling. The Wilcoxon signed ranks test considers both the magnitude

---

* This test is discussed in S. Siegel and N. Castellan, Jr., *Nonparametric Statistics for the Behavioral Sciences,* 2nd ed., McGraw-Hill, New York, 1988, pp. 103–111. It is also discussed in W. Daniel, *Applied Nonparametric Statistics,* 2nd ed., PWS-Kent, Boston, 1990, pp. 150–162.

of the difference scores and their direction, which makes it more powerful than the sign test. It is, however, less powerful than the t test for correlated groups. To illustrate this test, let's consider the following experiment.

AN EXPERIMENT

Promoting More Favorable Attitudes Toward Wildlife Conservation

A prominent ecological group is planning to mount an active campaign to increase wildlife conservation in their country. As part of the campaign, they plan to show a film designed to promote more favorable attitudes toward wildlife conservation. Before showing the film to the public at large, they want to evaluate its effects. A group of 10 subjects are randomly sampled and given a questionnaire that measures an individual's attitude toward wildlife conservation. Next, they are shown the film, after which they are again given the attitude questionnaire. The questionnaire has 50 possible points, and the higher the score, the more favorable is the attitude toward wildlife conservation. The results are shown in Table 18.5.

1. What is the alternative hypothesis? Use a nondirectional hypothesis.
2. What is the null hypothesis?
3. What do you conclude? Use $\alpha = 0.05_{2\,\text{tail}}$.

TABLE 18.5   Data and Solution for Wildlife Conservation Problem

Subject	Attitude Before	Attitude After	Difference	Rank of |Difference|	Signed Rank of Difference	Sum of Positive Ranks	Sum of Negative Ranks
1	40	44	4	4	4	4	
2	33	40	7	6	6	6	
3	36	49	13	10	10	10	
4	34	36	2	2	2	2	
5	40	39	−1	1	−1		1
6	31	40	9	8	8	8	
7	30	27	−3	3	−3		3
8	36	42	6	5	5	5	
9	24	35	11	9	9	9	
10	20	28	8	7	7	7	
				55		51	4

$$\frac{n(n+1)}{2} = \frac{10(11)}{2} = 55 \qquad\qquad T_{\text{obt}} = 4$$

From Table I, with $N = 10$ and $\alpha = 0.05_{2\,\text{tail}}$,

$$T_{\text{crit}} = 8$$

Since $T_{\text{obt}} < 8$, H_0 is rejected. The film appears to promote more favorable attitudes toward wildlife conservation.

SOLUTION

1. The alternative hypothesis is usually stated without specifying any population parameters. For this example, it states that the film affects attitudes toward wildlife conservation.

2. The null hypothesis is also usually stated without specifying any population parameters. For this example, it states that the film has no effect on attitudes toward wildlife conservation.

3. Conclusion, using $\alpha = 0.05_{2\,\text{tail}}$: As with all the other inference tests, the first step is to calculate the appropriate statistic. Since the data have been obtained from questionnaires, they are at least of ordinal scaling. To illustrate use of the Wilcoxon signed ranks test, we shall assume that the data meet the assumptions of this test (these will be discussed shortly). The statistic calculated by the Wilcoxon signed ranks test is T_{obt}. Determining T_{obt} involves four steps:

   a. Calculate the difference between each pair of scores.

   b. Rank the absolute values of the difference scores from the smallest to the largest.

   c. Assign to the resulting ranks the sign of the difference score whose absolute value yielded that rank.

   d. Compute the sum of the ranks separately for the positive and negative signed ranks. The lower sum is T_{obt}.

These four steps have been done with the data from the attitude questionnaire, and the resultant values have been entered in Table 18.5. Thus, the difference scores have been calculated and are shown in the fourth column of Table 18.5. The ranks of the absolute values of the difference scores are shown in the fifth column. Note that, as a check on whether the ranking has been done correctly, the sum of the unsigned ranks should equal $n(n + 1)/2$. In the present example, this sum should equal 55 [10(11)/2 = 55], which it does. Step **c** asks us to give each rank the sign of the difference score whose absolute value yielded that rank. This has been done in the sixth column. Thus, the ranks of 1 and 3 are assigned minus signs, and the rest are positive. The ranks of 1 and 3 received minus signs because their associated difference scores are negative. T_{obt} is determined by computing the sum of the positive ranks and the sum of the negative ranks. T_{obt} is the lower of the two sums. In this example, the sum of the positive ranks equals 51, and the sum of the negative ranks equals 4. Thus,

$$T_{\text{obt}} = 4$$

Note that often it is not necessary to compute both sums. Usually it is apparent by inspection which sum will be lower. The final step is to evaluate T_{obt}. Table I in Appendix D contains the critical values of T for various values of N. With $N = 10$ and $\alpha = 0.05_{2\,\text{tail}}$, from Table I,

$$T_{\text{crit}} = 8$$

With the Wilcoxon signed ranks test, as with the Mann–Whitney U test, the decision rule is

$$\text{If } T_{\text{obt}} \leq T_{\text{crit}}, \text{ reject } H_0.$$

Note that this is opposite to the rule we have been using for most of the other tests. Since $T_{obt} < 8$, we reject H_0 and conclude that the film does affect attitudes toward wildlife conservation. It appears to promote more favorable attitudes.

It is easy to see why the Wilcoxon signed ranks test is more powerful than the sign test but not as powerful as the t test for correlated groups. The Wilcoxon signed ranks test takes into account the magnitude of the difference scores, which makes it more powerful than the sign test. However, it only considers the rank order of the difference scores, not their actual magnitude, as does the t test. Therefore, the Wilcoxon signed ranks test is not as powerful as the t test.

Let's try another problem for practice.

PRACTICE PROBLEM 18.4

An investigator is interested in determining whether the difficulty of the material to be learned affects the anxiety level of college students. A random sample of 12 students is each given hard and easy learning tasks. Before doing each task, they are shown a few sample examples of the material to be learned. Then their anxiety level is assessed using an anxiety questionnaire. Thus, anxiety level is assessed before each learning task. The data are shown in Table 18.6. The higher the score, the greater is the anxiety level. What is the conclusion, using the Wilcoxon signed ranks test and $\alpha = 0.05_{2\,tail}$?

SOLUTION

The solution is shown in Table 18.6. Note that there are ties in some of the difference scores. Generally, there are two kinds of ties possible. First, the raw scores may be tied, yielding a difference score of 0. If this occurs, then these scores are disregarded, and the overall N is reduced by 1 for each 0 difference score. Ties can also occur in the difference scores, as in the present example. When this happens, the ranks of these scores are given a value equal to the mean of the tied ranks. This is the same procedure we followed for the Spearman rho correlation coefficient and for the Mann–Whitney U test. Thus, in this example, the two tied difference scores of 3 are assigned ranks of 2.5 $[(2 + 3)/2 = 2.5]$, and the tied difference scores of 10 receive the rank of 9.5. Otherwise, the solution is quite similar to that of the previous example.

TABLE 18.6

| Student No. | Anxiety | | Difference | Rank of \|Difference\| | Signed Rank of Difference | Sum of Positive Ranks | Sum of Negative Ranks |
	Hard Tasks	*Easy Tasks*					
1	48	40	8	7	7	7	
2	33	27	6	5	5	5	
3	46	34	12	11	11	11	
4	42	28	14	12	12	12	
5	40	30	10	9.5	9.5	9.5	
6	27	24	3	2.5	2.5	2.5	
7	31	33	−2	1	−1		1
8	42	39	3	2.5	2.5	2.5	
9	38	31	7	6	6	6	
10	34	39	−5	4	−4		4
11	38	29	9	8	8	8	
12	44	34	10	9.5	9.5	9.5	
				78		73	5

$$\frac{n(n+1)}{2} = \frac{12(13)}{2} = 78 \qquad\qquad T_{obt} = 5$$

From Table I, with $N = 12$ and $\alpha = 0.05_{2\ tail}$,

$$T_{crit} = 13$$

Since $T_{obt} < 13$, we reject H_0 and conclude that the difficulty of material does affect anxiety. It appears that more difficult material produces increased anxiety.

Assumptions of the Wilcoxon Signed Ranks Test

There are two assumptions underlying the Wilcoxon signed ranks test. First, the scores within each pair must be at least of ordinal measurement. Second, the difference scores must also have at least ordinal scaling. The second requirement arises because in computing T_{obt} we rank-order the difference scores. Thus, the magnitude of the difference scores must be at least ordinal so that they can be rank-ordered.

KRUSKAL–WALLIS TEST

The Kruskal–Wallis test is a nonparametric test that is used with an independent groups design employing k samples. It is used as a substitute for the parametric one-way ANOVA discussed in Chapter 15, when the assumptions of that test are seriously violated. The Kruskal–Wallis test does not assume population normality, nor homogeneity of variance, as does parametric ANOVA, and requires only ordinal scaling of the dependent variable. It is used when violations of population normality and/or homogeneity of variance are extreme, or when interval or ratio scaling are required and not met by the data. To understand this test, let's begin with an experiment.

AN EXPERIMENT

Evaluating Two Weight Reduction Programs

A health psychologist, employed by a large corporation, is interested in evaluating two weight reduction programs she is considering using with employees of her corporation. She conducts an experiment in which 18 obese employees are randomly assigned to three conditions, with 6 subjects per condition. The subjects in condition 1 are placed on a diet that reduces their daily caloric intake by 500 calories. The subjects in condition 2 receive the same restricted diet, but in addition are required to walk 2 miles each day. Condition 3 is a control condition, in which the subjects are asked to maintain their usual eating and exercise habits. The data presented in the table are the number of pounds lost by each subject over a 6-month period. A positive number indicates weight loss and a negative number, weight gain. Assume the data show that there is a strong violation of population normality such that the psychologist decides to analyze the data with the Kruskal–Wallis test, rather than using parametric ANOVA.

1 Diet		2 Diet + Exercise		3 Control	
Pounds Lost	*Rank*	*Pounds Lost*	*Rank*	*Pounds Lost*	*Rank*
2	5	12	12	8	9
15	14	9	10	3	6
7	8	20	16	−1	4
6	7	17	15	−3	2
10	11	28	17	−2	3
14	13	30	18	−8	1
$n_1 = 6$	$R_1 = 58$	$n_2 = 6$	$R_2 = 88$	$n_3 = 6$	$R_3 = 25$

a. What is the alternative hypothesis?
b. What is the null hypothesis?
c. What is the conclusion? Use $\alpha = 0.05$.

SOLUTION

a. Alternative hypothesis: As with parametric ANOVA, the alternative hypothesis states that at least one of the conditions affects weight loss differently than at least one of the other conditions. In the same manner as parametric ANOVA, each sample is considered a random sample from its own population set of scores. If there are k samples, there are k populations. In this example, $k = 3$. However, since this is a nonparametric test, Kruskal–Wallis makes no prediction about the population means μ_1, μ_2, or μ_3. It merely asserts that at least one of the population distributions is different from at least one of the other population distributions.

b. Null hypothesis: The samples are random samples from the same or identical population distributions. There is no prediction specifically regarding μ_1, μ_2, or μ_3.

c. Conclusion, using $\alpha = 0.05$: As usual, in evaluating H_0, we follow the two-step process: compute the appropriate statistic and then evaluate the statistic using its sampling distribution.

Step 1: **Compute the appropriate statistic.** The statistic we compute for the Kruskal–Wallis test is H_{obt}. The procedure is very much like computing U_{obt} for the Mann–Whitney U test. All of the scores are grouped together and rank-ordered, assigning the rank of 1 to the lowest score, 2 to the next to lowest, and N to the highest. When this is done, the ranks for each condition or sample are summed. These procedures have been carried out for the data of the present example and entered in the data table.

The sum of ranks for each group have been symbolized as R_1, R_2, and R_3, respectively. For these data, $R_1 = 58$, $R_2 = 88$, and $R_3 = 25$. The Kruskal–Wallis test assesses whether these sums of ranks are so different as to make it unreasonable to consider that they come from samples that were randomly selected from the same population. The larger the differences between the sums of the ranks of each sample, the less likely it is that the samples are from the same population.

The equation for computing H_{obt} is as follows:

$$H_{obt} = \left[\frac{12}{N(N+1)} \right] \left[\sum_{i=1}^{k} \frac{(R_i)^2}{n_i} \right] - 3(N+1)$$

$$= \left[\frac{12}{N(N+1)} \right] \left[\frac{R_1^2}{n_1} + \frac{R_2^2}{n_2} + \frac{R_3^2}{n_3} + \cdots + \frac{R_k^2}{n_k} \right] - 3(N+1)$$

where $\sum_{i=1}^{k} \frac{(R_i)^2}{n_i}$ tells us to square the sum of ranks for each sample, divide each squared value by the number of scores in the sample and sum over samples

k = number of samples or groups

n_i = number of scores in the ith sample

n_1 = number of scores in sample 1

n_2 = number of scores in sample 2

n_3 = number of scores in sample 3

n_k = number of scores in sample k

N = number of scores in all samples combined

R_i = sum of the ranks for the ith sample

R_1 = sum of the ranks for sample 1

R_2 = sum of the ranks for sample 2

R_3 = sum of the ranks for sample 3

R_k = sum of the ranks for sample k

Substituting the appropriate values from the data table into this equation, we obtain

$$H_{obt} = \left[\frac{12}{N(N+1)}\right]\left[\frac{(R_1)^2}{n_1} + \frac{(R_2)^2}{n_2} + \frac{(R_3)^2}{n_3}\right] - 3(N+1)$$

$$= \left[\frac{12}{18(18+1)}\right]\left[\frac{(58)^2}{6} + \frac{(88)^2}{6} + \frac{(25)^2}{6}\right] - 3(18+1)$$

$$= 68.61 - 57$$

$$= 11.61$$

Step 2: Evaluate the statistic. It can be shown that, if the number of scores in each sample is 5 or more, the sampling distribution of the statistic H is approximately the same as chi-square with df $= k - 1$. In the present experiment, df $= k - 1 = 3 - 1 = 2$. From Table H, with $\alpha = 0.05$, and df $= 2$,

$$H_{crit} = 5.991$$

As with parametric ANOVA, the Kruskal–Wallis test is a nondirectional test. The decision rule states that

If $H_{obt} \geq H_{crit}$, reject H_0.

If $H_{obt} < H_{crit}$, retain H_0.

Since $H_{obt} > 5.991$, we reject H_0. It appears that the conditions are not equal with regard to weight loss.

PRACTICE PROBLEM 18.5

A business consultant is doing research in the area of management training. There are two effective managerial styles: One is people-oriented and a second, task-oriented. Well-defined, static jobs are better served by the people-oriented managers and changing, newly created jobs by the task-oriented managers. The experiment being conducted investigates whether it is better to try to train managers to have both styles or whether it is better to match managers to jobs with no attempt to train in a second style. The managers for this experiment are 24 army officers, randomly selected from a large army base. The experiment involves three conditions. In condition 1, the subjects receive training in both managerial styles. After training is completed, these subjects are randomly assigned to new jobs without matching style and job. In condition 2, the subjects receive no additional training, but are assigned to jobs according to a match between their single managerial style and the job requirements. Condition 3 is a control condition, in which subjects receive no additional training and are assigned to new jobs, like those in condition 1, without matching. After they are in their new job assignments for 6 months, a performance rating is obtained on each officer.

The data follow. The higher the score, the better the performance. At the beginning of the experiment, there were 8 subjects in each condition. However, one of the subjects in condition 2 dropped out of the experiment midway into the experiment, and was not replaced. Assume the data do not meet the assumptions for the parametric one-way ANOVA.

Condition 1 Training		Condition 2 Matching		Condition 3 Control	
Score	*Rank*	*Score*	*Rank*	*Score*	*Rank*
65	8	90	21	55	3
84	16	83	15	82	14
87	19.5	76	12	71	10
53	2	87	19.5	60	6
70	9	92	22	52	1
85	17	86	18	81	13
56	4	93	23	73	11
63	7			57	5
$n_1 = 8$ $R_1 = 82.5$		$n_2 = 7$ $R_2 = 130.5$		$n_3 = 8$ $R_3 = 63$	

a. What is the alternative hypothesis?
b. What is the null hypothesis?
c. What is the conclusion? Use $\alpha = 0.05$.

SOLUTION

a. Alternative hypothesis: At least one of the conditions has a different effect on job performance than at least one of the other conditions. Therefore, at least one of the population distributions is different from one of the others.
b. Null hypothesis: The conditions have the same effect on job performance. Therefore, the samples are random samples from the same or identical population distributions.
c. Conclusion, using $\alpha = 0.05$:

Step 1: Compute the appropriate statistic.

$$H_{obt} = \left[\frac{12}{N(N+1)} \right] \left[\frac{(R_1)^2}{n_1} + \frac{(R_2)^2}{n_2} + \frac{(R_3)^2}{n_3} \right] - 3(N+1)$$

$$= \left[\frac{12}{23(23+1)} \right] \left[\frac{(82.5)^2}{8} + \frac{(130.5)^2}{7} + \frac{(63)^2}{8} \right] - 3(23+1)$$

$$= 82.17 - 72$$

$$= 10.17$$

Step 2: **Evaluate the statistic.** In the present experiment, df = $k - 1$ = 3 $-$ 1 = 2. From Table H, with α = 0.05, and df = 2,

$$H_{\text{crit}} = 5.991$$

Since $H_{\text{obt}} > 5.991$, we reject H_0. It appears that the conditions are not equal with regard to their effect on job performance.

Assumptions Underlying the Kruskal–Wallis Test

To use the Kruskal–Wallis test, the data must be of at least ordinal scaling. Additionally, there must be at least 5 scores in each sample to use the probabilities given in the table of chi-square.*

---

* To analyze data with fewer than 5 scores in a sample, see S. Siegel and N. Castellan, Jr., *Nonparametric Statistics for the Behavioral Sciences,* 2nd ed., McGraw-Hill, New York, 1988, pp. 206–212.

STATISTICS AND APPLIED SOCIAL RESEARCH—USEFUL OR "ABUSEFUL"?

A recent front-page article in *The Wall Street Journal* discussed the possible misuses of social science research in connection with federally mandated changes in the U.S. welfare system. Excerpts from the article are reproduced below.

Think Tanks Battle to Judge the Impact of Welfare Overhaul

Now that welfare overhaul is under way across the country, so is something else: an ideologically charged battle of the experts to label the various innovations as successes or failures.

Each side—liberal and conservative—fears the other will use early, and perhaps not totally reliable, results of surveys and studies of the impact of welfare reform to push a political agenda. And conservatives, along with other supporters of the law Congress recently passed to allow states to tinker with welfare, worry that they have the most to lose, because of the overwhelming presence of liberal scholars in the field of social science . . . .

Critics of the overhaul process make no bones that they intend to use research into the bill's effects to turn yesterday's angry taxpayers into tomorrow's friends of the downtrodden . . . .

The climate of ideological suspicions is exemplified by the California-based Kaiser Family Foundation's handling of a survey it financed of New Jersey welfare recipients. The study focused on the effect of New Jersey's controversial rule barring additional benefits for women who have children while on welfare. Last spring, shortly before Kaiser was to announce the results, it canceled a planned news conference. **Some critics suspect that Kaiser didn't like some of the answers it got, such as one showing that most recipients considered the "family cap" fair.** The Rutgers University researcher who conducted the survey says he has been forbidden to discuss it. Kaiser says the suspicions are unfounded and that methodological errors destroyed the survey's usefulness . . ..

The hopes and fears of both sides are embodied in one of the biggest private social-policy research projects ever undertaken: a five-year, $30-million study of welfare overhaul and other elements of the "New Federalism." The kick-off of the study will be announced today by the Washington-based Urban Institute.

The institute, founded three decades ago to examine the woes of the nation's cities, has assembled a politically balanced project staff and **promises to post "nonpartisan, reliable data" on the Internet for all sides to examine.** But memories are still fresh of the institute's prediction last year that the law would toss one million children into poverty, so even some of its top officials fret about how the research will be received in a political culture increasingly riven between opposing ideological camps.

"Everyone wants to attach a political label to everything that

comes out," says Isabel Sawhill, a former Clinton administration budget official . . .. **For people on both sides, "there's no such thing as unbiased information or apolitical studies anymore,"** she says.

Playing the Numbers

"Hopefully, the data we produce will be unassailable," says Anna Kondratas, a onetime Reagan administration aide who is the project's co-director. Yet she recalls a decade-old admonition from a Democratic congressman who didn't like her testimony about the food-stamp program: **"Everybody's entitled to his own statistics"** . . ..

In recent years, evaluations of high-profile social programs have often found them less beneficial than many political liberals had hoped. **"We are in desperate need to learn about what works,"** says Doug Nelson, president of the Annie E. Casey Foundation, the philanthropic organization that is the largest single backer of the Urban Institute study.

(continues)

WHAT IS THE TRUTH?

STATISTICS AND APPLIED SOCIAL RESEARCH—USEFUL OR "ABUSEFUL"? (continued)

Assessing the crazy quilt of state welfare innovations poses an immense challenge for researchers of any ideological stripe. With so many policies changing at once, all involving potential effects on employment, childbearing and family life, figuring out which questions to ask may be as difficult as finding the answers. **And the swirling crosscurrents over race, economics and values—the possible conflict between attempts to reduce illegitimate births and attempts to prevent increases in abortion, for example—make determinations of "success" or "failure" highly subjective.**

That is precisely why backers of the new law, which aims to reduce welfare spending by some $54 billion over seven years, are bracing for a flood of critical studies.

"Social scientists want to help children and families, and they believe the way to do that is to give them more benefits," says Ron Haskins, a former developmental psychologist at the University of North Carolina who now heads the staff of a House welfare subcommittee. **As a result, adds the Heritage Foundation's Robert Rector, most studies of the law's effect "will emphasize a bogus measure of material poverty" while shortchanging shifts in attitudes, values and behavior that may be hard to gauge and slower to manifest themselves . . . .**

That contentious atmosphere has embroiled the Kaiser Foundation's $90,000 survey of New Jersey welfare recipients. Ted Goertzel, a Rutgers University sociologist not involved in the survey, notes rumors that ideology played a role in Kaiser's decision not to release the data. "The recipients said they agreed with the family cap," he says. **"If you do research and don't like the results, are you obligated to release it? There's an ethical question here."**

I've included this article at the end of this chapter because chi-square is one of the major inference tests used in social science research. I think you will agree that this article poses some very interesting and important questions regarding applied research in this area. For example, in applied social science research, is it really true that "there is no such thing as unbiased information?" Was the Democratic congressman correct in stating, "Everybody's entitled to his own statistics"? Is it really true that "most studies of the law's effect 'will emphasize a bogus measure of material poverty, while shortchanging shifts in attitudes, values and behavior that may be hard to gauge and slower to manifest themselves' "? If so, how will we ever find out whether this and other social programs really work? Or doesn't it matter that we ever find out what social programs really work? Would it be better just to "pontificate" out of one's own private social values (Method of Authority), or per-haps just to cite individual case examples? Or do you agree with Doug Nelson that, "We are in desperate need to learn about what works"? If so, how can appropriate data be collected?

What do you think about the notion that social scientists with strong political views will go out and do research, deliberately biasing their research instruments so that the data will confirm their own political views? If it is true, does that mean we should stop funding such research? How about the ordinary layperson: How will she be able to intelligently interpret the results of such studies? Does this mean students should stop studying statistics, or quite the contrary, that students need to learn even more about statistics so that they will not be taken in by poor, pseudo, or biased research? Finally, what do you think about the situation where an organization conducts socially relevant research and the findings turn out to be against the interest of the organization? Does the organization have an obligation to inform the public? Is it unethical if it doesn't?

SUMMARY

In this chapter, we have discussed nonparametric statistics. Nonparametric inference tests depend considerably less on population characteristics than do parametric tests. The z, t, and F tests are examples of parametric tests; the sign test and the Mann–Whitney U test are examples of nonparametric tests. Parametric tests are used whenever possible because they are more powerful and versatile. However, when the assumptions of the parametric tests are violated, nonparametric tests are frequently employed.

One of the most frequently used inference tests for analyzing nominal data is the nonparametric test called chi-square (χ^2). It is appropriate for analyzing frequency data dealing with one or two variables. Chi-square essentially measures the discrepancy between the observed frequency (f_o) and the expected frequency (f_e) for each of the cells in a one-way or two-way table. In equation form, $\chi^2_{obt} = \Sigma (f_o - f_e)^2/f_e$, where the summation is over all the cells. In single-variable situations, the data are presented in a one-way table, and the various expected frequency values are determined on an *a priori* basis. In two-variable situations, the frequency data are presented in a contingency table, and we are interested in determining whether there is a relationship between the two variables. The null hypothesis states that there is no relationship—that the two variables are independent. The alternative hypothesis states that the two variables are related. The expected frequency for each cell is the frequency that would be expected if sampling is random from a population where the proportions for each category on one variable are equal for each category on the other variable. Since the population proportions are unknown, their expected values under the null hypothesis are estimated from the sample data, and the expected frequencies are calculated using these estimates.

The obtained value of χ^2 is evaluated by comparing it with χ^2_{crit}. If $\chi^2_{obt} \geq \chi^2_{crit}$, we reject the null hypothesis. The critical value of χ^2 is determined by the sampling distribution of χ^2 and the alpha level. The sampling distribution of χ^2 is a family of curves that vary with the degrees of freedom. In the one-variable experiment, df $= k - 1$. In the two-variable situation, df $= (r - 1)(c - 1)$. A basic assumption of χ^2 is that each subject can have only one entry in the table. A

second assumption is that the expected frequency in each cell must be of a certain minimum size. The use of χ^2 is not limited to nominal data, but regardless of the scaling, the data must finally be divided into mutually exclusive categories, and the cell entries must be frequencies.

The Wilcoxon matched-pairs signed ranks test is a nonparametric test that is used with a correlated groups design. The statistic calculated is T_{obt}. Determination of T_{obt} involves four steps: (1) finding the difference between each pair of scores, (2) ranking the absolute values of the difference scores, (3) assigning the appropriate sign to the ranks, and (4) separately summing the positive and negative ranks. T_{obt} is the lower sum. T_{obt} is evaluated by comparison with T_{crit}. If $T_{obt} \leq T_{crit}$, we reject H_0. The Wilcoxon signed ranks test requires that (1) the within-pair scores are at least of ordinal scaling and (2) the difference scores are also at least of ordinal scaling. This test serves as an alternate to the t test for correlated groups when the assumptions of the t test have not been met. It is more powerful than the sign test, but not as powerful as the t test.

The Kruskal–Wallis test is used as a substitute for one-way parametric ANOVA. It uses the independent groups design with k samples. The null hypothesis asserts that the k samples are random samples from the same or identical population distributions. No attempt is made to specifically test for population mean differences, as is the case with parametric ANOVA. The statistic computed is H_{obt}. If the number of scores in each sample is 5 or more, the sampling distribution of H_{obt} is close enough to that of chi-square to use the latter in determining H_{crit}. If $H_{obt} \geq H_{crit}$, H_0 is rejected. In computing H_{obt}, the scores of the k samples are combined and rank ordered, assigning 1 to the lowest score. The ranks are then summed for each sample. Kruskal–Wallis tests whether it is reasonable to consider the summed ranks for each sample to be due to random sampling from a single population set of scores. The greater the differences between the sum of ranks for each sample the less tenable is the null hypothesis. This test assumes that the dependent variable is measured on a scale that is of at least ordinal scaling. There must also be five or more scores in each sample to validly use the chi-square sampling distribution.

IMPORTANT TERMS

Chi-square (χ^2) (p. 424)
Contingency table (p. 430)
Expected frequency (f_e) (p. 425)

Kruskal–Wallis test (H) (p. 443)
Marginals (p. 430)
Observed frequency (f_o) (p. 425)

Wilcoxon matched-pairs signed
ranks test (T) (p. 440)

QUESTIONS AND PROBLEMS

1. Briefly identify or define the terms in the "Important Terms" section.
2. What is the underlying rationale for the determination of f_e in the two-variable experiment?
3. What are the assumptions underlying chi-square?
4. In situations involving more than 1 degree of freedom, the χ^2 test is nondirectional. Is this statement correct? Explain.
5. What distinguishes parametric from nonparametric tests? Explain, giving some examples.
6. Are parametric tests preferable to nonparametric tests? Explain.
7. When might we use a nonparametric test? Give an example.
8. Under what conditions might one use the Wilcoxon signed ranks test?
9. Compare the Wilcoxon signed ranks test to the sign test and the t test for correlated groups with regard to power. Explain any differences.
10. What are the assumptions of the Wilcoxon signed ranks test?
11. What are the assumptions underlying the Kruskal–Wallis test?
12. A researcher is interested in whether there really is a prevailing view that fat people are more jolly. A random sample of 80 individuals was asked the question, "Do you believe fat people are more jolly?" The following results were obtained:

Yes	No
44	36	80

Using $\alpha = 0.05$, what is your conclusion?

13. A study was conducted to determine whether big city and small town dwellers differed in their helpfulness to strangers. In this study, the investigators rang the doorbells of strangers living in New York City or small towns in the vicinity. They explained they had misplaced the address of a friend living in the neighborhood and asked to use the phone. The following data show the number of individuals who admitted or did not admit the strangers (the investigators) into their homes:

Helpfulness to Strangers

	Admitted Strangers into Their Home	Did Not Admit Strangers into Their Home	
Big city dweller	60	90	150
Small town dweller	70	30	100
	130	120	250

Do big city dwellers differ in their helpfulness to strangers? Use $\alpha = 0.05$ in making your decision.

14. Due to rampant inflation, the government is considering imposing wage and price controls. A government economist, interested in determining whether there is a relationship between occupation and attitude toward wage and price controls, collects the following data. The data show for each occupation the number of individuals in the sample who were *for* or *against* the controls:

Attitude Toward Wage and Price Controls

	For	Against	
Labor	90	60	150
Business	100	150	250
Professions	110	90	200
	300	300	600

Do these occupations differ regarding attitudes toward wage and price controls? Use $\alpha = 0.01$ in making your decision.

15. The head of the marketing division of a leading soap manufacturer must decide among four differently styled wrappings for the soap. To provide a data base for the decision, he has the soap placed in the different wrapping styles and distributed to five supermarkets. At the end of 2 weeks, he finds that the following amounts of soap were sold:

Wrapping A	Wrapping B	Wrapping C	Wrapping D	
90	98	130	82	400

Is there sufficient basis for making a decision among wrappings? If so, which should he pick? Use $\alpha = 0.05$.

16. A researcher believes that individuals in different occupations will show differences in their ability to be hypnotized. Six lawyers, six physicians, and six professional dancers are randomly selected for the experiment. A test of hypnotic susceptibility is administered to each. The results are shown here. The higher the score, the higher the hypnotizability. Assume the data violate the assumptions required for use of the F test, but are at least of ordinal scaling. Using $\alpha = 0.05$, what is your conclusion?

Condition 1 Lawyers	Condition 2 Physicians	Condition 3 Dancers
26	14	30
17	19	21
27	28	35
32	22	29
20	25	37
25	15	34

17. A professor of religious studies is interested in finding out whether there is a relationship between church attendance and educational level. Data are collected on a sample of individuals who only completed high school and on another sample who received a college education. The following are the resultant frequency data:

Church Attendance

	Attend Regularly	Do Not Attend Regularly	
High school	88	112	200
College	56	104	160
	144	216	360

What is your conclusion? Use $\alpha = 0.05$.

18. A coffee manufacturer advertises that, in a recent experiment in which their brand (brand A) was compared with the other four leading brands of coffee, more people preferred their brand to the other four. The data from the experiment are given at the top of the next column:

Coffee Brand

A	B	C	D	E	
60	45	52	43	50	250

Do you believe the ad to be misleading? Use $\alpha = 0.05$ in making your decision.

19. A study was conducted to determine if there is a relationship between the amount of contact white housewives have with blacks and changes in their attitudes toward blacks. In this study, the changes in attitude toward blacks were measured for white housewives who had moved into segregated public housing projects where there was little daily contact with blacks and for white housewives who had moved into fully integrated public housing projects where there was a great deal of contact. The following frequency data were recorded:

Attitude Toward Blacks

	Less Favorable	No Change	More Favorable	
Segregated housing proj.	9	42	24	75
Integrated housing proj.	7	46	72	125
	16	88	96	200

Based upon these data, what is your conclusion? Use $\alpha = 0.05$.

20. A psychologist investigates the hypothesis that birth order affects assertiveness. Her subjects are twenty young adults between 20 and 25 years of age. There are seven first-born, six second-born, and seven third-born subjects. Each subject is given an assertiveness test, with the following results. High scores indicate greater assertiveness. Assume the data are so nonnormally distributed that the F test can't be used, but that the data are at least of ordinal scaling. Use $\alpha = 0.01$ to evaluate the data. What is your conclusion?

Condition 1 First-born	Condition 2 Second-born	Condition 3 Third-born
18	18	7
8	12	19
4	3	2
21	24	30
28	22	18
32	1	5
10		14

21. An investigator believes that students who rank high in certain kinds of motives will behave differently in gambling situations. To investigate this hypothesis, the investigator randomly samples 50 students high in affiliation motivation, 50 students high in achievement motivation, and 50 students high in power motivation. The students are asked to play the game of roulette, and a record is kept of the bets they make. The data are then grouped into the number of subjects with each kind of motivation who make bets involving low, medium, and high risk. Low risk means they make bets involving low odds (even money or less), medium risk involves bets of medium odds (from 2 to 1 to 5 to 1), and high risk involves playing long shots (from 17 to 1 to 35 to 1). The following data are obtained:

Kind of Motive

	Affiliation	Achievement	Power	
Low risk	26	13	9	48
Med. risk	16	27	14	57
High risk	8	10	27	45
	50	50	50	150

Using $\alpha = 0.05$, is there a relationship between these different kinds of motives and gambling behavior? How do the groups differ?

22. A major oil company conducts an experiment to assess whether a film designed to tell the truth about, and also promote more favorable attitudes toward, large oil companies really does result in more favorable attitudes. Twelve individuals are run in a replicated measures design. In the "before" condition, each subject fills out a questionnaire designed to assess attitudes toward large oil companies. In the "after" condition, the subjects see the film, after which they fill out the questionnaire. The following scores were obtained. High scores indicate more favorable attitudes toward large oil companies.

Before	After
43	45
48	60
25	22
24	33
15	7
18	22
35	41
28	21
41	55
28	33
34	44
12	23

Analyze the data using the Wilcoxon signed ranks test with $\alpha = 0.05_{1\ tail}$. What do you conclude?

23. Refer to Chapter 14, Problem 16, p. 347. Assume the t test cannot be used because of an extreme violation of its normality assumption. Use the Wilcoxon signed ranks test to analyze the data. What do you conclude, using $\alpha = 0.05_{2\ tail}$?

24. Refer to Chapter 14, Problem 12, p. 345. Assume the data are so nonnormal as to invalidate use of the t test for correlated groups. Analyze the data with the Wilcoxon signed ranks test. What do you conclude, using $\alpha = 0.01_{2\,tail}$?

25. Refer to Chapter 15, Problem 17, p. 378. Assume the F test cannot be used because of an extreme violation of the normality assumption. Analyze the data with the Kruskal–Wallis test, using $\alpha = 0.05$.

26. A social psychologist is interested in whether there is a relationship between cohabitation before marriage and divorce. A random sample of 150 couples that were married in the past 10 years in a midwestern city were asked if they lived together prior to getting married and if their marriage was still intact. The following results were obtained.

	Divorced	Still Married	
Cohabited before marriage	58	42	100
Did not cohabit before marriage	18	32	50
	76	74	150

Using $\alpha = 0.05$, what do you conclude?

NOTES

18.1 When df = 1, directional alternative hypotheses can be tested with χ^2. With df = 1, $z_{obt} = \sqrt{\chi^2_{obt}}$. Therefore, we can convert χ^2_{obt} to z_{obt} and evaluate z_{obt} using z_{crit} for the appropriate one-tailed alpha level. Of course, the difference between f_o and f_e must be in the predicted direction to perform this test.

19 REVIEW OF INFERENTIAL STATISTICS

INTRODUCTION

We have covered a lot of material since we began our discussion of hypothesis testing with the sign test. We shall begin our review of this material with the most important terms and concepts pertaining to the general process of hypothesis testing. Then we shall discuss the general process itself. From there, we shall summarize the experimental designs and the inference tests used with each design. Since this material is very logical and interconnected, I hope this review will help bring closure and greater insight to the topic of inferential statistics.

TERMS AND CONCEPTS

Alternative Hypothesis (H_1)   The alternative hypothesis states that the differences in scores between conditions is due to the action of the independent variable. The alternative hypothesis may be nondirectional or directional. A nondirectional hypothesis states that the independent variable has an effect on the dependent variable but doesn't specify the direction of the effect. A directional hypothesis states the direction of the expected effect.

Null Hypothesis (H_0)   The null hypothesis is set up as the logical counterpart to the alternative hypothesis such that if the null hypothesis is false, the alternative hypothesis must be true. Conversely, if the null hypothesis is true, the alternative hypothesis must be false. The null hypothesis for a nondirectional alternative hypothesis is that the independent variable has no effect on the dependent variable. For a directional alternative hypothesis, the null hypothesis states that the independent variable does not have an effect in the direction specified.

Null-Hypothesis Population(s)   The null hypothesis population(s) is the set or sets of scores that would result if the experiment were done on the entire population and the independent variable had no effect. In a single sample design, it is the population with known μ. In a replicated measures design, it is the population of difference scores with $\mu_D = 0$ or $P = 0.50$. In an independent groups design, there are as many populations as there are groups, and the samples are random samples from populations where $\mu_1 = \mu_2 = \mu_3 = \mu_k$.

Sampling Distribution   The sampling distribution of a statistic gives all the values the statistic can take, along with the probability of getting that value if chance alone is responsible or if sampling is random from the null-hypothesis population(s). This distribution can be derived theoretically from basic probability, as we did with the sign test and Mann–Whitney U test, or empirically as with the z, t, and F tests. There are three steps involved in constructing the sampling distribution of a statistic using the empirical approach. First, all possible different samples of size N that can be formed from the population are determined. Second, the statistic for each of the samples is calculated. Finally, the probability of getting each value of the statistic is calculated under the assumption that sampling is random from the null-hypothesis population(s).

Critical Region for Rejection of H_0   The critical region for rejection of H_0 is the area under the curve that contains all the values of the statistic that will allow rejection of the null hypothesis. The critical value of a statistic is that value of the statistic that bounds the critical region. It is determined by the alpha level.

Alpha Level (α)   The alpha level is the threshold probability level against which the obtained probability is compared to determine the reasonableness of the null hypothesis. It also determines the critical region for rejection of the null hypothesis. Alpha is usually set at 0.05 or 0.01. The alpha level is set at the beginning of an experiment and limits the probability of making a Type I error.

Type I Error   A Type I error occurs when the null hypothesis is rejected and it is true.

Type II Error   A Type II error occurs when the null hypothesis is retained and it is false. Beta is equal to the probability of making a Type II error.

Power   The power of an experiment is equal to the probability of rejecting the null hypothesis if the independent variable has a real effect. It is useful to know the power of an experiment when designing the experiment and when interpreting nonsignificant results from an experiment that has already been conducted. Calculation of power involves two steps: (1) determining the sample outcomes that will allow rejection of the null hypothesis and (2) determining the probability of getting these outcomes under the assumed real effect of the independent variable. Power $= 1 - \beta$. Thus, as power increases, beta decreases. Power can be increased by increasing the number of subjects in the experiment, by increasing the magnitude of real effect of the independent variable, by decreasing the variability of the data through careful experimental control and proper experimental design, and by using the most sensitive inference test possible for the design and data.

 # PROCESS OF HYPOTHESIS TESTING

We have seen that in every experiment involving hypothesis testing there are two hypotheses that attempt to explain the data. They are the alternative hypothesis and the null hypothesis. In analyzing the data, we always evaluate the null hypothesis and indirectly conclude with regard to the alternative hypothesis. If H_0 can be rejected, then H_1 is accepted. If H_0 is not rejected, then H_1 is not accepted.

There are two steps involved in assessing the null hypothesis. First, we calculate the appropriate statistic, and second, we evaluate the statistic. To evaluate the statistic, we assume that the independent variable has no effect and that chance alone is responsible for the score differences between conditions. Another way of saying this is that we assume sampling is random from the null-hypothesis population(s). Then we calculate the probability of getting the obtained result or any more extreme under the previous assumption. This probability is one- or two-tailed, depending on whether the alternative hypothesis is directional or nondirectional. To calculate the obtained probability, we must know the sampling distribution of the statistic. If the obtained probability is equal to or less than the alpha level, we reject H_0. Alternatively, we determine whether the obtained statistic falls in the critical region for rejecting H_0. If it does, we reject the null hypothesis. Otherwise, H_0 remains a reasonable explanation, and we retain it.

If we reject H_0 and it is true, we have made a Type I error. The alpha level limits the probability of a Type I error. If we retain H_0 and it is false, we have made a Type II error. The power of the experiment determines the probability of making a Type II error. We have defined beta as the probability of making a Type II error. As power increases, beta decreases. By maintaining alpha sufficiently low and power sufficiently high, we achieve a high probability of making a correct decision when analyzing the data, no matter whether H_0 is true or false.

These statements apply to all experiments involving hypothesis testing. What varies from experiment to experiment is the inference test used and consequently the statistic calculated and evaluated. The inference test used will depend on the experimental design and the data collected.

SINGLE SAMPLE DESIGNS

With single sample experimental designs, one or more of the null-hypothesis population parameters (the mean and/or standard deviation) must be specified. Since it is not common to have this information, the single sample experiment occurs rather infrequently. The z and t tests are appropriate for this design. Both tests evaluate the effect of the independent variable on the mean ($\overline{X}_{obt}$) of the sample. For these tests, the nondirectional H_1 states that $\overline{X}_{obt}$ is a random sample from a population having a mean μ that is not equal to the mean of the null-hypothesis population. The corresponding H_0 states that the μ equals the mean of the null-hypothesis population. The directional H_1 states that $\overline{X}_{obt}$ is a random sample from a population where μ is greater or less than the mean of the null-hypothesis population, depending on the expected direction of the effect. Let's now review the z and t tests for single samples:

Test	Statistic Calculated	Decision Rule
z test for single samples	$z_{obt} = \dfrac{\overline{X}_{obt} - \mu}{\sigma/\sqrt{N}}$	If $|z_{obt}| \geq |z_{crit}|$, reject H_0.

General Comments   The z test is used in situations where both the mean and standard deviation of the null-hypothesis population can be specified. To evaluate H_0, we assume $\overline{X}_{obt}$ is a random sample from a population having a mean μ and standard deviation σ that are equal to the mean and standard deviation of the null-hypothesis population. The sampling distribution of $\overline{X}$ gives all the possible values of $\overline{X}$ for samples of size N and the probability of getting each value if sampling is random from the population with a mean μ and a standard deviation σ. The sampling distribution of $\overline{X}$ has a mean $\mu_{\overline{X}} = \mu$, has a standard deviation $\sigma_{\overline{X}} = \sigma/\sqrt{N}$, and is normally shaped if the population from which the sample was drawn is normal or if $N \geq 30$, provided the population does not differ greatly from normality.

We can assess H_0 by (1) converting $\overline{X}_{obt}$ to its z-transformed value (z_{obt}) and determining the probability of getting a value as extreme as or more extreme than z_{obt} if chance alone is operating or (2) calculating z_{obt} and comparing it with z_{crit}. It is easier to do the latter. The equation for z_{obt} is given on p. 459. The value of z_{obt} is evaluated by comparison with z_{crit}. The alpha level in conjunction with the sampling distribution of z determines the value of z_{crit}. The sampling distribution of z has a mean of 0 and a standard deviation of 1. If $\overline{X}_{obt}$ is normally distributed, then so is the corresponding z distribution, and z_{crit} can be determined from Table A in Appendix D. Thus, the z test requires that $N \geq 30$ or that the population of raw scores be normally distributed.

Test	Statistic Calculated	Decision Rule
t test for single samples	$t_{obt} = \dfrac{\overline{X}_{obt} - \mu}{s/\sqrt{N}}$ $t_{obt} = \dfrac{\overline{X}_{obt} - \mu}{\sqrt{\dfrac{SS}{N(N-1)}}}$	If $\|t_{obt}\| \geq \|t_{crit}\|$, reject H_0.

General Comments   The t test is used in situations where the mean of the null-hypothesis population can be specified and standard deviation is unknown. In testing H_0, we assume $\overline{X}_{obt}$ is a random sample from a population having a mean μ equal to the mean of the null-hypothesis population and an unknown standard deviation. The t test is very much like the z test, except that since σ is unknown, we estimate it with s. When s is substituted for σ in the equation for z_{obt}, the first equation given in the table for t_{obt} results. The second equation in the table is a computational equation for t_{obt} using the raw scores. To evaluate H_0, the value of t_{obt} is compared against t_{crit}, using the decision rule. The value of t_{crit} is determined by the alpha level and the sampling distribution of t. This distribution is a family of curves, shaped like the z distribution. The curves vary uniquely with degrees of freedom. The degrees of freedom for a statistic are equal to the number of scores that are free to vary in calculating the statistic. For the t test used with single samples, df $= N - 1$, because 1 degree of freedom is lost calculating s. The values of t_{crit} are found in Table D in Appendix D, using df and α. The t test has the same underlying assumptions as the z test. The population of raw scores should be normally distributed.

Test	Statistic Calculated	Decision Rule
t test for testing the significance of Pearson r	r_{obt}	If $|r_{\text{obt}}| \geq |r_{\text{crit}}|$, reject H_0.

General Comments   To determine whether a correlation exists in the population, we must test the significance of r_{obt}. This can be done using the t test. The resulting equation is

$$t_{\text{obt}} = \frac{r_{\text{obt}} - \rho}{\sqrt{\dfrac{1 - r_{\text{obt}}^2}{N - 2}}}$$

By substituting t_{crit} for t_{obt} in this equation, r_{crit} can be determined for any df and α level. Once r_{crit} is known, all we need to do is compare r_{obt} with r_{crit}. The decision rule is also given. The values of r_{crit} are found in Table E in Appendix D, using df and α. Degrees of freedom equal $N - 2$.

TWO CONDITION EXPERIMENTS: CORRELATED GROUPS DESIGN

This design is also called the *repeated measures* or *replicated measures* design. The essential feature of the design is that there are paired scores between the conditions, and the differences between the paired scores are analyzed. The paired scores can result from using the same subjects in each condition, from using identical twins, or from using subjects that have been matched in some other way. The most basic form of the design employs just two conditions, an experimental condition and a control condition. The two conditions are kept as alike as possible except for values of the independent variable, which are intentionally made different. We covered three tests for analyzing data from experiments of this design, the t test for correlated groups, the Wilcoxon matched-pairs signed-ranks test, and the sign test.

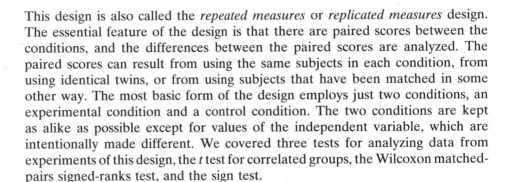

Test	Statistic Calculated	Decision Rule
t test for correlated groups	$t_{\text{obt}} = \dfrac{\bar{D}_{\text{obt}} - \mu_D}{\sqrt{\dfrac{SS_D}{N(N-1)}}}$	If $|t_{\text{obt}}| \geq |t_{\text{crit}}|$, reject H_0.

General Comments   The t test for correlated groups analyzes the effect of the independent variable on the mean of the sample difference scores ($\bar{D}_{\text{obt}}$). If the independent variable has no effect, then $\bar{D}_{\text{obt}}$ is a random sample from a population of difference scores having a mean $\mu_D = 0$ and unknown σ_D. This situation is the same as we encountered when using the t test for single samples (specifiable population mean but unknown standard deviation), except that we are dealing with difference scores rather than raw scores. Thus, the t test for correlated groups is identical to the t test for single samples, except that it evaluates difference scores instead of raw scores.

The nondirectional H_1 states that the independent variable has an effect, in which case $\bar{D}_{obt}$ is due to random sampling from a population of difference scores where $\mu_D \neq 0$. The directional H_1 specifies that $\mu_D > 0$ (for which H_0 states that $\mu_D \leq 0$) or $\mu_D < 0$ (for which H_0 states that $\mu_D \geq 0$). H_0 is tested by assuming $\bar{D}_{obt}$ is a random sample from a population of difference scores where $\mu_D = 0$.

The statistic calculated is t_{obt} (see the equation on p. 461), which is evaluated by comparing it with t_{crit}. The sampling distribution of t is the same as discussed in conjunction with the t test for single samples. The degrees of freedom are equal to $N - 1$, where $N =$ the number of difference scores. The values of t_{crit} are found in Table D, using df and α. The assumptions of this test are the same as those for the t test for single samples. This test is more sensitive than (1) the t test for independent groups when the correlation between the paired scores is high and (2) the Wilcoxon matched-pairs signed-ranks test and the sign test, which are also appropriate for the correlated groups design.

Test	Statistic Calculated	Decision Rule
Wilcoxon matched-pairs signed-ranks test	T_{obt}	If $T_{obt} \leq T_{crit}$, reject H_0.

General Comments   This is a nonparametric test that takes into account the magnitude and direction of the difference scores. It is therefore much more powerful than the sign test. Both the alternative and null hypotheses are usually stated without specifying population parameters. In analyzing the data, T_{obt} is calculated by (1) obtaining the difference score for each pair of scores, (2) rank-ordering the absolute values of the difference scores, (3) assigning the appropriate signs to the ranks, and (4) separately summing the positive and negative ranks. T_{obt} is the lower of the sums. T_{obt} is compared with T_{crit}. The values of T_{crit} are given in Table I in Appendix D, using N and α. The decision rule is shown in the table above. This test is recommended as an alternative to the t test for correlated groups when the assumptions of the t test are not met. The Wilcoxon signed-ranks test requires that the within-pair scores be at least of ordinal scaling and that the differences also be at least of ordinal scaling.

Test	Statistic Calculated	Decision Rule
Sign test	Number of P events in a sample of size N	If the one- or two-tailed p(number of P events) $\leq \alpha$, reject H_0.

General Comments   We used the sign test to introduce hypothesis testing, because it is a simple test to understand. It is not commonly used in practice, because it ignores the magnitude of the difference scores and considers only their direction.

In analyzing the data with the sign test, we determine the number of pluses in the sample and evaluate this statistic by using the binomial distribution. The binomial distribution is the appropriate sampling distribution for situations where (1) there is a series of N trials, (2) there are only two possible outcomes on each trial, (3) there is independence between trials, (4) the outcomes on each trial are mutually exclusive, and (5) the probability of each possible outcome on any trial stays the same from trial to trial. The binomial distribution is given by $(P + Q)^N$, where P is the probability of a plus on any trial and Q the probability of a minus. If the independent variable has no effect, then $P = Q = 0.50$. The nondirectional H_1 states that $P \neq Q \neq 0.50$. The directional H_1 specifies $P > 0.50$ or $P < 0.50$, depending on the expected direction of the effect.

H_0 is tested by assuming that the number of P events in the sample is due to random sampling from a population where $P = Q = 0.50$. The one- or two-tailed p(number of P events) is compared with the alpha level to evaluate H_0. This probability is found in Table B in Appendix D, using N, number of P events, and $P = 0.50$. Alternatively, given alpha, N, and the binomial distribution, we could have also determined the critical region for rejecting H_0 (as we did with the other statistics), in which case we would compare the obtained number of P events with the critical number of P events. To use the sign test, the data must be at least ordinal in scaling, and ties must be excluded from the analysis.

INDEPENDENT GROUPS DESIGN: TWO GROUPS

This design involves random sampling of subjects from the population and then random assignment of the subjects to each condition. There can be many conditions. The most basic form of the design uses two conditions, with each condition employing a different level of the independent variable. This design differs from the correlated groups design in that there is no basis for pairing scores between conditions. Analysis is performed separately on the raw scores of each sample, not on the difference scores. Both the t test and the Mann–Whitney U test are appropriate for this design.

Test	Statistic Calculated	Decision Rule
t test for independent groups	$t_{obt} = \dfrac{(\bar{X}_1 - \bar{X}_2) - \mu_{\bar{X}_1 - \bar{X}_2}}{\sqrt{\left(\dfrac{SS_1 + SS_2}{n_1 + n_2 - 2}\right)\left(\dfrac{1}{n_1} + \dfrac{1}{n_2}\right)}}$ When $n_1 = n_2$, $t_{obt} = \dfrac{(\bar{X}_1 - \bar{X}_2) - \mu_{\bar{X}_1 - \bar{X}_2}}{\sqrt{\dfrac{SS_1 + SS_2}{n(n-1)}}}$	If $|t_{obt}| \geq |t_{crit}|$, reject H_0.

General Comments   This test assumes that the independent variable affects the mean of the scores and not their standard deviation. The mean of each sample is calculated, and then the difference between sample means $(\overline{X}_1 - \overline{X}_2)$ is determined. The t test for independent groups analyzes the effect of the independent variable on $\overline{X}_1 - \overline{X}_2$. The sample value $\overline{X}_1$ is due to random sampling from a population having a mean μ_1 and a variance σ_1^2. The sample value $\overline{X}_2$ is due to random sampling from a population having a mean μ_2 and a variance σ_2^2. The variance of both populations is assumed equal $(\sigma_1^2 = \sigma_2^2 = \sigma^2)$.

The sampling distribution of $\overline{X}_1 - \overline{X}_2$ has the following characteristics: (1) It has a mean $\mu_{\overline{X}_1 - \overline{X}_2} = \mu_1 - \mu_2$, (2) it has a standard deviation $\sigma_{\overline{X}_1 - \overline{X}_2} = \sqrt{\sigma^2[(1/n_1) + (1/n_2)]}$, and (3) it is normally shaped if the population from which the samples have been taken is normal. If the independent variable has no effect, then $\mu_1 = \mu_2$. The nondirectional H_1 states that $\mu_1 \neq \mu_2$. The directional H_1 states that $\mu_2 > \mu_1$ or $\mu_1 < \mu_2$, depending on the expected direction of the effect.

To assess H_0, we assume that the independent variable has no effect, in which case $\mu_1 = \mu_2$ and $\mu_{\overline{X}_1 - \overline{X}_2} = 0$. To test H_0, we could calculate z_{obt}, but we need to know σ^2 for this calculation. Since σ^2 is unknown, we estimate it using a weighted estimate from both samples. The resulting statistic is t_{obt}. Two equations for calculating t_{obt} are given on p. 463. The first is a general equation, and the second can be used when the ns in the two samples are equal. The degrees of freedom associated with calculating t_{obt} for the independent groups design is $N - 2$. We calculate two variances in determining t_{obt}, and we lose 1 degree of freedom for each calculation. The sampling distribution of t is as described before. The value of t_{obt} is evaluated by comparing it with t_{crit} according to the decision rule given on p. 463. The values of t_{crit} are found in Table D, using df and α.

To use this test, the sampling distribution of $\mu_{\overline{X}_1 - \overline{X}_2}$ must be normally distributed. This means that the populations from which the samples were taken should be normally distributed.

In addition, to use the t test for independent groups, there should be homogeneity of variance. This test is considered robust with regard to violations of the normality and homogeneity of variance assumptions, provided $n_1 = n_2 \geq 30$. If there is a severe violation of an assumption, the Mann–Whitney U test serves as an alternate to the t test.

Test	Statistic Calculated	Decision Rule
Mann–Whitney U test	U_{obt} or U'_{obt}, where $$U_{\text{obt}} = n_1 n_2 + \frac{n_1(n_1 + 1)}{2} - R_1$$ $$U_{\text{obt}} = n_1 n_2 + \frac{n_2(n_2 + 1)}{2} - R_2$$	If $U_{\text{obt}} < U_{\text{crit}}$, reject H_0.

General Comments   The Mann–Whitney U test is a nonparametric test that analyzes the degree of separation between the samples. The less the separation, the more reasonable chance is as the underlying explanation. For any analysis, there are two values that indicate the degree of separation. They both indicate the same degree of separation. The lower value is called U_{obt}, and the higher value is called U'_{obt}. The lower the U_{obt} value, the greater the separation.

U_{obt} and U'_{obt} can be determined in two ways: (1) by counting *E*s lower than *C*s and *C*s lower than *E*s and (2) by using the equations given on p. 464. Since the equations are more general, we have used them most often. For any analysis, one of the equations will yield *U* and the other *U'*. However, which yields *U* and which *U'* depends on which group is labeled group 1 and which group 2. Since both *U* and *U'* are measures of the same degree of separation, it is only necessary to evaluate one of them.

To evaluate U_{obt}, it is compared with the *U* values given in Tables C.1–C.4. These values are the critical values of *U* (although we didn't call them by this name because we hadn't developed the concept yet). Naturally, these values depend on the sampling distribution of *U*. The decision rule for rejecting H_0 is given in the table.

The Mann–Whitney *U* test is appropriate for an independent groups design where the data are at least ordinal in scaling. It is a powerful test, often used in place of Student's *t* test when the data do not meet the assumptions of the *t* test.

MULTIGROUP EXPERIMENTS: PARAMETRIC ONE-WAY ANALYSIS OF VARIANCE, *F* TEST

Although a two group design is used fairly frequently in the behavioral sciences, it is more common to encounter experiments with three or more groups. Having more than two groups has two main advantages: (1) additional groups often clarify the interpretation of the results, and (2) additional groups allow many levels of the independent variable to be evaluated in one experiment. There is, however, one problem with doing multigroup experiments. Since there are many comparisons that can be made, we run the risk of an inflated Type I error probability when analyzing the data. The analysis of variance technique allows us to analyze the data without incurring this risk.

Test	Statistic Calculated	Decision Rule
Analysis of variance, the *F* test	$F_{obt} = \dfrac{s_B^2}{s_W^2}$	If $F_{obt} \geq F_{crit}$, reject H_0.

General Comments The parametric analysis of variance uses the *F* test to evaluate the data. In using this test, we calculate F_{obt}, which is fundamentally the ratio of two independent variance estimates of a population variance σ^2. The sampling distribution of *F* is composed of a family of positively skewed curves that vary with degrees of freedom. There are two values for degrees of freedom: one for the numerator and one for the denominator. The *F* distribution (1) is positively skewed, (2) has no negative values, and (3) has a median approximately equal to 1.

The parametric analysis of variance technique can be used with both the independent groups and the correlated groups designs. We have only considered the one-way ANOVA independent groups design. The technique allows the means of all the groups to be compared in one overall evaluation, thus avoiding the inflated Type I error probability that occurs when doing many individual comparisons. Essentially, the analysis of variance partitions the total variability of the data into two parts: the variability that exists within each group (the within-groups sum of squares) and the variability that exists between the groups (the between-groups sum of squares). Each sum of squares is used to form an independent estimate of the variance of the null-hypothesis populations, σ^2. Finally, an F ratio is calculated where the between-groups variance estimate is in the numerator and the within-groups variance estimate is in the denominator.

The steps and equations for calculating F_{obt} are as follows:

Step 1: Calculate the between-groups sum of squares, SS_B:

$$SS_B = \left[\frac{(\Sigma X_1)^2}{n_1} + \frac{(\Sigma X_2)^2}{n_2} + \frac{(\Sigma X_3)^2}{n_3} + \cdots + \frac{(\Sigma X_k)^2}{n_k} \right] - \frac{\left(\overset{\text{all scores}}{\Sigma X} \right)^2}{N}$$

Step 2: Calculate the within-groups sum of squares, SS_W:

$$SS_W = \overset{\text{all scores}}{\Sigma} X^2 - \left[\frac{(\Sigma X_1)^2}{n_1} + \frac{(\Sigma X_2)^2}{n_2} + \frac{(\Sigma X_3)^2}{n_3} + \cdots + \frac{(\Sigma X_k)^2}{n_k} \right]$$

Step 3: Calculate the total sum of squares, SS_T; check that $SS_T = SS_W + SS_B$:

$$SS_T = \overset{\text{all scores}}{\Sigma} X^2 - \frac{\left(\overset{\text{all scores}}{\Sigma X} \right)^2}{N}$$

Step 4: Calculate the degrees of freedom for each estimate:

$$df_B = k - 1$$
$$df_W = N - k$$
$$df_T = N - 1$$

Step 5: Calculate the between-groups variance estimate, s_B^2:

$$s_B^2 = \frac{SS_B}{df_B}$$

Step 6: Calculate the within-groups variance estimate, s_W^2:

$$s_W^2 = \frac{SS_W}{df_W}$$

Step 7: Calculate F_{obt}:

$$F_{obt} = \frac{s_B^2}{s_W^2}$$

The null hypothesis for the analysis of variance assumes that the independent variable has no effect and that the samples are random samples from populations where $\mu_1 = \mu_2 = \mu_3 = \mu_k$. Since the between-groups variance estimate increases with the effect of the independent variable and the within-groups variance estimate remains constant, the larger the F ratio, the more unreasonable the null hypothesis becomes. We evaluate F_{obt} by comparing it with F_{crit}. If $F_{obt} \geq F_{crit}$, we reject H_0 and conclude that at least one of the conditions differs from at least one of the other conditions. Note that the analysis of variance technique is nondirectional.

Multiple Comparisons   To determine which conditions differ from each other, *a priori* or *a posteriori* comparisons between pairs of groups are performed. *A priori comparisons* (also called *planned comparisons*) are appropriate when the comparisons have been planned in advance. No adjustment for multiple comparisons is made. Planned comparisons should be relatively few in number and should arise from the logic and meaning of the experiment. In doing the planned comparisons, we usually compare the means of the specified groups using the t test for independent groups. The value for t_{obt} is determined in the usual way, except we use s_W^2 from the analysis of variance in the denominator of the t equation. The t_{obt} value is compared with t_{crit}, using df_W and the alpha level and Table D to determine t_{crit}. The equations used for calculating t_{obt} are given as follows:

$$t_{obt} = \frac{\overline{X}_1 - \overline{X}_2}{\sqrt{s_W^2 \left(\frac{1}{n_1} + \frac{1}{n_2} \right)}}$$

If $n_1 = n_2$,

$$t_{obt} = \frac{\overline{X}_1 - \overline{X}_2}{\sqrt{2s_W^2/n}}$$

A posteriori, or *post hoc, comparisons* were not planned prior to conducting the experiment. They arise either after looking at the data or from assuming the "shotgun" approach of doing all possible mean comparisons in an attempt to gain as much information from the experiment as possible. For these reasons, comparisons made *post hoc* must correct for the increase in the probability of a Type I error that arises due to multiple comparisons. There are many techniques that do this. We have described Tukey's Honestly Significant Difference (HSD) test and the Newman–Keuls test.

Test	Statistic Calculated	Decision Rule
Tukey's HSD test	Q_{obt}, where $$Q_{obt} = \frac{\overline{X}_i - \overline{X}_j}{\sqrt{s_W^2/n}}$$	If $Q_{obt} \geq Q_{crit}$, reject H_0.

The HSD test is designed to compare all possible pairs of means while maintaining the Type I error rate for making the complete set of comparisons at α. The Q statistic is very much like the t statistic except it is always positive and it uses the Q distributions rather than the t distributions. The Q (Studentized range) distributions are derived by randomly taking k samples of equal n from the same population rather than just two samples as with the t distributions and determining the difference between the highest and lowest sample means. To use this test, we calculate Q_{obt} for the desired comparisons and compare Q_{obt} with Q_{crit}. The values of Q_{crit} are found in Table G in Appendix D, using k, df for s_W^2, and α. The decision rule is given above.

Test	Statistic Calculated	Decision Rule
Newman–Keuls test	Q_{obt}, where $$Q_{obt} = \frac{\overline{X}_i - \overline{X}_j}{\sqrt{s_W^2/n}}$$	If $Q_{obt} \geq Q_{crit}$, reject H_0.

General Comments   The Newman–Keuls test is also a *post hoc* test that allows us to make all possible pairwise comparisons among the sample means. The Newman–Keuls test is like the HSD test in that Q_{obt} is calculated and compared with Q_{crit} to evaluate H_0. However, it maintains the Type I error rate at α for each comparison rather than for the entire set of comparisons. It does this by changing the value of Q_{crit} for each comparison. The value of Q_{crit} for any given comparison is given by the sampling distribution of Q for the number of means that are encompassed by $\overline{X}_i$ and $\overline{X}_j$ after all the means have been rank-ordered. This number is symbolized by r. The specific values of Q_{crit} for any analysis are found in Table G, using r, df for s_W^2, and α.

The assumptions underlying the analysis of variance are the same as for the t test for independent groups. There are two assumptions: (1) the populations from which the samples were drawn should be normally distributed, and (2) there should be homogeneity of variance between the groups. The F test is robust with regard to violations of normality and homogeneity of variance, provided there are an equal number of subjects in each group and $n \geq 30$.

NONPARAMETRIC ONE-WAY ANALYSIS OF VARIANCE, KRUSKAL–WALLIS TEST

Test	Statistic Calculated	Decision Rule
Kruskal–Wallis	$H_{obt} = \left[\dfrac{12}{N(N+1)}\right]\left[\sum\limits_{i=1}^{k}\dfrac{(R_i)^2}{n_i}\right] - 3(N+1)$	If $H_{obt} > H_{crit}$, reject H_0.

General Comments   The Kruskal–Wallis test is a nonparametric test, appropriate for a k group, independent groups design. It is used as an alternate test to one-way parametric ANOVA, when the assumptions of that test are seriously violated. The Kruskal–Wallis test does not assume population normality and requires only ordinal scaling of the dependent variable. All the scores are grouped together and rank-ordered, assigning the rank of 1 to the lowest score, 2 to the next to lowest, and N to the highest. The ranks for each condition are then summed. The Kruskal–Wallis test assesses whether these sums of ranks are so different as to make it unreasonable to consider that they come from samples that were randomly selected from the same population.

TWO-WAY ANALYSIS OF VARIANCE, F TEST

The parametric two-way analysis of variance allows us to evaluate the effects of two variables and their interaction in one experiment. In the parametric two-way ANOVA, we partition the total sum of squares (SS_T) into four components: the within-cells sum of squares (SS_W), the row sum of squares (SS_R), the column sum of squares (SS_C), and the row $\times$ column sum of squares (SS_{RC}). When these sums of squares are divided by the appropriate degrees of freedom, they form four variance estimates: the within-cells variance estimate (s_W^2), the row variance estimate (s_R^2), the column variance estimate (s_C^2), and the row $\times$ column variance estimate (s_{RC}^2). The effect of each of the variables is determined by computing the appropriate F_{obt} value and comparing it with F_{crit}.

Test	Statistic Calculated	Decision Rule
Two-way analysis of variance, F test	$F_{obt} = \dfrac{s_R^2}{s_W^2}$ $F_{obt} = \dfrac{s_C^2}{s_W^2}$ $F_{obt} = \dfrac{s_{RC}^2}{s_W^2}$	If $F_{obt} > F_{crit}$, reject H_0.

The steps and equations for calculating the various F_{obt} values are as follows: Calculate F_{obt} for the main effects and interaction:

Step 1: Calculate the row sum of squares, SS_R:

$$SS_R = \left[\frac{\left(\overset{\text{row}}{\underset{1}{\sum}} X \right)^2 + \left(\overset{\text{row}}{\underset{2}{\sum}} X \right)^2 + \cdots + \left(\overset{\text{row}}{\underset{r}{\sum}} X \right)^2}{n_{\text{row}}} \right] - \frac{\left(\overset{\text{all scores}}{\sum} X \right)^2}{N}$$

Step 2: Calculate the column sum of squares, SS_C:

$$SS_C = \left[\frac{\left(\overset{\text{col.}}{\underset{1}{\sum}} X \right)^2 + \left(\overset{\text{col.}}{\underset{2}{\sum}} X \right)^2 + \cdots + \left(\overset{\text{col.}}{\underset{c}{\sum}} X \right)^2}{n_{\text{col.}}} \right] - \frac{\left(\overset{\text{all scores}}{\sum} X \right)^2}{N}$$

Step 3: Calculate the row × column sum of squares, SS_{RC}:

$$SS_{RC} = \left[\frac{\left(\overset{\text{cell}}{\underset{11}{\sum}} X \right)^2 + \left(\overset{\text{cell}}{\underset{12}{\sum}} X \right)^2 + \cdots + \left(\overset{\text{cell}}{\underset{rc}{\sum}} X \right)^2}{n_{\text{cell}}} \right] - \frac{\left(\overset{\text{all scores}}{\sum} X \right)^2}{N} - SS_R - SS_C$$

Step 4: Calculate the within-cells sum of squares, SS_W:

$$SS_W = \overset{\text{all scores}}{\sum} X^2 - \left[\frac{\left(\overset{\text{cell}}{\underset{11}{\sum}} X \right)^2 + \left(\overset{\text{cell}}{\underset{12}{\sum}} X \right)^2 + \cdots + \left(\overset{\text{cell}}{\underset{rc}{\sum}} X \right)^2}{n_{\text{cell}}} \right]$$

Step 5: Calculate the total sum of squares, SS_T, and check that $SS_T = SS_R + SS_C + SS_{RC} + SS_W$:

$$SS_T = \overset{\text{all scores}}{\sum} X^2 - \frac{\left(\overset{\text{all scores}}{\sum} X \right)^2}{N}$$

Step 6: Calculate the degrees of freedom for each variance estimate:

$$df_R = r - 1$$
$$df_C = c - 1$$
$$df_{RC} = (r - 1)(c - 1)$$
$$df_W = rc\,(n_{\text{cell}} - 1)$$
$$df_T = N - 1$$

Step 7: Calculate the variance estimates:

$$\text{Row variance estimate} = s_R{}^2 = \frac{SS_R}{df_R}$$

$$\text{Column variance estimate} = s_C{}^2 = \frac{SS_C}{df_C}$$

$$\text{Row} \times \text{column variance estimate} = s_{RC}{}^2 = \frac{SS_{RC}}{df_{RC}}$$

$$\text{Within-cells variance estimate} = s_W{}^2 = \frac{SS_W}{df_W}$$

Step 8: Calculate the F ratios:

For the row effect,

$$F_{obt} = \frac{s_R{}^2}{s_W{}^2}$$

For the column effect,

$$F_{obt} = \frac{s_C{}^2}{s_W{}^2}$$

For the row $\times$ column interaction effect,

$$F_{obt} = \frac{s_{RC}{}^2}{s_W{}^2}$$

Compare the F_{obt} values with F_{crit} and conclude.

ANALYZING NOMINAL DATA

You will recall that with nominal data, observations are grouped into several discrete, mutually exclusive categories, and one counts the frequency of occurrence in each category. The inference test most often used with nominal data is *chi-square*.

Test	Statistic Calculated	Decision Rule
Chi-square	$\chi^2_{obt} = \sum \frac{(f_o - f_e)^2}{f_e}$	If $\chi^2_{obt} \geq \chi^2_{crit}$, reject H_0.

General Comments   This test is appropriate for analyzing frequency data involving one or two variables. In the two variable situation, the frequency data are presented in a contingency table, and we test to see whether there is a relationship between the two variables. The null hypothesis states that there is no relation-ship—that the variables are independent. The alternative hypothesis states that the two variables are related.

Chi-square measures the discrepancy between the observed frequency (f_o) and the expected frequency (f_e) for each cell in the table and then sums across cells. The equation for χ^2_{obt} is given in the table. When the data involve two variables, the expected frequency for each cell is the frequency that would be expected if sampling is random from a population where the two variables are equal in proportions for each category. Since the population proportions are unknown, their expected values under H_0 are estimated from the sample data, and the expected frequencies are calculated using these estimates. The simplest way to determine f_e for each cell is to multiply the marginals for that cell and divide by N. If the data involve only one variable, the population proportions are determined on some *a priori* basis, e.g., equal population proportions for each category.

The obtained value of χ^2 is evaluated by comparing it with χ^2_{crit} according to the decision rule given on p. 471. The critical value of χ^2 is determined by the sampling distribution of χ^2 and the alpha level. The sampling distribution of χ^2 is a family of curves that varies with the degrees of freedom. In the one variable experiment, df = $k - 1$. In the two variable situation, df = $(r - 1)(c - 1)$. The values of χ^2_{crit} are found in Table H in Appendix D, using df and α.

Proper use of this test assumes that (1) each subject has only one entry in the table (no repeated measures on the same subjects), (2) if r or c is greater than 2, f_e for each cell should be at least 5, and (3) if the table is a 1×2 or 2×2 table, each f_e should be at least 10.

Chi-square can also be used with ordinal, interval, and ratio data. However, regardless of the actual scaling, to use χ^2, the data must be reduced to mutually exclusive categories and appropriate frequencies.

CHOOSING THE APPROPRIATE TEST

One of the important aspects of statistical inference is choosing which test to use for any experiment or problem. Up to now, it has been easy. We just used the test that we were studying for the particular chapter. However, in the review chapter, the situation is more challenging. Since we have covered many inference tests, we now have the opportunity to choose among them in deciding which to use. This, of course, is much more like the situation we face when doing research.

In choosing an inference test, the fundamental rule that we should follow is:

- *Use the most powerful test possible.*

To determine which tests are possible for a given experiment or problem, we must consider two factors, the measurement scale of the dependent variable, and the design of the experiment. Referring to the flowchart of Figure 19.1, the first question we ask is, "What is the level of measurement used for the dependent variable?" If it is nominal, the only inference test we've covered that is appropriate for nominal data is χ^2. Thus, if the data are nominal in scaling and the requirements of χ^2 (frequency data, large enough N, mutually exclusive categories, and independent observations) are met, then we should choose the χ^2 test. If the assumptions are not met, then we don't know what test to use, because it hasn't been covered in this introductory text. In the flow diagram, this regretta-

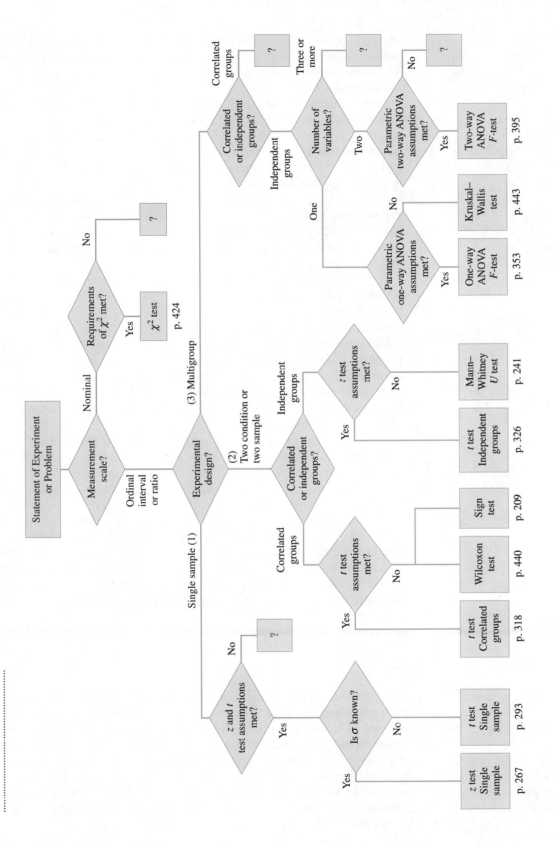

FIGURE 19.1   Decision flowchart for choosing the appropriate inference test

ble state-of-affairs is indicated by a "?". I hasten to reassure you, however, that the inference tests we've covered are the most commonly encountered ones, with the possible exception of very complicated experiments involving three or more variables.

If the data are not nominal, they must be ordinal, interval, or ratio in scaling. Having ruled out nominal data, we should next ask, "What is the experimental design?" The design used in the experiment limits the inference tests that we can use to analyze the data. We have covered three basic designs: single sample, two sample, or two condition, and multigroup experiments. If the design used is a single sample design (path 1 in Figure 19.1), the two tests we have covered for this design are the z test and the t test for single samples. If the data meet the assumptions for these tests, to decide which to use, we must ask the question, "Is σ known?" If the answer is "yes," then the appropriate test is the z test for single samples. If the answer is "no," then we must estimate σ, and use the t test for single samples.

If the experimental design is a two condition or two sample design (path 2), we need to determine whether it is a correlated or independent groups design. If it is correlated groups, and the assumptions of t are met, the appropriate test is the t test for correlated groups. Why? Because, if the assumptions are met, it is the most powerful test we can use for that design. If the assumptions are seriously violated, we should use an alternate test like the Wilcoxon (if its assumptions are met) or the sign test. If it is an independent groups design, and the assumptions of t are met, we should use the t test for independent groups. If the assumptions of t are seriously violated, we should use an alternate test, such as the Mann–Whitney U test.

If the experimental design is a multigroup design (path 3), we need to determine whether it is an independent or correlated groups design. In this text, we have covered multigroup experiments that use the independent groups design. If the experiment is multigroup, uses an independent groups design, involves one variable, and the assumptions of parametric ANOVA are met, the appropriate test is parametric one-way ANOVA (F test). If the assumptions are seriously violated, we should use its alternate, the Kruskal–Wallis test. If the design is a multigroup, independent groups design, involving two variables, and the data meet the assumptions of parametric two-way ANOVA, we would use parametric two-way ANOVA (F test) to analyze the data. We have not considered the more complex designs involving three or more variables.

QUESTIONS AND PROBLEMS

Note to the student: In the previous chapters covering inferential statistics, when you were asked to solve an end-of-chapter problem, there was no question about which inference test you would use—you would use the test covered in the chapter. For example, if you were doing a problem in Chapter 13, you knew you should use the t test for single samples, because that was the test the chapter covered. Now you have reached the elevated positon in which you know so much statistics that when solving a problem, there may be more than one inference test that could be used. Often both a parametric and nonparametric text may be possible. This is a new challenge. The rule to follow is to use the most sensitive test that the data will allow. For the problems in this chapter, always assume that the assumptions underlying the parametric test are met, unless the problem *explicitly* states otherwise.

1. Briefly define the following terms:

    Alternative hypothesis
    Null hypothesis
    Null-hypothesis population
    Sampling distribution
    Critical region for rejection of H_0
    Alpha level
    Type I error
    Type II error
    Power

2. Briefly describe the process of hypothesis testing. Be sure to include the terms listed in Question 1 in your discussion.

3. Why are sampling distributions important in hypothesis testing?

4. An educator conducts an experiment using an independent groups design to evaluate two methods of teaching third-grade spelling. The results are not significant, and the educator concludes that the two methods are equal. Is this conclusion sound? Assume that the study was properly designed and conducted; e.g., proper controls were present, sample size was reasonably large, proper statistics used, etc.

5. Why are parametric tests generally preferred over nonparametric tests?

6. List the factors that affect the power of an experiment and explain how they can be used to increase power.

7. What factors determine which inference test to use in analyzing the data of an experiment?

8. List the various experimental designs covered in this textbook. In addition, list the inference tests appropriate for each design in the order of their sensitivity.

9. What are the assumptions underlying each inference test?

10. What are the two steps followed in analyzing the data from any study involving hypothesis testing?

11. A new competitor in the scotch whiskey industry conducts an experiment to compare their scotch whiskey (called McPherson's Joy) to the other three leading brands. Two hundred scotch drinkers are randomly sampled from the scotch drinkers living in New York City. Each individual is asked to taste the four scotch whiskeys and pick the one they like the best. Of course, the whiskeys are unmarked, and the order in which they are tasted is balanced. The number of subjects that

preferred each brand is shown in the following table:

McPherson's Joy	Brand X	Brand Y	Brand Z	
58	52	48	42	200

a. What is the alternative hypothesis? Use a nondirectional hypothesis.
b. What is the null hypothesis?
c. Using $\alpha = 0.05$, what do you conclude?

12. A psychologist interested in animal learning conducts an experiment to determine the effect of ACTH (adrenocorticotropic hormone) on avoidance learning. Twenty 100-day-old male rats are randomly selected from the university vivarium for the experiment. Of the 20, 10 randomly chosen rats receive injections of ACTH 30 minutes before being placed in the avoidance situation. The other 10 receive placebo injections. The number of trials for each animal to learn the task is given below:

ACTH	Placebo
58	74
73	92
80	87
78	84
75	72
74	82
79	76
72	90
66	95
77	85

a. What is the nondirectional alternative hypothesis?
b. What is the null hypothesis?
c. Using $\alpha = 0.01_{2\ tail}$, what do you conclude?
d. What error may you have made by concluding as you did in part c?
e. To what population do these results apply?

13. A university nutritionist wonders whether the recent emphasis on eating a healthy diet has affected freshman students at her university. Con-

sequently, she conducts a study to determine whether the diet of freshman students currently enrolled contains less fat than that of previous freshmen. To determine their percentage of daily fat intake, 15 students in this year's freshman class keep a record of everything they eat for 7 days. The results show that for the 15 students, the mean percentage of daily fat intake is 37%, with a standard deviation of 12%. Records kept on a large number of freshman students from previous years show a mean percentage of daily fat intake of 40%, a standard deviation of 10.5%, and a normal distribution of scores.

a. Based on these data, is the daily fat intake of currently enrolled freshmen less than that of previous years? Use $\alpha = 0.05_{1\ tail}$.

b. If the actual mean daily fat intake of currently enrolled freshmen is 35%, what is the power of the experiment to detect this level of real effect?

c. If N is increased to 30, what is the power to detect a real mean daily fat intake of 35%?

d. If the nutritionist wants a power of 0.9000 to detect a real effect of at least 5 mean points below the established population norms, what N should she run?

14. A physiologist conducts an experiment designed to determine the effect of exogenous thyroxin (a hormone produced by the thyroid gland) on activity. Forty male rats are randomly assigned to four groups such that there are 10 rats per group. Each of the groups is injected with a different amount of thyroxin. Group 1 gets no thyroxin and merely receives saline solution. Group 2 receives a small amount, group 3 a moderate amount, and group 4 a high amount of thyroxin. After the injections, each animal is tested in an open field apparatus to measure its activity level. The open field apparatus is composed of a fairly large platform with sides around it to prevent the animal from leaving the platform. A grid configuration is painted on the surface of the platform such that the entire surface is covered with squares. To measure activity, the experimenter merely counts the number of squares that the animal has crossed during a fixed period of time. In the present experiment, each rat is tested in the open field apparatus for 10 minutes. The results are shown here; the scores are the number of squares crossed per minute.

Amount of Thyroxin

Zero, 1	Low, 2	Moderate, 3	High, 4
2	4	8	12
3	3	7	10
3	5	9	8
2	5	6	7
5	3	5	9
2	2	8	13
1	4	9	11
3	3	7	8
4	6	8	7
5	4	4	9

a. What is the overall null hypothesis?

b. Using $\alpha = 0.05$, what do you conclude?

c. Evaluate the *a priori* hypothesis that a high amount of exogenous thyroxin produces an effect on activity different from that of saline. Use $\alpha = 0.05_{2\ tail}$.

d. Use the Tukey HSD test with $\alpha = 0.05_{2\ tail}$ to compare all possible pairs of means. What do you conclude?

15. A study is conducted to determine whether dieting plus exercise is more effective in producing weight loss than dieting alone. Twelve pairs of matched subjects are run in the study. Subjects are matched on initial weight, initial level of exercise, age, and gender. One member of each pair is put on a diet for 3 months. The other member receives the same diet but, in addition, is put on a moderate exercise regime. The following scores indicate the weight loss in pounds over the 3-month period for each subject:

Pair	Diet Plus Exercise	Diet Alone
1	24	16
2	20	18
3	22	19
4	15	16
5	23	18
6	21	18
7	16	17
8	17	19
9	19	13
10	25	18
11	24	19
12	13	14

In answering the following questions, assume the data are very nonnormal so as to preclude using the appropriate parametric test.

a. What is the alternative hypothesis? Use a directional hypothesis.
b. What is the null hypothesis?
c. Using $\alpha = 0.05_{1 \text{ tail}}$, what do you conclude?

16. a. What other nonparametric test could you have used to analyze the data presented in Problem 15?
b. Use this test to analyze the data. What do you conclude with $\alpha = 0.05_{1 \text{ tail}}$?
c. Explain the difference between your conclusions for Problems 16b and 15c.

17. A researcher in human sexuality is interested in determining whether there is a relationship between sexual gender and time-of-day preference for having intercourse. A survey is conducted, and the results are shown in the following table; entries are the number of individuals who preferred morning or evening times:

Sexual Gender	Intercourse		
	Morning	Evening	
Male	36	24	60
Female	28	32	60
	64	56	120

a. What is the null hypothesis?
b. Using $\alpha = 0.05$, what do you conclude?

18. A psychologist is interested in whether the internal states of individuals affect their perceptions. Specifically, the psychologist wants to determine whether hunger influences perception. To test this hypothesis, she randomly divides 24 subjects into three groups of 8 subjects per group. The subjects are asked to describe "pictures" that they are shown on a screen. Actually, there are no pictures, just ambiguous shapes or forms. Hunger is manipulated through food deprivation. One group is shown the pictures 1 hour after eating, another group 4 hours after eating, and the last group 12 hours after eating. The number

of food-related objects reported by each subject is recorded. The following data are collected:

Food Deprivation		
1 hr	*4 hr*	*12 hr*
1	*2*	*3*
2	6	8
5	7	10
7	6	15
2	10	19
1	15	9
8	12	14
7	7	15
6	6	12

a. What is the overall null hypothesis?
b. What is your conclusion? Use $\alpha = 0.05$.
c. Using the Newman–Keuls test with $\alpha = 0.05_{2 \text{ tail}}$, do all possible *post hoc* comparisons between pairs of means. What is your conclusion?

19. An engineer working for a leading electronics firm claims to have invented a process for making longer-lasting TV picture tubes. Tests run on 24 picture tubes made with the new process show a mean life of 1,725 hours and a standard deviation of 85 hours. Tests run over the last 3 years on a very large number of TV picture tubes made with the old process show a mean life of 1,538 hours. Is the engineer correct in her claim? Use $\alpha = 0.01_{1 \text{ tail}}$ in making your decision.

20. In a study to determine the effect of alcohol on aggressiveness, 17 adult volunteers were randomly assigned to two groups, an experimental group and a control group. The subjects in the experimental group drank vodka disguised in orange juice, and the subjects in the control group drank only orange juice. After the drinks were finished, a test of aggressiveness was administered. The scores on page 478 were obtained; higher scores indicate greater aggressiveness:

Orange Juice	Vodka Plus Orange Juice
11	14
9	13
14	19
15	16
7	15
10	17
8	11
10	18
8	

a. What is the alternative hypothesis? Use a non-directional hypothesis.
b. What is the null hypothesis?
c. Using $\alpha = 0.05_{2\,tail}$, what is your conclusion?

21. The dean of admissions at a large university wonders how strong the relationship is between high school grades and college grades. During the 2 years that he has held this position, he has weighted high school grades heavily when deciding which students to admit to the university, yet he has never seen any data relating the two variables. Having a strong experimental background, he decides to conduct a study and find out for himself. He randomly samples 15 seniors from his university and obtains their high school and college grades. The following data are obtained:

Grades

Subject	High School	College
1	2.2	1.5
2	2.6	1.7
3	2.5	2.0
4	2.2	2.4
5	3.0	1.7
6	3.0	2.3
7	3.1	3.0
8	2.6	2.7
9	2.8	3.2
10	3.2	3.6
11	3.4	2.5
12	3.5	2.8
13	4.0	3.2
14	3.6	3.9
15	3.8	4.0

a. Compute r_{obt} for these data.
b. Is the correlation significant? Use $\alpha = 0.05_{2\,tail}$.
c. What proportion of the variability in college grades is accounted for by the high school grades?
d. Is the dean justified in weighting high school grades heavily when determining which students to admit to the university?

22. An experiment is conducted to evaluate the effect of smoking on heart rate. Ten subjects who smoke cigarettes are randomly selected for the experiment. Each subject serves in two conditions. In condition 1, the subject rests for an hour, after which heart rate is measured. In condition 2, the subject rests for an hour and then smokes two cigarettes. In condition 2, heart rate is measured after the subject has finished smoking the cigarettes. The data follow.

Heart Beats Per Minute

Subject	No Smoking, 1	Smoking, 2
1	72	76
2	80	84
3	68	75
4	74	73
5	80	86
6	85	88
7	86	84
8	78	80
9	68	72
10	67	70

a. What is the nondirectional alternative hypothesis?
b. What is the null hypothesis?
c. Using $\alpha = 0.05_{2\,tail}$, what do you conclude?

23. To meet the current oil crisis, the government must decide on a course of action to follow. There are two choices: (1) to allow the price of oil to rise or (2) to impose gasoline rationing. A survey is taken among individuals of various occupations to see whether there is a relationship between the occupations and the course of action which is favored. The results are shown in the following

3×2 table; cell entries are the number of individuals favoring the course of action which heads the cell:

	Course of Action		
Occupation	Oil Price Rise	Gasoline Rationing	
Business	180	120	300
Homemaker	135	165	300
Labor	152	148	300
	467	433	900

a. What is the null hypothesis?
b. Using $\alpha = 0.05$, what do you conclude?

24. You are interested in testing the hypothesis that adult men and women differ in logical reasoning ability. To do so, you randomly select 16 adults from the city in which you live and administer a logical reasoning test to them. A higher score indicates better logical reasoning ability. The following scores are obtained:

Men	Women
70	80
60	50
82	81
65	75
83	95
92	85
85	93
	75
	90

In answering the following questions, assume that the data violate the assumptions underlying the use of the appropriate parametric test and that you must analyze the data with a nonparametric test.
a. What is the null hypothesis?
b. Using $\alpha = 0.05_{2\,\text{tail}}$, what is your conclusion?

25. For her doctoral thesis, a graduate student in women's studies investigated the effects of stress on the menstrual cycle. Forty-two women were randomly sampled and run in a two condition replicated measures design. However, one of the women dropped out of the study. In the stress condition, the mean length of menstrual cycle for the remaining 41 women was 29 days with a standard deviation of 14 days. Based on these data, determine the 95% confidence interval for the population mean length of menstrual cycle when under stress.

26. A student believes that physical science professors are more authoritarian than social science professors. She conducts an experiment in which 6 physics, 6 psychology, and 6 sociology professors are randomly selected and given a questionnaire measuring authoritarianism. The results are shown here. The higher the score, the more authoritarian is the individual. Assume the data seriously violate normality assumptions. What do you conclude, using $\alpha = 0.05$?

Professors		
Physics	Psychology	Sociology
75	73	71
82	80	80
80	85	90
97	92	78
94	70	94
76	69	68

27. A sleep researcher is interested in determining whether taking naps can improve performance and, if so, whether it matters if the naps are taken in the afternoon or evening. Thirty undergraduates are randomly sampled and assigned to one of six conditions, napping in the afternoon or evening, resting in the afternoon or evening, or a normal activity control condition again in the afternoon or evening. There are 5 subjects in each condition. Each subject performs the activity appropriate for his or her assigned condition, after which a performance test is given. The higher the score, the better is the performance. The following results were obtained. What is

your conclusion? Use $\alpha = 0.05$, and assume the
data are from normally distributed populations.

Time of Day	Activity					
	Napping		Resting		Normal	
Afternoon	8	9	7	8	3	4
	7	5	6	4	5	5
	6		5		6	
Evening	6	5	5	3	4	2
	7	4	4	5	3	4
	6		4		3	

A | REVIEW OF PREREQUISITE MATHEMATICS

 ## INTRODUCTION

In this appendix, we shall present a review of some basic mathematical skills that we believe are important as background for an introductory course in statistics. This appendix is intended to be a review of material that you have already learned but which may be a little "rusty" from disuse. For students who have been away from mathematics for many years and who feel unsure of their mathematical background, we recommend the following books: H. M. Walker, *Mathematics Essential for Elementary Statistics* (rev. ed., Holt, New York, 1951) or A. J. Washington, *Arithmetic and Beginning Algebra* (Addison-Wesley, Reading, 1984).

Algebraic Symbols

Symbol	Explanation
$>$	Is greater than
$5 > 4$	5 is greater than 4.
$X > 10$	X is greater than 10.
$a > b$	a is greater than b.
$<$	Is less than
$7 < 9$	7 is less than 9.
$X < 12$	X is less than 12.
$a < b$	a is less than b.
$2 < X < 20$	X is greater than 2 and less than 20, or the value of X lies between 2 and 20.
$X < 2$ or $X > 20$	X is less than 2 or greater than 20, or the value of X lies outside the interval of 2 to 20.
$\geq$	Is equal to or greater than
$X \geq 3$	X is equal to or greater than 3.
$a \geq b$	a is equal to or greater than b.
$\leq$	Is equal to or less than
$X \leq 5$	X is equal to or less than 5.
$a \leq b$	a is equal to or less than b.
$\neq$	Is not equal to
$3 \neq 5$	3 is not equal to 5.
$X \neq 8$	X is not equal to 8.
$a \neq b$	a is not equal to b.
$\lvert X \rvert$	The absolute value of X; the absolute value of X equals the magnitude of X irrespective of its sign.
$\lvert +7 \rvert$	The absolute value of 7; $\lvert +7 \rvert = 7$.
$\lvert -5 \rvert$	The absolute value of -5; $\lvert -5 \rvert = 5$.

Arithmetic Operations

Operation	Example
1. *Addition of two positive numbers:* To add two positive numbers, sum their absolute values and give the result a plus sign.	$2 + 8 = 10$
2. *Addition of two negative numbers:* To add two negative numbers, sum their absolute values and give the result a minus sign.	$-3 + (-4) = -7$
3. *Addition of two numbers with opposite signs:* To add two numbers with opposite signs, find the difference between their absolute values and give the number the sign of the larger absolute value.	$16 + (-10) = 6$ $3 + (-14) = -11$
4. *Subtraction of one number from another:* To subtract one number from another, change the sign of the number to be subtracted and proceed as in addition (operations 1, 2, or 3).	$16 - 4 = 16 + (-4) = 12$ $5 - 8 = 5 + (-8) = -3$ $9 - (-6) = 9 + (+6) = 15$ $-3 - 5 = -3 + (-5) = -8$
5. *Multiplying a series of numbers:* a. When multiplying a series of numbers, the result is positive if there are an even number of negative values in the series.	$2(-5)(-6)(3) = 180$ $-3(-7)(-2)(-1) = 42$ $-a(-b) = ab$
b. When multiplying a series of numbers, the result is negative if there are an odd number of negative values in the series.	$4(-5)(2) = -40$ $-8(-2)(-5)(3) = -240$ $-a(-b)(-c) = -abc$
6. *Dividing a series of numbers:* a. When dividing a series of numbers, the result is positive if there are an even number of negative values.	$\dfrac{-4}{-8} = \dfrac{1}{2}$ $\dfrac{-3(-4)(2)}{6} = 4$ $\dfrac{-a}{-b} = \dfrac{a}{b}$
b. When dividing a series of numbers, the result is negative if there are an odd number of negative values.	$\dfrac{-2}{5} = -0.40$ $\dfrac{-3(-2)}{-4} = -1.5$ $\dfrac{-a}{b} = -\dfrac{a}{b}$

Rules Governing the Order of Arithmetic Operations

Rule	Example
1. The order in which numbers are added does not change the result.	$6 + 4 + 11 = 4 + 6 + 11 = 11 + 6 + 4 = 21$ $6 + (-3) + 2 = -3 + 6 + 2 = 2 + 6 + (-3) = 5$
2. The order in which numbers are multiplied does not change the result.	$3 \times 5 \times 8 = 8 \times 5 \times 3 = 5 \times 8 \times 3 = 120$
3. If both multiplication and addition or subtraction are specified, the multiplication should be performed first unless parentheses or brackets indicate otherwise.	$4 \times 5 + 2 = 20 + 2 = 22$ $6 \times (14 - 12) \times 3 = 6 \times 2 \times 3 = 36$ $6 \times (4 + 3) \times 2 = 6 \times 7 \times 2 = 84$
4. If both division and addition or subtraction are specified, the division should be performed first unless parentheses or brackets indicate otherwise.	$12 \div 4 + 2 = 3 + 2 = 5$ $12 \div (4 + 2) = 12 \div 6 = 2$ $12 \div 4 - 2 = 3 - 2 = 1$ $12 \div (4 - 2) = 12 \div 2 = 6$

Rules Governing Parentheses and Brackets

Rule	Example
1. Parentheses and brackets indicate that whatever is shown within them is to be treated as a single number.	$(2 + 8)(6 - 3 + 2) = 10(5) = 50$
2. Where there are parentheses contained within brackets, perform the operations contained within the parentheses first.	$[(4)(6 - 3 + 2) + 6][2] = [(4)(5) + 6][2]$ $= [20 + 6][2] = [26][2] = 52$
3. When it is inconvenient to reduce whatever is contained within the parentheses to a single number, the parentheses may be removed as follows:	
a. If a positive sign precedes the parentheses, remove the parentheses without changing the sign of any number it contained.	$1 + (3 + 5 - 2) = 1 + 3 + 5 - 2 = 7$
b. If a negative sign precedes the parentheses, remove the parentheses and change the signs of the numbers it contained.	$4 - (6 - 2 - 1) = 4 - 6 + 2 + 1 = 1$
c. If a multiplier exists outside the parentheses, all the terms within the parentheses must be multiplied by the multiplier.	$3(2 + 3 - 4) = 6 + 9 - 12 = 3$ $a(b + c + d) = ab + ac + ad$
d. The product of two sums is found by multiplying each element of one sum by the elements of the other sum.	$(a + b)(c + d) = ac + ad + bc + bd$
e. If whatever is contained within the parentheses is operated on in any way, always do the operation first, before combining with other terms.	$5 + 4(3 + 1) + 6 = 5 + 4(4) + 6 = 5 + 16 + 6 = 27$ $4 + (3 + 1)/2 = 4 + 4/2 = 4 + 2 = 6$ $1 + (3 + 2)^2 = 1 + (5)^2 = 1 + 25 = 26$

Fractions

Operation	Example
1. *Addition of fractions:* To add two fractions, (1) find the least common denominator, (2) express each fraction in terms of the least common denominator, and (3) add the numerators and divide the sum by the common denominator.	$\frac{1}{3} + \frac{1}{2} = \frac{2}{6} + \frac{3}{6} = \frac{5}{6}$ $\frac{a}{b} + \frac{c}{d} = \frac{ad}{bd} + \frac{cb}{bd} = \frac{ad+cb}{bd}$
2. *Multiplication of fractions:* To multiply two fractions, multiply the numerators together and divide by the product of the denominators.	$\frac{2}{5}(\frac{3}{7}) = \frac{6}{35}$ $\frac{a}{b}\left(\frac{c}{d}\right) = \frac{ac}{bd}$
3. *Changing a fraction into its decimal equivalent:* To convert a fraction into a decimal, perform the indicated division, rounding to the required number of digits (two digits in the example shown).	$\frac{3}{7} = 0.429 = 0.43$
4. *Changing a decimal into a percentage:* To convert a decimal fraction into a percentage, multiply the decimal fraction by 100.	$0.43 \times 100 = 43\%$
5. *Multiplying an integer by a fraction:* To multiply an integer by a fraction, multiply the integer by the numerator of the fraction and divide the product by the denominator.	$\frac{2}{5}(4) = \frac{8}{5} = 1.60$
6. *Cancellation:* When multiplying several fractions together, identical factors in the numerator and denominator may be canceled.	$\frac{\cancel{5}}{\cancel{12}}\left(\frac{\cancel{4}}{\cancel{9}}\right)\left(\frac{\cancel{3}}{\cancel{10}}\right) = \frac{1}{18}$

Exponents

Operation	Example
1. *Multiplying a number by itself 2 times*	$(4)^2 = 4(4) = 16$ $a^2 = aa$
2. *Multiplying a number by itself 3 times*	$(4)^3 = 4(4)(4) = 64$ $a^3 = aaa$
3. *Multiplying a number by itself N times*	$4^N = \overbrace{4(4)(4)\cdots(4)}^{N}$
4. *Multiplying two exponential quantities having the same base:* The product of two exponential quantities having the same base is the base raised to the sum of the exponents.	$(2)^2(2)^4 = (2)^{2+4} = (2)^6$ $a^N a^P = a^{N+P}$
5. *Dividing two exponential quantities having the same base:* The quotient of two exponential quantities having the same base is the base raised to an exponent equal to the exponent of the quantity in the numerator minus the exponent of the quantity in the denominator.	$\dfrac{(2)^4}{(2)^2} = (2)^2$ $\dfrac{a^N}{a^P} = a^{N-P}$
6. *Raising a base to a negative exponent:* A base raised to a negative exponent is equal to 1 divided by the base raised to the positive value of the exponent.	$(2)^{-3} = \dfrac{1}{(2)^3}$ $a^{-N} = \dfrac{1}{a^N}$

Factoring   When factoring an algebraic expression, we try to reduce the expression to the simplest components that when multiplied together yield the original expression.

Example	Explanation
$ab + ac + ad = a(b + c + d)$	We factored out *a* from each item.
$abc - 2ab = ab(c - 2)$	We factored out *ab* from both terms.
$a^2 + 2ab + b^2 = (a + b)^2$	This expression can be reduced to $a + b$ times itself.

SOLVING EQUATIONS WITH ONE UNKNOWN

When solving equations with one unknown, the basic idea is to isolate the unknown on one side of the equation and reduce the other side to its smallest possible value. In so doing, we make use of the principle that the equation remains an equality if whatever we do to one side of the equation, we also do the same to the other side. Thus, for example, the equation remains an equality if we add the same number to both sides. In solving the equation, we alter the equation by adding, subtracting, multiplying dividing, squaring, etc., so as to isolate the unknown. This is permissible as long as we do the same operation to both sides of the equation, thus maintaining the equality. The following examples illustrate many of the operations commonly used to solve equations having one unknown. In each of the examples, we shall be solving the equation for Y.

Example	**Explanation**
$Y + 5 = 2$ $Y = 2 - 5$ $= -3$	To isolate Y, we subtract 5 from both sides of the equation.
$Y - 4 = 6$ $Y = 6 + 4$ $= 10$	To isolate Y, we add 4 to both sides of the equation.
$\dfrac{Y}{2} = 8$ $Y = 8(2)$ $= 16$	To isolate Y, we multiply both sides of the equation by 2.
$3Y = 7$ $Y = \frac{7}{3}$ $= 2.33$	To isolate Y, we divide both sides of the equation by 3.
$6 = \dfrac{Y - 3}{2}$ $12 = Y - 3$ $3 + 12 = Y$ $15 = Y$ $Y = 15$	To isolate Y, (1) multiply both sides by 2, and (2) add 3 to both sides.
$3 = \dfrac{2}{Y}$ $3Y = 2$ $Y = \frac{2}{3}$	To isolate Y, (1) multiply both sides by Y, and (2) divide both sides by 3.
$4(Y + 1) = 3$ $Y + 1 = \frac{3}{4}$ $Y = \frac{3}{4} - 1$ $= -\frac{1}{4}$	To isolate Y, (1) divide both sides by 4, and (2) subtract 1 from both sides.

$$\frac{4}{Y+2} = 8$$

To isolate Y,
  (1) take the reciprocal of both sides,

$$\frac{Y+2}{4} = \frac{1}{8}$$

  (2) multiply both sides by 4, and

$$Y + 2 = \tfrac{4}{8}$$

  (3) subtract 2 from both sides.

$$Y = \tfrac{4}{8} - 2$$

$$= -1\tfrac{1}{2}$$

$$2Y + 4 = 10$$

To isolate Y,

$$2Y = 10 - 4$$

  (1) subtract 4 from both sides, and

$$Y = \frac{10 - 4}{2}$$

  (2) divide both sides by 2.

$$= 3$$

 ## LINEAR INTERPOLATION

Linear interpolation is often necessary when looking up values in a table. For example, suppose we wanted to find the square root of 96.5 using a table that only has the square root of 96 and 97 but not 96.5, as shown here:

Number	Square Root
96	9.7980
97	9.8489

Looking in the column headed by Number, we note that there is no value corresponding to 96.5. The closest values are 96 and 97. From the table, we can see that the square root of 96 is 9.7980 and that the square root of 97 is 9.8489. Obviously, the square root of 96.5 must lie between 9.7980 and 9.8489. Using *linear* interpolation, we assume there is a linear relationship between the number and its square root, and we use this linear relationship to approximate the square root of numbers not given in the table. Since 96.5 is halfway between 96 and 97, using linear interpolation, we would expect the square root of 96.5 to lie halfway between 9.7980 and 9.8489. If we let X equal the square root of 96.5, then

$$X = 9.7980 + 0.5(9.8489 - 9.7980) = 9.8234$$

Linear interpolation for $\sqrt{96.5}$

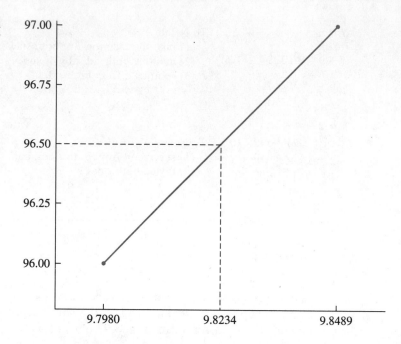

The relationship is shown graphically in Figure A.1. Although it wasn't made explicit, the computed value for X was the result of setting up the following proportions and solving for X:

$$\frac{96.5 - 96}{97 - 96} = \frac{X - 9.7980}{9.8489 - 9.7980}$$

Number	Square Root
96	9.7980
96.5	X
97	9.8489

B | EQUATIONS

Listed here are the computational equations used in this textbook. The page number refers to the page where the equation first appeared.

Equation	Description	Equation First Occurs on Page:
$\displaystyle\sum_{i=1}^{N} X_i = X_1 + X_2 + X_3 + \cdots + X_N$	*summation*	21
$\text{cum } \% = \dfrac{\text{cum } f}{N} \times 100$	*cumulative percentage*	42
$\text{Percentile point} = X_L + (i/f_i)(\text{cum } f_P - \text{cum } f_L)$	*equation for computing percentile point*	45
$\text{Percentile rank} = \dfrac{\text{cum } f_L + (f_i/i)(X - X_L)}{N} \times 100$	*equation for computing percentile rank*	47
$\overline{X} = \dfrac{\Sigma X_i}{N}$	*mean of a sample*	62
$\mu = \dfrac{\Sigma X_i}{N}$	*mean of a population of raw scores*	62
$\overline{X}_{\text{overall}} = \dfrac{n_1 \overline{X}_1 + n_2 \overline{X}_2 + \cdots + n_k \overline{X}_k}{n_1 + n_2 + \cdots + n_k}$	*overall mean of several groups*	65
$\text{Mdn} = X_L + (i/f_i)(\text{cum } f_P - \text{cum } f_L)$	*median of a distribution*	67
$\text{Range} = \text{Highest score} - \text{Lowest score}$	*range of a distribution*	71
$X - \overline{X}$	*deviation score for sample data*	71
$X - \mu$	*deviation score for population data*	71

491

Equation	Description	Equation First Occurs on Page:
$s = \sqrt{\dfrac{SS}{N-1}} = \sqrt{\dfrac{\Sigma\,(X-\bar{X})^2}{N-1}}$	*standard deviation of a sample of raw scores*	73
$\sigma = \sqrt{\dfrac{SS_{\text{pop}}}{N}} = \sqrt{\dfrac{\Sigma\,(X-\mu)^2}{N}}$	*standard deviation of a population of raw scores*	73
$SS = \Sigma\,X^2 - \dfrac{(\Sigma\,X)^2}{N}$	*sum of squares of a sample*	74
$s^2 = \dfrac{SS}{N-1}$	*variance of a sample of raw scores*	77
$\sigma^2 = \dfrac{SS_{\text{pop}}}{N}$	*variance of a population of raw scores*	77
$z = \dfrac{X-\mu}{\sigma}$	*z score for population data*	84
$z = \dfrac{X-\bar{X}}{s}$	*z score for sample data*	84
$X = \mu + \sigma z$	*equation for finding a population raw score from its z score*	93
$Y = bX + a$	*equation of a straight line*	101
$b = \text{Slope} = \dfrac{\Delta Y}{\Delta X} = \dfrac{Y_2 - Y_1}{X_2 - X_1}$	*slope of a straight line*	101
$r = \dfrac{\Sigma\,z_X z_Y}{N-1}$	*equation for Pearson r using z scores*	110
$r = \dfrac{\Sigma\,XY - \dfrac{(\Sigma\,X)(\Sigma\,Y)}{N}}{\sqrt{\left[\Sigma\,X^2 - \dfrac{(\Sigma\,X)^2}{N}\right]\left[\Sigma\,Y^2 - \dfrac{(\Sigma\,Y)^2}{N}\right]}}$	*computational equation for Pearson r*	110
$r_s = 1 - \dfrac{6\,\Sigma\,D_i^2}{N^3 - N}$	*computational equation for Spearman rho*	117
$Y' = b_Y X + a_Y$	*linear regression equation for predicting Y given X*	130
$b_Y = \dfrac{\Sigma\,XY - \dfrac{(\Sigma\,X)(\Sigma\,Y)}{N}}{\Sigma\,X^2 - \dfrac{(\Sigma\,X)^2}{N}}$	*regression constant b for predicting Y given X, computational equation with raw scores*	130
$a_Y = \bar{Y} - b_Y \bar{X}$	*regression constant a for predicting Y given X*	130

Equation	Description	Equation First Occurs on Page:
$X' = b_X Y + a_X$	*linear regression equation for predicting X given Y*	137
$b_X = \dfrac{\Sigma XY - \dfrac{(\Sigma X)(\Sigma Y)}{N}}{\Sigma Y^2 - \dfrac{(\Sigma Y)^2}{N}}$	*regression constant b for predicting X given Y, computational equation with raw scores*	137
$a_X = \bar{X} - b_X \bar{Y}$	*regression constant a for predicting X given Y*	137
$s_{Y\mid X} = \sqrt{\dfrac{SS_Y - \dfrac{[\Sigma XY - (\Sigma X)(\Sigma Y)/N]^2}{SS_X}}{N-2}}$	*computational equation for the standard error of estimate when predicting Y given X*	140
$b_Y = r\dfrac{s_Y}{s_X}$	*equation relating r to the b_Y regression constant*	144
$b_X = r\dfrac{s_X}{s_Y}$	*equation relating r to the b_X regression constant*	144
$R^2 = \dfrac{r_{YX_1}{}^2 + r_{YX_2}{}^2 - 2r_{YX_1} r_{YX_2} r_{X_1 X_2}}{1 - r_{X_1 X_2}{}^2}$	*equation for computing the squared multiple correlation*	148
$p(A) = \dfrac{\text{Number of events classifiable as } A}{\text{Total number of possible events}}$	a priori *probability*	159
$p(A) = \dfrac{\text{Number of times } A \text{ has occurred}}{\text{Total number of occurrences}}$	a posteriori *probability*	160
$p(A \text{ or } B) = p(A) + p(B) - p(A \text{ and } B)$	*addition rule for two events, general equation*	161
$p(A \text{ or } B) = p(A) + p(B)$	*addition rule when A and B are mutually exclusive*	163
$p(A \text{ or } B \text{ or } C \text{ or} \ldots \text{or } Z) = p(A) + p(B) + p(C) + \cdots + p(Z)$	*addition rule with more than two mutually exclusive events*	164
$p(A) + p(B) + p(C) + \cdots + p(Z) = 1.00$	*when events are exhaustive and mutually exclusive*	166
$P + Q = 1.00$	*when two events are exhaustive and mutually exclusive*	166
$p(A \text{ and } B) = p(A)p(B\mid A)$	*multiplication rule with two events, general equation*	166

Equation	Description	Equation First Occurs on Page:
$p(A \text{ and } B) = 0$	*multiplication rule with mutually exclusive events*	167
$p(A \text{ and } B) = p(A)p(B)$	*multiplication rule with independent events*	167
$p(A \text{ and } B \text{ and } C \text{ and } \dots \text{ and } Z) = p(A)p(B)p(C)\cdots p(Z)$	*multiplication rule with more than two independent events*	173
$p(A \text{ and } B) = p(A)p(B\mid A)$	*multiplication rule with dependent events*	174
$p(A \text{ and } B \text{ and } C \text{ and } \dots \text{ and } Z)$ $= p(A)p(B\mid A)p(C\mid AB)\cdots p(Z\mid ABC\dots)$	*multiplication rule with more than two dependent events*	177
$p(A) = \dfrac{\text{Area under curve corresponding to } A}{\text{Total area under curve}}$	*probability of A with a continuous variable*	181
$(P + Q)^N$	*binomial expansion*	194
Number of Q events $= N -$ Number of P events	*relationship between number of Q events, number of P events, and N*	201
Beta $= 1 -$ Power	*relationship between beta and power*	230
$U_{obt} = n_1 n_2 + \dfrac{n_1(n_1 + 1)}{2} - R_1$	*general equation for calculating U_{obt} or U'_{obt}*	246
$U_{obt} = n_1 n_2 + \dfrac{n_2(n_2 + 1)}{2} - R_2$	*general equation for calculating U_{obt} or U'_{obt}*	246
$\mu_{\bar{X}} = \mu$	*mean of the sampling distribution of the mean*	269
$\sigma_{\bar{X}} = \dfrac{\sigma}{\sqrt{N}}$	*standard deviation of the sampling distribution of the mean or standard error of the mean*	269
$z_{obt} = \dfrac{\bar{X}_{obt} - \mu}{\sigma_{\bar{X}}}$	*z transformation for $\bar{X}_{obt}$*	276
$z_{obt} = \dfrac{\bar{X}_{obt} - \mu}{\sigma/\sqrt{N}}$	*z transformation for $\bar{X}_{obt}$*	278
$N = \left[\dfrac{\sigma(z_{crit} - z_{obt})}{\mu_{real} - \mu_{null}}\right]^2$	*determining N for a specified power*	286

Equation	Description	Equation First Occurs on Page:
$t_{obt} = \dfrac{\overline{X}_{obt} - \mu}{s/\sqrt{N}}$	equation for calculating the t statistic	293
$t_{obt} = \dfrac{\overline{X}_{obt} - \mu}{s_{\overline{X}}}$	equation for calculating the t statistic	293
$s_{\overline{X}} = \dfrac{s}{\sqrt{N}}$	estimated standard error of the mean	294
$df = N - 1$	degrees of freedom for t test (single sample)	296
$t_{obt} = \dfrac{\overline{X}_{obt} - \mu}{\sqrt{\dfrac{SS}{N(N-1)}}}$	equation for calculating the t statistic from raw scores	299
$\mu_{lower} = \overline{X}_{obt} - s_{\overline{X}} t_{0.025}$	lower limit for the 95% confidence interval	306
$\mu_{upper} = \overline{X}_{obt} + s_{\overline{X}} t_{0.025}$	upper limit for the 95% confidence interval	306
$\mu_{lower} = \overline{X}_{obt} - s_{\overline{X}} t_{0.005}$	lower limit for the 99% confidence interval	307
$\mu_{upper} = \overline{X}_{obt} + s_{\overline{X}} t_{0.005}$	upper limit for the 99% confidence interval	307
$\mu_{lower} = \overline{X}_{obt} - s_{\overline{X}} t_{crit}$	general equation for the lower limit of the confidence interval	307
$\mu_{upper} = \overline{X}_{obt} + s_{\overline{X}} t_{crit}$	general equation for the upper limit of the confidence interval	307
$t_{obt} = \dfrac{r_{obt} - p}{s_r}$	t test for testing the significance of r	310
$t_{obt} = \dfrac{r_{obt}}{\sqrt{\dfrac{1 - r_{obt}^2}{N-2}}}$	t test for testing the significance of r	310
$t_{obt} = \dfrac{\overline{D}_{obt} - \mu_D}{s_D/\sqrt{N}}$	t test for correlated groups	319
$t_{obt} = \dfrac{\overline{D}_{obt} - \mu_D}{\sqrt{\dfrac{SS_D}{N(N-1)}}}$	t test for correlated groups	319

Equation	Description	Equation First Occurs on Page:
$SS_D = \Sigma D^2 - \dfrac{(\Sigma D)^2}{N}$	sum of squares of the difference scores	319
$\sigma_{\bar{X}_1 - \bar{X}_2} = \sqrt{\sigma^2\left(\dfrac{1}{n_1} + \dfrac{1}{n_2}\right)}$	standard deviation of the difference between sample means	329
$\mu_{\bar{X}_1 - \bar{X}_2} = \mu_1 - \mu_2$	mean of the difference between sample means	329
$t_{obt} = \dfrac{(\bar{X}_1 - \bar{X}_2) - \mu_{\bar{X}_1 - \bar{X}_2}}{\sqrt{\left(\dfrac{SS_1 + SS_2}{n_1 + n_2 - 2}\right)\left(\dfrac{1}{n_1} + \dfrac{1}{n_2}\right)}}$	computational equation for t_{obt}, independent groups design	332
$t_{obt} = \dfrac{\bar{X}_1 - \bar{X}_2}{\sqrt{\left(\dfrac{SS_1 + SS_2}{n_1 + n_2 - 2}\right)\left(\dfrac{1}{n_1} + \dfrac{1}{n_2}\right)}}$	computational equation for t_{obt} assuming the independent variable has no effect	332
$t_{obt} = \dfrac{\bar{X}_1 - \bar{X}_2}{\sqrt{\dfrac{SS_1 + SS_2}{n(n-1)}}}$	computational equation for t_{obt} when $n_1 = n_2$	334
$SS_1 = \Sigma X_1^2 - \dfrac{(\Sigma X_1)^2}{n_1}$	sum of squares for group 1	336
$SS_2 = \Sigma X_2^2 - \dfrac{(\Sigma X_2)^2}{n_2}$	sum of squares for group 2	336
$F = \dfrac{\text{Variance estimate 1 of } \sigma^2}{\text{Variance estimate 2 of } \sigma^2}$	basic definition of F	351
$F_{obt} = \dfrac{s_B^2}{s_W^2}$	F equation for the analysis of variance	355
$MS_B = s_B^2$	mean square between	355
$MS_W = s_W^2$	mean square within mean square error	355
$s_W^2 = \dfrac{SS_W}{N - k}$	within-groups variance estimate	357
$SS_W = \sum^{\text{all scores}} X^2 - \left[\dfrac{(\Sigma X_1)^2}{n_1} + \dfrac{(\Sigma X_2)^2}{n_2} + \dfrac{(\Sigma X_3)^2}{n_3} + \cdots + \dfrac{(\Sigma X_k)^2}{n_k}\right]$	within-groups sum of squares, computational equation	357
$df_W = N - k$	degrees of freedom for the within-groups variance estimate	357

Equation	Description	Equation First Occurs on Page:
$$s_B^2 = \frac{SS_B}{k-1}$$	between-groups variance estimate	358
$$SS_B = \left[\frac{(\Sigma X_1)^2}{n_1} + \frac{(\Sigma X_2)^2}{n_2} + \frac{(\Sigma X_3)^2}{n_3} + \cdots + \frac{(\Sigma X_k)^2}{n_k}\right] - \frac{\left(\overset{\text{all scores}}{\sum} X\right)^2}{N}$$	between-groups sum of squares, computational equation	358
$$\text{df}_B = k - 1$$	degrees of freedom for the between-groups variance estimate	358
$$SS_T = SS_W + SS_B$$	equation for checking SS_W and SS_B	361
$$SS_T = \overset{\text{all scores}}{\sum} X^2 - \frac{\left(\overset{\text{all scores}}{\sum} X\right)^2}{N}$$	equation for calculating the total variability	361
$$t_{\text{obt}} = \frac{\overline{X}_1 - \overline{X}_2}{\sqrt{s_W^2\left(\frac{1}{n_1} + \frac{1}{n_2}\right)}}$$	t equation for a priori comparisons, general equation	382
$$t_{\text{obt}} = \frac{\overline{X}_1 - \overline{X}_2}{\sqrt{2s_W^2/n}}$$	t equation for a priori comparisons with equal n in the two groups	382
$$Q_{\text{obt}} = \frac{\overline{X}_i - \overline{X}_j}{\sqrt{s_W^2/n}}$$	equation for calculating Q_{obt}	385
$$s_W^2 = \frac{SS_W}{\text{df}_W}$$	equation for within-cells variance estimate	396
$$SS_W = \overset{\text{all scores}}{\sum} X^2 - \left[\frac{\left(\overset{\text{cell}11}{\sum} X\right)^2 + \left(\overset{\text{cell}12}{\sum} X\right)^2 + \cdots + \left(\overset{\text{cell}rc}{\sum} X\right)^2}{n_{\text{cell}}}\right]$$	within-cells sum of squares, computational equation	398
$$\text{df}_W = rc(n-1)$$	within-cells degrees of freedom	398
$$s_R^2 = \frac{SS_R}{\text{df}_R}$$	equation for the row variance estimate	399
$$SS_R = \left[\frac{\left(\overset{\text{row}1}{\sum} X\right)^2 + \left(\overset{\text{row}2}{\sum} X\right)^2 + \cdots + \left(\overset{\text{row}r}{\sum} X\right)^2}{n_{\text{row}}}\right] - \frac{\left(\overset{\text{all scores}}{\sum} X\right)^2}{N}$$	computational equation for the row sum of squares	399

Equation	Description	Equation First Occurs on Page:
$\mathrm{df}_R = r - 1$	*row degrees of freedom*	399
$s_C^2 = \dfrac{SS_C}{\mathrm{df}_C}$	*column variance estimate*	400
$\mathrm{df}_C = c - 1$	*column degrees of freedom*	400
$SS_C = \left[\dfrac{\left(\overset{\text{col.}}{\underset{1}{\sum}} X\right)^2 + \left(\overset{\text{col.}}{\underset{2}{\sum}} X\right)^2 + \cdots + \left(\overset{\text{col.}}{\underset{c}{\sum}} X\right)^2}{n_{\text{col.}}} \right] - \dfrac{\left(\overset{\text{all}}{\overset{\text{scores}}{\sum}} X\right)^2}{N}$	*computational equation for the column sum of squares*	401
$s_{RC}^2 = \dfrac{SS_{RC}}{\mathrm{df}_{RC}}$	*row × column variance estimate*	401
$SS_{RC} = \left[\dfrac{\left(\overset{\text{cell}}{\underset{11}{\sum}} X\right)^2 + \left(\overset{\text{cell}}{\underset{12}{\sum}} X\right)^2 + \cdots + \left(\overset{\text{cell}}{\underset{rc}{\sum}} X\right)^2}{n_{\text{cell}}} \right] - \dfrac{\left(\overset{\text{all}}{\overset{\text{scores}}{\sum}} X\right)^2}{N}$ $- SS_R - SS_C$	*computational equation for the row × column sum of squares*	402
$\mathrm{df}_{RC} = (r - 1)(c - 1)$	*row × column degrees of freedom*	402
$\chi_{\text{obt}}^2 = \Sigma \dfrac{(f_o - f_e)^2}{f_e}$	*equation for calculating χ_{obt}^2*	425
$H_{\text{obt}} = \left[\dfrac{12}{N(N + 1)} \right] \left[\sum_{i=1}^{k} \dfrac{(R_i)^2}{n_i} \right] - 3(N + 1)$	*equation for computing H_{obt}*	445

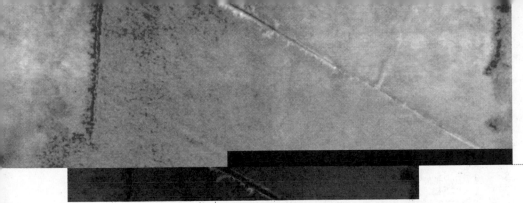

C | ANSWERS TO END-OF-CHAPTER QUESTIONS AND PROBLEMS

CHAPTER 1

6. b. (1) functioning of the hypothalamus; (2) daily food intake; (3) the 30 rats selected for the experiment; (4) all rats living in the university vivarium at the time of the experiment; (5) the daily food intake of each animal during the 2-week period after recovery; (6) the mean daily food intake of each group   c. (1) the arrangement of typing keys; (2) typing speed; (3) the 20 secretarial trainees who were in the experiment; (4) all secretarial trainees enrolled in the business school at the time of the experiment; (5) the typing speed scores of each trainee obtained at the end of training; (6) the mean typing speed of each group;   d. (1) methods of treating depression; (2) degree or amount of depression; (3) the 60 depressed students; (4) the undergraduate body at a large university at the time of the experiment; (5) depression scores of the 60 depressed students; (6) mean of the depression scores of each treatment

7. a. constant   b. constant   c. variable   d. variable   e. constant   f. variable   g. variable   h. constant

8. a. descriptive statistics   b. descriptive statistics   c. descriptive statistics   d. inferential statistics   e. inferential statistics   f. descriptive statistics

9. a. The sample scores are the 20 scores given. The population scores are the 213 scores which would have resulted if the number of drinks during "happy hour" were measured from all of the bars.   c. The sample scores are the 25 lengths measured. The population scores are the 600 lengths that would be obtained if all 600 blanks were measured.

CHAPTER 2

2. a. continuous   b. discrete   c. discrete   d. continuous   e. discrete   f. continuous   g. continuous   h. continuous

3. a. ratio   b. nominal   c. interval   d. ordinal   e. ordinal   f. ratio   g. ratio   h. interval   i. ordinal

4. No, ratios are not legitimate on an interval scale. We need an absolute zero point to perform ratios. Since an ordinal scale does not have an absolute zero point, the ratio of the absolute values represented by 30 and 60 will not be $\frac{1}{2}$.

5. a. 18   b. 21.1   d. 590

6. a. 14.54   c. 37.84   d. 46.50   e. 52.46   f. 25.49

7. a. 9.5–10.5   b. 2.45–2.55   d. 2.005–2.015   e. 5.2315–5.2325

8. a. 25   b. 35   d. 101

9. a. $X_1 = 250$, $X_2 = 378$, $X_3 = 451$, $X_4 = 275$, $X_5 = 225$, $X_6 = 430$, $X_7 = 325$, $X_8 = 334$ b. 2668

10. a. $\sum_{i=1}^{N} X_i$ b. $\sum_{i=1}^{3} X_i$ c. $\sum_{i=2}^{4} X_i$ d. $\sum_{i=2}^{5} X_i^2$

11. a. 1.4   b. 23.2   c. 100.8   d. 41.7   e. 35.3

12. For 5b: $\sum X^2 = 104.45$ and $(\sum X)^2 = 445.21$; for 5c: $\sum X^2 = 3434$ and $(\sum X)^2 = 21{,}904$

13. a. 34   b. 14   d. 6.5

CHAPTER 3

4. a.

Score	f	Score	f	Score	f	Score	f
98	1	84	2	70	2	57	2
97	0	83	3	69	2	56	1
96	0	82	4	68	4	55	2
95	0	81	3	67	2	54	0
94	2	80	0	66	1	53	0
93	2	79	2	65	1	52	0
92	1	78	5	64	3	51	0
91	2	77	2	63	1	50	0
90	2	76	4	62	2	49	2
89	1	75	3	61	1	48	0
88	1	74	1	60	1	47	0
87	2	73	4	59	0	46	0
86	0	72	6	58	0	45	1
85	4	71	3				

b.

Class Interval	Real Limits	f
96–99	95.5–99.5	1
92–95	91.5–95.5	5
88–91	87.5–91.5	6
84–87	83.5–87.5	8
80–83	79.5–83.5	10
76–79	75.5–79.5	13
72–75	71.5–75.5	14
68–71	67.5–71.5	11
64–67	63.5–67.5	7
60–63	59.5–63.5	5
56–59	55.5–59.5	3
52–55	51.5–55.5	2
48–51	47.5–51.5	2
44–47	43.5–47.5	1
		88

5.

Class Interval	f	Relative f	Cumulative f	Cumulative %
96–99	1	0.01	88	100.00
92–95	5	0.06	87	98.86
88–91	6	0.07	82	93.18
84–87	8	0.09	76	86.36
80–83	10	0.11	68	77.27
76–79	13	0.15	58	65.91
72–75	14	0.16	45	51.14
68–71	11	0.12	31	35.23
64–67	7	0.08	20	22.73
60–63	5	0.06	13	14.77
56–59	3	0.03	8	9.09
52–55	2	0.02	5	5.68
48–51	2	0.02	3	3.41
44–47	1	0.01	1	1.14
	88	1.00		

6. a. 82.70   **b.** 72.70

7. a. 70.17   **c.** 85.23

9. a.

Class Interval	f
60–64	1
55–59	1
50–54	2
45–49	2
40–44	4
35–39	5
30–34	7
25–29	12
20–24	17
15–19	16
10–14	8
5–9	3
	78

10.

Class Interval	f	Relative f	Cumulative f
60–64	1	0.01	78
55–59	1	0.01	77
50–54	2	0.03	76
45–49	2	0.03	74
40–44	4	0.05	72
35–39	5	0.06	68
30–34	7	0.09	63
25–29	12	0.15	56
20–24	17	0.22	44
15–19	16	0.21	27
10–14	8	0.10	11
5–9	3	0.04	3
	78	1.00	

11. a. 23.03

12. a. 88.72   **b.** 67.18

15. a. 3.30   **b.** 2.09

16. 65.62

17. a. 34.38   **b.** 33.59   **c.** Some accuracy is lost when grouping scores because the grouped scores analysis assumes the scores are evenly distributed throughout the interval.

CHAPTER 4

14. a. $\overline{X} = 3.56$, Mdn = 3, mode = 2
   c. $\overline{X} = 3.03$, Mdn = 2.70, no mode

15. a. $\overline{X}_{orig.} = 4.00$, $\overline{X}_{new} = 6.00$, $\overline{X}_{new} = \overline{X}_{orig.} + a$
    b. $\overline{X}_{orig.} = 4.00$, $\overline{X}_{new} = 2.00$, $\overline{X}_{new} = \overline{X}_{orig.} - a$
    c. $\overline{X}_{orig.} = 4.00$, $\overline{X}_{new} = 8.00$, $\overline{X}_{new} = a\overline{X}_{orig.}$
    d. $\overline{X}_{orig.} = 4.00$, $\overline{X}_{new} = 2.00$, $\overline{X}_{new} = \overline{X}_{orig.}/a$

16. a. 72.00   b. 72

17. a. 68.83   b. 64.5

18. a. 2.93   b. 2.8

19. a. the mean, because there are no extreme scores
    b. the mean, again because there are no extreme scores
    c. the median, because the distribution contains an extreme score (25)

20. a. positively skewed   b. negatively skewed
    c. symmetrical

21. a. $\overline{X} = 2.54$ hours per day
    b. Mdn = 2.7 hours per day
    c. mode = 0 hours per day

23. a. $\overline{X} = 15.44$ b. Mdn = 14   c. There is no mode.

24. 197.44

25. a. range = 6, $s = 2.04$, $s^2 = 4.14$
    c. range = 9.1, $s = 3.64$, $s^2 = 13.24$

27. 4.00 minutes

28. 7.17 months

29. a. $s_{orig.} = 2.12$, $s_{new} = 2.12$, $s_{new} = s_{orig.}$
    b. $s_{orig.} = 2.12$, $s_{new} = 2.12$, $s_{new} = s_{orig.}$
    c. $s_{orig.} = 2.12$, $s_{new} = 4.24$, $s_{new} = as_{orig.}$
    d. $s_{orig.} = 2.12$, $s_{new} = 1.06$, $s_{new} = s_{orig.}/a$

30. a. 4.50   b. 5   c. 7   d. 7   e. 2.67   f. 7.14

31. Distribution b is most variable, followed by distribution a and then distribution c. For distribution b, $s = 11.37$; for distribution a, $s = 3.16$; and for distribution c, $s = 0$.

32. a. $s = 1.86$ b. 11.96. Because the standard deviation is sensitive to extreme scores and 35 is an extreme score

33. a. $\overline{X} = 7.90$ b. 8   c. 8   d. 15   e. 4.89
    f. 23.88

34. a. $\overline{X} + a$ b. $\overline{X} - a$ c. $a\overline{X}$ d. $\overline{X}/a$

35. a. s stays the same.   b. s stays the same.
    c. s is multiplied by a.   d. s is divided by a.

36. a. 2.67, 3.33, 6.33, 4.67, 6.67, 4.67, 3.00, 4.00, 7.67, 6.67   b. 3.00, 2.00, 8.00, 6.00, 6.00, 4.00, 2.00, 3.00, 8.00, 7.00   c. Expect more variability in the medians.   d. s(medians) = 2.38, s(means) = 1.76

37. All the scores must have the same value.

CHAPTER 5

6. a.

Raw Score	z Score
10	−1.41
12	−0.94
16	0.00
18	0.47
19	0.71
21	1.18

b. mean = 0.00, standard deviation = 1.00

7. a.

Raw Score	z Score
10	−1.55
12	−1.03
16	0.00
18	0.52
19	0.77
21	1.29

b. mean = 0.00, standard deviation = 1.00

8. a. 1.14   b. −0.86   c. 1.86   d. 0.00   e. 1.00   f. −1.00

9. a. 50.00%   b. 15.87%   c. 6.18%   d. 2.02%
   e. 0.07%   f. 32.64%

10. a. 34.13%   b. 34.13%   c. 49.04%   d. 49.87%
    e. 0.00%   f. 25.17%   g. 26.73%

11. a. 0.00   b. 1.96   c. 1.64   d. 0.52
    e. −0.84   f. −1.28

13. a. statistics   b. 92.07

14. a. 3.75%   b. 99.81%   c. 98.54%   d. 12.97%
    e. 3.95 kilograms   f. 3.64   g. 13,623

15. a. 95.99%   b. 99.45%   c. 15.87%   d. 4.36%   e. 50.00%

16. a. 16.85%   b. 0.99%   c. 59.87%   d. 97.50%   e. 50.00%

17. a. 51.57%   b. 34.71%   c. 23.28%

18. a.

Distance	z Score
30	−1.88
31	−1.50
32	−1.13
33	−0.75
34	−0.38
35	0.00
36	0.38
37	0.75
38	1.13

b. and c.

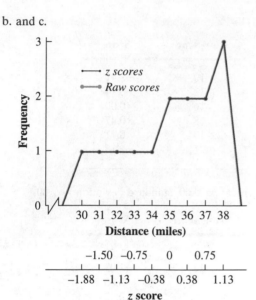

d. The z distribution is not normally shaped. The z distribution takes the same shape as the distribution of raw scores. In this problem, the raw scores are not normally shaped. Therefore, the z distribution will not be normally shaped either.   e. mean = 0.00, standard deviation = 1.00

19. a. 92.36%   b. 55.04%   c. 96.78%   d. $99.04

20. business

21. Rebecca did better on exam 2. Maurice did better on exam 1.

CHAPTER 6

3. a. linear, perfect positive   b. curvilinear, perfect   c. linear, imperfect negative   d. curvilinear, imperfect   e. linear, perfect negative   f. linear, imperfect positive

13. a. for set A, $r = 1.00$; for set B, $r = 0.11$; for set C, $r = -1.00$.   b. same value as in part **a**.   c. The r values are the same.   d. The r values remain the same.   e. The r values do not change if a constant is subtracted from the raw scores or if the raw scores are divided by a constant. The value of r does not change when the scale is altered by adding or subtracting a constant to it, nor does r change if the scale is transformed by multiplying or dividing by a constant.

15. b. $r = 0.68$ c. 0.18. Decreasing the range produced a decrease in r.   d. $r^2 = 0.46$. If illness is causally related to smoking, r^2 allows us to evaluate how important a factor smoking is in producing illness.

16. b. $r = 0.98$.   c. Yes, this is a reliable test because $r^2 = 0.95$. Almost all of the variability of the scores on the second administration can be accounted for by the scores on the first administration.

17. a. $r = 0.85$ b. $r_s = 0.86$

18. a. $r_s = 0.85$ b. For the paper and pencil test and psychiatrist A, $r_s = 0.73$; for the paper and pencil test and psychiatrist B, $r_s = 0.79$.

19. b. $r = 0.59$ d. $r = 0.91$ e. Yes, test 2, because r^2 accounts for 82.4% of the variability in work performance. Although there are no doubt other factors operating, this test appears to offer a good adjunct to the interview. Test 1 does not do nearly as well.

CHAPTER 7

10. No. A scatter plot of the paired scores reveals that there is a perfect relationship between length of left index finger and weight. Thus, Mr. Clairvoyant can exactly predict my weight having measured the length of my left index finger.
    b. $Y' = 7.5X + 37$ c. 79.75

11. b. The relationship is negative, imperfect, and linear.   c. $r = -0.69$ d. $Y' = -1.429X + 125.883$; negative, because the relationship is negative.   f. 71.60   g. 10.87

12. a. $Y' = 10.828X + 11.660$ b. $196,000   c. Technically, the relationship holds only within the range of the base data. It may be that if a lot more money is spent, the relationship would change such that no additional profit or even loss is the result. Of course, the manager could experiment by "testing the waters," e.g., by spending $25,000 on advertising, to see whether the relationship still holds at that level.

13. a. $Y' = -1.212X + 131.77$ b. $130.96   c. 17.31

14. $R^2 = 85.3\%$, $r^2 = 82.4\%$; using test 1 doesn't seem worth the extra work.

CHAPTER 8

8. a, c, e

9. a, c, d, e

10. a, c

11. a. 0.0192   b. 0.0769   c. 0.3077   d. 0.5385

13. a. 0.4000   b. 0.0640   c. 0.0350   d. 0.2818

14. a. 0.4000   b. 0.0491   c. 0.0409   d. 0.2702

15. a. 0.0029   b. 0.0087   c. 0.0554

16. 0.0001

17. 0.0238

18. a. 0.0687   b. 0.5581   c. 0.2005

19. 0.1479

20. 0.00000001

21. a. 0.1429   b. 0.1648   c. 0.0116

23. a. 0.0301   b. 0.6201   c. 0.0301

24. a. 0.0146   b. 0.9233   c. 0.0344

25. a. 0.0764   b. 0.4983   c. 0.0021

26. a. 0.0400   b. 0.1200   c. 0.0400

CHAPTER 9

3. 0.90

4. a. 0.0369   b. $6P^5Q$ c. 0.0369 (The answers are the same.)

5. a. 0.0161   b. 0.0031   c. 0.0192   d. 0.0384

6. a. 0.0407   b. 0.0475   c. 0.0475

7. a. 0.1369   b. 0.2060   c. 0.2060

8. a. 0.0020   b. 0.0899   c. 0.1798

10. 0.3487

11. 0.0037

13. 0.0039

14. a. 0.0001   b. 0.0113

15. a. 32   b. 37

16. a. 0.0001   b. 0.4437   c. 0.1382   d. 0.5563

17. a. 0.0039   b. 0.3164   c. 0.6836

18. a. 0.0576   b. 0.0001   c. 0.0100   d. 0.8059

19. a. 0.0098   b. 1

CHAPTER 10

13. a. The alternative hypothesis states that the new teaching method increases the amount learned.   b. The null hypothesis states that the two methods are equal in the amount of material learned or the old method does better.   c. $p(14$ or more pluses$) = 0.0577$. Since $0.0577 > 0.05$, you retain H_0. You cannot conclude that the new method is better.   d. You may be making a Type II error, retaining H_0 if it is false. e. The results apply to the eighth-grade students in the school district at the time of the experiment.

14. a. The alternative hypothesis states that increases in the level of angiotensin II will produce change in thirst level.   b. The null hypothesis states that increases in the level of angiotensin II will not have any effect on thirst.   c. $p(0, 1, 2, 14, 15,$ or 16 pluses$) = 0.0040$. Since $0.0040 < 0.05$, you reject H_0. Increases in the level of angiotensin II appear to increase thirst. d. You may be making a Type I error, rejecting H_0 if it is true.   e. The results apply to the rats living in the vivarium of the drug company at the time of the experiment.

15. a. The alternative hypothesis states that using Very Bright toothpaste instead of Brand X results in brighter teeth.   b. The null hypothesis states that Very Bright and Brand X toothpastes are equal in their brightening effects, or Brand X is better.   c. $p(7$ or more pluses$) = 0.1719$. Since $0.1719 > 0.05$, you retain H_0. You cannot conclude that Very Bright is better. d. You may be making a Type II error, retaining H_0 if it is false.   e. The results apply to the employees of the Pasadena plant at the time of the experiment.

16. a. The alternative hypothesis states that acupuncture affects pain tolerance.   b. The null hypothesis states that acupuncture has no effect on pain tolerance. c. $p(0, 1, 2, 3, 12, 13, 14,$ or 15 pluses$) = 0.0352$. Since $0.0352 < 0.05$, you reject H_0 and conclude that acupuncture affects pain tolerance. It appears to increase pain tolerance.   d. You may have made a Type I error, rejecting H_0 if it is true.   e. The conclusion applies to the large pool of university undergraduate volunteers.

CHAPTER 11

6. a. The alternative hypothesis states that treatment A and treatment B differ in their effect on the dependent variable.   b. The null hypothesis states that treatment A and treatment B are equal in their effect on the dependent variable.   c. $U_{obt} = 9$, $U'_{obt} = 47$. Since $U_{obt} < 10$, you reject H_0 and conclude that treatment A and treatment B differ significantly in their effect on the dependent variable.

8. a. The alternative hypothesis states that men and women differ in abstract reasoning ability.   b. The null hypothesis states that men and women are equal in abstract reasoning ability.   c. $U_{obt} = 26$, $U'_{obt} = 38$. Since $U_{obt} > 13$, you retain H_0. You cannot conclude on the basis of this experiment that men and women differ in reasoning ability.   d. You may be making a Type II error.   e. The results apply to the adult men and women living in your hometown.

9. a. The alternative hypothesis states that FSH increases the singing rate in captive male cotingas.   b. The null hypothesis states that FSH does not increase the singing rate of captive male cotingas.   c. $U_{obt} = 15.5$, $U'_{obt} = 64.5$. Since $U_{obt} < 20$, you reject H_0 and conclude that FSH appears to increase the singing rate of male cotingas.

10. a. The alternative hypothesis states that right-handed and left-handed people differ in spatial ability. b. The null hypothesis states that right-handed people and left-handed people are equal in spatial ability. c. $U_{obt} = 20.5$, $U'_{obt} = 69.5$. Since $U_{obt} > 20$, you retain H_0. You cannot conclude that right-handed and left-handed people differ in spatial ability.

11. a. The alternative hypothesis states that hypnosis is more effective than the standard treatment in reducing test anxiety.   b. The null hypothesis states that hypnosis and the standard treatment are equal in reducing test anxiety.   c. $U_{obt} = 31$, $U'_{obt} = 90$. Since $U_{obt} < 34$, you reject H_0 and conclude that hypnosis is more effective than the standard treatment in reducing test anxiety.

CHAPTER 12

16. a. The sampling distribution of the mean is given here:

$(\bar{X})$	$p(\bar{X})$
7.0	0.04
6.5	0.08
6.0	0.12
5.5	0.16
5.0	0.20
4.5	0.16
4.0	0.12
3.5	0.08
3.0	0.04

b. From the population raw scores, $\mu = 5.00$, and from the 25 sample means, $\mu_{\bar{X}} = 5.00$. Therefore, $\mu_{\bar{X}} = \mu$.
c. From the 25 sample means, $\sigma_{\bar{X}} = 1.00$. From the population raw scores, $\sigma = 1.41$. Thus, $\sigma_{\bar{X}} = \sigma/\sqrt{N} = 1.41/\sqrt{2} = 1.41/1.41 = 1.00$.

17. a. The distribution is normally shaped; $\mu_{\bar{X}} = 80$, $\sigma_{\bar{X}} = 2.00$.   b. The distribution is normally shaped; $\mu_{\bar{X}} = 80$, $\sigma_{\bar{X}} = 1.35$.   c. The distribution is normally shaped; $\mu_{\bar{X}} = 80$, $\sigma_{\bar{X}} = 1.13$.   d. As N increases, $\mu_{\bar{X}}$ stays the same, but $\sigma_{\bar{X}}$ decreases.

18. $z_{obt} = 3.16$, and $z_{crit} = \pm1.96$. Since $|z_{obt}| > 1.96$, we reject H_0. It is not reasonable to consider the sample a random sample from a population with $\mu = 60$ and $\sigma = 10$.

19. a. $z_{obt} = -2.05$, and $z_{crit} = -2.33$. Since $|z_{obt}| < 2.33$, we can't reject the hypothesis that the sample is a random sample from a population with $\mu = 22$ and $\sigma = 8$.   b. Power = 0.1685   c. Power = 0.5675   d. $N = 161$ (actually gives a power = 0.7995)

21. $z_{obt} = 4.03$, and $z_{crit} = 1.645$. Since $|z_{obt}| > 1.645$, reject H_0 and conclude that this year's class is superior to the previous ones.

22. $z_{obt} = 1.56$, and $z_{crit} = 1.645$. Since $|z_{obt}| < 1.645$, retain H_0. We cannot conclude that the new engine saves gas.

23. a. Power = 0.5871   b. Power = 0.9750   c. $N = 91$ (rounded to nearest integer)

24. $z_{obt} = 4.35$, and $z_{crit} = 1.645$. Since $|z_{obt}| > 1.645$, reject H_0 and conclude that exercise appears to slow down the "aging" process, at least as measured by maximum oxygen consumption.

CHAPTER 13

10. $t_{obt} = -1.37$, and t_{crit} with 29 df = ±2.756. Since $|t_{obt}| < 2.756$, we retain H_0. It is reasonable to consider the sample a random sample from a population with $\mu = 85$.

11. $t_{obt} = 3.08$ and t_{crit} with 28 df = 2.467. Since $|t_{obt}| > 2.467$, we can reject H_0, which specifies that the sample is a random sample from a population with a mean = 72. Therefore, we can accept the hypothesis that the sample is a random sample from a population with a mean > 72.

12. $t_{obt} = 2.08$, and t_{crit} with 21 df = 1.721. Since $|t_{obt}| > 1.721$, we reject H_0. It is not reasonable to consider the sample a random sample from a normal population with $\mu = 38$.

13.

	95%	99%
a.	21.68–28.32	20.39–29.61
c.	28.28–32.92	27.45–33.75
d.	22.76–27.24	21.98–28.02

Increasing N decreases the width of confidence interval.

14. $t_{obt} = -1.57$, and $t_{crit} = \pm2.093$. Since $|t_{obt}| < 2.093$, you fail to reject H_0 and therefore cannot conclude that the student's technique shortens the duration of stay. The difference in conclusions is due to the greater sensitivity of the z test.

16. a. 18.34–21.66   b. 17.79–22.21.

17. a. $z_{obt} = 3.96$, and $z_{crit} = 1.645$. Since $|z_{obt}| > 1.645$, we reject H_0 and conclude that the amount of smoking in women appears to have increased in recent years. The professor was correct.   b. $t_{obt} = 3.67$, and $t_{crit} = 1.658$. Reject H_0; same conclusion as in part a.

18. $t_{obt} = 1.65$, and $t_{crit} = 1.796$. Since $|t_{obt}| < 1.796$, we retain H_0. We cannot conclude that middle-age men employed by the corporation have become fatter.

19. $t_{obt} = 0.98$, and $t_{crit} = \pm2.262$. Since $|t_{obt}| < 2.262$, we retain H_0. From these data, we cannot conclude that the graduates of the local business school get higher salaries for their first jobs than the national average.

20. a. 109.09–140.91   b. 103.14–146.86

21. $r = 0.63$. Reject H_0 since $r_{crit} = 0.5760$.

22. $r = 0.98$. Reject H_0 since $r_{crit} = 0.6319$.

23. a. $r_{obt} = 0.630$ b. $r_{crit} = 0.7067$. Retain H_0; correlation is not significant. No, ρ may actually differ from 0, and power may be too low to detect it.   c. $r_{crit} = 0.4438$. Reject H_0; correlation is significant. Power is greater with $N = 20$.

24. $r_{crit} = 0.5139$. Reject H_0; correlation is significant.

CHAPTER 14

13. a. The alternative hypothesis states that memory for pictures is superior to memory for words. $\mu_1 > \mu_2$.
b. The null hypothesis states that memory for pictures is not superior to memory for words. $\mu_1 \leq \mu_2$.
c. $t_{obt} = 1.86$, and $t_{crit} = 1.761$. Since $|t_{obt}| > 1.761$, you reject H_0 and conclude that memory for pictures is superior to memory for words.

15. $t_{obt} = 4.10$, and $t_{crit} = \pm 2.145$. Since $|t_{obt}| > 2.145$, you reject H_0 and conclude that newspaper advertising really does make a difference. It appears to increase cosmetics sales.

16. a. $t_{obt} = 4.09$, and $t_{crit} = \pm 2.306$. Since $|t_{obt}| > 2.306$, you reject H_0 and conclude that biofeedback training reduces tension headaches.   b. If the sampling distribution of $\bar{D}$ is not normally distributed, you cannot use the t test. However, you can use the sign test, since it does not assume anything about the shape of the scores. By using the sign test, $p(0, 1, 8, \text{ or } 9 \text{ pluses}) = 0.0392$. Since $0.0392 < 0.05$, you reject H_0, as before.

17. $t_{obt} = 3.50$, and $t_{crit} = \pm 2.306$. Since $|t_{obt}| > 2.306$, we reject H_0 and conclude that other factors such as attention, etc., have an effect on tension headaches. They appear to decrease them.

18. $t_{obt} = 2.83$, and $t_{crit} = \pm 2.120$. Since $|t_{obt}| > 2.120$, you reject H_0 and conclude that the decrease obtained with biofeedback training cannot be attributed solely to other factors, such as attention. The biofeedback training itself has an effect on tension headaches. It appears to decrease them.

19. $t_{obt} = 0.60$, and $t_{crit} = \pm 2.228$. Since $|t_{obt}| < 2.228$, we retain H_0. Based on these data, we cannot conclude that hiring part-time workers instead of full-time workers will affect productivity.

20. a. $t_{obt} = -2.83$, and $t_{crit} = \pm 2.131$. Since $|t_{obt}| > 2.131$, you reject H_0 and conclude that the clinician was right. Depression interferes with sleep.   b. If the populations from which the samples were taken are very nonnormal, you may decide the t test is not valid. However, you can use the Mann–Whitney U test, since it doesn't

depend on the shape of the population scores. $U_{obt} = 10.5$. Since $U_{obt} < 15$, you reject H_0, as before.

21. $t_{obt} = 2.11$ and $t_{crit} = \pm 2.201$. Since $|t_{obt}| < 2.201$, you retain H_0. You cannot conclude that early exposure to schooling affects IQ.

CHAPTER 15

13. a. $F_{crit} = 3.63$ b. $F_{crit} = 2.86$ c. $F_{crit} = 4.38$

14. $F_{obt} = 8.64$, and t_{obt} from Practice Problem 14.4 $= -2.94$. Therefore, $F = t^2$.

15. a.

Source	SS	df	s^2	F_{obt}
Between	1253.68	3	417.89	3.40
Within	3762.72	36	104.52	
Total	5016.40	39		

b. 4   c. 10   d. 2.86   e. Yes, the effect is significant.

16.

Source	SS	df	s^2	F_{obt}
Between	23.444	2	11.722	4.77
Within	36.833	15	2.456	
Total	60.278	17		

$F_{crit} = 3.68$. Since $F_{obt} > 3.68$, you reject H_0 and conclude that at least one of the cereals differs in sugar content.

18.

Source	SS	df	s^2	F_{obt}
Between	108.333	3	36.111	5.40
Within	133.667	20	6.683	
Total	242.000	23		

$F_{crit} = 3.10$. Since $F_{obt} > 3.10$, we reject H_0 and conclude that age affects memory.

19.

Source	SS	df	s^2	F_{obt}
Between	100.167	2	50.084	4.87
Within	92.500	9	10.278	
Total	192.667	11		

$F_{crit} = 4.26$. Since $F_{obt} > 4.26$, we reject H_0 and conclude that the batteries of at least one manufacturer differ regarding useful life.

20.

Source	SS	df	s^2	F_{obt}
Between	221.6	3	73.867	22.77
Within	116.8	36	3.244	
Total	338.4	39		

$F_{crit} = 2.86$. Since $F_{obt} > 2.86$, we reject H_0 and conclude that hormone X affects sexual behavior.

21. a. The treatments are equally effective; $\mu_1 = \mu_2 = \mu_3 = \mu_4$

b.

Source	SS	df	s^2	F_{obt}
Between	762.88	3	254.29	19.92
Within	574.90	36	15.97	
Total	1337.78			

$F_{obt} = 15.92$, $F_{crit} = 2.86$; reject H_0, affirm H_1.

CHAPTER 16

6. a.

	Breakfast Cereal		
Condition	A	B	C
$\overline{X}$	3.000	5.333	5.500
$\overline{X}_i - \overline{X}_j$		2.333	2.500
			0.167
Q_{obt}		3.65	3.91
			0.26
$Q_{crit} = 3.67$			

Reject H_0 for the comparison between cereals A and C. Cereal A is significantly lower in sugar content than cereal C. Retain H_0 for all other comparisons.

b.

	Breakfast Cereal		
Condition	A	B	C
$\overline{X}$	3.000	5.333	5.500
$\overline{X}_i - \overline{X}_j$		2.333	2.500
			0.167
Q_{obt}		3.65	3.91
			0.26
Q_{crit}		3.01	3.67
			3.01

Reject H_0 for the comparison of cereals B and C. Cereal A is significantly lower in sugar content than cereals B and C. Retain H_0 for the comparison between cereals B and C. We cannot conclude that they differ in sugar content.

8. a. $t_{obt} = 3.13$, and $t_{crit} = \pm 2.086$. Reject H_0 and conclude that the 60-year-old group is significantly different from the 30-year-old group.

b.

Condition	60 yr	30 yr	50 yr	40 yr
$\overline{X}$	8.833	13.500	13.667	14.000
$\overline{X}_i - \overline{X}_j$		4.667	4.834	5.167
			0.167	0.500
				0.333
Q_{obt}		4.42	4.58	4.90
			0.16	0.47
				0.32
Q_{crit}		2.95	3.58	3.96
			2.95	3.58
				2.95

Reject H_0 for all comparisons involving the 60-year-old group. This age group is significantly different from each of the other age groups. Retain H_0 for all other comparisons. It appears that memory begins to deteriorate somewhere between the ages of 50 and 60.

9.

	Battery		
Condition	C	B	A
$\overline{X}$	49.5	49.75	55.75
$\overline{X}_i - \overline{X}_j$		0.25	6.25
			6.00
Q_{obt}		0.16	3.90
			3.74
$Q_{crit} = 3.95$			

Reject H_0 for the comparisons between the batteries of manufacturer A and the other two manufacturers. Retain H_0 for the comparison between the batteries of manufacturers B and C. The batteries of manufacturer A have significantly longer life. On this basis, you recommend them over the batteries made by manufacturers B and C.

10. a. $t_{obt} = 7.20$, and $t_{crit} = \pm 2.029$. Reject H_0 and conclude that concentration 3 of hormone X significantly increases the number of matings.

b.

Condition	Concentration of Hormone X			
	0	1	2	3
$\overline{X}$	5.6	5.8	8.4	11.4
$\overline{X}_i - \overline{X}_j$		0.2	2.8	5.8
			2.6	5.6
				3.0
Q_{obt}		0.35	4.92	10.18
			4.56	9.83
				5.27
Q_{crit}		2.87	3.46	3.81
			2.87	3.46
				2.87

Reject H_0 for all comparisons except between the placebo and concentration 1. Thus, increasing the concentration of hormone X increases the number of matings. The failure to find a significant effect for concentration 1 was probably due to low power to detect a difference for this low level of concentration. Alternatively, there may be a threshold that must be exceeded before the hormone becomes effective.

c.

Condition	Concentration of Hormone X			
	0	1	2	3
$\overline{X}$	5.6	5.8	8.4	11.4
$\overline{X}_i - \overline{X}_j$		0.2	2.8	5.8
			2.6	5.6
				3.0
Q_{obt}		0.35	4.92	10.18
			4.56	9.83
				5.27
$Q_{crit} = 3.81$				

Same conclusion as in part **b**.

11.

Condition	Treatments			
	1	2	3	4
$\overline{X}$	22.80	11.60	16.00	20.90
$\overline{X}_i - \overline{X}_j$		11.20	6.80	1.90
			4.40	9.30
				4.90
Q_{obt}		8.86	5.38	1.50
			3.48	7.36
				3.88
$Q_{crit} = 3.79$				

Reject H_0 for cognitive restructuring (2) and assertiveness training (3) versus the placebo control group (1). Retain H_0 for exercise/nutrition (4) versus the placebo group (1). Reject H_0 for cognitive restructuring (2) and assertiveness training (3) versus exercise/nutrition (4). Retain H_0 for cognitive restructuring (2) versus assertiveness training (3).

CHAPTER 17

8. a. The different types of intervening material have the same effect on recall. $\mu_{a_1} = \mu_{a_2}$. The different amounts of repetition have the same effect on recall. $\mu_{b_1} = \mu_{b_2} = \mu_{b_3}$. There is no interaction between the number of repetitions and the type of intervening material in their effects on recall. With any main effects removed, $\mu_{a_1b_1} = \mu_{a_1b_2} = \mu_{a_1b_3} = \mu_{a_2b_1} = \mu_{a_2b_2} = \mu_{a_2b_3}$.

b.

Source	SS	df	s^2	F_{obt}	F_{crit}
Rows (intervening material)	121.000	1	121.000	41.41	4.17
Columns (number of repetitions)	176.167	2	88.084	30.14	3.32
Rows × columns	35.166	2	17.583	6.02	3.32
Within-cells	87.667	30	2.922		
Total	420.000	35			

Since in all cases $F_{obt} > F_{crit}$, H_0 is rejected for both main effects and the interaction effect. From the pattern of cell means it is apparent that (1) increasing the number of repetitions increases recall, (2) using nonsense syllable pairs for intervening material decreases recall, and (3) there is an interaction such that the lower the number of repetitions, the greater is the difference in effect between the two types of material.

10. a. For the concentrations administered, previous use of Drowson has no effect on its effectiveness. $\mu_{a_1} = \mu_{a_2}$. There is no difference between the placebo and the minimum recommended dosage of Drowson in their effects on insomnia. $\mu_{b_1} = \mu_{b_2}$. There is no interaction between the previous use of Drowson and the effect on insomnia of the two concentrations of Drowson. With any main effects removed, $\mu_{a_1b_1} = \mu_{a_1b_2} = \mu_{a_2b_1} = \mu_{a_2b_2}$.

b.

Source	SS	df	s^2	F_{obt}	F_{crit}
Rows (previous use)	639.031	1	639.031	11.63	4.20
Columns (concentration)	979.031	1	979.031	17.82	4.20
Rows × columns	472.782	1	472.782	8.61	4.20
Within-cells	1538.125	28	54.933		
Total	3628.969	31			

Since $F_{obt} > F_{crit}$ for each comparison, H_0 is rejected for both main effects and the interaction effect. From the pattern of cell means, it is apparent that Drowson promotes faster sleep onset in subjects who have had no previous use of the drug. However, the effect, if any, is much lower in chronic users, indicating that a tolerance to Drowson develops with chronic use.

CHAPTER 18

12. $\chi^2_{obt} = 0.80$. Since $\chi^2_{crit} = 3.841$, we retain H_0. These data do not support the hypothesis that the prevailing view is fat people are more jolly.

13. $\chi^2_{obt} = 21.63$. Since $\chi^2_{crit} = 3.841$, we reject H_0 and conclude that big city and small town dwellers differ in their helpfulness to strangers. Converting the frequencies to proportions, we can see that the small town dwellers were more helpful.

15. $\chi^2_{obt} = 13.28$, and $\chi^2_{crit} = 7.815$. Since $\chi^2_{obt} > 7.815$, we reject H_0 and conclude that the wrappings differ in their effect on sales. The manager should choose wrapping C.

16. $H_{obt} = 7.34$. Since $H_{crit} = 5.991$, we reject H_0 and conclude that at least one of the professions differs from at least one of the others.

17. $\chi^2_{obt} = 3.00$. Since $\chi^2_{crit} = 3.841$, we retain H_0. Based on these data, we cannot conclude that church attendance and educational level are related.

18. $\chi^2_{obt} = 3.56$, and $\chi^2_{crit} = 9.488$. Since $\chi^2_{obt} < 9.488$, we retain H_0. Yes, the advertising is misleading because the data do not show a significant difference among brands.

19. $\chi^2_{obt} = 12.73$, and $\chi^2_{crit} = 5.991$. Since $\chi^2_{obt} > 5.991$, we reject H_0 and conclude that there is a relationship between the amount of contact white housewives have with blacks and changes in their attitudes toward blacks. The contact in the integrated housing projects appears to have had a positive effect on the attitude of the white housewives.

20. $H_{obt} = 1.82$. Since $H_{crit} = 9.2110$, we must retain H_0. We cannot conclude that birth order affects assertiveness.

21. $\chi^2_{obt} = 29.57$, and $\chi^2_{crit} = 9.488$. Since $\chi^2_{obt} > 9.488$, we reject H_0. There is a relationship between gambling behavior and the different motives. Those high in power motivation appear to take the high risks more often. Most of the subjects with high power motivation placed high-risk bets, whereas the majority of those high in achievement motivation opted for medium-risk bets and the majority of those high in affiliation motivation chose the low-risk bets. These results are consistent with the views that (1) people with high power motivation will take high risks to achieve the attention and status that accompany such risk, (2) people high in achievement motivation will take medium risks to maximize the probability of having a sense of personal accomplishment, and (3) people with high affiliation motivation will take low risks to avoid competition and maximize the sense of belongingness.

22. $T_{obt} = 15$, and $T_{crit} = 17$. Since $T_{obt} < 17$, you reject H_0 and conclude that the film promotes more favorable attitudes toward major oil companies.

23. $T_{obt} = 1$, and $T_{crit} = 5$. Since $T_{obt} < 5$, you reject H_0 and conclude that biofeedback to relax frontalis muscle affects tension headaches. It appears to decrease them.

24. $T_{obt} = 14$, and $T_{crit} = 3$. Since $T_{obt} > 3$, you retain H_0. You cannot conclude that the pill affects blood pressure.

25. $H_{obt} = 10.16$, $H_{crit} = 5.991$. Reject H_0; affirm H_1. Sleep deprivation has an effect on the ability to maintain sustained attention.

26. $\chi^2_{obt} = 6.454$, $\chi^2_{crit} = 3.841$. Reject H_0; affirm H_1. There is a relationship between cohabitation and divorce. There is a significantly higher proportion of divorced couples among those that cohabited before marriage than among those that did not cohabit before marriage.

CHAPTER 19

11. a. The alternative hypothesis states that the four brands of scotch whiskey are not equal in preference among the scotch drinkers in New York City. b. The null hypothesis states that the four brands of scotch whiskey are equal in preference among the scotch drinkers in New York City. c. $\chi^2_{obt} = 2.72$, $\chi^2_{crit} = 7.815$. Since $\chi^2_{obt} < 7.815$, retain H_0. You cannot conclude that the scotch drinkers in New York City differ in their preference for the four brands of scotch whiskey.

12. a. The alternative hypothesis states that ACTH affects avoidance learning. $\mu_1 \neq \mu_2$. b. The null hypothesis states that ACTH has no effect on avoidance learning. $\mu_1 = \mu_2$. c. $t_{obt} = -3.24$, $t_{crit} = \pm 2.878$. Since $|t_{obt}| > 2.878$, reject H_0 and conclude that ACTH has an effect on avoidance learning. It appears to facilitate avoidance learning. d. You may be making a Type I er-

ror. The null hypothesis may be true and it has been rejected.   e. These results apply to the 100-day-old male rats living in the university vivarium at the time the sample was selected.   f. Yes, the Mann–Whitney U test   g. $U_{obt} = 17$, $U_{crit} = 16$; retain H_0. Mann–Whitney U test is not as powerful as the t test.

14. a. Exogenous thyroxin has no effect on activity. Therefore, $\mu_1 = \mu_2 = \mu_3 = \mu_4$.

b.

Source	SS	df	s^2	F_{obt}
Between	260.9	3	86.967	33.96
Within	92.2	36	2.561	
Total	353.1	39		

$F_{crit} = 2.86$. Since $F_{obt} > 2.86$, you reject H_0 and conclude that exogenous thyroxin affects activity level.   c. $t_{obt} = 8.94$, $t_{crit} = \pm 2.029$. Reject H_0 and conclude that there is a significant difference between high amounts of exogenous thyroxin and saline on activity level. Exogenous thyroxin appears to increase activity level.

d.

	Amount of Thyroxin			
Condition	*Zero,* 1	*Low,* 2	*Moderate,* 3	*High,* 4
$\overline{X}$	3.0	3.9	7.1	9.4
$\overline{X}_i - \overline{X}_j$		0.9	4.1	6.4
			3.2	5.5
Q_{obt}		1.78	8.10	12.65
			6.32	10.87
				4.54
$Q_{crit} = 3.81$				

Reject H_0 for all comparisons except between groups 1 and 2. Increases in the amount of exogenous thyroxin produce significantly higher levels of activity. The failure to find a significant difference between groups 1 and 2 is probably due to low power. Alternatively, there may be a threshold that must be exceeded before exogenous thyroxin becomes effective.

15. a. The alternative hypothesis states that dieting plus exercise is more effective in producing weight loss than dieting alone.   b. The null hypothesis states that dieting plus exercise is not more effective in producing weight loss than dieting alone.   c. Since it is not valid to use the t test for correlated groups, the next most sensitive test is the Wilcoxon signed ranks test. $T_{obt} = 10.5$, $T_{crit} = 17$. Since $T_{obt} < 17$, reject H_0 and conclude that dieting plus exercise is more effective than dieting alone in producing weight loss.

16. a. sign test   b. $p(8, 9, 10, 11, \text{ or } 12 \text{ pluses}) = 0.1937$. Since the obtained probability is greater than alpha, you retain H_0.   c. The Wilcoxon signed ranks test is more powerful than the sign test.

17. a. The null hypothesis states that there is no relationship between gender and time-of-day preference for having intercourse.   b. $\chi^2_{obt} = 2.14$, $\chi^2_{crit} = 3.841$. Since $\chi^2_{obt} < 3.841$, retain H_0. You cannot conclude that there is a relationship between gender and time-of-day preference for having intercourse.

18. a. The null hypothesis states that food deprivation (hunger) has no effect on the number of food-related objects reported. Therefore, $\mu_1 = \mu_2 = \mu_3$.

b.

Source	SS	df	s^2	F_{obt}
Between	256.083	2	128.042	11.85
Within	226.875	21	10.804	
Total	482.958	23		

$F_{crit} = 3.47$. Since $F_{obt} > 3.47$, reject H_0 and conclude that food deprivation has an effect on the number of food-related objects reported.

c.

	Food Deprivation		
Condition	*1 hr,* 1	*4 hr,* 2	*12 hr,* 3
$\overline{X}$	4.750	8.625	12.750
$\overline{X}_i - \overline{X}_j$		3.875	8.000
			4.125
Q_{obt}		3.33	6.88
			3.55
Q_{crit}		2.94	3.57
			2.94

Reject H_0 for all comparisons. All three conditions differ significantly from each other. Increasing the number of hours from eating results in an increase in the number of food-related objects reported.

19. $t_{obt} = 10.78$, $t_{crit} = 2.500$. Since $|t_{obt}| > 2.500$, reject H_0. Yes, the engineer is correct in her opinion. The new process results in significantly longer life for TV picture tubes.

20. a. The alternative hypothesis states that alcohol has an effect on aggressiveness, $\mu_1 \neq \mu_2$.   b. The null hypothesis states that alcohol has no effect on aggressiveness. $\mu_1 = \mu_2$.   c. $t_{obt} = -3.93$, $t_{crit} = \pm 2.131$. Since $|t_{obt}| > 2.131$, reject H_0 and conclude that alcohol has an effect on aggressiveness. It appears to increase aggressiveness.   d. Yes, the Mann–Whitney U

test   e. $U_{obt} = 6.5$, $U_{crit} = 15$. Reject H_0, affirm H_1. This is the same conclusion reached previously with the t test.

21. a. $r_{obt} = 0.70$.   b. $r_{crit} = 0.5139$. Since $|r_{obt}| > 0.5139$, reject H_0. The correlation is significant.   c. $r^2 = 0.48$.   d. Yes, although it is clear that there is still a lot of variability unaccounted for.

22. a. The alternative hypothesis states that smoking affects heart rate. $\mu_D \neq 0$.   b. The null hypothesis states that smoking has no effect on heart rate. $\mu_D = 0$.   c. $t_{obt} = -3.40$, $t_{crit} = \pm 2.262$. Since $|t_{obt}| > 2.262$, reject H_0 and conclude that smoking affects heart rate. It appears to increase heart rate.

23. a. The null hypothesis states that there is no relationship between the occupations and the course of action that is favored.   b. $\chi^2_{obt} = 13.79$, $\chi^2_{crit} = 5.991$. Since $\chi^2_{obt} > 5.991$, reject H_0 and conclude that there is a relationship between the occupations and the course of action that is favored. From the proportions shown in the sample, business is in favor of letting the oil price rise, the homemakers favor gasoline rationing, and labor is fairly evenly divided.

24. a. The null hypothesis states that men and women do not differ in logical reasoning ability.   b. $U_{obt} = 25.5$, $U'_{obt} = 37.5$, and $U_{crit} = 12$. Since $U_{obt} > 12$, retain H_0. You cannot conclude that men and women differ in logical reasoning ability.   c. You may be making a Type II error. The null hypothesis may be false and you retained it.

25. 24.58–33.42

26. $H_{obt} = 1.46$. Since $H_{crit} = 5.991$, we must retain H_0. We cannot conclude that physical science professors are more authoritarian than social science professors.

27.

Source	SS	df	s^2	F_{obt}	F_{crit}
Rows (activity)	17.633	1	17.633	11.76*	4.26
Columns (time)	28.800	2	14.400	9.60*	3.40
Rows × columns	0.267	2	0.133	0.09	3.40
Within-cells	36.000	24	1.500		
Total	82.700	29			

* Since $F_{obt} > F_{crit}$ for the rows and columns effects, we reject H_0 for the main effects. We must retain H_0 for the interaction effect. It appears that performance is affected differently by at least one of the activity conditions and by the time of day when it is conducted. It appears that napping and afternoon produce superior performance.

D TABLES

Acknowledgments

 ACKNOWLEDGMENTS

The tables contained in this appendix have been adapted with permission from the following sources:

Table A R. Clarke, A. Coladarch, and J. Caffrey, *Statistical Reasoning and Procedures,* Charles E. Merrill Publishers, Columbus, Ohio, 1965, Appendix 2.

Table B R. S. Burington and D. C. May, *Handbook of Probability and Statistics with Tables,* 2nd ed., McGraw-Hill Book Company, New York, 1970.

Table C H. B. Mann and D. R. Whitney, "On a Test of Whether One of Two Random Variables Is Stochastically Larger Than the Other," *Annals of Mathematical Statistics,* 18 (1947), 50–60, and D. Auble, "Extended Tables for the Mann–Whitney Statistic," *Bulletin of the Institute of Educational Research at Indiana University, 1,* No. 2 (1953), as used in Runyon and Haber, *Fundamentals of Behavioral Statistics,* 3rd ed., Addison-Wesley Publishing Company, Inc., Reading, Mass., 1976.

Table D   Fisher and Yates, *Statistical Tables for Biological, Agricultural, and Medical Research,* Longman Group Ltd., London (previously published by Oliver & Boyd Ltd., Edinburgh), 1974, Table III.

Table E   Fisher and Yates, *Statistical Tables for Biological, Agricultural, and Medical Research,* Longman Group Ltd., London (previously published by Oliver & Boyd Ltd., Edinburgh), 1974, Table VII.

Table F   G. W. Snedecor, *Statistical Methods,* 5th ed., Iowa State University Press, Ames, 1956.

Table G   E. S. Pearson and H. O. Hartley, eds., *Biometrika Tables for Statisticians,* Vol. 1, 3rd ed., New York, Cambridge University Press, 1966, Table 29.

Table H   Fisher and Yates, *Statistical Tables for Biological, Agricultural, and Medical Research,* Longman Group Ltd., London (previously published by Oliver & Boyd Ltd., Edinburgh), 1974, Table IV.

Table I   F. Wilcoxon, S. Katte, and R. A. Wilcox, *Critical Values and Probability Levels for the Wilcoxon Rank Sum Test and the Wilcoxon Signed Rank Test,* American Cyanamid Co., New York, 1963, and F. Wilcoxon and R. A. Wilcox, *Some Rapid Approximate Statistical Procedures,* Lederle Laboratories, New York, 1964, as used in Runyon and Haber, *Fundamentals of Behavioral Statistics,* 3rd ed., Addison-Wesley Publishing Company, Inc., Reading, Mass., 1976

Table J   RAND Corporation, *A Million Random Digits,* Free Press of Glencoe, Glencoe, Ill., 1955.

Table A

Areas Under the
Normal Curve

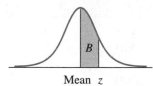

Mean z

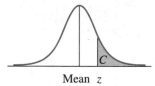

Mean z

Column A gives the positive *z*
score.

Column B gives the area between the mean and *z*. Since
the curve is symmetrical, areas
for negative *z* scores are the
same as for positive ones.

Column C gives the area that
is beyond *z*.

z (A)	Area Between Mean and z (B)	Area Beyond z (C)	z (A)	Area Between Mean and z (B)	Area Beyond z (C)
0.00	.0000	.5000	0.38	.1480	.3520
0.01	.0040	.4960	0.39	.1517	.3483
0.02	.0080	.4920			
0.03	.0120	.4880	0.40	.1554	.3446
0.04	.0160	.4840	0.41	.1591	.3409
			0.42	.1628	.3372
0.05	.0199	.4801	0.43	.1664	.3336
0.06	.0239	.4761	0.44	.1700	.3300
0.07	.0279	.4721			
0.08	.0319	.4681	0.45	.1736	.3264
0.09	.0359	.4641	0.46	.1772	.3228
			0.47	.1808	.3192
0.10	.0398	.4602	0.48	.1844	.3156
0.11	.0438	.4562	0.49	.1879	.3121
0.12	.0478	.4522			
0.13	.0517	.4483	0.50	.1915	.3085
0.14	.0557	.4443	0.51	.1950	.3050
			0.52	.1985	.3015
0.15	.0596	.4404	0.53	.2019	.2981
0.16	.0636	.4364	0.54	.2054	.2946
0.17	.0675	.4325			
0.18	.0714	.4286	0.55	.2088	.2912
0.19	.0753	.4247	0.56	.2123	.2877
			0.57	.2157	.2843
0.20	.0793	.4207	0.58	.2190	.2810
0.21	.0832	.4168	0.59	.2224	.2776
0.22	.0871	.4129			
0.23	.0910	.4090	0.60	.2257	.2743
0.24	.0948	.4052	0.61	.2291	.2709
			0.62	.2324	.2676
0.25	.0987	.4013	0.63	.2357	.2643
0.26	.1026	.3974	0.64	.2389	.2611
0.27	.1064	.3936			
0.28	.1103	.3897	0.65	.2422	.2578
0.29	.1141	.3859	0.66	.2454	.2546
			0.67	.2486	.2514
0.30	.1179	.3821	0.68	.2517	.2483
0.31	.1217	.3783	0.69	.2549	.2451
0.32	.1255	.3745			
0.33	.1293	.3707	0.70	.2580	.2420
0.34	.1331	.3669	0.71	.2611	.2389
			0.72	.2642	.2358
0.35	.1368	.3632	0.73	.2673	.2327
0.36	.1406	.3594	0.74	.2704	.2296
0.37	.1443	.3557			

TABLE A (Continued)

Areas Under the
Normal Curve

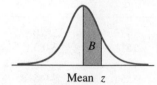

Mean z

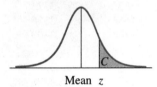

Mean z

Column A gives the positive z
score.

Column B gives the area between the mean and z. Since
the curve is symmetrical, areas
for negative z scores are the
same as for positive ones.

Column C gives the area that
is beyond z.

z / A	Area Between Mean and z / B	Area Beyond z / C	z / A	Area Between Mean and z / B	Area Beyond z / C
0.75	.2734	.2266	1.08	.3599	.1401
0.76	.2764	.2236	1.09	.3621	.1379
0.77	.2794	.2206			
0.78	.2823	.2177	1.10	.3643	.1357
0.79	.2852	.2148	1.11	.3665	.1335
			1.12	.3686	.1314
0.80	.2881	.2119	1.13	.3708	.1292
0.81	.2910	.2090	1.14	.3729	.1271
0.82	.2939	.2061			
0.83	.2967	.2033	1.15	.3749	.1251
0.84	.2995	.2005	1.16	.3770	.1230
			1.17	.3790	.1210
0.85	.3023	.1977	1.18	.3810	.1190
0.86	.3051	.1949	1.19	.3830	.1170
0.87	.3078	.1922			
0.88	.3106	.1894	1.20	.3849	.1151
0.89	.3133	.1867	1.21	.3869	.1131
			1.22	.3888	.1112
0.90	.3159	.1841	1.23	.3907	.1093
0.91	.3186	.1814	1.24	.3925	.1075
0.92	.3212	.1788			
0.93	.3238	.1762	1.25	.3944	.1056
0.94	.3264	.1736	1.26	.3962	.1038
			1.27	.3980	.1020
0.95	.3289	.1711	1.28	.3997	.1003
0.96	.3315	.1685	1.29	.4015	.0985
0.97	.3340	.1660			
0.98	.3365	.1635	1.30	.4032	.0968
0.99	.3389	.1611	1.31	.4049	.0951
			1.32	.4066	.0934
1.00	.3413	.1587	1.33	.4082	.0918
1.01	.3438	.1562	1.34	.4099	.0901
1.02	.3461	.1539			
1.03	.3485	.1515	1.35	.4115	.0885
1.04	.3508	.1492	1.36	.4131	.0869
			1.37	.4147	.0853
1.05	.3531	.1469	1.38	.4162	.0838
1.06	.3554	.1446	1.39	.4177	.0823
1.07	.3577	.1423			

TABLE A (*Continued*)

Areas Under the
Normal Curve

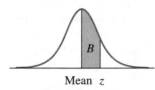

Mean z

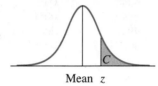

Mean z

Column A gives the positive z score.

Column B gives the area between the mean and z. Since the curve is symmetrical, areas for negative z scores are the same as for positive ones.

Column C gives the area that is beyond z.

z A	Area Between Mean and z B	Area Beyond z C	z A	Area Between Mean and z B	Area Beyond z C
1.40	.4192	.0808	1.78	.4625	.0375
1.41	.4207	.0793	1.79	.4633	.0367
1.42	.4222	.0778			
1.43	.4236	.0764	1.80	.4641	.0359
1.44	.4251	.0749	1.81	.4649	.0351
			1.82	.4656	.0344
1.45	.4265	.0735	1.83	.4664	.0336
1.46	.4279	.0721	1.84	.4671	.0329
1.47	.4292	.0708			
1.48	.4306	.0694	1.85	.4678	.0322
1.49	.4319	.0681	1.86	.4686	.0314
			1.87	.4693	.0307
1.50	.4332	.0668	1.88	.4699	.0301
1.51	.4345	.0655	1.89	.4706	.0294
1.52	.4357	.0643			
1.53	.4370	.0630	1.90	.4713	.0287
1.54	.4382	.0618	1.91	.4719	.0281
			1.92	.4726	.0274
1.55	.4394	.0606	1.93	.4732	.0268
1.56	.4406	.0594	1.94	.4738	.0262
1.57	.4418	.0582			
1.58	.4429	.0571	1.95	.4744	.0256
1.59	.4441	.0559	1.96	.4750	.0250
			1.97	.4756	.0244
1.60	.4452	.0548	1.98	.4761	.0239
1.61	.4463	.0537	1.99	.4767	.0233
1.62	.4474	.0526			
1.63	.4484	.0516	2.00	.4772	.0228
1.64	.4495	.0505	2.01	.4778	.0222
			2.02	.4783	.0217
1.65	.4505	.0495	2.03	.4788	.0212
1.66	.4515	.0485	2.04	.4793	.0207
1.67	.4525	.0475			
1.68	.4535	.0465	2.05	.4798	.0202
1.69	.4545	.0455	2.06	.4803	.0197
			2.07	.4808	.0192
1.70	.4554	.0446	2.08	.4812	.0188
1.71	.4564	.0436	2.09	.4817	.0183
1.72	.4573	.0427			
1.73	.4582	.0418	2.10	.4821	.0179
1.74	.4591	.0409	2.11	.4826	.0174
			2.12	.4830	.0170
1.75	.4599	.0401	2.13	.4834	.0166
1.76	.4608	.0392	2.14	.4838	.0162
1.77	.4616	.0384			

TABLE A (*Continued*)

Areas Under the
Normal Curve

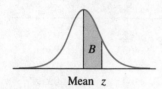

Mean z

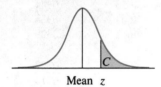

Mean z

Column A gives the positive z score.

Column B gives the area between the mean and z. Since the curve is symmetrical, areas for negative z scores are the same as for positive ones.

Column C gives the area that is beyond z.

z A	Area Between Mean and z B	Area Beyond z C	z A	Area Between Mean and z B	Area Beyond z C
2.15	.4842	.0158	2.52	.4941	.0059
2.16	.4846	.0154	2.53	.4943	.0057
2.17	.4850	.0150	2.54	.4945	.0055
2.18	.4854	.0146			
2.19	.4857	.0143	2.55	.4946	.0054
			2.56	.4948	.0052
2.20	.4861	.0139	2.57	.4949	.0051
2.21	.4864	.0136	2.58	.4951	.0049
2.22	.4868	.0132	2.59	.4952	.0048
2.23	.4871	.0129			
2.24	.4875	.0125	2.60	.4953	.0047
			2.61	.4955	.0045
2.25	.4878	.0122	2.62	.4956	.0044
2.26	.4881	.0119	2.63	.4957	.0043
2.27	.4884	.0116	2.64	.4959	.0041
2.28	.4887	.0113			
2.29	.4890	.0110	2.65	.4960	.0040
			2.66	.4961	.0039
2.30	.4893	.0107	2.67	.4962	.0038
2.31	.4896	.0104	2.68	.4963	.0037
2.32	.4898	.0102	2.69	.4964	.0036
2.33	.4901	.0099			
2.34	.4904	.0096	2.70	.4965	.0035
			2.71	.4966	.0034
2.35	.4906	.0094	2.72	.4967	.0033
2.36	.4909	.0091	2.73	.4968	.0032
2.37	.4911	.0089	2.74	.4969	.0031
2.38	.4913	.0087			
2.39	.4916	.0084	2.75	.4970	.0030
			2.76	.4971	.0029
2.40	.4918	.0082	2.77	.4972	.0028
2.41	.4920	.0080	2.78	.4973	.0027
2.42	.4922	.0078	2.79	.4974	.0026
2.43	.4925	.0075			
2.44	.4927	.0073	2.80	.4974	.0026
			2.81	.4975	.0025
			2.82	.4976	.0024
2.45	.4929	.0071	2.83	.4977	.0023
2.46	.4931	.0069	2.84	.4977	.0023
2.47	.4932	.0068			
2.48	.4934	.0066	2.85	.4978	.0022
2.49	.4936	.0064	2.86	.4979	.0021
			2.87	.4979	.0021
2.50	.4938	.0062	2.88	.4980	.0020
2.51	.4940	.0060	2.89	.4981	.0019

TABLE A (*Continued*)

Areas Under the
Normal Curve

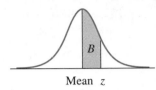

Mean   z

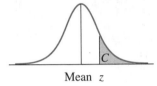

Mean   z

Column A gives the positive z
score.

Column B gives the area between the mean and z. Since
the curve is symmetrical, areas
for negative z scores are the
same as for positive ones.

Column C gives the area that
is beyond z.

z	Area Between Mean and z	Area Beyond z	z	Area Between Mean and z	Area Beyond z
A	B	C	A	B	C
2.90	.4981	.0019	3.10	.4990	.0010
2.91	.4982	.0018	3.11	.4991	.0009
2.92	.4982	.0018	3.12	.4991	.0009
2.93	.4983	.0017	3.13	.4991	.0009
2.94	.4984	.0016	3.14	.4992	.0008
2.95	.4984	.0016	3.15	.4992	.0008
2.96	.4985	.0015	3.16	.4992	.0008
2.97	.4985	.0015	3.17	.4992	.0008
2.98	.4986	.0014	3.18	.4993	.0007
2.99	.4986	.0014	3.19	.4993	.0007
3.00	.4987	.0013	3.20	.4993	.0007
3.01	.4987	.0013	3.21	.4993	.0007
3.02	.4987	.0013	3.22	.4994	.0006
3.03	.4988	.0012	3.23	.4994	.0006
3.04	.4988	.0012	3.24	.4994	.0006
3.05	.4989	.0011	3.30	.4995	.0005
3.06	.4989	.0011	3.40	.4997	.0003
3.07	.4989	.0011	3.50	.4998	.0002
3.08	.4990	.0010	3.60	.4998	.0002
3.09	.4990	.0010	3.70	.4999	.0001

TABLE B

Binomial Distribution

N	No. of P or Q Events	.05	.10	.15	.20	.25	.30	.35	.40	.45	.50
1	0	.9500	.9000	.8500	.8000	.7500	.7000	.6500	.6000	.5500	.5000
	1	.0500	.1000	.1500	.2000	.2500	.3000	.3500	.4000	.4500	.5000
2	0	.9025	.8100	.7225	.6400	.5625	.4900	.4225	.3600	.3025	.2500
	1	.0950	.1800	.2550	.3200	.3750	.4200	.4550	.4800	.4950	.5000
	2	.0025	.0100	.0225	.0400	.0625	.0900	.1225	.1600	.2025	.2500
3	0	.8574	.7290	.6141	.5120	.4219	.3430	.2746	.2160	.1664	.1250
	1	.1354	.2430	.3251	.3840	.4219	.4410	.4436	.4320	.4084	.3750
	2	.0071	.0270	.0574	.0960	.1406	.1890	.2389	.2880	.3341	.3750
	3	.0001	.0010	.0034	.0080	.0156	.0270	.0429	.0640	.0911	.1250
4	0	.8145	.6561	.5220	.4096	.3164	.2401	.1785	.1296	.0915	.0625
	1	.1715	.2916	.3685	.4096	.4219	.4116	.3845	.3456	.2995	.2500
	2	.0135	.0486	.0975	.1536	.2109	.2646	.3105	.3456	.3675	.3750
	3	.0005	.0036	.0115	.0256	.0469	.0756	.1115	.1536	.2005	.2500
	4	.0000	.0001	.0005	.0016	.0039	.0081	.0150	.0256	.0410	.0625
5	0	.7738	.5905	.4437	.3277	.2373	.1681	.1160	.0778	.0503	.0312
	1	.2036	.3280	.3915	.4096	.3955	.3602	.3124	.2592	.2059	.1562
	2	.0214	.0729	.1382	.2048	.2637	.3087	.3364	.3456	.3369	.3125
	3	.0011	.0081	.0244	.0512	.0879	.1323	.1811	.2304	.2757	.3125
	4	.0000	.0004	.0022	.0064	.0146	.0284	.0488	.0768	.1128	.1562
	5	.0000	.0000	.0001	.0003	.0010	.0024	.0053	.0102	.0185	.0312
6	0	.7351	.5314	.3771	.2621	.1780	.1176	.0754	.0467	.0277	.0156
	1	.2321	.3543	.3993	.3932	.3560	.3025	.2437	.1866	.1359	.0938
	2	.0305	.0984	.1762	.2458	.2966	.3241	.3280	.3110	.2780	.2344
	3	.0021	.0146	.0415	.0819	.1318	.1852	.2355	.2765	.3032	.3125
	4	.0001	.0012	.0055	.0154	.0330	.0595	.0951	.1382	.1861	.2344
	5	.0000	.0001	.0004	.0015	.0044	.0102	.0205	.0369	.0609	.0938
	6	.0000	.0000	.0000	.0001	.0002	.0007	.0018	.0041	.0083	.0156
7	0	.6983	.4783	.3206	.2097	.1335	.0824	.0490	.0280	.0152	.0078
	1	.2573	.3720	.3960	.3670	.3115	.2471	.1848	.1306	.0872	.0547
	2	.0406	.1240	.2097	.2753	.3115	.3177	.2985	.2613	.2140	.1641
	3	.0036	.0230	.0617	.1147	.1730	.2269	.2679	.2903	.2918	.2734
	4	.0002	.0026	.0109	.0287	.0577	.0972	.1442	.1935	.2388	.2734
	5	.0000	.0002	.0012	.0043	.0115	.0250	.0466	.0774	.1172	.1641
	6	.0000	.0000	.0001	.0004	.0013	.0036	.0084	.0172	.0320	.0547
	7	.0000	.0000	.0000	.0000	.0001	.0002	.0006	.0016	.0037	.0078
8	0	.6634	.4305	.2725	.1678	.1001	.0576	.0319	.0168	.0084	.0039
	1	.2793	.3826	.3847	.3355	.2670	.1977	.1373	.0896	.0548	.0312
	2	.0515	.1488	.2376	.2936	.3115	.2965	.2587	.2090	.1569	.1094
	3	.0054	.0331	.0839	.1468	.2076	.2541	.2786	.2787	.2568	.2188
	4	.0004	.0046	.0185	.0459	.0865	.1361	.1875	.2322	.2627	.2734
	5	.0000	.0004	.0026	.0092	.0231	.0467	.0808	.1239	.1719	.2188
	6	.0000	.0000	.0002	.0011	.0038	.0100	.0217	.0413	.0703	.1094
	7	.0000	.0000	.0000	.0001	.0004	.0012	.0033	.0079	.0164	.0312
	8	.0000	.0000	.0000	.0000	.0000	.0001	.0002	.0007	.0017	.0039

P or *Q*

TABLE B (Continued)

Binomial Distribution

N	No. of P or Q Events	P or Q									
		.05	.10	.15	.20	.25	.30	.35	.40	.45	.50
9	0	.6302	.3874	.2316	.1342	.0751	.0404	.0277	.0101	.0046	.0020
	1	.2985	.3874	.3679	.3020	.2253	.1556	.1004	.0605	.0339	.0176
	2	.0629	.1722	.2597	.3020	.3003	.2668	.2162	.1612	.1110	.0703
	3	.0077	.0446	.1069	.1762	.2336	.2668	.2716	.2508	.2119	.1641
	4	.0006	.0074	.0283	.0661	.1168	.1715	.2194	.2508	.2600	.2461
	5	.0000	.0008	.0050	.0165	.0389	.0735	.1181	.1672	.2128	.2461
	6	.0000	.0001	.0006	.0028	.0087	.0210	.0424	.0743	.1160	.1641
	7	.0000	.0000	.0000	.0003	.0012	.0039	.0098	.0212	.0407	.0703
	8	.0000	.0000	.0000	.0000	.0001	.0004	.0013	.0035	.0083	.0176
	9	.0000	.0000	.0000	.0000	.0000	.0000	.0001	.0003	.0008	.0020
10	0	.5987	.3487	.1969	.1074	.0563	.0282	.0135	.0060	.0025	.0010
	1	.3151	.3874	.3474	.2684	.1877	.1211	.0725	.0403	.0207	.0098
	2	.0746	.1937	.2759	.3020	.2816	.2335	.1757	.1209	.0763	.0439
	3	.0105	.0574	.1298	.2013	.2503	.2668	.2522	.2150	.1665	.1172
	4	.0010	.0112	.0401	.0881	.1460	.2001	.2377	.2508	.2384	.2051
	5	.0001	.0015	.0085	.0264	.0584	.1029	.1536	.2007	.2340	.2461
	6	.0000	.0001	.0012	.0055	.0162	.0368	.0689	.1115	.1596	.2051
	7	.0000	.0000	.0001	.0008	.0031	.0090	.0212	.0425	.0746	.1172
	8	.0000	.0000	.0000	.0001	.0004	.0014	.0043	.0106	.0229	.0439
	9	.0000	.0000	.0000	.0000	.0000	.0001	.0005	.0016	.0042	.0098
	10	.0000	.0000	.0000	.0000	.0000	.0000	.0000	.0001	.0003	.0010
11	0	.5688	.3138	.1673	.0859	.0422	.0198	.0088	.0036	.0014	.0005
	1	.3293	.3835	.3248	.2362	.1549	.0932	.0518	.0266	.0125	.0054
	2	.0867	.2131	.2866	.2953	.2581	.1998	.1395	.0887	.0513	.0269
	3	.0137	.0710	.1517	.2215	.2581	.2568	.2254	.1774	.1259	.0806
	4	.0014	.0158	.0536	.1107	.1721	.2201	.2428	.2365	.2060	.1611
	5	.0001	.0025	.0132	.0388	.0803	.1231	.1830	.2207	.2360	.2256
	6	.0000	.0003	.0023	.0097	.0268	.0566	.0985	.1471	.1931	.2256
	7	.0000	.0000	.0003	.0017	.0064	.0173	.0379	.0701	.1128	.1611
	8	.0000	.0000	.0000	.0002	.0011	.0037	.0102	.0234	.0462	.0806
	9	.0000	.0000	.0000	.0000	.0001	.0005	.0018	.0052	.0126	.0269
	10	.0000	.0000	.0000	.0000	.0000	.0000	.0002	.0007	.0021	.0054
	11	.0000	.0000	.0000	.0000	.0000	.0000	.0000	.0000	.0002	.0005
12	0	.5404	.2824	.1422	.0687	.0317	.0138	.0057	.0022	.0008	.0002
	1	.3413	.3766	.3012	.2062	.1267	.0712	.0368	.0174	.0075	.0029
	2	.0988	.2301	.2924	.2835	.2323	.1678	.1088	.0639	.0339	.0161
	3	.0173	.0852	.1720	.2362	.2581	.2397	.1954	.1419	.0923	.0537
	4	.0021	.0213	.0683	.1329	.1936	.2311	.2367	.2128	.1700	.1208
	5	.0002	.0038	.0193	.0532	.1032	.1585	.2039	.2270	.2225	.1934
	6	.0000	.0005	.0040	.0155	.0401	.0792	.1281	.1766	.2124	.2256
	7	.0000	.0000	.0006	.0033	.0115	.0291	.0591	.1009	.1489	.1934
	8	.0000	.0000	.0001	.0005	.0024	.0078	.0199	.0420	.0762	.1208
	9	.0000	.0000	.0000	.0001	.0004	.0015	.0048	.0125	.0277	.0537
	10	.0000	.0000	.0000	.0000	.0000	.0002	.0008	.0025	.0068	.0161
	11	.0000	.0000	.0000	.0000	.0000	.0000	.0001	.0003	.0010	.0029
	12	.0000	.0000	.0000	.0000	.0000	.0000	.0000	.0000	.0001	.0002

Binomial Distribution

N	No. of P or Q Events	.05	.10	.15	.20	.25	.30	.35	.40	.45	.50
13	0	.5133	.2542	.1209	.0550	.0238	.0097	.0037	.0013	.0004	.0001
	1	.3512	.3672	.2774	.1787	.1029	.0540	.0259	.0113	.0045	.0016
	2	.1109	.2448	.2937	.2680	.2059	.1388	.0836	.0453	.0220	.0095
	3	.0214	.0997	.1900	.2457	.2517	.2181	.1651	.1107	.0660	.0349
	4	.0028	.0277	.0838	.1535	.2097	.2337	.2222	.1845	.1350	.0873
	5	.0003	.0055	.0266	.0691	.1258	.1803	.2154	.2214	.1989	.1571
	6	.0000	.0008	.0063	.0230	.0559	.1030	.1546	.1968	.2169	.2095
	7	.0000	.0001	.0011	.0058	.0186	.0442	.0833	.1312	.1775	.2095
	8	.0000	.0000	.0001	.0011	.0047	.0142	.0336	.0656	.1089	.1571
	9	.0000	.0000	.0000	.0001	.0009	.0034	.0101	.0243	.0495	.0873
	10	.0000	.0000	.0000	.0000	.0001	.0006	.0022	.0065	.0162	.0349
	11	.0000	.0000	.0000	.0000	.0000	.0001	.0003	.0012	.0036	.0095
	12	.0000	.0000	.0000	.0000	.0000	.0000	.0000	.0001	.0005	.0016
	13	.0000	.0000	.0000	.0000	.0000	.0000	.0000	.0000	.0000	.0001
14	0	.4877	.2288	.1028	.0440	.0178	.0068	.0024	.0008	.0002	.0001
	1	.3593	.3559	.2539	.1539	.0832	.0407	.0181	.0073	.0027	.0009
	2	.1229	.2570	.2912	.2501	.1802	.1134	.0634	.0317	.0141	.0056
	3	.0259	.1142	.2056	.2501	.2402	.1943	.1366	.0845	.0462	.0222
	4	.0037	.0349	.0998	.1720	.2202	.2290	.2022	.1549	.1040	.0611
	5	.0004	.0078	.0352	.0860	.1468	.1963	.2178	.2066	.1701	.1222
	6	.0000	.0013	.0093	.0322	.0734	.1262	.1759	.2066	.2088	.1833
	7	.0000	.0002	.0019	.0092	.0280	.0618	.1082	.1574	.1952	.2095
	8	.0000	.0000	.0003	.0020	.0082	.0232	.0510	.0918	.1398	.1833
	9	.0000	.0000	.0000	.0003	.0018	.0066	.0183	.0408	.0762	.1222
	10	.0000	.0000	.0000	.0000	.0003	.0014	.0049	.0136	.0312	.0611
	11	.0000	.0000	.0000	.0000	.0000	.0002	.0010	.0033	.0093	.0222
	12	.0000	.0000	.0000	.0000	.0000	.0000	.0001	.0005	.0019	.0056
	13	.0000	.0000	.0000	.0000	.0000	.0000	.0000	.0001	.0002	.0009
	14	.0000	.0000	.0000	.0000	.0000	.0000	.0000	.0000	.0000	.0001
15	0	.4633	.2059	.0874	.0352	.0134	.0047	.0016	.0005	.0001	.0000
	1	.3658	.3432	.2312	.1319	.0668	.0305	.0126	.0047	.0016	.0005
	2	.1348	.2669	.2856	.2309	.1559	.0916	.0476	.0219	.0090	.0032
	3	.0307	.1285	.2184	.2501	.2252	.1700	.1110	.0634	.0318	.0139
	4	.0049	.0428	.1156	.1876	.2252	.2186	.1792	.1268	.0780	.0417
	5	.0006	.0105	.0449	.1032	.1651	.2061	.2123	.1859	.1404	.0916
	6	.0000	.0019	.0132	.0430	.0917	.1472	.1906	.2066	.1914	.1527
	7	.0000	.0003	.0030	.0138	.0393	.0811	.1319	.1771	.2013	.1964
	8	.0000	.0000	.0005	.0035	.0131	.0348	.0710	.1181	.1647	.1964
	9	.0000	.0000	.0001	.0007	.0034	.0116	.0298	.0612	.1048	.1527
	10	.0000	.0000	.0000	.0001	.0007	.0030	.0096	.0245	.0515	.0916
	11	.0000	.0000	.0000	.0000	.0001	.0006	.0024	.0074	.0191	.0417
	12	.0000	.0000	.0000	.0000	.0000	.0001	.0004	.0016	.0052	.0139
	13	.0000	.0000	.0000	.0000	.0000	.0000	.0001	.0003	.0010	.0032
	14	.0000	.0000	.0000	.0000	.0000	.0000	.0000	.0000	.0001	.0005
	15	.0000	.0000	.0000	.0000	.0000	.0000	.0000	.0000	.0000	.0000
16	0	.4401	.1853	.0743	.0281	.0100	.0033	.0010	.0003	.0001	.0000
	1	.3706	.3294	.2097	.1126	.0535	.0228	.0087	.0030	.0009	.0002
	2	.1463	.2745	.2775	.2111	.1336	.0732	.0353	.0150	.0056	.0018

N	No. of P or Q Events	.05	.10	.15	.20	.25	.30	.35	.40	.45	.50
16	3	.0359	.1423	.2285	.2463	.2079	.1465	.0888	.0468	.0215	.0085
	4	.0061	.0514	.1311	.2001	.2252	.2040	.1553	.1014	.0572	.0278
	5	.0008	.0137	.0555	.1201	.1802	.2099	.2008	.1623	.1123	.0667
	6	.0001	.0028	.0180	.0550	.1101	.1649	.1982	.1983	.1684	.1222
	7	.0000	.0004	.0045	.0197	.0524	.1010	.1524	.1889	.1969	.1746
	8	.0000	.0001	.0009	.0055	.0197	.0487	.0923	.1417	.1812	.1964
	9	.0000	.0000	.0001	.0012	.0058	.0185	.0442	.0840	.1318	.1746
	10	.0000	.0000	.0000	.0002	.0014	.0056	.0167	.0392	.0755	.1222
	11	.0000	.0000	.0000	.0000	.0002	.0013	.0049	.0142	.0337	.0667
	12	.0000	.0000	.0000	.0000	.0000	.0002	.0011	.0040	.0115	.0278
	13	.0000	.0000	.0000	.0000	.0000	.0000	.0002	.0008	.0029	.0085
	14	.0000	.0000	.0000	.0000	.0000	.0000	.0000	.0001	.0005	.0018
	15	.0000	.0000	.0000	.0000	.0000	.0000	.0000	.0000	.0001	.0002
	16	.0000	.0000	.0000	.0000	.0000	.0000	.0000	.0000	.0000	.0000
17	0	.4181	.1668	.0631	.0225	.0075	.0023	.0007	.0002	.0000	.0000
	1	.3741	.3150	.1893	.0957	.0426	.0169	.0060	.0019	.0005	.0001
	2	.1575	.2800	.2673	.1914	.1136	.0581	.0260	.0102	.0035	.0010
	3	.0415	.1556	.2359	.2393	.1893	.1245	.0701	.0341	.0144	.0052
	4	.0076	.0605	.1457	.2093	.2209	.1868	.1320	.0796	.0411	.0182
	5	.0010	.0175	.0668	.1361	.1914	.2081	.1849	.1379	.0875	.0472
	6	.0001	.0039	.0236	.0680	.1276	.1784	.1991	.1839	.1432	.0944
	7	.0000	.0007	.0065	.0267	.0668	.1201	.1685	.1927	.1841	.1484
	8	.0000	.0001	.0014	.0084	.0279	.0644	.1143	.1606	.1883	.1855
	9	.0000	.0000	.0003	.0021	.0093	.0276	.0611	.1070	.1540	.1855
	10	.0000	.0000	.0000	.0004	.0025	.0095	.0263	.0571	.1008	.1484
	11	.0000	.0000	.0000	.0001	.0005	.0026	.0090	.0242	.0525	.0944
	12	.0000	.0000	.0000	.0000	.0001	.0006	.0024	.0081	.0215	.0472
	13	.0000	.0000	.0000	.0000	.0000	.0001	.0005	.0021	.0068	.0182
	14	.0000	.0000	.0000	.0000	.0000	.0000	.0001	.0004	.0016	.0052
	15	.0000	.0000	.0000	.0000	.0000	.0000	.0000	.0001	.0003	.0010
	16	.0000	.0000	.0000	.0000	.0000	.0000	.0000	.0000	.0000	.0001
	17	.0000	.0000	.0000	.0000	.0000	.0000	.0000	.0000	.0000	.0000
18	0	.3972	.1501	.0536	.0180	.0056	.0016	.0004	.0001	.0000	.0000
	1	.3763	.3002	.1704	.0811	.0338	.0126	.0042	.0012	.0003	.0001
	2	.1683	.2835	.2556	.1723	.0958	.0458	.0190	.0069	.0022	.0006
	3	.0473	.1680	.2406	.2297	.1704	.1046	.0547	.0246	.0095	.0031
	4	.0093	.0070	.1592	.2153	.2130	.1681	.1104	.0614	.0291	.0117
	5	.0014	.0218	.0787	.1507	.1988	.2017	.1664	.1146	.0666	.0327
	6	.0002	.0052	.0310	.0816	.1436	.1873	.1941	.1655	.1181	.0708
	7	.0000	.0010	.0091	.0350	.0820	.1376	.1792	.1892	.1657	.1214
	8	.0000	.0002	.0022	.0120	.0376	.0811	.1327	.1734	.1864	.1669
	9	.0000	.0000	.0004	.0033	.0139	.0386	.0794	.1284	.1694	.1855
	10	.0000	.0000	.0001	.0008	.0042	.0149	.0385	.0771	.1248	.1669
	11	.0000	.0000	.0000	.0001	.0010	.0046	.0151	.0374	.0742	.1214
	12	.0000	.0000	.0000	.0000	.0002	.0012	.0047	.0145	.0354	.0708
	13	.0000	.0000	.0000	.0000	.0000	.0002	.0012	.0045	.0134	.0327
	14	.0000	.0000	.0000	.0000	.0000	.0000	.0002	.0011	.0039	.0117

TABLE B (*Continued*)

Binomial Distribution

N	No. of P or Q Events	P or Q									
		.05	.10	.15	.20	.25	.30	.35	.40	.45	.50
18	15	.0000	.0000	.0000	.0000	.0000	.0000	.0000	.0002	.0009	.0031
	16	.0000	.0000	.0000	.0000	.0000	.0000	.0000	.0000	.0001	.0006
	17	.0000	.0000	.0000	.0000	.0000	.0000	.0000	.0000	.0000	.0001
	18	.0000	.0000	.0000	.0000	.0000	.0000	.0000	.0000	.0000	.0000
19	0	.3774	.1351	.0456	.0144	.0042	.0011	.0003	.0001	.0000	.0000
	1	.3774	.2852	.1529	.0685	.0268	.0093	.0029	.0008	.0002	.0000
	2	.1787	.2852	.2428	.1540	.0803	.0358	.0138	.0046	.0013	.0003
	3	.0533	.1796	.2428	.2182	.1517	.0869	.0422	.0175	.0062	.0018
	4	.0112	.0798	.1714	.2182	.2023	.1491	.0909	.0467	.0203	.0074
	5	.0018	.0266	.0907	.1636	.2023	.1916	.1468	.0933	.0497	.0222
	6	.0002	.0069	.0374	.0955	.1574	.1916	.1844	.1451	.0949	.0518
	7	.0000	.0014	.0122	.0443	.0974	.1525	.1844	.1797	.1443	.0961
	8	.0000	.0002	.0032	.0166	.0487	.0981	.1489	.1797	.1771	.1442
	9	.0000	.0000	.0007	.0051	.0198	.0514	.0980	.1464	.1771	.1762
	10	.0000	.0000	.0001	.0013	.0066	.0220	.0528	.0976	.1449	.1762
	11	.0000	.0000	.0000	.0003	.0018	.0077	.0233	.0532	.0970	.1442
	12	.0000	.0000	.0000	.0000	.0004	.0022	.0083	.0237	.0529	.0961
	13	.0000	.0000	.0000	.0000	.0001	.0005	.0024	.0085	.0233	.0518
	14	.0000	.0000	.0000	.0000	.0000	.0001	.0006	.0024	.0082	.0222
	15	.0000	.0000	.0000	.0000	.0000	.0000	.0001	.0005	.0022	.0074
	16	.0000	.0000	.0000	.0000	.0000	.0000	.0000	.0001	.0005	.0018
	17	.0000	.0000	.0000	.0000	.0000	.0000	.0000	.0000	.0001	.0003
	18	.0000	.0000	.0000	.0000	.0000	.0000	.0000	.0000	.0000	.0000
	19	.0000	.0000	.0000	.0000	.0000	.0000	.0000	.0000	.0000	.0000
20	0	.3585	.1216	.0388	.0115	.0032	.0008	.0002	.0000	.0000	.0000
	1	.3774	.2702	.1368	.0576	.0211	.0068	.0020	.0005	.0001	.0000
	2	.1887	.2852	.2293	.1369	.0669	.0278	.0100	.0031	.0008	.0002
	3	.0596	.1901	.2428	.2054	.1339	.0716	.0323	.0123	.0040	.0011
	4	.0133	.0898	.1821	.2182	.1897	.1304	.0738	.0350	.0139	.0046
	5	.0022	.0319	.1028	.1746	.2023	.1789	.1272	.0746	.0365	.0148
	6	.0003	.0089	.0454	.1091	.1686	.1916	.1712	.1244	.0746	.0370
	7	.0000	.0020	.0160	.0545	.1124	.1643	.1844	.1659	.1221	.0739
	8	.0000	.0004	.0046	.0222	.0609	.1144	.1614	.1797	.1623	.1201
	9	.0000	.0001	.0011	.0074	.0271	.0654	.1158	.1597	.1771	.1602
	10	.0000	.0000	.0002	.0020	.0099	.0308	.0686	.1171	.1593	.1762
	11	.0000	.0000	.0000	.0005	.0030	.0120	.0336	.0710	.1185	.1602
	12	.0000	.0000	.0000	.0001	.0008	.0039	.0136	.0355	.0727	.1201
	13	.0000	.0000	.0000	.0000	.0002	.0010	.0045	.0146	.0366	.0739
	14	.0000	.0000	.0000	.0000	.0000	.0002	.0012	.0049	.0150	.0370
	15	.0000	.0000	.0000	.0000	.0000	.0000	.0003	.0013	.0049	.0148
	16	.0000	.0000	.0000	.0000	.0000	.0000	.0000	.0003	.0013	.0046
	17	.0000	.0000	.0000	.0000	.0000	.0000	.0000	.0000	.0002	.0011
	18	.0000	.0000	.0000	.0000	.0000	.0000	.0000	.0000	.0000	.0002
	19	.0000	.0000	.0000	.0000	.0000	.0000	.0000	.0000	.0000	.0000
	20	.0000	.0000	.0000	.0000	.0000	.0000	.0000	.0000	.0000	.0000

TABLE C.1

Critical Values of U and U' for a One-Tailed Test at $\alpha = 0.005$ or a Two-Tailed Test at $\alpha = 0.01$

*To be significant for any given n_1 and n_2: U_{obt} must be equal to or **less than** the value shown in the table. U'_{obt} must be equal to or **greater than** the value shown in the table.*

Each cell shows the critical value of U (top) and U' (bottom, underlined).

n_2 \ n_1	1	2	3	4	5	6	7	8	9	10	11	12	13	14	15	16	17	18	19	20
1	—	—	—	—	—	—	—	—	—	—	—	—	—	—	—	—	—	—	—	—
2	—	—	—	—	—	—	—	—	—	—	—	—	—	—	—	—	—	—	0/38	0/40
3	—	—	—	—	—	—	—	—	0/27	0/30	0/33	1/35	1/38	1/41	2/43	2/46	2/49	2/52	3/54	3/57
4	—	—	—	—	—	0/24	0/28	1/31	1/35	2/38	2/42	3/45	3/49	4/52	5/55	5/59	6/62	6/66	7/69	8/72
5	—	—	—	—	0/25	1/29	1/34	2/38	3/42	4/46	5/50	6/54	7/58	7/63	8/67	9/71	10/75	11/79	12/83	13/87
6	—	—	—	0/24	1/29	2/34	3/39	4/44	5/49	6/54	7/59	9/63	10/68	11/73	12/78	13/83	15/87	16/92	17/97	18/102
7	—	—	—	0/28	1/34	3/39	4/45	6/50	7/56	9/61	10/67	12/72	13/78	15/83	16/89	18/94	19/100	21/105	22/111	24/116
8	—	—	—	1/31	2/38	4/44	6/50	7/57	9/63	11/69	13/75	15/81	17/87	18/94	20/100	22/106	24/112	26/118	28/124	30/130
9	—	—	0/27	1/35	3/42	5/49	7/56	9/63	11/70	13/77	16/83	18/90	20/97	22/104	24/111	27/117	29/124	31/131	33/138	36/144
10	—	—	0/30	2/38	4/46	6/54	9/61	11/69	13/77	16/84	18/92	21/99	24/106	26/114	29/121	31/129	34/136	37/143	39/151	42/158
11	—	—	0/33	2/42	5/50	7/59	10/67	13/75	16/83	18/92	21/100	24/108	27/116	30/124	33/132	36/140	39/148	42/156	45/164	48/172
12	—	—	1/35	3/45	6/54	9/63	12/72	15/81	18/90	21/99	24/108	27/117	31/125	34/134	37/143	41/151	44/160	47/169	51/177	54/186
13	—	—	1/38	3/49	7/58	10/68	13/78	17/87	20/97	24/106	27/116	31/125	34/125	38/144	42/153	45/163	49/172	53/181	56/191	60/200
14	—	—	1/41	4/52	7/63	11/73	15/83	18/94	22/104	26/114	30/124	34/134	38/144	42/154	46/164	50/174	54/184	58/194	63/203	67/213
15	—	—	2/43	5/55	8/67	12/78	16/89	20/100	24/111	29/121	33/132	37/143	42/153	46/164	51/174	55/185	60/195	64/206	69/216	73/227
16	—	—	2/46	5/59	9/71	13/83	18/94	22/106	27/117	31/129	36/140	41/151	45/163	50/174	55/185	60/196	65/207	70/218	74/230	79/241
17	—	—	2/49	6/62	10/75	15/87	19/100	24/112	29/124	34/148	39/148	44/160	49/172	54/184	60/195	65/207	70/219	75/231	81/242	86/254
18	—	—	2/52	6/66	11/79	16/92	21/105	26/118	31/131	37/143	42/156	47/169	53/181	58/194	64/206	70/218	75/231	81/243	87/255	92/268
19	—	0/38	3/54	7/69	12/83	17/97	22/111	28/124	33/138	39/151	45/164	51/177	56/191	63/203	69/216	74/230	81/242	87/255	93/268	99/281
20	—	0/40	3/57	8/72	13/87	18/102	24/116	30/130	36/144	42/158	48/172	54/186	60/200	67/213	73/227	79/241	86/254	92/268	99/281	105/295

Dashes in the body of the table indicate that no decision is possible at the stated level of significance.

TABLE C.2

Critical Values of U and U' for a One-Tailed Test at $\alpha = 0.01$ or a Two-Tailed Test at $\alpha = 0.02$

*To be significant for any given n_1 and n_2: U_{obt} must be equal to or **less than** the value shown in the table. U'_{obt} must be equal to or **greater than** the value shown in the table.*

Each cell shows U (top) over U' (underlined, bottom).

$n_2 \backslash n_1$	1	2	3	4	5	6	7	8	9	10	11	12	13	14	15	16	17	18	19	20
1	—	—	—	—	—	—	—	—	—	—	—	—	—	—	—	—	—	—	—	—
2	—	—	—	—	—	—	—	—	—	—	—	—	0/_26_	0/_28_	0/_30_	0/_32_	0/_34_	0/_36_	1/_37_	1/_39_
3	—	—	—	—	—	—	0/_21_	0/_24_	1/_26_	1/_29_	1/_32_	2/_34_	2/_37_	2/_40_	3/_42_	3/_45_	4/_47_	4/_50_	4/_52_	5/_55_
4	—	—	—	—	0/_20_	1/_23_	1/_27_	2/_30_	3/_33_	3/_37_	4/_40_	5/_43_	5/_47_	6/_50_	7/_53_	7/_57_	8/_60_	9/_63_	9/_67_	10/_70_
5	—	—	—	0/_20_	1/_24_	2/_28_	3/_32_	4/_36_	5/_40_	6/_44_	7/_48_	8/_52_	9/_56_	10/_60_	11/_64_	12/_68_	13/_72_	14/_76_	15/_80_	16/_84_
6	—	—	—	1/_23_	2/_28_	3/_33_	4/_38_	6/_42_	7/_47_	8/_52_	9/_57_	11/_61_	12/_66_	13/_71_	15/_75_	16/_80_	18/_84_	19/_89_	20/_94_	22/_93_
7	—	—	0/_21_	1/_27_	3/_32_	4/_38_	6/_43_	7/_49_	9/_54_	11/_59_	12/_65_	14/_70_	16/_75_	17/_81_	19/_86_	21/_91_	23/_96_	24/_102_	26/_107_	28/_112_
8	—	—	0/_24_	2/_30_	4/_36_	6/_42_	7/_49_	9/_55_	11/_61_	13/_67_	15/_73_	17/_79_	20/_84_	22/_90_	24/_96_	26/_102_	28/_108_	30/_114_	32/_120_	34/_126_
9	—	—	1/_26_	3/_33_	5/_40_	7/_47_	9/_54_	11/_61_	14/_67_	16/_74_	18/_81_	21/_87_	23/_94_	26/_100_	28/_107_	31/_113_	33/_120_	36/_126_	38/_133_	40/_140_
10	—	—	1/_29_	3/_37_	6/_44_	8/_52_	11/_59_	13/_67_	16/_74_	19/_81_	22/_88_	24/_96_	27/_103_	30/_110_	33/_117_	36/_124_	38/_132_	41/_139_	44/_146_	47/_153_
11	—	—	1/_32_	4/_40_	7/_48_	9/_57_	12/_65_	15/_73_	18/_81_	22/_88_	25/_96_	28/_104_	31/_112_	34/_120_	37/_128_	41/_135_	44/_143_	47/_151_	50/_159_	53/_167_
12	—	—	2/_34_	5/_43_	8/_52_	11/_61_	14/_70_	17/_79_	21/_87_	24/_96_	28/_104_	31/_113_	35/_121_	38/_130_	42/_138_	46/_146_	49/_155_	53/_163_	56/_172_	60/_180_
13	—	0/_26_	2/_37_	5/_47_	9/_56_	12/_66_	16/_75_	20/_84_	23/_94_	27/_103_	31/_112_	35/_121_	39/_130_	43/_139_	47/_148_	51/_157_	55/_166_	59/_175_	63/_184_	67/_193_
14	—	0/_28_	2/_40_	6/_50_	10/_60_	13/_71_	17/_81_	22/_90_	26/_100_	30/_110_	34/_120_	38/_130_	43/_139_	47/_149_	51/_159_	56/_168_	60/_178_	65/_187_	69/_197_	73/_207_
15	—	0/_30_	3/_42_	7/_53_	11/_64_	15/_75_	19/_86_	24/_96_	28/_107_	33/_117_	37/_128_	42/_138_	47/_148_	51/_159_	56/_169_	61/_179_	66/_189_	70/_200_	75/_210_	80/_220_
16	—	0/_32_	3/_45_	7/_57_	12/_68_	16/_80_	21/_91_	26/_102_	31/_113_	36/_124_	41/_135_	46/_146_	51/_157_	56/_168_	61/_179_	66/_190_	71/_201_	76/_212_	82/_222_	87/_233_
17	—	0/_34_	4/_47_	8/_60_	13/_72_	18/_84_	23/_96_	28/_108_	33/_120_	38/_132_	44/_143_	49/_155_	55/_166_	60/_178_	66/_189_	71/_201_	77/_212_	82/_224_	88/_234_	93/_247_
18	—	0/_36_	4/_50_	9/_63_	14/_76_	19/_89_	24/_102_	30/_114_	36/_126_	41/_139_	47/_151_	53/_163_	59/_175_	65/_187_	70/_200_	76/_212_	82/_224_	88/_236_	94/_248_	100/_260_
19	—	1/_37_	4/_53_	9/_67_	15/_80_	20/_94_	26/_107_	32/_120_	38/_133_	44/_146_	50/_159_	56/_172_	63/_184_	69/_197_	75/_210_	82/_222_	88/_235_	94/_248_	101/_260_	107/_273_
20	—	1/_39_	5/_55_	10/_70_	16/_84_	22/_98_	28/_112_	34/_126_	40/_140_	47/_153_	53/_167_	60/_180_	67/_193_	73/_207_	80/_220_	87/_233_	93/_247_	100/_260_	107/_273_	114/_286_

Dashes in the body of the table indicate that no decision is possible at the stated level of significance.

TABLE C.3

Critical Values of U and U' for a One-Tailed Test at $\alpha = 0.025$ or a Two-Tailed Test at $\alpha = 0.05$

*To be significant for any given n_1 and n_2: U_{obt} must be equal to or **less than** the value shown in the table. U'_{obt} must be equal to or **greater than** the value shown in the table.*

n_2 \ n_1	1	2	3	4	5	6	7	8	9	10	11	12	13	14	15	16	17	18	19	20
1	—	—	—	—	—	—	—	—	—	—	—	—	—	—	—	—	—	—	—	—
2	—	—	—	—	—	—	—	0	0	0	0	1	1	1	1	1	2	2	2	2
								16	18	20	22	23	25	27	29	31	32	34	36	38
3	—	—	—	—	0	1	1	2	2	3	3	4	4	5	5	6	6	7	7	8
					15	17	20	22	25	27	30	32	35	37	40	42	45	47	50	52
4	—	—	—	0	1	2	3	4	4	5	6	7	8	9	10	11	11	12	13	13
				16	19	22	25	28	32	35	38	41	44	47	50	53	57	60	63	67
5	—	—	0	1	2	3	5	6	7	8	9	11	12	13	14	15	17	18	19	20
			15	19	23	27	30	34	38	42	46	49	53	57	61	65	68	72	76	80
6	—	—	1	2	3	5	6	8	10	11	13	14	16	17	19	21	22	24	25	27
			17	22	27	31	36	40	44	49	53	58	62	67	71	75	80	84	89	93
7	—	—	1	3	5	6	8	10	12	14	16	18	20	22	24	26	28	30	32	34
			20	25	30	36	41	46	51	56	61	66	71	76	81	86	91	96	101	106
8	—	0	2	4	6	8	10	13	15	17	19	22	24	26	29	31	34	36	38	41
		16	22	28	34	40	46	51	57	63	69	74	80	86	91	97	102	108	114	119
9	—	0	2	4	7	10	12	15	17	20	23	26	28	31	34	37	39	42	45	48
		18	25	32	38	44	51	57	64	70	76	82	89	95	101	107	114	120	126	132
10	—	0	3	5	8	11	14	17	20	23	26	29	33	36	39	42	45	48	52	55
		20	27	35	42	49	56	63	70	77	84	91	97	104	111	118	125	132	138	145
11	—	0	3	6	9	13	16	19	23	26	30	33	37	40	44	47	51	55	58	62
		22	30	38	46	53	61	69	76	84	91	99	106	114	121	129	136	143	151	158
12	—	1	4	7	11	14	18	22	26	29	33	37	41	45	49	53	57	61	65	69
		23	32	41	49	58	66	74	82	91	99	107	115	123	131	139	147	155	163	171
13	—	1	4	8	12	16	20	24	28	33	37	41	45	50	54	59	63	67	72	76
		25	35	44	53	62	71	80	89	97	106	115	124	132	141	149	158	167	175	184
14	—	1	5	9	13	17	22	26	31	36	40	45	50	55	59	64	69	74	78	83
		27	37	47	57	67	76	86	95	104	114	123	132	141	151	160	169	178	188	197
15	—	1	5	10	14	19	24	29	34	39	44	49	54	59	64	70	75	80	85	90
		29	40	50	61	71	81	91	101	111	121	131	141	151	161	170	180	190	200	210
16	—	1	6	11	15	21	26	31	37	42	47	53	59	64	70	75	81	86	92	98
		31	42	53	65	75	86	97	107	118	129	139	149	160	170	181	191	202	212	222
17	—	2	6	11	17	22	28	34	39	45	51	57	63	69	75	81	87	93	99	105
		32	45	57	68	80	91	102	114	125	136	147	158	169	180	191	202	213	224	235
18	—	2	7	12	18	24	30	36	42	48	55	61	67	74	80	86	93	99	106	112
		34	47	60	72	84	96	108	120	132	143	155	167	178	190	202	213	225	236	248
19	—	2	7	13	19	25	32	38	45	52	58	65	72	78	85	92	99	106	113	119
		36	50	63	76	89	101	114	126	138	151	163	175	188	200	212	224	236	248	261
20	—	2	8	13	20	27	34	41	48	55	62	69	76	83	90	98	105	112	119	127
		38	52	67	80	93	106	119	132	145	158	171	184	197	210	222	235	248	261	273

Dashes in the body of the table indicate that no decision is possible at the stated level of significance.

TABLE C.4

Critical Values of U and U' for a One-Tailed Test at $\alpha = 0.05$ or a Two-Tailed Test at $\alpha = 0.10$

*To be significant for any given n_1 and n_2: U_{obt} must be equal to or **less than** the value shown in the table. U'_{obt} must be equal to or **greater than** the value shown in the table.*

Each cell is shown as U / U' (where U' is the underlined value).

n_2 \ n_1	1	2	3	4	5	6	7	8	9	10	11	12	13	14	15	16	17	18	19	20
1	—	—	—	—	—	—	—	—	—	—	—	—	—	—	—	—	—	—	0/19	0/20
2	—	—	—	—	0/10	0/12	0/14	1/15	1/17	1/19	1/21	2/22	2/24	2/26	3/27	3/29	3/31	4/32	4/34	4/36
3	—	—	0/9	0/12	1/14	2/16	2/19	3/21	3/24	4/26	5/28	5/31	6/33	7/35	7/38	8/40	9/42	9/45	10/47	11/49
4	—	—	0/12	1/15	2/18	3/21	4/24	5/27	6/30	7/33	8/36	9/39	10/42	11/45	12/48	14/50	15/53	16/56	17/59	18/62
5	—	0/10	1/14	2/18	4/21	5/25	6/29	8/32	9/36	11/39	12/43	13/47	15/50	16/54	18/57	19/61	20/65	22/68	23/72	25/75
6	—	0/12	2/16	3/21	5/25	7/29	8/34	10/38	12/42	14/46	16/50	17/55	19/59	21/63	23/67	25/71	26/76	28/80	30/84	32/88
7	—	0/14	2/19	4/24	6/29	8/34	11/38	13/43	15/48	17/53	19/58	21/63	24/67	26/72	28/77	30/82	33/86	35/91	37/96	39/101
8	—	1/15	3/21	5/27	8/32	10/38	13/43	15/49	18/54	20/60	23/65	26/70	28/76	31/81	33/87	36/92	39/97	41/103	44/108	47/113
9	—	1/17	3/24	6/30	9/36	12/42	15/48	18/54	21/60	24/66	27/72	30/78	33/84	36/90	39/96	42/102	45/108	48/114	51/120	54/126
10	—	1/19	4/26	7/33	11/39	14/46	17/53	20/60	24/66	27/73	31/79	34/86	37/93	41/99	44/106	48/112	51/119	55/125	58/132	62/138
11	—	1/21	5/28	8/36	12/43	16/50	19/58	23/65	27/72	31/79	34/87	38/94	42/101	46/108	50/115	54/122	57/130	61/137	65/144	69/151
12	—	2/22	5/31	9/39	13/47	17/55	21/63	26/70	30/78	34/86	38/94	42/102	47/109	51/117	55/125	60/132	64/140	68/148	72/156	77/163
13	—	2/24	6/33	10/42	15/50	19/59	24/67	28/76	33/84	37/93	42/101	47/109	51/118	56/126	61/134	65/143	70/151	75/159	80/167	84/176
14	—	2/26	7/35	11/45	16/54	21/63	26/72	31/81	36/90	41/99	46/108	51/117	56/126	61/135	66/144	71/153	77/161	82/170	87/179	92/188
15	—	3/27	7/38	12/48	18/57	23/67	28/77	33/87	39/96	44/106	50/115	55/125	61/134	66/144	72/153	77/163	83/172	88/182	94/191	100/200
16	—	3/29	8/40	14/50	19/61	25/71	30/82	36/92	42/102	48/112	54/122	60/132	65/143	71/153	77/163	83/173	89/183	95/193	101/203	107/213
17	—	3/31	9/42	15/53	20/65	26/76	33/86	39/97	45/108	51/119	57/130	64/140	70/151	77/161	83/172	89/183	96/193	102/204	109/214	115/225
18	—	4/32	9/45	16/56	22/68	28/80	35/91	41/103	48/114	55/123	61/137	68/148	75/159	82/170	88/182	95/193	102/204	109/215	116/226	123/237
19	0/19	4/34	10/47	17/59	23/72	30/84	37/96	44/108	51/120	58/132	65/144	72/156	80/167	87/179	94/191	101/203	109/214	116/226	123/238	130/250
20	0/20	4/36	11/49	18/62	25/75	32/88	39/101	47/113	54/126	62/138	69/151	77/163	84/176	92/188	100/200	107/213	115/225	123/237	130/250	138/262

Dashes in the body of the table indicate that no decision is possible at the stated level of significance.

TABLE D

Critical Values of Student's
t Distribution

The values listed in the table are the critical values of t for the specified degrees of freedom (left column) and the alpha level (column heading). For two-tailed alpha levels, t_{crit} is both + and −. To be significant, $|t_{obt}| \geq |t_{crit}|$.

df	Level of Significance for One-Tailed Test					
	.10	.05	.025	.01	.005	.0005
	Level of Significance for Two-Tailed Test					
	.20	.10	.05	.02	.01	.001
1	3.078	6.314	12.706	31.821	63.657	636.619
2	1.886	2.920	4.303	6.965	9.925	31.598
3	1.638	2.353	3.182	4.541	5.841	12.941
4	1.533	2.132	2.776	3.747	4.604	8.610
5	1.476	2.015	2.571	3.365	4.032	6.859
6	1.440	1.943	2.447	3.143	3.707	5.959
7	1.415	1.895	2.365	2.998	3.499	5.405
8	1.397	1.860	2.306	2.986	3.355	5.041
9	1.383	1.833	2.262	2.821	3.250	4.781
10	1.372	1.812	2.228	2.764	3.169	4.587
11	1.363	1.796	2.201	2.718	3.106	4.437
12	1.356	1.782	2.179	2.681	3.055	4.318
13	1.350	1.771	2.160	2.650	3.012	4.221
14	1.345	1.761	2.145	2.624	2.977	4.140
15	1.341	1.753	2.131	2.602	2.947	4.073
16	1.337	1.746	2.120	2.583	2.921	4.015
17	1.333	1.740	2.110	2.567	2.898	3.965
18	1.330	1.734	2.101	2.552	2.878	3.922
19	1.328	1.729	2.093	2.539	2.861	3.883
20	1.325	1.725	2.086	2.528	2.845	3.850
21	1.323	1.721	2.080	2.518	2.831	3.819
22	1.321	1.717	2.074	2.508	2.819	3.792
23	1.319	1.714	2.069	2.500	2.807	3.767
24	1.318	1.711	2.064	2.492	2.797	3.745
25	1.316	1.708	2.060	2.485	2.787	3.725
26	1.315	1.706	2.056	2.479	2.779	3.707
27	1.314	1.703	2.052	2.473	2.771	3.690
28	1.313	1.701	2.048	2.467	2.763	3.674
29	1.311	1.699	2.045	2.462	2.756	3.659
30	1.310	1.697	2.042	2.457	2.750	3.646
40	1.303	1.684	2.021	2.423	2.704	3.551
60	1.296	1.671	2.000	2.390	2.660	3.460
120	1.289	1.658	1.980	2.358	2.617	3.373
∞	1.282	1.645	1.960	2.326	2.576	3.291

TABLE E

Critical Values of
Pearson r

The values listed in the table are the critical values of r for the specified degrees of freedom (left column) and the alpha level (column heading). For two-tailed alpha levels, r_{crit} is both + and −. To be significant, $|r_{obt}| \geq |r_{crit}|$.

$df = N - 2$	Level of Significance for One-Tailed Test				
	.05	.025	.01	.005	.0005
	Level of Significance for Two-Tailed Test				
	.10	.05	.02	.01	.001
1	.9877	.9969	.9995	.9999	1.0000
2	.9000	.9500	.9800	.9900	.9990
3	.8054	.8783	.9343	.9587	.9912
4	.7293	.8114	.8822	.9172	.9741
5	.6694	.7545	.8329	.8745	.9507
6	.6215	.7067	.7887	.8343	.9249
7	.5822	.6664	.7498	.7977	.8982
8	.5494	.6319	.7155	.7646	.8721
9	.5214	.6021	.6851	.7348	.8471
10	.4973	.5760	.6581	.7079	.8233
11	.4762	.5529	.6339	.6835	.8010
12	.4575	.5324	.6120	.6614	.7800
13	.4409	.5139	.5923	.6411	.7603
14	.4259	.4973	.5742	.6226	.7420
15	.4124	.4821	.5577	.6055	.7246
16	.4000	.4683	.5425	.5897	.7084
17	.3887	.4555	.5285	.5751	.6932
18	.3783	.4438	.5155	.5614	.6787
19	.3687	.4329	.5034	.5487	.6652
20	.3598	.4227	.4921	.5368	.6524
25	.3233	.3809	.4451	.4869	.5974
30	.2960	.3494	.4093	.4487	.5541
35	.2746	.3246	.3810	.4182	.5189
40	.2573	.3044	.3578	.3932	.4896
45	.2428	.2875	.3384	.3721	.4648
50	.2306	.2732	.3218	.3541	.4433
60	.2108	.2500	.2948	.3248	.4078
70	.1954	.2319	.2737	.3017	.3799
80	.1829	.2172	.2565	.2830	.3568
90	.1726	.2050	.2422	.2673	.3375
100	.1638	.1946	.2301	.2540	.3211

TABLE F Critical Values of the F Distribution for $\alpha = 0.05$ (Roman Type) and $\alpha = 0.01$ (**Boldface Type**)

The values listed in the table are the critical values of F for the degrees of freedom of the numerator of the F ratio (column headings) and the degrees of freedom of the denominator of the F ratio (row headings). To be significant, $F_{obt} \geq F_{crit}$.

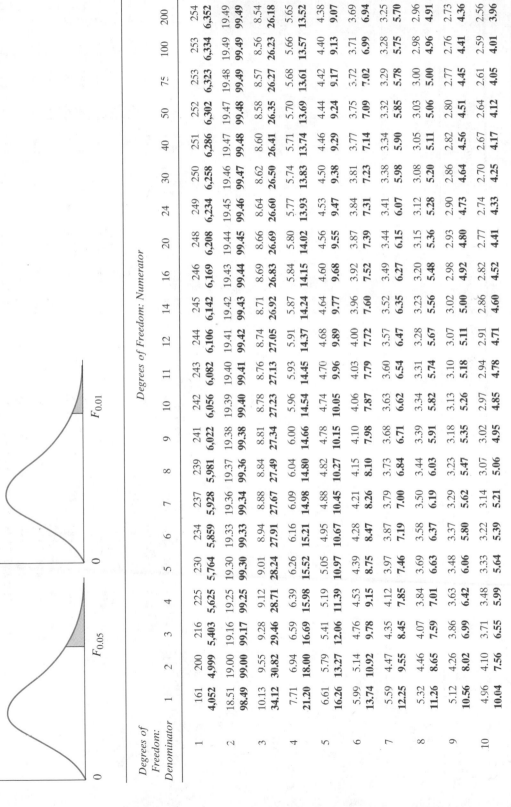

$F_{0.05}$

$F_{0.01}$

Degrees of Freedom: Numerator

Degrees of Freedom: Denominator	1	2	3	4	5	6	7	8	9	10	11	12	14	16	20	24	30	40	50	75	100	200	500	∞
1	161	200	216	225	230	234	237	239	241	242	243	244	245	246	248	249	250	251	252	253	253	254	254	254
	4,052	**4,999**	**5,403**	**5,625**	**5,764**	**5,859**	**5,928**	**5,981**	**6,022**	**6,056**	**6,082**	**6,106**	**6,142**	**6,169**	**6,208**	**6,234**	**6,258**	**6,286**	**6,302**	**6,323**	**6,334**	**6,352**	**6,361**	**6,366**
2	18.51	19.00	19.16	19.25	19.30	19.33	19.36	19.37	19.38	19.39	19.40	19.41	19.42	19.43	19.44	19.45	19.46	19.47	19.47	19.48	19.49	19.49	19.50	19.50
	98.49	**99.00**	**99.17**	**99.25**	**99.30**	**99.33**	**99.34**	**99.36**	**99.38**	**99.40**	**99.41**	**99.42**	**99.43**	**99.44**	**99.45**	**99.46**	**99.47**	**99.48**	**99.48**	**99.49**	**99.49**	**99.49**	**99.50**	**99.50**
3	10.13	9.55	9.28	9.12	9.01	8.94	8.88	8.84	8.81	8.78	8.76	8.74	8.71	8.69	8.66	8.64	8.62	8.60	8.58	8.57	8.56	8.54	8.54	8.53
	34.12	**30.82**	**29.46**	**28.71**	**28.24**	**27.91**	**27.67**	**27.49**	**27.34**	**27.23**	**27.13**	**27.05**	**26.92**	**26.83**	**26.69**	**26.60**	**26.50**	**26.41**	**26.35**	**26.27**	**26.23**	**26.18**	**26.14**	**26.12**
4	7.71	6.94	6.59	6.39	6.26	6.16	6.09	6.04	6.00	5.96	5.93	5.91	5.87	5.84	5.80	5.77	5.74	5.71	5.70	5.68	5.66	5.65	5.64	5.63
	21.20	**18.00**	**16.69**	**15.98**	**15.52**	**15.21**	**14.98**	**14.80**	**14.66**	**14.54**	**14.45**	**14.37**	**14.24**	**14.15**	**14.02**	**13.93**	**13.83**	**13.74**	**13.69**	**13.61**	**13.57**	**13.52**	**13.48**	**13.46**
5	6.61	5.79	5.41	5.19	5.05	4.95	4.88	4.82	4.78	4.74	4.70	4.68	4.64	4.60	4.56	4.53	4.50	4.46	4.44	4.42	4.40	4.38	4.37	4.36
	16.26	**13.27**	**12.06**	**11.39**	**10.97**	**10.67**	**10.45**	**10.27**	**10.15**	**10.05**	**9.96**	**9.89**	**9.77**	**9.68**	**9.55**	**9.47**	**9.38**	**9.29**	**9.24**	**9.17**	**9.13**	**9.07**	**9.04**	**9.02**
6	5.99	5.14	4.76	4.53	4.39	4.28	4.21	4.15	4.10	4.06	4.03	4.00	3.96	3.92	3.87	3.84	3.81	3.77	3.75	3.72	3.71	3.69	3.68	3.67
	13.74	**10.92**	**9.78**	**9.15**	**8.75**	**8.47**	**8.26**	**8.10**	**7.98**	**7.87**	**7.79**	**7.72**	**7.60**	**7.52**	**7.39**	**7.31**	**7.23**	**7.14**	**7.09**	**7.02**	**6.99**	**6.94**	**6.90**	**6.88**
7	5.59	4.47	4.35	4.12	3.97	3.87	3.79	3.73	3.68	3.63	3.60	3.57	3.52	3.49	3.44	3.41	3.38	3.34	3.32	3.29	3.28	3.25	3.24	3.23
	12.25	**9.55**	**8.45**	**7.85**	**7.46**	**7.19**	**7.00**	**6.84**	**6.71**	**6.62**	**6.54**	**6.47**	**6.35**	**6.27**	**6.15**	**6.07**	**5.98**	**5.90**	**5.85**	**5.78**	**5.75**	**5.70**	**5.67**	**5.65**
8	5.32	4.46	4.07	3.84	3.69	3.58	3.50	3.44	3.39	3.34	3.31	3.28	3.23	3.20	3.15	3.12	3.08	3.05	3.03	3.00	2.98	2.96	2.94	2.93
	11.26	**8.65**	**7.59**	**7.01**	**6.63**	**6.37**	**6.19**	**6.03**	**5.91**	**5.82**	**5.74**	**5.67**	**5.56**	**5.48**	**5.36**	**5.28**	**5.20**	**5.11**	**5.06**	**5.00**	**4.96**	**4.91**	**4.88**	**4.86**
9	5.12	4.26	3.86	3.63	3.48	3.37	3.29	3.23	3.18	3.13	3.10	3.07	3.02	2.98	2.93	2.90	2.86	2.82	2.80	2.77	2.76	2.73	2.72	2.71
	10.56	**8.02**	**6.99**	**6.42**	**6.06**	**5.80**	**5.62**	**5.47**	**5.35**	**5.26**	**5.18**	**5.11**	**5.00**	**4.92**	**4.80**	**4.73**	**4.64**	**4.56**	**4.51**	**4.45**	**4.41**	**4.36**	**4.33**	**4.31**
10	4.96	4.10	3.71	3.48	3.33	3.22	3.14	3.07	3.02	2.97	2.94	2.91	2.86	2.82	2.77	2.74	2.70	2.67	2.64	2.61	2.59	2.56	2.55	2.54
	10.04	**7.56**	**6.55**	**5.99**	**5.64**	**5.39**	**5.21**	**5.06**	**4.95**	**4.85**	**4.78**	**4.71**	**4.60**	**4.52**	**4.41**	**4.33**	**4.25**	**4.17**	**4.12**	**4.05**	**4.01**	**3.96**	**3.93**	**3.91**

The values listed in the table are the critical values of F for the degrees of freedom of the numerator of the F ratio (column headings) and the degrees of freedom of the denominator of the F ratio (row headings). To be significant, $F_{obt} \geq F_{crit}$.

Degrees of Freedom: Numerator

Degrees of Freedom: Denominator	1	2	3	4	5	6	7	8	9	10	11	12	14	16	20	24	30	40	50	75	100	200	500	∞
11	4.84	3.98	3.59	3.36	3.20	3.09	3.01	2.95	2.90	2.86	2.82	2.79	2.74	2.70	2.65	2.61	2.57	2.53	2.50	2.47	2.45	2.42	2.41	2.40
	9.65	**7.20**	**6.22**	**5.67**	**5.32**	**5.07**	**4.88**	**4.74**	**4.63**	**4.54**	**4.46**	**4.40**	**4.29**	**4.21**	**4.10**	**4.02**	**3.94**	**3.86**	**3.80**	**3.74**	**3.70**	**3.66**	**3.62**	**3.60**
12	4.75	3.88	3.49	3.26	3.11	3.00	2.92	2.85	2.80	2.76	2.72	2.69	2.64	2.60	2.54	2.50	2.46	2.42	2.40	2.36	2.35	2.32	2.31	2.30
	9.33	**6.93**	**5.95**	**5.41**	**5.06**	**4.82**	**4.65**	**4.50**	**4.39**	**4.30**	**4.22**	**4.16**	**4.05**	**3.98**	**3.86**	**3.78**	**3.70**	**3.61**	**3.56**	**3.49**	**3.46**	**3.41**	**3.38**	**3.36**
13	4.67	3.80	3.41	3.18	3.02	2.92	2.84	2.77	2.72	2.67	2.63	2.60	2.55	2.51	2.46	2.42	2.38	2.34	2.32	2.28	2.26	2.24	2.22	2.21
	9.07	**6.70**	**5.74**	**5.20**	**4.86**	**4.62**	**4.44**	**4.30**	**4.19**	**4.10**	**4.02**	**3.96**	**3.85**	**3.78**	**3.67**	**3.59**	**3.51**	**3.42**	**3.37**	**3.30**	**3.27**	**3.21**	**3.18**	**3.16**
14	4.60	3.74	3.34	3.11	2.96	2.85	2.77	2.70	2.65	2.60	2.56	2.53	2.48	2.44	2.39	2.35	2.31	2.27	2.24	2.21	2.19	2.16	2.14	2.13
	8.86	**6.51**	**5.56**	**5.03**	**4.69**	**4.46**	**4.28**	**4.14**	**4.03**	**3.94**	**3.86**	**3.80**	**3.70**	**3.62**	**3.51**	**3.43**	**3.34**	**3.26**	**3.21**	**3.14**	**3.11**	**3.06**	**3.02**	**3.00**
15	4.54	3.68	3.29	3.06	2.90	2.79	2.70	2.64	2.59	2.55	2.51	2.48	2.43	2.39	2.33	2.29	2.25	2.21	2.18	2.15	2.12	2.10	2.08	2.07
	8.68	**6.36**	**5.42**	**4.89**	**4.56**	**4.32**	**4.14**	**4.00**	**3.89**	**3.80**	**3.73**	**3.67**	**3.56**	**3.48**	**3.36**	**3.29**	**3.20**	**3.12**	**3.07**	**3.00**	**2.97**	**2.92**	**2.89**	**2.87**
16	4.49	3.63	3.24	3.01	2.85	2.74	2.66	2.59	2.54	2.49	2.45	2.42	2.37	2.33	2.28	2.24	2.20	2.16	2.13	2.09	2.07	2.04	2.02	2.01
	8.53	**6.23**	**5.29**	**4.77**	**4.44**	**4.20**	**4.03**	**3.89**	**3.78**	**3.69**	**3.61**	**3.55**	**3.45**	**3.37**	**3.25**	**3.18**	**3.10**	**3.01**	**2.96**	**2.89**	**2.86**	**2.80**	**2.77**	**2.75**
17	4.45	3.59	3.20	2.96	2.81	2.70	2.62	2.55	2.50	2.45	2.41	2.38	2.33	2.29	2.23	2.19	2.15	2.11	2.08	2.04	2.02	1.99	1.97	1.96
	8.40	**6.11**	**5.18**	**4.67**	**4.34**	**4.10**	**3.93**	**3.79**	**3.68**	**3.59**	**3.52**	**3.45**	**3.35**	**3.27**	**3.16**	**3.08**	**3.00**	**2.92**	**2.86**	**2.79**	**2.76**	**2.70**	**2.67**	**2.65**
18	4.41	3.55	3.16	2.93	2.77	2.66	2.58	2.51	2.46	2.41	2.37	2.34	2.29	2.25	2.19	2.15	2.11	2.07	2.04	2.00	1.98	1.95	1.93	1.92
	8.28	**6.01**	**5.09**	**4.58**	**4.25**	**4.01**	**3.85**	**3.71**	**3.60**	**3.51**	**3.44**	**3.37**	**3.27**	**3.19**	**3.07**	**3.00**	**2.91**	**2.83**	**2.78**	**2.71**	**2.68**	**2.62**	**2.59**	**2.57**
19	4.38	3.52	3.13	2.90	2.74	2.63	2.55	2.48	2.43	2.38	2.34	2.31	2.26	2.21	2.15	2.11	2.07	2.02	2.00	1.96	1.94	1.91	1.90	1.88
	8.18	**5.93**	**5.01**	**4.50**	**4.17**	**3.94**	**3.77**	**3.63**	**3.52**	**3.43**	**3.36**	**3.30**	**3.19**	**3.12**	**3.00**	**2.92**	**2.84**	**2.76**	**2.70**	**2.63**	**2.60**	**2.54**	**2.51**	**2.49**
20	4.35	3.49	3.10	2.87	2.71	2.60	2.52	2.45	2.40	2.35	2.31	2.28	2.23	2.18	2.12	2.08	2.04	1.99	1.96	1.92	1.90	1.87	1.85	1.84
	8.10	**5.85**	**4.94**	**4.43**	**4.10**	**3.87**	**3.71**	**3.56**	**3.45**	**3.37**	**3.30**	**3.23**	**3.13**	**3.05**	**2.94**	**2.86**	**2.77**	**2.69**	**2.63**	**2.56**	**2.53**	**2.47**	**2.44**	**2.42**
21	4.32	3.47	3.07	2.84	2.68	2.57	2.49	2.42	2.37	2.32	2.28	2.25	2.20	2.15	2.09	2.05	2.00	1.96	1.93	1.89	1.87	1.84	1.82	1.81
	8.02	**5.78**	**4.87**	**4.37**	**4.04**	**3.81**	**3.65**	**3.51**	**3.40**	**3.31**	**3.24**	**3.17**	**3.07**	**2.99**	**2.88**	**2.80**	**2.72**	**2.63**	**2.58**	**2.51**	**2.47**	**2.42**	**2.38**	**2.36**
22	4.30	3.44	3.05	2.82	2.66	2.55	2.47	2.40	2.35	2.30	2.26	2.23	2.18	2.13	2.07	2.03	1.98	1.93	1.91	1.87	1.84	1.81	1.80	1.78
	7.94	**5.72**	**4.82**	**4.31**	**3.99**	**3.76**	**3.59**	**3.45**	**3.35**	**3.26**	**3.18**	**3.12**	**3.02**	**2.94**	**2.83**	**2.75**	**2.67**	**2.58**	**2.53**	**2.46**	**2.42**	**2.37**	**2.33**	**2.31**
23	4.28	3.42	3.03	2.80	2.64	2.53	2.45	2.38	2.32	2.28	2.24	2.20	2.14	2.10	2.04	2.00	1.96	1.91	1.88	1.84	1.82	1.79	1.77	1.76
	7.88	**5.66**	**4.76**	**4.26**	**3.94**	**3.71**	**3.54**	**3.41**	**3.30**	**3.21**	**3.14**	**3.07**	**2.97**	**2.89**	**2.78**	**2.70**	**2.62**	**2.53**	**2.48**	**2.41**	**2.37**	**2.32**	**2.28**	**2.26**
24	4.26	3.40	3.01	2.78	2.62	2.51	2.43	2.36	2.30	2.26	2.22	2.18	2.13	2.09	2.02	1.98	1.94	1.89	1.86	1.82	1.80	1.76	1.74	1.73
	7.82	**5.61**	**4.72**	**4.22**	**3.90**	**3.67**	**3.50**	**3.36**	**3.25**	**3.17**	**3.09**	**3.03**	**2.93**	**2.85**	**2.74**	**2.66**	**2.58**	**2.49**	**2.44**	**2.36**	**2.33**	**2.27**	**2.23**	**2.21**
25	4.24	3.38	2.99	2.76	2.60	2.49	2.41	2.34	2.28	2.24	2.20	2.16	2.11	2.06	2.00	1.96	1.92	1.87	1.84	1.80	1.77	1.74	1.72	1.71
	7.77	**5.57**	**4.68**	**4.18**	**3.86**	**3.63**	**3.46**	**3.32**	**3.21**	**3.13**	**3.05**	**2.99**	**2.89**	**2.81**	**2.70**	**2.62**	**2.54**	**2.45**	**2.40**	**2.32**	**2.29**	**2.23**	**2.19**	**2.17**

Degrees of Freedom: Numerator

Degrees of Freedom: Denominator	1	2	3	4	5	6	7	8	9	10	11	12	14	16	20	24	30	40	50	75	100	200	500	∞
26	4.22 / **7.72**	3.37 / **5.53**	2.98 / **4.64**	2.74 / **4.14**	2.59 / **3.82**	2.47 / **3.59**	2.39 / **3.42**	2.32 / **3.29**	2.27 / **3.17**	2.22 / **3.09**	2.18 / **3.02**	2.15 / **2.96**	2.10 / **2.86**	2.05 / **2.77**	1.99 / **2.66**	1.95 / **2.58**	1.90 / **2.50**	1.85 / **2.41**	1.82 / **2.36**	1.78 / **2.28**	1.76 / **2.25**	1.72 / **2.19**	1.70 / **2.15**	1.69 / **2.13**
27	4.21 / **7.68**	3.35 / **5.49**	2.96 / **4.60**	2.73 / **4.11**	2.57 / **3.79**	2.46 / **3.56**	2.37 / **3.39**	2.30 / **3.26**	2.25 / **3.14**	2.20 / **3.06**	2.16 / **2.98**	2.13 / **2.93**	2.08 / **2.83**	2.03 / **2.74**	1.97 / **2.63**	1.93 / **2.55**	1.88 / **2.47**	1.84 / **2.38**	1.80 / **2.33**	1.76 / **2.25**	1.74 / **2.21**	1.71 / **2.16**	1.68 / **2.12**	1.67 / **2.10**
28	4.20 / **7.64**	3.34 / **5.45**	2.95 / **4.57**	2.71 / **4.07**	2.56 / **3.76**	2.44 / **3.53**	2.36 / **3.36**	2.29 / **3.23**	2.24 / **3.11**	2.19 / **3.03**	2.15 / **2.95**	2.12 / **2.90**	2.06 / **2.80**	2.02 / **2.71**	1.96 / **2.60**	1.91 / **2.52**	1.87 / **2.44**	1.81 / **2.35**	1.78 / **2.30**	1.75 / **2.22**	1.72 / **2.18**	1.69 / **2.13**	1.67 / **2.09**	1.65 / **2.06**
29	4.18 / **7.60**	3.33 / **5.42**	2.93 / **4.54**	2.70 / **4.04**	2.54 / **3.73**	2.43 / **3.50**	2.35 / **3.33**	2.28 / **3.20**	2.22 / **3.08**	2.18 / **3.00**	2.14 / **2.92**	2.10 / **2.87**	2.05 / **2.77**	2.00 / **2.68**	1.94 / **2.57**	1.90 / **2.49**	1.85 / **2.41**	1.80 / **2.32**	1.77 / **2.27**	1.73 / **2.19**	1.71 / **2.15**	1.68 / **2.10**	1.65 / **2.06**	1.64 / **2.03**
30	4.17 / **7.56**	3.32 / **5.39**	2.92 / **4.51**	2.69 / **4.02**	2.53 / **3.70**	2.42 / **3.47**	2.34 / **3.30**	2.27 / **3.17**	2.21 / **3.06**	2.16 / **2.98**	2.12 / **2.90**	2.09 / **2.84**	2.04 / **2.74**	1.99 / **2.66**	1.93 / **2.55**	1.89 / **2.47**	1.84 / **2.38**	1.79 / **2.29**	1.76 / **2.24**	1.72 / **2.16**	1.69 / **2.13**	1.66 / **2.07**	1.64 / **2.03**	1.62 / **2.01**
32	4.15 / **7.50**	3.30 / **5.34**	2.90 / **4.46**	2.67 / **3.97**	2.51 / **3.66**	2.40 / **3.42**	2.32 / **3.25**	2.25 / **3.12**	2.19 / **3.01**	2.14 / **2.94**	2.10 / **2.86**	2.07 / **2.80**	2.02 / **2.70**	1.97 / **2.62**	1.91 / **2.51**	1.86 / **2.42**	1.82 / **2.34**	1.76 / **2.25**	1.74 / **2.20**	1.69 / **2.12**	1.67 / **2.08**	1.64 / **2.02**	1.61 / **1.98**	1.59 / **1.96**
34	4.13 / **7.44**	3.28 / **5.29**	2.88 / **4.42**	2.65 / **3.93**	2.49 / **3.61**	2.38 / **3.38**	2.30 / **3.21**	2.23 / **3.08**	2.17 / **2.97**	2.12 / **2.89**	2.08 / **2.82**	2.05 / **2.76**	2.00 / **2.66**	1.95 / **2.58**	1.89 / **2.47**	1.84 / **2.38**	1.80 / **2.30**	1.74 / **2.21**	1.71 / **2.15**	1.67 / **2.08**	1.64 / **2.04**	1.61 / **1.98**	1.59 / **1.94**	1.57 / **1.91**
36	4.11 / **7.39**	3.26 / **5.25**	2.86 / **4.38**	2.63 / **3.89**	2.48 / **3.58**	2.36 / **3.35**	2.28 / **3.18**	2.21 / **3.04**	2.15 / **2.94**	2.10 / **2.86**	2.06 / **2.78**	2.03 / **2.72**	1.98 / **2.62**	1.93 / **2.54**	1.87 / **2.43**	1.82 / **2.35**	1.78 / **2.26**	1.72 / **2.17**	1.69 / **2.12**	1.65 / **2.04**	1.62 / **2.00**	1.59 / **1.94**	1.56 / **1.90**	1.55 / **1.87**
38	4.10 / **7.35**	3.25 / **5.21**	2.85 / **4.34**	2.62 / **3.86**	2.46 / **3.54**	2.35 / **3.32**	2.26 / **3.15**	2.19 / **3.02**	2.14 / **2.91**	2.09 / **2.82**	2.05 / **2.75**	2.02 / **2.69**	1.96 / **2.59**	1.92 / **2.51**	1.85 / **2.40**	1.80 / **2.32**	1.76 / **2.22**	1.71 / **2.14**	1.67 / **2.08**	1.63 / **2.00**	1.60 / **1.97**	1.57 / **1.90**	1.54 / **1.86**	1.53 / **1.84**
40	4.08 / **7.31**	3.23 / **5.18**	2.84 / **4.31**	2.61 / **3.83**	2.45 / **3.51**	2.34 / **3.29**	2.25 / **3.12**	2.18 / **2.99**	2.12 / **2.88**	2.07 / **2.80**	2.04 / **2.73**	2.00 / **2.66**	1.95 / **2.56**	1.90 / **2.49**	1.84 / **2.37**	1.79 / **2.29**	1.74 / **2.20**	1.69 / **2.11**	1.66 / **2.05**	1.61 / **1.97**	1.59 / **1.94**	1.55 / **1.88**	1.53 / **1.84**	1.51 / **1.81**
42	4.07 / **7.27**	3.22 / **5.15**	2.83 / **4.29**	2.59 / **3.80**	2.44 / **3.49**	2.32 / **3.26**	2.24 / **3.10**	2.17 / **2.96**	2.11 / **2.86**	2.06 / **2.77**	2.02 / **2.70**	1.99 / **2.64**	1.94 / **2.54**	1.89 / **2.46**	1.82 / **2.35**	1.78 / **2.26**	1.73 / **2.17**	1.68 / **2.08**	1.64 / **2.02**	1.60 / **1.94**	1.57 / **1.91**	1.54 / **1.85**	1.51 / **1.80**	1.49 / **1.78**
44	4.06 / **7.24**	3.21 / **5.12**	2.82 / **4.26**	2.58 / **3.78**	2.43 / **3.46**	2.31 / **3.24**	2.23 / **3.07**	2.16 / **2.94**	2.10 / **2.84**	2.05 / **2.75**	2.01 / **2.68**	1.98 / **2.62**	1.92 / **2.52**	1.88 / **2.44**	1.81 / **2.32**	1.76 / **2.24**	1.72 / **2.15**	1.66 / **2.06**	1.63 / **2.00**	1.58 / **1.92**	1.56 / **1.88**	1.52 / **1.82**	1.50 / **1.78**	1.48 / **1.75**
46	4.05 / **7.21**	3.20 / **5.10**	2.81 / **4.24**	2.57 / **3.76**	2.42 / **3.44**	2.30 / **3.22**	2.22 / **3.05**	2.14 / **2.92**	2.09 / **2.82**	2.04 / **2.73**	2.00 / **2.66**	1.97 / **2.60**	1.91 / **2.50**	1.87 / **2.42**	1.80 / **2.30**	1.75 / **2.22**	1.71 / **2.13**	1.65 / **2.04**	1.62 / **1.98**	1.57 / **1.90**	1.54 / **1.86**	1.51 / **1.80**	1.48 / **1.76**	1.46 / **1.72**
48	4.04 / **7.19**	3.19 / **5.08**	2.80 / **4.22**	2.56 / **3.74**	2.41 / **3.42**	2.30 / **3.20**	2.21 / **3.04**	2.14 / **2.90**	2.08 / **2.80**	2.03 / **2.71**	1.99 / **2.64**	1.96 / **2.58**	1.90 / **2.48**	1.86 / **2.40**	1.79 / **2.28**	1.74 / **2.20**	1.70 / **2.11**	1.64 / **2.02**	1.61 / **1.96**	1.56 / **1.88**	1.53 / **1.84**	1.50 / **1.78**	1.47 / **1.73**	1.45 / **1.70**
50	4.03 / **7.17**	3.18 / **5.06**	2.79 / **4.20**	2.56 / **3.72**	2.40 / **3.41**	2.29 / **3.18**	2.20 / **3.02**	2.13 / **2.88**	2.07 / **2.78**	2.02 / **2.70**	1.98 / **2.62**	1.95 / **2.56**	1.90 / **2.46**	1.85 / **2.39**	1.78 / **2.26**	1.74 / **2.18**	1.69 / **2.10**	1.63 / **2.00**	1.60 / **1.94**	1.55 / **1.86**	1.52 / **1.82**	1.48 / **1.76**	1.46 / **1.71**	1.44 / **1.68**
55	4.02 / **7.12**	3.17 / **5.01**	2.78 / **4.16**	2.54 / **3.68**	2.38 / **3.37**	2.27 / **3.15**	2.18 / **2.98**	2.11 / **2.85**	2.05 / **2.75**	2.00 / **2.66**	1.97 / **2.59**	1.93 / **2.53**	1.88 / **2.43**	1.83 / **2.35**	1.76 / **2.23**	1.72 / **2.15**	1.67 / **2.06**	1.61 / **1.96**	1.58 / **1.90**	1.52 / **1.82**	1.50 / **1.78**	1.46 / **1.71**	1.43 / **1.66**	1.41 / **1.64**

TABLE F (*Continued*)

The values listed in the table are the critical values of F for the degrees of freedom of the numerator of the F ratio (column headings) and the degrees of freedom of the denominator of the F ratio (row headings). To be significant, $F_{obt} \geq F_{crit}$.

Degrees of Freedom: Denominator	Degrees of Freedom: Numerator																							
	1	2	3	4	5	6	7	8	9	10	11	12	14	16	20	24	30	40	50	75	100	200	500	∞
60	4.00 **7.08**	3.15 **4.98**	2.76 **4.13**	2.52 **3.65**	2.37 **3.34**	2.25 **3.12**	2.17 **2.95**	2.10 **2.82**	2.04 **2.72**	1.99 **2.63**	1.95 **2.56**	1.92 **2.50**	1.86 **2.40**	1.81 **2.32**	1.75 **2.20**	1.70 **2.12**	1.65 **2.03**	1.59 **1.93**	1.56 **1.87**	1.50 **1.79**	1.48 **1.74**	1.44 **1.68**	1.41 **1.63**	1.39 **1.60**
65	3.99 **7.04**	3.14 **4.95**	2.75 **4.10**	2.51 **3.62**	2.36 **3.31**	2.24 **3.09**	2.15 **2.93**	2.08 **2.79**	2.02 **2.70**	1.98 **2.61**	1.94 **2.54**	1.90 **2.47**	1.85 **2.37**	1.80 **2.30**	1.73 **2.18**	1.68 **2.09**	1.63 **2.00**	1.57 **1.90**	1.54 **1.84**	1.49 **1.76**	1.46 **1.71**	1.42 **1.64**	1.39 **1.60**	1.37 **1.56**
70	3.98 **7.01**	3.13 **4.92**	2.74 **4.08**	2.50 **3.60**	2.35 **3.29**	2.23 **3.07**	2.14 **2.91**	2.07 **2.77**	2.01 **2.67**	1.97 **2.59**	1.93 **2.51**	1.89 **2.45**	1.84 **2.35**	1.79 **2.28**	1.72 **2.15**	1.67 **2.07**	1.62 **1.98**	1.56 **1.88**	1.53 **1.82**	1.47 **1.74**	1.45 **1.69**	1.40 **1.62**	1.37 **1.56**	1.35 **1.53**
80	3.96 **6.96**	3.11 **4.88**	2.72 **4.04**	2.48 **3.56**	2.33 **3.25**	2.21 **3.04**	2.12 **2.87**	2.05 **2.74**	1.99 **2.64**	1.95 **2.55**	1.91 **2.48**	1.88 **2.41**	1.82 **2.32**	1.77 **2.24**	1.70 **2.11**	1.65 **2.03**	1.60 **1.94**	1.54 **1.84**	1.51 **1.78**	1.45 **1.70**	1.42 **1.65**	1.38 **1.57**	1.35 **1.52**	1.32 **1.49**
100	3.94 **6.90**	3.09 **4.82**	2.70 **3.98**	2.46 **3.51**	2.30 **3.20**	2.19 **2.99**	2.10 **2.82**	2.03 **2.69**	1.97 **2.59**	1.92 **2.51**	1.88 **2.43**	1.85 **2.36**	1.79 **2.26**	1.75 **2.19**	1.68 **2.06**	1.63 **1.98**	1.57 **1.89**	1.51 **1.79**	1.48 **1.73**	1.42 **1.64**	1.39 **1.59**	1.34 **1.51**	1.30 **1.46**	1.28 **1.43**
125	3.92 **6.84**	3.07 **4.78**	2.68 **3.94**	2.44 **3.47**	2.29 **3.17**	2.17 **2.95**	2.08 **2.79**	2.01 **2.65**	1.95 **2.56**	1.90 **2.47**	1.86 **2.40**	1.83 **2.33**	1.77 **2.23**	1.72 **2.15**	1.65 **2.03**	1.60 **1.94**	1.55 **1.85**	1.49 **1.75**	1.45 **1.68**	1.39 **1.59**	1.36 **1.54**	1.31 **1.46**	1.27 **1.40**	1.25 **1.37**
150	3.91 **6.81**	3.06 **4.75**	2.67 **3.91**	2.43 **3.44**	2.27 **3.14**	2.16 **2.92**	2.07 **2.76**	2.00 **2.62**	1.94 **2.53**	1.89 **2.44**	1.85 **2.37**	1.82 **2.30**	1.76 **2.20**	1.71 **2.12**	1.64 **2.00**	1.59 **1.91**	1.54 **1.83**	1.47 **1.72**	1.44 **1.66**	1.37 **1.56**	1.34 **1.51**	1.29 **1.43**	1.25 **1.37**	1.22 **1.33**
200	3.89 **6.76**	3.04 **4.71**	2.65 **3.88**	2.41 **3.41**	2.26 **3.11**	2.14 **2.90**	2.05 **2.73**	1.98 **2.60**	1.92 **2.50**	1.87 **2.41**	1.83 **2.34**	1.80 **2.28**	1.74 **2.17**	1.69 **2.09**	1.62 **1.97**	1.57 **1.88**	1.52 **1.79**	1.45 **1.69**	1.42 **1.62**	1.35 **1.53**	1.32 **1.48**	1.26 **1.39**	1.22 **1.33**	1.19 **1.28**
400	3.86 **6.70**	3.02 **4.66**	2.62 **3.83**	2.39 **3.36**	2.23 **3.06**	2.12 **2.85**	2.03 **2.69**	1.96 **2.55**	1.90 **2.46**	1.85 **2.37**	1.81 **2.29**	1.78 **2.23**	1.72 **2.12**	1.67 **2.04**	1.60 **1.92**	1.54 **1.84**	1.49 **1.74**	1.42 **1.64**	1.38 **1.57**	1.32 **1.47**	1.28 **1.42**	1.22 **1.32**	1.16 **1.24**	1.13 **1.19**
1000	3.85 **6.66**	3.00 **4.62**	2.61 **3.80**	2.38 **3.34**	2.22 **3.04**	2.10 **2.82**	2.02 **2.66**	1.95 **2.53**	1.89 **2.43**	1.84 **2.34**	1.80 **2.26**	1.76 **2.20**	1.70 **2.09**	1.65 **2.01**	1.58 **1.89**	1.53 **1.81**	1.47 **1.71**	1.41 **1.61**	1.36 **1.54**	1.30 **1.44**	1.26 **1.38**	1.19 **1.28**	1.13 **1.19**	1.08 **1.11**
∞	3.84 **6.64**	2.99 **4.60**	2.60 **3.78**	2.37 **3.32**	2.21 **3.02**	2.09 **2.80**	2.01 **2.64**	1.94 **2.51**	1.88 **2.41**	1.83 **2.32**	1.79 **2.24**	1.75 **2.18**	1.69 **2.07**	1.64 **1.99**	1.57 **1.87**	1.52 **1.79**	1.46 **1.69**	1.40 **1.59**	1.35 **1.52**	1.28 **1.41**	1.24 **1.36**	1.17 **1.25**	1.11 **1.15**	1.00 **1.00**

The values listed in the table are the critical values of Q for $\alpha = 0.05$ *and 0.01, as a function of degrees of freedom of* s_W^2 *and k (the number of means) or r (the number of means encompassed by* $\overline{X}_i$ *and* $\overline{X}_j$, *after the means have been rank-ordered). To be significant,* $Q_{obt} \geq Q_{crit}$.

s_W^2 df	α	\multicolumn{10}{c}{k (Number of Means) or r (Number of Means Encompassed by $\overline{X}_i$ and $\overline{X}_j$)}
		2	3	4	5	6	7	8	9	10	11
5	.05	3.64	4.60	5.22	5.67	6.03	6.33	6.58	6.80	6.99	7.17
	.01	5.70	6.98	7.80	8.42	8.91	9.32	9.67	9.97	10.24	10.48
6	.05	3.46	4.34	4.90	5.30	5.63	5.90	6.12	6.32	6.49	6.65
	.01	5.24	6.33	7.03	7.56	7.97	8.32	8.61	8.87	9.10	9.30
7	.05	3.34	4.16	4.68	5.06	5.36	5.61	5.82	6.00	6.16	6.30
	.01	4.95	5.92	6.54	7.01	7.37	7.68	7.94	8.17	8.37	8.55
8	.05	3.26	4.04	4.53	4.89	5.17	5.40	5.60	5.77	5.92	6.05
	.01	4.75	5.64	6.20	6.62	6.96	7.24	7.47	7.68	7.86	8.03
9	.05	3.20	3.95	4.41	4.76	5.02	5.24	5.43	5.59	5.74	5.87
	.01	4.60	5.43	5.96	6.35	6.66	6.91	7.13	7.33	7.49	7.65
10	.05	3.15	3.88	4.33	4.65	4.91	5.12	5.30	5.46	5.60	5.72
	.01	4.48	5.27	5.77	6.14	6.43	6.67	6.87	7.05	7.21	7.36
11	.05	3.11	3.82	4.26	4.57	4.82	5.03	5.20	5.35	5.49	5.61
	.01	4.39	5.15	5.62	5.97	6.25	6.48	6.67	6.84	6.99	7.13
12	.05	3.08	3.77	4.20	4.51	4.75	4.95	5.12	5.27	5.39	5.51
	.01	4.32	5.05	5.50	5.84	6.10	6.32	6.51	6.67	6.81	6.94
13	.05	3.06	3.73	4.15	4.45	4.69	4.88	5.05	5.19	5.32	5.43
	.01	4.26	4.96	5.40	5.73	5.98	6.19	6.37	6.53	6.67	6.79
14	.05	3.03	3.70	4.11	4.41	4.64	4.83	4.99	5.13	5.25	5.36
	.01	4.21	4.89	5.32	5.63	5.88	6.08	6.26	6.41	6.54	6.66
15	.05	3.01	3.67	4.08	4.37	4.59	4.78	4.94	5.08	5.20	5.31
	.01	4.17	4.84	5.25	5.56	5.80	5.99	6.16	6.31	6.44	6.55
16	.05	3.00	3.65	4.05	4.33	4.56	4.74	4.90	5.03	5.15	5.26
	.01	4.13	4.79	5.19	5.49	5.72	5.92	6.08	6.22	6.35	6.46
17	.05	2.98	3.63	4.02	4.30	4.52	4.70	4.86	4.99	5.11	5.21
	.01	4.10	4.74	5.14	5.43	5.66	5.85	6.01	6.15	6.27	6.38
18	.05	2.97	3.61	4.00	4.28	4.49	4.67	4.82	4.96	5.07	5.17
	.01	4.07	4.70	5.09	5.38	5.60	5.79	5.94	6.08	6.20	6.31
19	.05	2.96	3.59	3.98	4.25	4.47	4.65	4.79	4.92	5.04	5.14
	.01	4.05	4.67	5.05	5.33	5.55	5.73	4.89	6.02	6.14	6.25
20	.05	2.95	3.58	3.96	4.23	4.45	4.62	4.77	4.90	5.01	5.11
	.01	4.02	4.64	5.02	5.29	5.51	5.69	5.84	5.97	6.09	6.19
24	.05	2.92	3.53	3.90	4.17	4.37	4.54	4.68	4.81	4.92	5.01
	.01	3.96	4.55	4.91	5.17	5.37	5.54	5.69	5.81	5.92	6.02
30	.05	2.89	3.49	3.85	4.10	4.30	4.46	4.60	4.72	4.82	4.92
	.01	3.89	4.45	4.80	5.05	5.24	5.40	5.54	5.65	5.76	5.85
40	.05	2.86	3.44	3.79	4.04	4.23	4.39	4.52	4.63	4.73	4.82
	.01	3.82	4.37	4.70	4.93	5.11	5.26	5.39	5.50	5.60	5.69
60	.05	2.83	3.40	3.74	3.98	4.16	4.31	4.44	4.55	4.65	4.73
	.01	3.76	4.28	4.59	4.82	4.99	5.13	5.25	5.36	5.45	5.53
120	.05	2.80	3.36	3.68	3.92	4.10	4.24	4.36	4.47	4.56	4.64
	.01	3.70	4.20	4.50	4.71	4.87	5.01	5.12	5.21	5.30	5.37
∞	.05	2.77	3.31	3.63	3.86	4.03	4.17	4.29	4.39	4.47	4.55
	.01	3.64	4.12	4.40	4.60	4.76	4.88	4.99	5.08	5.16	5.23

TABLE H Chi-Square (χ^2) Distribution

The first column (df) locates each χ^2 distribution. The other columns give the proportion of area under the χ^2 distribution that is above the tabled value of χ^2. The χ^2 values under the column headings of .05 and .01 are the critical values of χ^2 for $\alpha = 0.05$ and 0.01. To be significant, $\chi^2_{obt} \geq \chi^2_{crit}$.

Degrees of freedom df	P = .99	.98	.95	.90	.80	.70	.50	.30	.20	.10	.05	.02	.01
1	.000157	.000628	.00393	.0158	.0642	.148	.455	1.074	1.642	2.706	3.841	5.412	6.635
2	.0201	.0404	.103	.211	.446	.713	1.386	2.408	3.219	4.605	5.991	7.824	9.210
3	.115	.185	.352	.584	1.005	1.424	2.366	3.665	4.642	6.251	7.815	9.837	11.341
4	.297	.429	.711	1.064	1.649	2.195	3.357	4.878	5.989	7.779	9.488	11.668	13.277
5	.554	.752	1.145	1.610	2.343	3.000	4.351	6.064	7.289	9.236	11.070	13.388	15.086
6	.872	1.134	1.635	2.204	3.070	3.828	5.348	7.231	8.558	10.645	12.592	15.033	16.812
7	1.239	1.564	2.167	2.833	3.822	4.671	6.346	8.383	9.803	12.017	14.067	16.622	18.475
8	1.646	2.032	2.733	3.490	4.594	5.527	7.344	9.524	11.030	13.362	15.507	18.168	20.090
9	2.088	2.532	3.325	4.168	5.380	6.393	8.343	10.656	12.242	14.684	16.919	19.679	21.666
10	2.558	3.059	3.940	4.865	6.179	7.267	9.342	11.781	13.442	15.987	18.307	21.161	23.209
11	3.053	3.609	4.575	5.578	6.989	8.148	10.341	12.899	14.631	17.275	19.675	22.618	24.725
12	3.571	4.178	5.226	6.304	7.807	9.034	11.340	14.011	15.812	18.549	21.026	24.054	26.217
13	4.107	4.765	5.892	7.042	8.634	9.926	12.340	15.119	16.985	19.812	22.362	25.472	27.688
14	4.660	5.368	6.571	7.790	9.467	10.821	13.339	16.222	18.151	21.064	23.685	26.873	29.141
15	5.229	5.985	7.261	8.547	10.307	11.721	14.339	17.322	19.311	22.307	24.996	28.259	30.578
16	5.812	6.614	7.962	9.312	11.152	12.624	15.338	18.418	20.465	23.542	26.296	29.633	32.000
17	6.408	7.255	8.672	10.085	12.002	13.531	16.338	19.511	21.615	24.769	27.587	30.995	33.409
18	7.015	7.906	9.390	10.865	12.857	14.440	17.338	20.601	22.760	25.989	28.869	32.346	34.805
19	7.633	8.567	10.117	11.651	13.716	15.352	18.338	21.689	23.900	27.204	30.144	33.687	36.191
20	8.260	9.237	10.851	12.443	14.578	16.266	19.337	22.775	25.038	28.412	31.410	35.020	37.566
21	8.897	9.915	11.591	13.240	15.445	17.182	20.337	23.858	26.171	29.615	32.671	36.343	38.932
22	9.542	10.600	12.338	14.041	16.314	18.101	21.337	24.939	27.301	30.813	33.924	37.659	40.289
23	10.196	11.293	13.091	14.848	17.187	19.021	22.337	26.018	28.429	32.007	35.172	38.968	41.638
24	10.856	11.992	13.848	15.659	18.062	19.943	23.337	27.096	29.553	33.196	36.415	40.270	42.980
25	11.524	12.697	14.611	16.473	18.940	20.867	24.337	28.172	30.675	34.382	37.652	41.566	44.314
26	12.198	13.409	15.379	17.292	19.820	21.792	25.336	29.246	31.795	35.563	38.885	42.856	45.642
27	12.879	14.125	16.151	18.114	20.703	22.719	26.336	30.319	32.912	36.741	40.113	44.140	46.963
28	13.565	14.847	16.928	18.939	21.588	23.647	27.336	31.391	34.027	37.916	41.337	45.419	48.278
29	14.256	15.574	17.708	19.768	22.475	24.577	28.336	32.461	35.139	39.087	42.557	46.693	49.588
30	14.953	16.306	18.493	20.599	23.364	25.508	29.336	33.530	36.250	40.256	43.773	47.962	50.892

TABLE I   Critical Values of T for Wilcoxon Signed Ranks Test

The values listed in the table are the critical values of T for the specified N (left column) and alpha level (column heading). To be significant, $T_{obt} \leq T_{crit}$.

	Level of Significance for One-Tailed Test					Level of Significance for One-Tailed Test			
	.05	.025	.01	.005		.05	.025	.01	.005
	Level of Significance for Two-Tailed Test					Level of Significance for Two-Tailed Test			
N	.10	.05	.02	.01	N	.10	.05	.02	.01
5	0	—	—	—	28	130	116	101	91
6	2	0	—	—	29	140	126	110	100
7	3	2	0	—	30	151	137	120	109
8	5	3	1	0	31	163	147	130	118
9	8	5	3	1	32	175	159	140	128
10	10	8	5	3	33	187	170	151	138
11	13	10	7	5	34	200	182	162	148
12	17	13	9	7	35	213	195	173	159
13	21	17	12	9	36	227	208	185	171
14	25	21	15	12	37	241	221	198	182
15	30	25	19	15	38	256	235	211	194
16	35	29	23	19	39	271	249	224	207
17	41	34	27	23	40	286	264	238	220
18	47	40	32	27	41	302	279	252	233
19	53	46	37	32	42	319	294	266	247
20	60	52	43	37	43	336	310	281	261
21	67	58	49	42	44	353	327	296	276
22	75	65	55	48	45	371	343	312	291
23	83	73	62	54	46	389	361	328	307
24	91	81	69	61	47	407	378	345	322
25	100	89	76	68	48	426	396	362	339
26	110	98	84	75	49	446	415	379	355
27	119	107	92	83	50	466	434	397	373

TABLE J

Random Numbers

	1	2	3	4	5	6	7	8	9
1	32942	95416	42339	59045	26693	49057	87496	20624	14819
2	07410	99859	83828	21409	29094	65114	36701	25762	12827
3	59981	68155	45673	76210	58219	45738	29550	24736	09574
4	46251	25437	69654	99716	11563	08803	86027	51867	12116
5	65558	51904	93123	27887	53138	21488	09095	78777	71240
6	99187	19258	86421	16401	19397	83297	40111	49326	81686
7	35641	00301	16096	34775	21562	97983	45040	19200	16383
8	14031	00936	81518	48440	02218	04756	19506	60695	88494
9	60677	15076	92554	26042	23472	69869	62877	19584	39576
10	66314	05212	67859	89356	20056	30648	87349	20389	53805
11	20416	87410	75646	64176	82752	63606	37011	57346	69512
12	28701	56992	70423	62415	40807	98086	58850	28968	45297
13	74579	33844	33426	07570	00728	07079	19322	56325	84819
14	62615	52342	82968	75540	80045	53069	20665	21282	07768
15	93945	06293	22879	08161	01442	75071	21427	94842	26210
16	75689	76131	96837	67450	44511	50424	82848	41975	71663
17	02921	16919	35424	93209	52133	87327	95897	65171	20376
18	14295	34969	14216	03191	61647	30296	66667	10101	63203
19	05303	91109	82403	40312	62191	67023	90073	83205	71344
20	57071	90357	12901	08899	91039	67251	28701	03846	94589
21	78471	57741	13599	84390	32146	00871	09354	22745	65806
22	89242	79337	59293	47481	07740	43345	25716	70020	54005
23	14955	59592	97035	80430	87220	06392	79028	57123	52872
24	42446	41880	37415	47472	04513	49494	08860	08038	43624
25	18534	22346	54556	17558	73689	14894	05030	19561	56517
26	39284	33737	42512	86411	23753	29690	26096	81361	93099
27	33922	37329	89911	55876	28379	81031	22058	21487	54613
28	78355	54013	50774	30666	61205	42574	47773	36027	27174
29	08845	99145	94316	88974	29828	97069	90327	61842	29604
30	01769	71825	55957	98271	02784	66731	40311	88495	18821
31	17639	38284	59478	90409	21997	56199	30068	82800	69692
32	05851	58653	99949	63505	40409	85551	90729	64938	52403
33	42396	40112	11469	03476	03328	84238	26570	51790	42122
34	13318	14192	98167	75631	74141	22369	36757	89117	54998
35	60571	54786	26281	01855	30706	66578	32019	65884	58485
36	09531	81853	59334	70929	03544	18510	89541	13555	21168
37	72865	16829	86542	00396	20363	13010	69645	49608	54738
38	56324	31093	77924	28622	83543	28912	15059	80192	83964
39	78192	21626	91399	07235	07104	73652	64425	85149	75409
40	64666	34767	97298	92708	01994	53188	78476	07804	62404
41	82201	75694	02808	65983	74373	66693	13094	74183	73020
42	15360	73776	40914	85190	54278	99054	62944	47351	89098
43	68142	67957	70896	37983	20487	95350	16371	03426	13895
44	19138	31200	30616	14639	44406	44236	57360	81644	94761
45	28155	03521	36415	78452	92359	81091	56513	88321	97910
46	87971	29031	51780	27376	81056	86155	55488	50590	74514
47	58147	68841	53625	02059	75223	16783	19272	61994	71090
48	18875	52809	70594	41649	32935	26430	82096	01605	65846
49	75109	56474	74111	31966	29969	70093	98901	84550	25769
50	35983	03742	76822	12073	59463	84420	15868	99505	11426

TABLE J (*Continued*)

Random Numbers

	1	2	3	4	5	6	7	8	9
51	12651	61646	11769	75109	86996	97669	25757	32535	07122
52	81769	74436	02630	72310	45049	18029	07469	42341	98173
53	36737	98863	77240	76251	00654	64688	09343	70278	67331
54	82861	54371	76610	94934	72748	44124	05610	53750	95938
55	21325	15732	24127	37431	09723	63529	73977	95218	96074
56	74146	47887	62463	23045	41490	07954	22597	60012	98866
57	90759	64410	54179	66075	61051	75385	51378	08360	95946
58	55683	98078	02238	91540	21219	17720	87817	41705	95785
59	79686	17969	76061	83748	55920	83612	41540	86492	06447
60	70333	00201	86201	69716	78185	62154	77930	67663	29529
61	14042	53536	07779	04157	41172	36473	42123	43929	50533
62	59911	08256	06596	48416	69770	68797	56080	14223	59199
63	62368	62623	62742	14891	39247	52242	98832	69533	91174
64	57529	97751	54976	48957	74599	08759	78494	52785	68526
65	15469	90574	78033	66885	13936	42117	71831	22961	94225
66	18625	23674	53850	32827	81647	80820	00420	63555	74489
67	74626	68394	88562	70745	23701	45630	65891	58220	35442
68	11119	16519	27384	90199	79210	76965	99546	30323	31664
69	41101	17336	48951	53674	17880	45260	08575	49321	36191
70	32123	91576	84221	78902	82010	30847	62329	63898	23268
71	26091	68409	69704	82267	14751	13151	93115	01437	56945
72	67680	79790	48462	59278	44185	29616	76531	19589	83139
73	15184	19260	14073	07026	25264	08388	27182	22557	61501
74	58010	45039	57181	10238	36874	28546	37444	80824	63981
75	56425	53996	86245	32623	78858	08143	60377	42925	42815
76	82630	84066	13592	60642	17904	99718	63432	88642	37858
77	14927	40909	23900	48761	44860	92467	31742	87142	03607
78	23740	22505	07489	85986	74420	21744	97711	36648	35620
79	32990	97446	03711	63824	07953	85965	87089	11687	92414
80	05310	24058	91946	78437	34365	82469	12430	84754	19354
81	21839	39937	27534	88913	49055	19218	47712	67677	51889
82	08833	42549	93981	94051	28382	83725	72643	64233	97252
83	58336	11139	47479	00931	91560	95372	97642	33856	54825
84	62032	91144	75478	47431	52726	30289	42411	91886	51818
85	45171	30557	53116	04118	58301	24375	65609	85810	18620
86	91611	62656	60128	35609	63698	78356	50682	22505	01692
87	55472	63819	86314	49174	93582	73604	78614	78849	23096
88	18573	09729	74091	53994	10970	86557	65661	41854	26037
89	60866	02955	90288	82136	83644	94455	06560	78029	98768
90	45043	55608	82767	60890	74646	79485	13619	98868	40857
91	17831	09737	79473	75945	28394	79334	70577	38048	03607
92	40137	03981	07585	18128	11178	32601	27994	05641	22600
93	77776	31343	14576	97706	16039	47517	43300	59080	80392
94	69605	44104	40103	95635	05635	81673	68657	09559	23510
95	19916	52934	26499	09821	87331	80993	61299	36979	73599
96	02606	58552	07678	56619	65325	30705	99582	53390	46357
97	65183	73160	87131	35530	47946	09854	18080	02321	05809
98	10740	98914	44916	11322	89717	88189	30143	52687	19420
99	98642	89822	71691	51573	83666	61642	46683	33761	47542
100	60139	25601	93663	25547	02654	94829	48672	28736	84994

GLOSSARY

Alpha level A probability level set by an investigator at the beginning of an experiment to limit the probability of making a Type I error. **(p. 216)**

Arithmetic mean The sum of the scores divided by the number of scores. In equation form,

$$\overline{X} = \Sigma \frac{X_i}{N} = \frac{X_1 + X_2 + X_3 + \cdots + X_N}{N}$$

mean of a sample

or $\quad \mu = \Sigma \frac{X_i}{N} = \frac{X_1 + X_2 + X_3 + \cdots + X_N}{N}$

mean of a population set of scores

where $\quad X_1, \ldots, X_N$ = raw scores
$\overline{X}$ (read "X bar") = mean of a sample set of scores
μ (read "mew") = mean of a population set of scores
Σ (read "sigma") = summation sign
N = number of scores

(p. 62)

Binomial distribution A probability distribution that results when five conditions are met: (1) there is a series of N trials; (2) on each trial there are only two possible outcomes; (3) on each trial, the two possible outcomes are mutually exclusive; (4) there is independence between the outcomes of each trial; and (5) the probability of each possible outcome on any trial stays the same from trial to trial. The binomial distribution gives each possible outcome of the N trials and the probability of getting each of these outcomes. **(p. 191)**

Comparison-wise error rate The probability of making a Type I error for any of the possible comparisons in an experiment. **(p. 384)**

Confidence interval A range of values that probably contains the population value. **(p. 304)**

Confidence limits The values that state the boundaries of the confidence interval. **(p. 304)**

Contingency table A two-way table showing the contingency between two variables where the variables have been classified into mutually exclusive categories and the cell entries are frequencies. **(p. 430)**

Continuous variable A variable that theoretically can have an infinite number of values between adjacent units on the scale. **(p. 28)**

Correlation coefficient A quantitative expression of the magnitude and direction of a relationship. **(p. 106)**

Critical region for rejection of the null hypothesis The area under the curve that contains all the values of the statistic that allow rejection of the null hypothesis. **(p. 277)**

Critical value of a statistic The value of the statistic that bounds the critical region. **(p. 277)**

Cumulative frequency distribution The number of scores that fall below the upper real limit of each interval. **(p. 42)**

Cumulative percentage distribution The percentage of scores that fall below the upper real limit of each interval. **(p. 42)**

Data The measurements that are made on the subjects of an experiment. **(p. 7)**

Degrees of freedom (df) The number of scores that are free to vary in calculating a statistic. **(p. 295)**

Dependent variable The variable in an experiment that an investigator measures to determine the effect of the independent variable. **(p. 6)**

Descriptive statistics Techniques that are used to describe or characterize the obtained sample data. **(p. 10)**

Deviation score The distance of the raw score from the mean of its distribution. **(p. 71)**

Discrete variable A variable for which no values are possible between adjacent units on the scale. **(p. 28)**

Exhaustive set of events A set that includes all of the possible events. **(p. 165)**

Experiment-wise error rate The probability of making one or more Type I errors for the full set of possible comparisons in an experiment. **(p. 384)**

Factorial experiment An experiment in which the effects of two or more factors are assessed and the treatments used are combinations of the levels of the factors. **(p. 371)**

Frequency distribution A listing of score values and their frequency of occurrence. **(p. 35)**

Imperfect relationship A positive or negative relationship for which all of the points do not fall on the line. **(p. 104)**

Independence of two events The occurrence of one event has no effect on the probability of occurrence of the other. **(p. 167)**

Independent variable The variable in an experiment that is systematically manipulated by an investigator. **(p. 6)**

Inferential statistics Techniques that use the obtained sample data to infer to populations. **(p. 10)**

Interaction effect The result observed when the effect of one factor is not the same at all levels of the other factor. **(p. 372)**

Least-squares regression line The prediction line that minimizes $\Sigma (Y - Y')^2$. **(p. 129)**

Linear relationship A relationship between two variables that can be most accurately represented by a straight line. **(p. 100)**

Main effects The effect of factor A (averaged over the levels of factor B) and the effect of factor B (averaged over the levels of factor A). **(p. 372)**

Median (Mdn) The scale value below which 50% of the scores fall. **(p. 67)**

Mode The most frequent score in the distribution. **(p. 69)**

Mutually exclusive events Two events that cannot occur together; that is, the occurrence of one precludes the occurrence of the other. **(p. 162)**

Negative relationship An inverse relationship between two variables. **(p. 103)**

Negatively skewed curve A curve on which most of the scores occur at the higher values, and the curve tails off toward the lower end of the horizontal axis. **(p. 55)**

Null-hypothesis population An actual or theoretical set of population scores that would result if the experiment were done on the entire population and the independent variable had no effect; it is used to test the validity of the null hypothesis. **(p. 265)**

Parameter A number calculated on population data that quantifies a characteristic of the population. **(p. 7)**

Pearson r A measure of the extent to which paired scores occupy the same or opposite positions within their own distributions. **(p. 108)**

Percentile The value on the measurement scale below which a specified percentage of the scores in the distribution fall. **(p. 43)**

Percentile point *See* Percentile.

Percentile rank (of a score) The percentage of scores with values lower than the score in question. **(p. 47)**

Perfect relationship A positive or negative relationship for which all of the points fall on the line. **(p. 104)**

Population The complete set of individuals, objects, or scores that an investigator is interested in studying. **(p. 6)**

Positive relationship A direct relationship between two variables. **(p. 103)**

Positively skewed curve A curve on which most of the scores occur at the lower values of the horizontal axis, and the curve tails off toward the higher end. **(p. 55)**

Power The probability that the results of an experiment will allow rejection of the null hypothesis if the independent variable has a real effect. **(p. 227)**

Probability of occurrence of A or B The probability of occurrence of A plus the probability of occurrence of B minus the probability of occurrence of both A and B. **(p. 161)**

Probability of occurrence of both A and B The probability of occurrence of A times the probability of occurrence of B given that A has occurred. **(p. 166)**

Random sample A sample selected from the population by a process that assures that (1) each possible sample of a given size has an equal chance of being selected and (2) all the members of the population have an equal chance of being selected into the sample. **(p. 156)**

Range The difference between the highest and lowest scores in the distribution. **(p. 71)**

Real limits of a continuous variable Those values that are above and below the recorded value by one-half of the smallest measuring unit of the scale. **(p. 29)**

Regression A topic that considers using the relationship between two or more variables for prediction. **(p. 127)**

Relative frequency distribution The proportion of the total number of scores that occur in each interval. **(p. 41)**

Sample A subset of the population. **(p. 6)**

Sampling distribution of F All the possible F values along with the $p(F)$ for each value, assuming sampling is random from the population. **(p. 351)**

Sampling distribution of t A probability distribution of the t values that would occur if all possible different samples of a fixed size N were drawn from the null-hypothesis population. It gives (1) all the possible different t values for samples of size N and (2) the probability of getting each value if sampling is random from the null-hypothesis population. **(p. 295)**

Sampling distribution of the mean A listing of all the values the mean can take, along with the probability of getting each value if sampling is random from the null-hypothesis population. **(p. 268)**

Sampling distribution of a statistic A listing of (1) all the values that the statistic can take and (2) the probability of getting each value under the assumption that it results from chance alone, or if sampling is random from the null-hypothesis population. **(pp. 263, 266)**

Sampling with replacement A method of sampling in which each member of the population selected for the sample is returned to the population before the next member is selected. **(p. 158)**

Sampling without replacement A method of sampling in which the members of the sample are not returned to the population prior to selecting subsequent members. **(p. 159)**

Scatter plot A graph of paired X and Y values. **(p. 100)**

Skewed curve A curve whose two sides do not coincide if the curve is folded in half, that is, a curve that is not symmetrical. **(p. 55)**

Statistic A number calculated on sample data that quantifies a characteristic of the sample. **(p. 7)**

Symmetrical curve A curve whose two sides coincide if the curve is folded in half. **(p. 55)**

Type I error A decision to reject the null hypothesis when the null hypothesis is true. **(p. 215)**

Type II error A decision to retain the null hypothesis when the null hypothesis is false. **(p. 215)**

Variable Any property or characteristic of some event, object, or person that may have different values at different times, depending on the conditions. **(p. 6)**

z score A transformed score that designates how many standard deviation units the corresponding raw score is above or below the mean. **(p. 84)**

INDEX

A

Absolute zero point, 23–25
Addition rule, 161–165
    computational equation for, 161, 163, 165–166
    multiplication rule and, 178–180
Advertising
    use of statistics in, 12–13, 232–233, 375
Alpha (α) level
    beta and, 230
    decision process and, 213
    defined, 216, 458
    power and, 230, 282, 287–288
    Type I error and, 215–217
    Type II error and, 217
Alternative hypothesis, 212
    defined, 457
Analysis of variance (ANOVA), 10, 27, 351–376, 423. *See also F test*
    F distribution, 351–352
    multiple comparisons and, 418
    null hypothesis in, 354–356
    one-way, 354–356, 395
        assumptions underlying, 370–371
        between-groups variance estimate, 355, 357–358
        F ratio, 358–363
        logic of, 363–367
        sum of squares between groups, 355, 358
        sum of squares within groups, 355–357
        use of, 465–469
        within-groups variance estimate, 355–357
    t test and, 368–370

Analysis of variance (ANOVA) (*continued*)
    two-way, 371–374, 395–418
        assumptions underlying, 418
        column sum of squares, 395, 400–401
        column variance estimate, 374, 400–401
        F ratios, 374, 402–406
        interpretations for, 406–418
        row sum of squares, 395, 399
        row variance estimate, 374, 398–399
        row $\times$ column sum of squares, 401–402
        row $\times$ column variance estimate, 374, 401–402
        use of, 469–471
        within-cells sum of squares, 395–398
        within-cells variance estimate, 374, 396–398
A posteriori comparisons, 383–384, 467
A priori comparisons, 381–383, 467
Arithmetic mean. *See* Mean
Authority, as method of knowing, 3
Averaged over the columns, 399

B

Bar graph, 51
Bell-shaped curve, 54–55, 70
Beta (β). *See also* Type II error
    alpha level and, 216–217, 230
    calculation of power and, 227, 229–230
    power and, 227, 229–230, 282
Between-groups sum of squares, 355, 357–358

Between-groups variance estimate, 355, 357–358
    computational equation for, 358
    independent variable, 363
Beveridge, W. I. B., 4–5
Biased sample, 157
Bimodal distribution, 69–70
Binomial distribution, 191–205, 264. *See also* Sign test
    binomial expansion and, 194–196
    defined, 191
    in hypothesis testing, 213–214
    illustration of, 191–194
    table of associated probabilities, 197–205
    tables for, 518
Binomial expansion, 194–196
    computational equation for, 194
    defined, 194
Biomedical Computer Programs—P series (BMDP), 11

C

Causation, correlation and, 120
Central Limit Theorem, 268, 270, 272
Central tendency, 10
    defined, 61
    mean, 62–66
    measures of, 61–70
    median, 66–69
    mode, 69
    symmetry, 69–70
Chi-square, 424–439
    assumptions underlying, 439
    computation of, 425–426, 431–433
    contingency table and, 430–435